RENEWABLE ENERGY
AND SUSTAINABILITY
———

RENEWABLE ENERGY AND SUSTAINABILITY
Prospects in the Developing Economies

Edited by

IMRAN KHAN

Department of Electrical and Electronic Engineering, Jashore University of Science and Technology, Jashore, Bangladesh

ELSEVIER

Elsevier

Radarweg 29, PO Box 211, 1000 AE Amsterdam, Netherlands
The Boulevard, Langford Lane, Kidlington, Oxford OX5 1GB, United Kingdom
50 Hampshire Street, 5th Floor, Cambridge, MA 02139, United States

Copyright © 2022 Elsevier Inc. All rights reserved.

No part of this publication may be reproduced or transmitted in any form or by any means, electronic or mechanical, including photocopying, recording, or any information storage and retrieval system, without permission in writing from the publisher. Details on how to seek permission, further information about the Publisher's permissions policies and our arrangements with organizations such as the Copyright Clearance Center and the Copyright Licensing Agency, can be found at our website: www.elsevier.com/permissions.

This book and the individual contributions contained in it are protected under copyright by the Publisher (other than as may be noted herein).

Notices

Knowledge and best practice in this field are constantly changing. As new research and experience broaden our understanding, changes in research methods, professional practices, or medical treatment may become necessary.

Practitioners and researchers must always rely on their own experience and knowledge in evaluating and using any information, methods, compounds, or experiments described herein. In using such information or methods they should be mindful of their own safety and the safety of others, including parties for whom they have a professional responsibility.

To the fullest extent of the law, neither the Publisher nor the authors, contributors, or editors, assume any liability for any injury and/or damage to persons or property as a matter of products liability, negligence or otherwise, or from any use or operation of any methods, products, instructions, or ideas contained in the material herein.

ISBN: 978-0-323-88668-0

For Information on all Elsevier publications visit our website at
https://www.elsevier.com/books-and-journals

Publisher: Charlotte Cockle
Acquisitions Editor: Peter Adamson
Editorial Project Manager: Mica Ella Ortega
Production Project Manager: Nirmala Arumugam
Cover Designer: Mark Rogers

Typeset by Aptara, New Delhi, India

Dedication

To one of the great Bangladeshi scientists

Professor Jamal Nazrul Islam

He is an inspiration to me towards research and patriotism.

&

To my wife, **K. N. Queen**, and my son **Reeyan Khan** for their patience and continual support and encouragement during the project.

Why should one bother about the ultimate fate of the universe? One answer to this question is similar to the answer to the question about climbing Mount Everest: because the problem exists. It is in the nature of the human mind to seek incessantly new frontiers of knowledge to explore.

Professor Jamal Nazrul Islam, The Ultimate Fate of The Universe

Contents

Contributors xv
About the editor xvii
Preface xix
Acknowledgment xxi

1. Sustainability—Concept and its application in the energy sector 1

Imran Khan and Md. Sahabuddin

1.1 Introduction 1
1.2 The concept of sustainability 4
1.3 Sustainability assessment methods 5
1.4 Sustainability assessment: An application 10
    1.4.1 The TOPSIS method 10
    1.4.2 Result and analysis 16
1.5 Conclusion 17
Self-evaluation questions 18
References 18

2. Prospects of bioelectricity in south Asian developing countries—A sustainable solution for future electricity 23

Pobitra Halder, Ibrahim Gbolahan Hakeem, Savankumar Patel, Shaheen Shah, Hafijur Khan and Kalpit Shah

2.1 Introduction 23
2.2 Electricity generation mix in south Asian developing countries 26
2.3 Why bioelectricity? 29
2.4 Technologies for bioelectricity generation 31
    2.4.1 Microbial fuel cell 33
    2.4.2 Anaerobic digestion-based technology 34
    2.4.3 Direct combustion/gasification-based technology 39
2.5 Bioelectricity prospects and technological status in south Asian developing countries 40
    2.5.1 Biomass resources potential 40
    2.5.2 Bioelectricity generation: Current status and future mandates 42
    2.5.3 Biomass-fired plant for combined heat and power: Opportunities and challenges in developing countries 46
2.6 Concluding remarks 49
Self-evaluation questions 50
References 50

3. Environmental, social, and economic impacts of renewable energy sources 57

Zobaidul Kabir, Nahid Sultana and Imran Khan

3.1 Introduction 57
3.2 Global scenario of renewable energy production 60
3.3 Impacts of renewable energy 62
    3.3.1 Solar energy 62
    3.3.2 Hydro energy 65
    3.3.3 Wind energy 67
    3.3.4 Bioenergy 69
    3.3.5 Solid waste 72
3.4 Discussion 73
    3.4.1 Future impacts in developing countries 77
3.5 Conclusion 79
Self-evaluation questions 79
References 80

4. Application of solar photovoltaic for enhanced electricity access and sustainable development in developing countries 85

Majbaul Alam

4.1 Introduction 85
    4.1.1 Solar energy: Global and developing world context 85
4.2 Solar technology 86
    4.2.1 Solar photovoltaic 86
    4.2.2 Concentrating solar power (CSP) 87
4.3 Solar energy for developing countries 88
    4.3.1 Present status 92
    4.3.2 Future prospects and challenges 94

4.4 Contribution to sustainability 100
4.5 Case study: Kenya 102
4.6 Conclusions 104
Self-evaluation questions 105
References 105

5. Hydropower–Basics and its role in achieving energy sustainability for the developing economies 107

Arun Kumar

5.1 Introduction 107
5.2 Historical background of hydropower development 108
5.3 Present status of hydropower development 108
5.4 Hydropower basics 111
5.5 Services from hydropower development 114
5.6 Required surveys and investigations 114
5.7 Hydrology 116
5.8 Cumulative impact assessment of hydropower plants 116
5.9 Climate change impact on hydropower potential 117
5.10 Technology 117
5.11 Components of hydropower plant 118
    5.11.1 Civil works components 119
    5.11.2 Hydroelectrical equipment 122
5.12 Present day operation strategy 125
5.13 Renovation, modernization, and upgrading 125
5.14 Hydropower industry 126
5.15 Cost of hydropower 126
5.16 Achieving energy sustainability through transboundary hydropower cooperation 127
5.17 Contribution of hydropower to sustainability in developing countries 128
5.18 Possible multiplier effects of hydropower projects 130
5.19 Conclusion 130
Self-evaluation questions 131
References 131

6. Wind energy and its link to sustainability in developing countries 135

Mahfuz Kabir, Navya Sree BN, Krishna J. Khatod, Vikrant P. Katekar and Sandip S. Deshmukh

6.1 Introduction 135
6.2 Foundational content of wind energy 136
    6.2.1 Historical development of wind energy 136
    6.2.2 Classification of wind turbines 138
    6.2.3 Thermodynamics of wind energy conversion systems 141
    6.2.4 Design of a wind turbine 144
6.3 Available technologies 145
    6.3.1 Conventional wind energy conversion system 145
    6.3.2 Advanced wind energy conversion system 146
6.4 Current status: Global and in developing countries 147
    6.4.1 Global status of wind energy 148
    6.4.2 Power generation 150
    6.4.3 Technology manufacturing and supply chain 151
    6.4.4 Cost of wind energy 153
    6.4.5 Employment generation 153
    6.4.6 Future outlook 155
6.5 Wind power in developing countries 157
    6.5.1 Wind power in selected developing countries 157
6.6 Contribution to sustainability 163
    6.6.1 Wind energy and its link to sustainability 163
    6.6.2 Impacts of wind energy 167
6.7 Conclusion 170
6.8 Self-evaluation questions and numerical problems 170
    6.8.1 Short questions 170
    6.8.2 Long questions 171
    6.8.3 Numerical problems 171
Acknowledgment 172
References 172

7. Substituting coal with renewable biomass for electricity production using co-gasification technique: A short-term sustainable pathway for developing countries 179

M. Shahabuddin and Sankar Bhattacharya

7.1 Introduction 179
7.2 Status of Coal and biomass as energy sources 181

7.3 Power and energy from coal and biomass in Bangladesh 181
7.4 Greenhouse gas emission 182
7.5 Why co-gasification? 183
7.6 Selection of biomass for the co-gasification with coal 186
7.7 Gasification technologies 186
    7.7.1 Commercial gasifiers 188
7.8 Present status of co-gasification of biomass and coal 191
7.9 Challenges of co-gasification 194
7.9 Conclusion 196
Self-evaluation questions 197
References 197

8. Waste-to-energy (WtE): A potential renewable source for future electricity generation in the developing world 203

Zobaidul Kabir, Mahfuz Kabir and Nigar Sultana

8.1 Introduction 203
8.2 Waste management tools and techniques 205
8.3 Available WtE generation technologies 206
    8.3.1 Sanitary landfilling 206
    8.3.2 Gasification 206
    8.3.3 Incineration 208
    8.3.4 Pyrolysis 209
    8.3.5 Anaerobic digestion 211
8.4 Present status of MSW in developing countries 211
    8.4.1 Collection of MSW 211
    8.4.2 Composition of waste 212
    8.4.3 Disposal of MSW 213
    8.4.4 WtE technology 214
8.5 Future prospect in the developing world 215
8.6 Contribution to sustainability in developing countries 216
8.7 Case study: Bangladesh 219
    8.7.1 Generation and composition of MSW 219
    8.7.2 WtE and institutional policy 220
    8.7.3 Potential of WtE technologies in Bangladesh 220
8.8 Conclusion 222
Self-evaluation questions 223
References 223

9. Geothermal energy in developing countries–The dilemma between renewable and nonrenewable 227

Nurdan Yildirim and Emin Selahattin Umdu

9.1 Introduction 227
9.2 Available technologies 230
    9.2.1 Geothermal power generation 230
    9.2.2 Direct use of geothermal energy 233
    9.2.3 Engineered (enhanced) geothermal systems or hot dry rock systems 237
    9.2.4 Combined (hybrid) geothermal systems 237
9.3 Current status 237
9.4 The dilemma: Renewable or nonrenewable 243
9.5 Contribution to sustainability 246
9.6 Future prospect 248
9.7 Thermodynamic analysis of geothermal energy systems 251
9.8 Case study 253
9.9 Self-evaluation questions and numerical problems 256
    9.9.1 Numerical problems 256
Nomenclature 257
References 258

10. Ocean renewable energy and its prospect for developing economies 263

Mahfuz Kabir, M.S. Chowdhury, Nigar Sultana, M.S. Jamal and Kuaanan Techato

10.1 Introduction 263
10.2 Ocean renewable energy conversion 264
    10.2.1 Tidal energy 264
    10.2.2 Wave energy 265
    10.2.3 Ocean thermal gradient energy 269
    10.2.4 Salinity gradient energy (SGE) 272
10.3 Present status 275
10.4 ORE in developing economies 279
    10.4.1 ORE projects in developing economies 282
10.5 Contribution of ORE to sustainability 284
10.6 ORE in selected developing economies 286
    10.6.1 India 286
    10.6.2 Bangladesh 287
    10.6.3 Indonesia 288

10.7 Way forward 289
10.8 Conclusion 291
Self-evaluation questions 292
Numerical problems 292
References 292

11. Hydrogen energy–Potential in developing countries 299

Minhaj Uddin Monir, Azrina Abd Aziz, Mohammad Tofayal Ahmed and Md. Yeasir Hasan

11.1 Introduction 299
11.2 Energy sectors in developing countries 301
11.3 Sources of hydrogen production 303
    11.3.1 Fossil fuel source 304
    11.3.2 Biomass and algae source 305
    11.3.3 Microbial source 306
11.4 Technologies for hydrogen production 307
    11.4.1 Thermal process 307
    11.4.2 Electrolytic process 311
    11.4.3 Photolytic process 312
    11.4.4 Fermentation process 313
11.5 Current status of hydrogen production 315
11.6 Challenges of hydrogen conversion 317
11.7 Hydrogen potentiality in developing countries 317
11.8 Conclusions 319
Self-evaluation questions 320
References 321

12. The role of demand-side management in sustainable energy sector development 325

Samuel Gyamfi, Felix Amankwah Diawuo, Emmanuel Yeboah Asuamah and Emmanuel Effah

12.1 Introduction 325
12.2 Concept of demand-side management 326
12.3 Basic techniques in DSM 327
    12.3.1 Flexible load techniques 332
    12.3.2 Flexible storage techniques 332
    12.3.3 Demand-side generation techniques 333
12.4 DSM contribution to sustainability 335
12.5 Case studies 336
    12.5.1 Case study 1: DSM in Ghana 336
    12.5.2 Case study 2: DSM in South Africa 338
    12.5.3 Case study 3: DSM in China 339
    12.5.4 Case study 4: DSM in India 340
12.6 Future directions of DSM 340
12.7 Conclusion 341
Self-evaluation questions 341
References 342

13. The role of energy storage technologies for sustainability in developing countries 347

Md Momtazur Rahman, Imran Khan and Kamal Alameh

13.1 Introduction 347
    13.1.1 Role of energy storage technologies in energy transitions 349
13.2 Classification of energy storage technologies 349
    13.2.1 Mechanical energy storage 349
    13.2.2 Thermal energy storage 355
    13.2.3 Electrochemical energy storage 357
    13.2.4 Electromagnetic energy storage 361
13.3 Progress and challenges of energy storage technologies in developing countries 362
13.4 Sustainability evaluation of energy storage technologies in developing countries 366
13.5 The case of China 367
    13.5.1 Summary and benefits of the energy storage project 367
13.6 Challenges and policy implications 367
13.7 Conclusion 369
13.8 Self-evaluation 370
    13.8.1 Questions 370
    13.8.2 Numerical problems 370
References 371

14. Climate change, sustainability, and renewable energy in developing economies 377

Mahfuz Kabir, Zobaidul Kabir and Nigar Sultana

14.1 Introduction 377
14.2 Linkage between climate change, energy, and sustainability 379
14.3 Growth of electricity demand and climate change: Impact on developing economies 382

- 14.3.1 Electricity consumption in developing economies and climate change 382
- 14.3.2 Production of renewable energy in developing economies 384
- 14.4 Mitigating climate change: Transition from nonrenewable to renewable 395
- 14.5 Opportunities and challenges in adopting renewable energy technology 398
- 14.6 Way forward: R&D, market, and fiscal and monetary policy support 402
  - 14.6.1 Investment in renewable energy capacity 402
  - 14.6.2 Research and development 404
  - 14.6.3 Fiscal policy 405
  - 14.6.4 Monetary policy 406
- 14.7 Conclusion 407
- Self-evaluation questions 408
- References 408

Index 415

Contributors

Imran Khan Department of Electrical and Electronic Engineering, Jashore University of Science and Technology, Jashore, Bangladesh

Md. Sahabuddin Department of Electrical and Electronic Engineering, Jashore University of Science and Technology, Jashore, Bangladesh

Pobitra Halder School of Engineering, RMIT University, VIC, Australia; Department of Industrial and Production Engineering, Jashore University of Science and Technology, Jashore, Bangladesh

Ibrahim Gbolahan Hakeem School of Engineering, RMIT University, VIC, Australia

Savankumar Patel School of Engineering, RMIT University, VIC, Australia

Shaheen Shah Department of Oil and Gas Engineering, Memorial University of Newfoundland, Newfoundland, Canada

Hafijur Khan University of Chinese Academy of Sciences, Beijing, China

Kalpit Shah School of Engineering, RMIT University, VIC, Australia

Zobaidul Kabir School of Environmental and Life Sciences, University of Newcastle, Ourimbah Campus, Australia

Nahid Sultana University of Southern Queensland, School of Business, Toowoomba, QLD, Australia

Majbaul Alam Sustainable Energy Research Group, Energy and Climate Change Division, University of Southampton, United Kingdom

Arun Kumar Department of Hydro and Renewable Energy, Indian Institute of Technology Roorkee, Uttarakhand, India

Mahfuz Kabir Bangladesh Institute of International and Strategic Studies (BIISS), Dhaka, Bangladesh

Navya Sree BN Department of Mechanical Engineering, Birla Institute of Technology & Science, Pilani, Hyderabad, India

Krishna J. Khatod Department of Mechanical Engineering, Birla Institute of Technology & Science, Pilani, Hyderabad, India

Vikrant P. Katekar Department of Mechanical Engineering, S. B. Jain Institute of Technology, Management and Research, Nagpur, Maharashtra, India

Sandip S. Deshmukh Department of Mechanical Engineering, Birla Institute of Technology & Science, Pilani, Hyderabad, India

M. Shahabuddin Department of Chemical Engineering, Monash University, Clayton, Australia; Department of Mechanical Engineering and Product Design Engineering, Swinburne University of Technology, Hawthorn, Australia

Sankar Bhattacharya Department of Chemical Engineering, Monash University, Clayton, Australia

Nigar Sultana Knowledge for Development Management (K4DM) Project, UNDP, Bangladesh

Nurdan Yildirim Mechanical Engineering Department, Yasar University, Bornova, İzmir-Turkey

Emin Selahattin Umdu Energy Systems Engineering Department, Yasar University, Bornova, İzmir-Turkey

M.S. Chowdhury Faculty of Environmental Management, Prince of Songkla University, Songkhla, Thailand

M.S. Jamal Institute of Fuel Research and Development (IFRD), Bangladesh Council of Scientific and Industrial Research (BCSIR), Dhaka, Bangladesh

Kuaanan Techato Faculty of Environmental Management, Prince of Songkla University, Songkhla, Thailand

Minhaj Uddin Monir Department of Petroleum and Mining Engineering, Jashore University of Science and Technology, Jashore, Bangladesh

Azrina Abd Aziz Faculty of Civil Engineering Technology, Universiti Malaysia Pahang, Gambang, Malaysia

Mohammad Tofayal Ahmed Department of Petroleum and Mining Engineering, Jashore University of Science and Technology, Jashore, Bangladesh

Md. Yeasir Hasan Department of Petroleum and Mining Engineering, Jashore University of Science and Technology, Jashore, Bangladesh

Samuel Gyamfi Regional Centre for Energy and Environmental Sustainability (RCEES), University of Energy and Natural Resources (UENR), Sunyani, Ghana; Department of Energy and Petroleum Engineering, School of Engineering, University of Energy and Natural Resources (UENR), Sunyani, Ghana

Felix Amankwah Diawuo Regional Centre for Energy and Environmental Sustainability (RCEES), University of Energy and Natural Resources (UENR), Sunyani, Ghana; Department of Energy and Petroleum Engineering, School of Engineering, University of Energy and Natural Resources (UENR), Sunyani, Ghana

Emmanuel Yeboah Asuamah Regional Centre for Energy and Environmental Sustainability (RCEES), University of Energy and Natural Resources (UENR), Sunyani, Ghana; Department of Energy and Petroleum Engineering, School of Engineering, University of Energy and Natural Resources (UENR), Sunyani, Ghana

Emmanuel Effah Regional Centre for Energy and Environmental Sustainability (RCEES), University of Energy and Natural Resources (UENR), Sunyani, Ghana; Department of Energy and Petroleum Engineering, School of Engineering, University of Energy and Natural Resources (UENR), Sunyani, Ghana

Md Momtazur Rahman School of Science, Edith Cowan University, Joondalup Drive, WA, Australia

Kamal Alameh School of Science, Edith Cowan University, Joondalup Drive, WA, Australia

About the editor

Imran Khan received his Ph.D. from the Centre for Sustainability and Department of Physics at the University of Otago, New Zealand. By background, he is an engineer. He obtained his first degree in Electronics and Communication Engineering (ECE) from Khulna University of Engineering and Technology, Bangladesh. He also completed his M.Sc. Engineering in ECE from the same university; part of his M.Sc. course work was completed from the University of Ottawa, Canada. Currently, Dr. Khan is serving as a Faculty Member in the Department of Electrical and Electronic Engineering (EEE), Jashore University of Science and Technology, Bangladesh. He leads the Energy Research Laboratory in the same university. He has more than 10 years of teaching and research experience. He has served several administrative positions, including Chairman of the EEE Department, Director of the ICT Cell. Dr. Khan has performed as one of the Research Cell Advisors to the same university. At present, he serves as an Editorial Board Member for the journal SN Applied Sciences published by Springer Nature and Guest Editor for an Engineering Topical Collection "Sustainable Energy Trends in the Developing Economies." He is also serving as a special issue Editor for the Sustainability journal published by MDPI. He has served as a reviewer for several reputed journals such as Renewable & Sustainable Energy Reviews; Energy Conversion and Management; Energy, Sustainability and Society; Sustainability and has received an "Outstanding Reviewer" award from journals such as Applied Energy, Journal of Cleaner Production, and Renewable Energy. Dr. Khan is a multidisciplinary researcher focusing on energy security challenges and transitions toward a low-carbon and sustainable future. He is interested in exploring the interactions between technology, society, the environment, and the economy about energy access and security. He is also interested in investigating demand-side management opportunities in the electricity sector of developing countries. Dr. Khan has authored and coauthored numerous research articles in reputed refereed journals and conference proceedings, in addition to several book chapters published by reputed international publishers.

Preface

Sustainable energy is a multidisciplinary concept that has received much attention due to global warming and adverse climate change-related issues, mainly in developed countries. Although this is equally important for the developing nations where access to basic electricity is yet to reach its total population, it has received limited focus due to many different factors such as technical, economic, environmental, and social. All these factors are well discussed in the literature for developed countries. Nevertheless, very little is known about the developing world. For example, how will the clean and sustainable energy technologies be adopted and deployed for the new electricity generation expansion plan? What would be the economic challenges for this new clean technology adoption? What would be the impacts on environmental and social issues? This book will help identify these challenges; consequently, this will underpin sustainable electricity generation expansion plans in these countries.

Many developing countries have potential renewable energy resources, but they are unable to utilize them due to sound policymaking. Hence, this book might be a good source of information for energy-related policymaking for developing economies. Overall, this book could serve as a potential source of information for energy sustainability issues in developing countries, which is scarce in the existing literature.

This book addresses all the issues mentioned earlier through a multidisciplinary approach and relevant case studies. The key features of this book are:

1. This book is explicitly focused on the sustainable energy status of developing economies.
2. Renewable energy potential and its link to sustainable electricity generation in developing countries are explored.
3. A multidisciplinary approach is employed to assess the energy sustainability issues in the developing world.

Renewable energy technology deployment in developing economies faces diverse challenges, and it varies from one nation to another. In a nutshell, the existing literature provides little or limited knowledge about the sustainability issues in the energy sectors of the developing world. Thus, the primary purpose of this book is to close this gap in the literature so that students, researchers, energy engineers, and policymakers can find relevant information in one place.

Editor
Imran Khan, PhD

Dept. of Electrical and Electronic Engineering
Jashore University of Science and Technology
E-mail: i.khan@just.edu.bd, ikr_ece@yahoo.com

Acknowledgment

This book would not have been possible without the valued cooperation of many people and organizations. I would like to express my special appreciation and thanks to those who so generously contributed directly or indirectly to work presented in this book. Special appreciations are given to all chapter reviewers around the globe for their expert opinions in evaluating the chapters' scientific quality. Special mention goes to my supervisors, *Associate Professor Michael Jack* and *Professor Janet Stephenson* from the University of Otago, New Zealand. I am grateful to them for always encouraging my research and allowing me to grow as a researcher. Special appreciation is given to the Jashore University of Science and Technology, Bangladesh, and the Department of Electrical and Electronic Engineering of the same university for supporting me during the project.

And finally, last but by no means least, thanks to my family for supporting me spiritually throughout the project.

CHAPTER 1

Sustainability—Concept and its application in the energy sector

Imran Khan and Md. Sahabuddin

Department of Electrical and Electronic Engineering, Jashore University of Science and Technology, Jashore, Bangladesh

1.1 Introduction

Sustainability is a concept that frequently evolves, making it difficult to reach a final definition. A precise definition of sustainability is not only difficult but also a challenge as it involves multiple factors. Similarly, "sustainable development, a concept that emerged in the context of a growing awareness of an imminent ecological crisis, seems to have been one of the driving forces of world history in the period around the end of the 20th century" (Du Pisani, 2006). The concept of sustainability and sustainable development is discussed in (Mensah, 2019). In brief, the history of sustainable energy development adoption by the world community is illustrated in Fig. 1.1.

In 1987, sustainability was defined by the United Nations Brundtland Commission as "meeting the needs of the present without compromising the ability of future generations to meet their own needs" (UNBC, 1987). In addition, the definition of sustainability or sustainable development is defined in a number of ways in the literature. A list of these common definitions is illustrated in Table 1.1.

Defining 'energy sustainability' has not gained global acceptance in the literature. Energy sustainability is concerned with the availability of reliable, adequate, and affordable energy in conformity with economic, environmental, and social needs. In general, any form of energy that is able to meet present demands without placing any threats to depletion of its sources could be a sustainable energy source. That is, a sustainable energy source would be providing energy in an environmentally friendly, socially acceptable, and most economic means.

Predominantly, due to negative climate change, use of sustainable energy is crucial for every nation. There are many important benefits of using sustainable energy, including:

- Sustainable energy would eventually eliminate dependence on fossil fuels, which is responsible for huge global greenhouse gas (GHG) emissions.

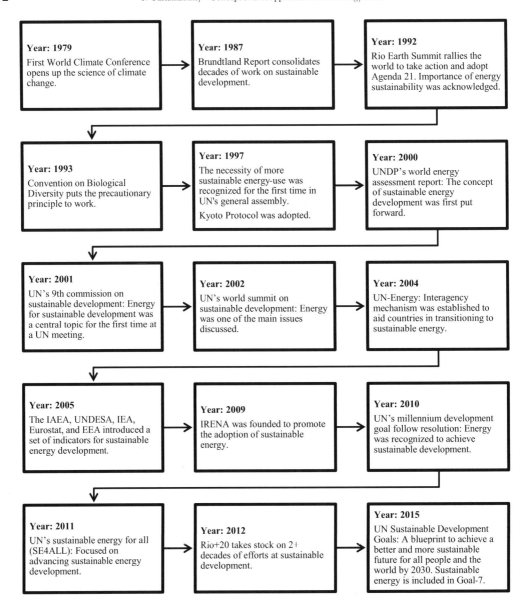

FIGURE 1.1 History of "sustainable energy development" concept adoption by the world community (Data source: UAlberta, 2012; Gunnarsdottir et al., 2021).

- Sustainable energy sources will never be depleted as they are produced naturally.
- During energy production from sustainable energy sources, almost zero GHGs are emitted, thus aiding the environment and public health.
- As the cost of sustainable energy production is decreasing rapidly, it will eventually be a more economic option for the global population.

TABLE 1.1 Common definitions of sustainability or sustainable development in the literature.

Definition	Source
Development that meets the needs of the present without compromising the ability of future generations to meet their own needs.	UNBC (1987)
The creation of a social and economic system that guarantees support for the following aims: increase in the real income, the improvement of the level of education, and the improvement in the populations' health and in the general quality of life.	Pearce et al. (1989)
The improvement in the population's quality of life while taking into consideration the ecosystem's regenerating capacity.	Catton (1986); Ciegis et al. (2009)
The process of increasing the spectrum of alternatives allowing individuals and communities to realize their aspirations and potential in the long perspective, at the same time maintaining the regeneration ability in economic, social, and ecological systems.	Munasinghe (1993)
Sustainable development is the society's development that creates the possibility for achieving overall wellbeing for the present and the future generations through combining environmental, economic, and social aims of the society without exceeding the allowable limits of the effect on the environment.	Ciegis et al. (2009)
Sustainable development, sustainable growth, and sustainable use have been used interchangeably, as if their meanings were the same. They are not. Sustainable growth is a contradiction in terms: nothing physical can grow indefinitely. Sustainable use is only applicable to renewable resources. Sustainable development is used in this strategy to mean: improving the quality of human life whilst living within the carrying capacity of the ecosystems.	IUCN, UNEP and WWF (1991)
Development is about realizing resource potential, Sustainable development of renewable natural resources implies respecting limits to the development process, even though these limits are adjustable by technology. The sustainability of technology may be judged by whether it increases production, but retains it other environmental and other limits	Holdgate and Synge (1993)
Sustainable development is concerned with the development of a society where the costs of development are not transferred to future generations, or at least an attempt is made to compensate for such costs.	Pearce (1993)
"Sustainability" should be interpreted purely as a technical characteristic of any project, program or development path, not as implying any moral injunction or over-riding criterion of choice	Beckerman (1994)
Sustainable development refers to making progress toward an economic system that uses natural resources in ways that do not deplete their capital or otherwise compromise their availability to future generations of people. In this sense, the present human economy is obviously non-sustainable because it involves rapid economic growth that is achieved by vigorously mining both non-renewable and potentially renewable resources.	Freedman (2018)
Sustainability is the process of living within the limits of available physical, natural and social resources in ways that allow the living systems in which humans are embedded to thrive in perpetuity.	UAlberta (2012)

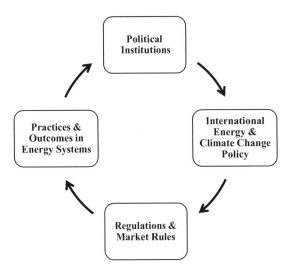

FIGURE 1.2 Interconnection for sustainable energy transition governing process.

Both national and global energy systems, in particular electricity generation systems, require a transition toward sustainable energy. In governing this sustainable transition, a number of interconnected processes should be followed (Kuzemko et al., 2016). This is illustrated in Fig. 1.2. For instance, political institutions must be involved in this process as they are the policymaking authorities for national energy policy development. To pay particular attention to sustainable energy sector development, international energy and climate change policy should be involved in the process. In accordance with international rules and regulations in relation to sustainable energy sector transition, related national energy rules, and regulations should be revised. The revised regulations must be applied to the energy systems, whose practices and outcomes should be monitored. However, if this does not work efficiently, political institutions should take related measures.

The aim of this chapter is threefold: first, to introduce the concept of sustainability, its history of development and definition. Second, to highlight the many different multi-criteria decision analysis/making (MCDA/M) methods of sustainability assessment. Third, to apply sustainability assessment for the renewable and fossil fueled electricity generation technologies, along with a case study.

1.2 The concept of sustainability

Sustainability is a multidisciplinary concept, referring to a system that maintains an effective balance between three criteria: economy, environment, and society. Of these criteria, the environment has limited resources and is able to deal with certain negative impacts from its surroundings; society must have a role in meeting present demands and ensuring that future generations' needs can be fulfilled; the economy is "a social domain that emphasizes the practices, discourses and material expressions associated with the production, use and management of resources" (James et al., 2015; Schönsleben et al., 2010). A system that has components from economy and environment but none or a little involvement from society

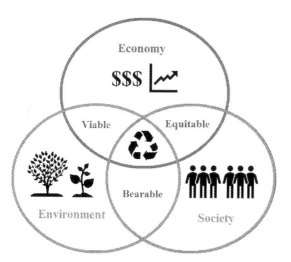

FIGURE 1.3 Sustainability concept. The common place (the recycling symbol in the figure) where all the three circles overlapped with each other represents a sustainable system.

makes the system a "viable" one, whereas dominant components from environment and society would make the system a "bearable" one. On the other hand, an "equitable" system might be developed with predominant components from economy and society. A sustainable system is only achievable if and only if an equal share of these three aspects can be confirmed, as shown in Fig. 1.3.

The majority of studies in the literature used these three criteria of sustainability for sustainability assessment of any energy systems (Khan, 2020c; Khan and Kabir, 2020; Khan, 2019; May and Brennan, 2006). However, numerous studies also consider other criteria along with these three for sustainability assessment of energy system. For example, an early study took into account four criteria for the sustainability ranking of four renewable energy generation technologies, including economic factors such as price, environmental (e.g., emissions, land use, water consumption), social and technical criteria (e.g., efficiency) (Evans et al., 2009). Similarly, a renewable energy ranking was conducted for the Algerian electricity system using sociopolitical criteria along with environmental and economic ones (Haddad et al., 2017). Klein and Whalley (2015) considered technical criterion in addition to economic, environmental, and social criteria for a sustainability comparison of US electricity options.

In summary, the three most common criteria for sustainability assessment of any energy system that were found in the literature are environment, economy, and society. However, the other most frequently found factor, which might be merged with the economic criterion as indicated in (Khan, 2019; 2020c) was technical criterion.

1.3 Sustainability assessment methods

There are several types of impact assessment procedures for a system and they can be conducted with many different levels, targets, and timing. For instance, the type of impact assessment could be for social, environmental, economic purposes, or integrated. Sustainability

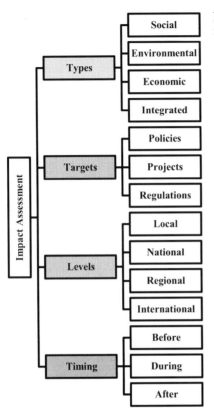

FIGURE 1.4 Impact assessment types, its targets, levels, and timing.

assessment falls under the integrated type, as it considers three different pillars, that is, society, economy, and environment. The targets for impact assessment might be set for policymaking, regulations, or for any specific project. In respect of level, it might be set for local, regional, national, or international. For example, in Khan (2020c), the author assessed the sustainability at regional (south Asia growth quadrangle) level. On the other hand, the sustainability of the power generation expansion plan was conducted at the national level for Bangladesh in Khan (2019). The timing for the impact assessment could be before, during or after the project or event (Khan, 2020a). However, these are general types, targets, levels, and times for impact assessments could differ, based on application such as in construction sector, energy sector. All these are summarized in Fig. 1.4.

Sustainable impact assessment must include long-term investment, effects, and impacts; an equal balance between social, economic, and environmental factors; and ensuring trade-offs between the three domains (social, economic, and environmental). There are a few basic steps that need to be followed for sustainability assessment, depicted in Fig. 1.5.

Before conducting sustainability assessment of a system, it is necessary to identify its relevance to sustainability assessment, that is, whether sustainability assessment is necessary for that system or not. If it is found that there is a true need for assessment, its scope must be

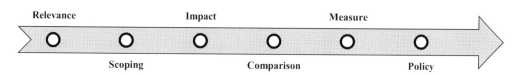

FIGURE 1.5 Basic steps for sustainability assessment.

identified. For instance, what would be the procedure, and to what extent it would be assessed, and which tool would be appropriate for this assessment. It must assess the economic, social, and environmental impacts of the system for both short- and long-term. A comparison should be made with the available similar systems and a trade-off must be proposed. If the assessed system is found not to be sustainable, what measures need to be conducted to make the system sustainable must be indicated. Based-on the obtained assessment results, related policy measures should be determined.

A suitable method or tool must be identified (in the scoping step, see Fig. 1.5) to conduct the sustainability assessment. The multi-criteria decision analysis/making (MCDA/M) is one very effective method used for the sustainability assessment for the energy systems. Numerous studies in the literature employed MCDA as the main method of sustainability assessment for the energy system, see for example, (Khan, 2020c, 2020b; Khan and Kabir, 2020; Khan, 2019; Lee and Chang, 2018; Siksnelyte et al., 2018; Ren and Dong, 2018; Shaaban et al., 2018; Atilgan and Azapagic, 2017a).

Some of the frequently employed methods for sustainability assessments in the literature are as follows:

- Weighted Sum Method (WSM)
- Weighted Product Method (WPM)
- Weighted Aggregated Sum Product Assessment (WASPAS)
- Analytic Hierarchy Process (AHP)
- Analytic Network Process (ANP)
- Technique for Order Preference by Similarity to Ideal Solution (TOPSIS)
- Preference Ranking Organization METHod for Enrichment of Evaluations (PROMETHEE)
- ELimination Et Coix Traduisant la REalite (ELECTRE)
- VlseKriterijumska Optimizacija I Kompromisno Resenje (VIKOR)
- COmplex PRoportional ASsessment (COPRAS)
- Additive Ratio Assessment (ARAS)
- Stepwise Weight Assessment Ratio Analysis (SWARA)

A comparative analysis between common MCDA methods can be found in Sahabuddin and Khan (2021). There are some challenges that should be met in applying any one of the MCDA methods. For instance, paying equal attention to each of the sustainability pillars (economic, environmental, and social) considering the longer-term impact, assessing qualitative and quantitative sustainability indicators at the same time, selecting appropriate sustainability indicators for the three pillars, and identifying the optimal solution among the three pillars of sustainability.

In summary, a few general steps that should be followed to conduct sustainability assessment for any system are:

- Identification of the target and its level, such as, for national policymaking, local project.
- To check the sustainability relevance for the targeted goal.
- Identifying proper sustainability indicators from the three pillars of sustainability.
- Selecting proper tools and methods of analysis.
- Assessing the impacts.
- Sorting the most, least, and optimal sustainable solutions from the alternatives.

Numerous studies in the literature have conducted sustainability assessment either for the electricity generation system or electricity generation technologies, often both. In these assessments, the three pillars of sustainability were predominantly followed. Often, they used different criteria along with the major three, that is, economic, social, and environmental criteria. For instance, in an earlier study, experts' designed experiments to evaluate the sustainability of ten renewable electricity generation technologies, and concentrated solar power was found to be the most sustainable one, followed by hydro (Dombi, Kuti and Balogh, 2014). The least sustainable generation technology was found to be biomass plant. In those assessments the authors considered seven attribute values of the technologies: GHG emission, land demand, other harmful ecological impacts, costs, levelized cost of electricity, new job creation, and impact on local income.

Maxim (2014) assessed the sustainability of different electricity generation technologies considering the following sustainability criteria: socio-political, economic, environmental, and technological. Large hydro followed by small hydro and onshore wind was found to be the most sustainable electricity generation technologies, whereas coal was the least sustainable technology.

Similarly, under four broad categories, out of 72 sustainability indicators available in the literature Shaaban and Scheffran (2017) identified 13 core indicators that might be used for sustainability assessment of the power generation sector in Egypt. All these indicators, along with others those are frequently found in the literature, are broadly categorized by four criteria, namely, social, environmental, economic, and technical, and are illustrated in Fig. 1.6. Of these four criteria, most often economic and technical criteria are merged together and termed economic criteria (shown in the dashed line in Fig. 1.6).

According to Iddrisu and Bhattacharyya (2015), economic criteria "evaluate whether the energy supply is cost effective and affordable," environmental criteria "aim to reduce the negative impact of energy use on society (and environment) and to extend the positive ones," social criteria "assesses the distributional effect of energy on the society," and technical criteria "is the supply side of the cycle that captures the ability of the energy supply system to meet the present and future needs of society reliably, efficiently and from clean sources. Technically, the supply system consists of the physical infrastructure that defines the configuration of the system and hence the expected output; and the resources inputs" (Iddrisu and Bhattacharyya, 2015).

Importantly, during indicator selection some general guidelines should be followed (*Reprinted from* Shortall and Davidsdottir (2017), *with permission from Elsevier*):

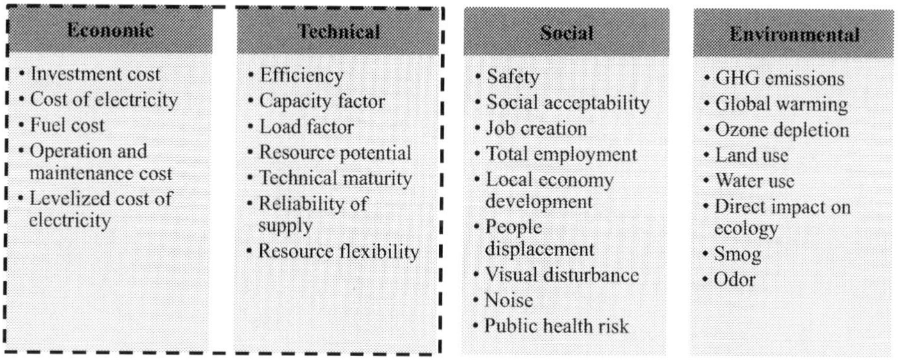

1.3 Sustainability assessment methods

Economic	Technical	Social	Environmental
• Investment cost • Cost of electricity • Fuel cost • Operation and maintenance cost • Levelized cost of electricity	• Efficiency • Capacity factor • Load factor • Resource potential • Technical maturity • Reliability of supply • Resource flexibility	• Safety • Social acceptability • Job creation • Total employment • Local economy development • People displacement • Visual disturbance • Noise • Public health risk	• GHG emissions • Global warming • Ozone depletion • Land use • Water use • Direct impact on ecology • Smog • Odor

FIGURE 1.6 Core sustainability indicators identified in the literature. Most often economic and technical criteria are merged together and termed as an economic criterion (shown in dashed line). *Sources: (Kabayo et al., 2019; Lee and Chang, 2018; Akber et al., 2017; Shaaban and Scheffran, 2017; Khan, 2019, 2020c; Santos et al., 2017; Bentsen et al., 2019; Yilan et al., 2020; Stamford and Azapagic, 2014; Aryanpur et al., 2019; Buchmayr et al., 2021; Garni et al., 2016; Roinioti and Koroneos, 2019; Atilgan and Azapagic, 2017b; Shortall and Davidsdottir, 2017).*

A good indicator must:

- Have policy relevance and utility for users;
- Match the interests of the target audience;
- Provide a representative picture of environmental conditions, pressures on the environment or society's responses;
- Be simple, accessible, easy to interpret and able to show trends over time;
- Invite action (reading further, investigate, ask questions, do something);
- Go with an explanation of causes behind the trends;
- Be comparable with other indicators that describe similar areas, sectors or activities;
- Be responsive to changes in the environment and related human activities;
- Provide a basis for international comparisons;
- Be either national in scope or applicable to regional environmental issues of national significance;
- Have a threshold or reference value against which to compare it so that users are able to assess the significance of the values associated with it;
- Be theoretically well founded in technical and scientific terms;
- Be based on international standards and international consensus about its validity;
- Lend itself to being linked to economic models, forecasting and information systems.

Furthermore, the data required to support the indicator should be:

- Readily available or made available at a reasonable cost/benefit ratio;
- Adequately documented and of known quality;
- Updated at regular intervals in accordance with reliable procedures.

TABLE 1.2 Sustainability indicators used for this analysis.

Economic	Environmental	Social
Efficiency (Eff.)	Greenhouse Gas Emission (GHGE)	Public Health Risk (PHR)
Capacity Factor (CF)	Land Use (LU)	New Job Creation (NJC)
Load Factor (LF)	Water Consumption (WC)	Local Economy Development (LED)
Levelized Cost of Electricity (LCOE)	Ecological Impact (EI)	Social Acceptance (SA)

1.4 Sustainability assessment: An application

To assess the sustainability of electricity generation system an example case is demonstrated in this section employing a MCDA method, TOPSIS. The generation technologies considered here are renewable (hydro, solar, geothermal, wind, and biomass), nonrenewable (oil, gas, and coal), and nuclear, that is, nine technologies. For indicator selection under the three pillars of sustainability, 12 core indicators are selected for this purpose as listed in Table 1.2. Although efficiency, capacity factor, and load factor can be categorized under technical criteria, in terms of the three basic pillars namely economic, environmental, and social, these indicators can be categorized in the economic category. There are a number of reasons for this. For instance, if the efficiency of any plant needs to be increased that involves technological improvement, this requires considerable funding. Hence, this falls into economic category. The same is true for the other two (capacity factor, load factor).

Out of the 12 indicators in Table 1.2, seven are quantitative and 5 are qualitative in nature. For instance, all the social indicators and the ecological impact in environmental criterion are qualitative. To interpret these qualitative indicators into quantitative values a scale of 1 to 5 was used from lowest (1) to very high (5). To assign values according to the scale, the opinions of seven experts were collected. The experts were electrical engineers in the electrical industries in Bangladesh such as Bangladesh Rural Electrification Board (BREB), West Zone Power Distribution Company Limited (WZPDCL), Northwest Power Generation Company Limited (NWPGCL), and Ashuganj Power Station Company Limited (APSCL) with 2 to 5 years of experience. According to the importance of the five indicators, the experts assigned weight to each of the indicators. From the survey, the average value was calculated for each of the qualitative indicator as listed in Table 1.3. Economic and environmental indicators' values were collected from existing literature and listed in Tables 1.4 and 1.5, respectively.

1.4.1 The TOPSIS method

The TOPSIS method measures the alternative's distance from the ideal best and worst values. The optimal alternative has the shortest distance from the best value and the longest distance from the worst value. This method has been employed to rank the electricity generation technologies by considering three sustainability criteria; each criterion is subdivided into several indicators as listed in Table 1.2. The steps are explained as follows (Sahabuddin and Khan, 2021):

TABLE 1.3 Qualitative indicators' value obtained from expert's opinion.

Generation technologies	Social indicators*				Environmental indicator*
	PHR	NJC	LED	SA	EI
Oil	3.29	2.57	2.43	2.00	3.43
Gas	1.86	3.29	3.43	3.29	2.00
Coal	4.57	3.50	3.43	1.29	4.43
Nuclear	4.15	4.00	4.00	3.72	2.00
Hydro	1.00	3.00	2.86	4.72	2.00
Solar	1.00	2.29	2.43	4.43	1.14
Geothermal	1.29	2.72	2.00	3.14	2.00
Wind	1.00	1.57	1.86	4.00	1.14
Biomass	1.29	2.00	2.14	3.86	1.14

*For these indicators, see Table 1.2.

TABLE 1.4 Economic indicators (quantitative) and their values.

Generation technologies	Economic indicators*				References
	Eff. (%)	CF (%)	LF (%)	LCOE (USD/kWh)	
Oil	39.37	13	36.42	0.054	(Khan, 2020c; Neill and Hashemi, 2018; BPDB, 2020; BHE, 2010; Richter, 2012)
Gas	35.39	27.9	43	0.048	
Coal	40	58.4	53.5	0.042	
Nuclear	33	73.8	62.9	0.043	
Hydro	90	31.7	37.5	0.051	
Solar	12	10.2	11.2	0.242	
Geothermal	35	73	95	0.064	
Wind	38	32.3	33.5	0.056	
Biomass	30	58	62	0.054	

*For these indicators, see Table 1.2.

(i) Formation of decision matrix

In this step, using Eq. (1.1), a decision matrix is formed and is shown in Table 1.6.

$$X = (x_{ij})_{m \times n} \tag{1.1}$$

Where:

x_{ij}: value of each indicator under specific criterion
i: number of row

TABLE 1.5 Environmental indicators (quantitative) and their values.

Generation technologies	Environmental indicators*			References
	GHGE (gCO₂-e/kWh)	LU (Sq.km/TWh)	WC (L/MWhr)	
Oil	705	44.7	7	(Khan, 2020c; Schlömer et al., 2014; Larsen and Drews, 2019; Richter, 2021)
Gas	400	18.6	12	
Coal	900	9.7	12	
Nuclear	12	2.4	11	
Hydro	24	54	3	
Solar	41	36.9	6	
Geothermal	38	7.5	7	
Wind	11	72.1	7	
Biomass	230	543	7	

*For these indicators, see Table 1.2.

j: number of column
m: number of alternatives
n: number of criteria

(ii) *Weighted normalized decision matrix*

The normalized decision matrix is obtained using Eq. (1.2) and values are listed in Table 1.7.

$$A_{ij} = \frac{x_{ij}}{\sqrt{\sum_{i=1}^{m} x_{ij}^2}} \tag{1.2}$$

The weighted normalized decision matrix is calculated using Eq. (1.3) and values are listed in Table 1.8.

$$D_{ij} = A_{ij} \times w_j, \quad i = 1, 2, \ldots, m; \quad j = 1, 2, \ldots, n \tag{1.3}$$

Where, w_j = weight of the jth criterion. Here, equal weights are considered for the indicators.

(iii) *Determination of ideal best and ideal worst*

$$\text{For beneficial indicator: ideal best, } p_j = max_j D_{ij} \tag{1.4}$$

$$\text{For non-beneficial indicator: ideal best, } p_j = min_j D_{ij} \tag{1.5}$$

$$\text{For beneficial indicator: ideal worst, } q_j = min_j D_{ij} \tag{1.6}$$

$$\text{For non-beneficial indicator: ideal worst, } q_j = max_j D_{ij} \tag{1.7}$$

These calculated ideal best and ideal worst values are listed in Table 1.9.

TABLE 1.6 Decision matrix.

Generation technologies	Economic criterion			Environmental criterion				Social criterion			
	Eff. (%)	CF (%)	LF (%)	LCOE (USD/kWh)	GHGE (gCO$_2$-e/kWh)	LU (Sq.km/TWh)	WC (L/MWhr)	EI	PHR	NJC	LED	SA
Oil	39.37	13.00	36.42	0.054	705	44.7	7	3.43	3.29	2.57	2.43	2.00
Gas	35.39	27.90	43.00	0.048	400	18.6	12	2.00	1.86	3.29	3.43	3.29
Coal	40.00	58.40	53.50	0.042	900	9.7	12	4.43	4.57	3.50	3.43	1.29
Nuclear	33.00	73.80	62.90	0.043	12	2.4	11	2.00	4.15	4.00	4.00	3.72
Hydro	90.00	31.70	37.50	0.051	24	54.0	3	2.00	1.00	3.00	2.86	4.72
Solar	12.00	10.20	11.20	0.242	41	36.9	6	1.14	1.00	2.29	2.43	4.43
Geothermal	35.00	73.00	95.00	0.064	38	7.5	7	2.00	1.29	2.72	2.00	3.14
Wind	38.00	32.30	33.50	0.056	11	72.1	7	1.14	1.00	1.57	1.86	4.00
Biomass	30.00	58.00	62.00	0.054	230	543.0	7	1.14	1.29	2.00	2.14	3.86

*For the indicators, see Table 1.2.

TABLE 1.7 Normalized decision matrix.

Technologies	Eff.	CF	LF	LCOE	GHGE	LU	WC	EI	PHR	NJC	LED	SA
Oil	0.2993	0.0905	0.2280	0.1907	0.5711	0.0807	0.2746	0.4789	0.4293	0.2992	0.2873	0.1882
Gas	0.2690	0.1942	0.2692	0.1696	0.3240	0.0336	0.4707	0.2793	0.2427	0.3830	0.4055	0.3095
Coal	0.3041	0.4064	0.3349	0.1484	0.7291	0.0175	0.4707	0.6186	0.5964	0.4075	0.4055	0.1214
Nuclear	0.2509	0.5136	0.3938	0.1519	0.0097	0.0043	0.4315	0.2793	0.5416	0.4657	0.4729	0.3500
Hydro	0.6842	0.2206	0.2348	0.1801	0.0194	0.0975	0.1177	0.2793	0.1305	0.3492	0.3381	0.4440
Solar	0.0912	0.0710	0.0701	0.8548	0.0332	0.0666	0.2353	0.1592	0.1305	0.2666	0.2873	0.4168
Geothermal	0.2661	0.5080	0.5947	0.2261	0.0308	0.0135	0.2746	0.2793	0.1683	0.3167	0.2364	0.2954
Wind	0.2889	0.2248	0.2097	0.1978	0.0089	0.1302	0.2746	0.1592	0.1305	0.1828	0.2199	0.3763
Biomass	0.2281	0.4036	0.3881	0.1907	0.1863	0.9803	0.2746	0.1592	0.1683	0.2328	0.2530	0.3631

TABLE 1.8 Weighted normalized decision matrix.

Technologies	Eff.	CF	LF	LCOE	GHGE	LU	WC	EI	PHR	NJC	LED	SA
Oil	0.0748	0.0226	0.0570	0.0477	0.1428	0.0202	0.0686	0.1197	0.1073	0.0748	0.0718	0.0470
Gas	0.0673	0.0485	0.0673	0.0424	0.0810	0.0084	0.1177	0.0698	0.0607	0.0958	0.1014	0.0774
Coal	0.0760	0.1016	0.0837	0.0371	0.1823	0.0044	0.1177	0.1546	0.1491	0.1019	0.1014	0.0303
Nuclear	0.0627	0.1284	0.0984	0.0380	0.0024	0.0011	0.1079	0.0698	0.1354	0.1164	0.1182	0.0875
Hydro	0.1710	0.0551	0.0587	0.0450	0.0049	0.0244	0.0294	0.0698	0.0326	0.0873	0.0845	0.1110
Solar	0.0228	0.0177	0.0175	0.2137	0.0083	0.0167	0.0588	0.0398	0.0326	0.0666	0.0718	0.1042
Geothermal	0.0665	0.1270	0.1487	0.0565	0.0077	0.0034	0.0686	0.0698	0.0421	0.0792	0.0591	0.0739
Wind	0.0722	0.0562	0.0524	0.0495	0.0022	0.0325	0.0686	0.0398	0.0326	0.0457	0.0550	0.0941
Biomass	0.0570	0.1009	0.0970	0.0477	0.0466	0.2451	0.0686	0.0398	0.0421	0.0582	0.0632	0.0908

TABLE 1.9 Ideal best (IB) and ideal worst (IW) values.

Type	Eff.	CF	LF	LCOE	GHGE	LU	WC	EI	PHR	NJC	LED	SA
I.B.	0.1710	0.1284	0.1487	0.0371	0.0022	0.0011	0.0294	0.0398	0.0326	0.0457	0.1182	0.1110
I.W.	0.0228	0.0177	0.0175	0.2137	0.1823	0.2451	0.1177	0.1546	0.1491	0.1164	0.0550	0.0303

(iv) *Euclidian distance from ideal best and worst*

$$\text{Euclidian distance from ideal best}, d_i^+ = \sqrt{\sum_{j=1}^{n} (D_{ij} - p_j)^2} \quad (1.8)$$

$$\text{Euclidian distance from ideal worst}, d_i^- = \sqrt{\sum_{j=1}^{n} (D_{ij} - q_j)^2} \quad (1.9)$$

1.4 Sustainability assessment: An application

TABLE 1.10 Euclidean distance (ED) from ideal best (IB) and ideal worst (IW), and performance score (PS).

Technologies	Economic criterion			Environmental criterion			Social criterion		
	EDIB	EDIW	PS	EDIB	EDIW	PS	EDIB	EDIW	PS
Oil	0.1702	0.1785	0.5119	0.1675	0.2361	0.5851	0.1126	0.0636	0.3608
Gas	0.1543	0.1864	0.5472	0.1223	0.2710	0.6891	0.0686	0.1123	0.6207
Coal	0.1182	0.2132	0.6433	0.2311	0.2407	0.5102	0.1533	0.0486	0.2408
Nuclear	0.1194	0.2264	0.6547	0.0840	0.3149	0.7894	0.1269	0.0863	0.4048
Hydro	0.1163	0.2313	0.6655	0.0381	0.3085	0.8901	0.0535	0.1476	0.7338
Solar	0.2874	0.0000	0.0000	0.0338	0.3148	0.9029	0.0514	0.1476	0.7418
Geothermal	0.1063	0.2361	0.6895	0.0497	0.3138	0.8632	0.0780	0.1214	0.6089
Wind	0.1562	0.1792	0.5343	0.0503	0.3052	0.8586	0.0655	0.1504	0.6968
Biomass	0.1286	0.2637	0.6722	0.2511	0.1844	0.4235	0.0606	0.1362	0.6920

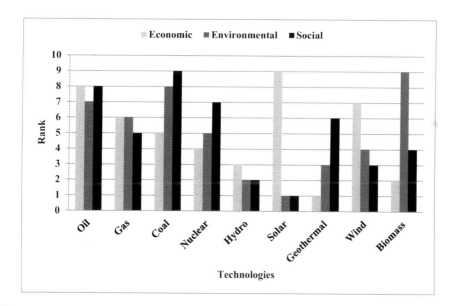

FIGURE 1.7 Ranking of technologies.

(v) *Determination of performance score and ranking*

$$\text{Performance score}, p_i = \frac{d_i^-}{d_i^+ + d_i^-} \quad (1.10)$$

The alternatives having the highest performance score is given first rank and successively the others. All these values are listed in Table 1.10. The ranks are depicted in Fig. 1.7.

1.4.2 Result and analysis

The application of the TOPSIS method to assess the sustainability of nine most common electricity generation technologies show that hydro is the most sustainable compared with any other technologies. Although solar achieved the best rank for environmental and social criteria, the economic indicators are not in favor of sustainable development yet. The next sustainable technology was found to be geothermal, followed by wind. On the other hand, coal was found to be the least sustainable technology, followed by oil.

In terms of each criterion, the most economical technology was found to be geothermal followed by biomass and hydro. Contrarily, the least economic technology was solar followed by oil. In environmental criterion, solar occupied a more environmentally friendly position, followed by hydro and geothermal. Instead of coal, biomass was found to be the least environment friendly electricity generation technology. In relation to the social criterion, solar and hydro were the first and second socially accepted technologies, respectively, whereas, the least socially accepted technology was coal, followed by oil.

Importantly, the results obtained might vary depending on the selection of the indicators under different criteria. It is crucial to select the proper and adequate number of indicators for sustainability assessment. Therefore, it is recommended to follow an acceptable framework or method for indicator selection. For example, "what-why/how-whom" is one such framework that helps to select proper sustainability indicators under different sustainability criteria (Khan, 2021). This framework predominantly requires few answers while selecting the indicators (Nathan and Reddy, 2011). First, "What is the main objective of the sustainability assessment" and "What are the indicators that are closely associated with the objective." Second, "Why the selected indicators are important" and "Are these indicators sufficient to serve the purpose." Are there any alternatives for the selected indicators? Finally, who are the individuals for whom these indicators are selected; for instance, are they technical personnel or policymakers?

Sensitivity analysis of the obtained results is another critical parameter for sustainability assessment of any system. Sensitivity analysis reveals the robustness of the applied MCDA method. The primary goal of sensitivity analysis is to check the influence of input factors changes on the final result of the MCDA analysis (Sahabuddin and Khan, 2021). The sensitivity analysis underpins the decision-makers to understand the uncertainties, advantages, limitations, and complete scope of the model. In particular, it ensures detailed insight into the problem and the solution obtained from the model. Importantly, the decision makers could realize the sensitivity of the optimum solution by observing changes in the input parameters and related consequences at the outputs.

In the MCDA method, the sensitivity analysis could be conducted either by changing the values of input indicators or assigned weights of the indicators. Most of the previous studies conducted the sensitivity analysis by changing the criteria weights, see for example Štreimikienė et al. (2016), Ribeiro et al. (2013), Chatzimouratidis and Pilavachi (2009). On the other hand, changing the values of input parameters was also used for sensitivity analysis in the literature, such as Häyhä et al. (2011), Strantzali (2017).

For this analysis, both weight and values were changed to conduct the sensitivity analysis. For the former, 25%, 50%, and 100% more weights were assigned for one of the environmental parameters, that is, greenhouse gas emission. On the other hand, for value changes, minimum and maximum values of efficiency for the relevant technologies were used and this was an

TABLE 1.11 Efficiencies of different technologies used for the sensitivity analysis.

Generation technologies	Efficiency (in %)		
	Base value	Minimum	Maximum
Oil	39.37	35	42
Gas	35.39	33	55
Coal	40	38	48
Nuclear	33	30	38
Hydro	90	85	92
Solar	12	10	20
Geothermal	35	15	37
Wind	38	30	45
Biomass	30	27	32

economic criterion. In both cases, the results were compared with the base case. The minimum and maximum values for the efficiencies that were used for sensitivity analysis are shown in Table 1.11.

Note that the changes were only conducted for the GHG emission factor, and thus the result was only observed for environmental criteria. Similarly, the result of sensitivity analysis for value changes was only observed for economic criteria as the value of efficiencies was only changed.

The sensitivity analysis revealed that a drastic rank change was only observed for the value change of input parameters, for instance, the rank of hydro and geothermal. For hydro, the rank change was about 66% for both minimum and maximum values (c.f., Fig. 1.8(a)). Similarly, for the minimum efficiency of the geothermal technology, the rank downgraded. The economic rank of four technologies was changed when minimum values were used. Whereas, three technologies' rank were changed when the maximum values were used.

In terms of weight change, no rank changes were observed at the output for 25% increase in weight. When 50% weight was changed, two technologies, namely coal and biomass, the environmental ranks were changed from the base case. Out of nine technologies, the environmental ranks of five were changed when 100% weight was increased (c.f., Fig. 1.8(b)).

1.5 Conclusion

Globally, the energy sector, particularly electricity generation is responsible for a huge amount of emission of GHGs, which contributes to negative climate change. Sustainability is a multidisciplinary concept that receives much attention nowadays due to this climate change. In this chapter, the concept of sustainability and its definitions were explored along with its history of development. The methods that are frequently used in the literature to assess the

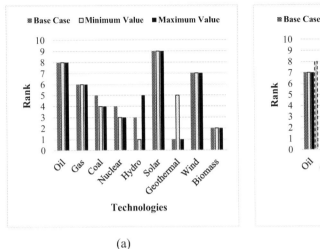
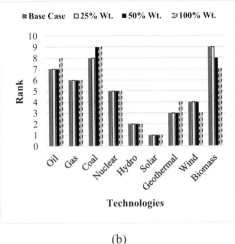

FIGURE 1.8 Results of sensitivity analysis for input parameters' (a) value change (for efficiency), and (b) weight (Wt.) change (for GHG emission).

sustainability of electricity generation systems are highlighted. The parameters in selecting effective indicators for different sustainability criteria are also discussed. As an example of how to apply a sustainability assessment method, the TOPSIS method was employed to assess the sustainability of nine electricity generation technologies and the results obtained were explained.

Self-evaluation questions

1. What do you mean by sustainability and sustainable development?
2. Define energy sustainability.
3. Explain the concept of sustainability.
4. Explain the basic steps for sustainability assessment.
5. What are the characteristics that must be satisfied to be an appropriate indicator under any criterion of the sustainability?
6. Conduct a literature survey and apply a multicriteria decision analysis method (other than TOPSIS) to assess the sustainability of the common renewable and nonrenewable electricity generation technologies. You can use the indicators used in this chapter.

References

Akber, M.Z., Thaheem, M.J., Arshad, H., 2017. Life cycle sustainability assessment of electricity generation in Pakistan: policy regime for a sustainable energy mix. Energy Policy 111, 111–126. doi:10.1016/j.enpol.2017.09.022.

Aryanpur, V., et al., 2019. An overview of energy planning in Iran and transition pathways towards sustainable electricity supply sector. Renew. Sustain. Energy Rev. 112, 58–74. doi:10.1016/j.rser.2019.05.047.

Atilgan, B., Azapagic, A., 2017a. Energy challenges for Turkey: identifying sustainable options for future electricity generation up to 2050. Sustain. Product. Consum. 12, 234–254. doi:10.1016/j.spc.2017.02.001.

References

Atilgan, B., Azapagic, A., 2017b. Energy challenges for Turkey: identifying sustainable options for future electricity generation up to 2050. Sustain. Product. Consum. 12, 234–254. doi:10.1016/j.spc.2017.02.001.

Beckerman, W., 1994. "Sustainable development": is it a useful concept? Environ. Values 3 (3), 191–209. Available at http://www.environmentandsociety.org/node/5516.

Bentsen, N.S., et al., 2019. Dynamic sustainability assessment of heat and electricity production based on agricultural crop residues in Denmark. J. Cleaner Prod. 213, 491–507. doi:10.1016/j.jclepro.2018.12.194.

BHE, 2010. The Efficiency of Power Plants of Different Types, Bright Hub Engineering. Available at: https://www.brighthubengineering.com/power-plants/72369-compare-the-efficiency-of-different-power-plants/. (Accessed: 16 May 2021).

BPDB, 2020. Annual Report 2019-2020. Dhaka, Bangladesh. Available at: https://www.bpdb.gov.bd/bpdb_new/index.php/site/new_annual_reports. (Accessed: 16 April 2021).

Buchmayr, A., et al., 2021. The path to sustainable energy supply systems: proposal of an integrative sustainability assessment framework. Renew. Sustain. Energy Rev. 138, 110666. doi:10.1016/j.rser.2020.110666.

Catton, W.R., 1986. Carrying Capacity and the Limits to Freedom. In: XI World Congress of Sociology. New Delhi doi:10.4236/sm.2012.21004.

Chatzimouratidis, A.I., Pilavachi, P.A., 2009. Technological, economic and sustainability evaluation of power plants using the Analytic Hierarchy Process. Energy Policy 37, 778–787. doi:10.1016/j.enpol.2008.11.021.

Ciegis, R., Ramanauskiene, J., Martinkus, B., 2009. The concept of sustainable development and its use for sustainability scenarios. Eng Econ 2, 28–37. Available at https://citeseerx.ist.psu.edu/viewdoc/download?doi=10.1.1.491.527&rep=rep1&type=pdf.

Dombi, M., Kuti, I., Balogh, P., 2014. Sustainability assessment of renewable power and heat generation technologies. Energy Policy 67, 264–271.

Evans, A., Strezov, V., Evans, T.J., 2009. Assessment of sustainability indicators for renewable energy technologies. Renewable Sustainable Energy Rev. 13, 1082–1088. doi:10.1016/j.rser.2008.03.008.

Freedman, B., 2018. Chapter 12: Resources and sustainable development. Environmental Science: A Canadian perspective, 6th ed. Dalhousie University Libraries, Halifax, Canada. Available at https://ecampusontario.pressbooks.pub/environmentalscience/chapter/chapter-12-resources-and-sustainable-development/.

Garni, H. Al et al., 2016. 'A multicriteria decision making approach for evaluating renewable power generation sources in Saudi Arabia', 16, pp. 137–150. doi: 10.1016/j.seta.2016.05.006.

Gunnarsdottir, I., et al., 2021. Sustainable energy development: history of the concept and emerging themes. Renewable Sustainable Energy Rev. 141 (110770), 1–17. doi:10.1016/j.rser.2021.110770.

Haddad, B., Liazid, A., Ferreira, P., 2017. A multi-criteria approach to rank renewables for the Algerian electricity system. Renewable Energy 107, 462–472. doi:10.1016/j.renene.2017.01.035.

Häyhä, T., Franzese, P.P., Ulgiati, S., 2011. Economic and environmental performance of electricity production in Finland: a multicriteria assessment framework. Ecol. Modell. 223, 81–90. doi:10.1016/j.ecolmodel.2011.10.013.

Holdgate, M.W., Synge, H., 1993. The future of IUCN. The World Conservation Union: Proceedings of a Symposium Held on the Occasion of the Inauguration of the New IUCN Headquarters. IUCN, Gland, Switzerland, p. 166. Available at: https://books.google.com.bd/books?id=d15kmZIO6VYC&source=gbs_navlinks_s&redir_esc=y.

Iddrisu, I., Bhattacharyya, S.C., 2015. Sustainable energy development index: a multi-dimensional indicator for measuring sustainable energy development. Renew. Sustain. Energy Rev. 50, 513–530. doi:10.1016/j.rser.2015.05.032.

IUCN, UNEP and WWF, 1991. Caring for the Earth. A strategy for Sustainable Living, 1st ed.1st edn. IUCN UNEP WWF, Gland, Switzerland Mining Survey. Available at: https://portals.iucn.org/library/efiles/documents/cfe-003.pdf.

James, P. et al., 2015. Urban Sustainability in Theory and Practice: Circles of sustainability. 1st edn. Edited by P. James et al. New York: Routledge. Available at: https://www.routledge.com/Urban-Sustainability-in-Theory-and-Practice-Circles-of-sustainability/James/p/book/9781138025738.

Kabayo, J., et al., 2019. Life-cycle sustainability assessment of key electricity generation systems in Portugal. Energy 176, 131–142. doi:10.1016/j.energy.2019.03.166.

Khan, I., 2019. Power generation expansion plan and sustainability in a developing country: a multi-criteria decision analysis. J. Cleaner Prod. 220, 707–720. doi:10.1016/J.JCLEPRO.2019.02.161.

Khan, I., 2020a. Critiquing social impact assessments: ornamentation or reality in the Bangladeshi electricity infrastructure sector? Energy Res. Soc. Sci. 60 (101339), 1–8. doi:10.1016/j.erss.2019.101339.

Khan, I., 2020b. Data and method for assessing the sustainability of electricity generation sectors in the south Asia growth quadrangle. Data Brief 28 (104808), 1–8. doi:10.1016/j.dib.2019.104808.

Khan, I., 2020c. Sustainability challenges for the south Asia growth quadrangle: a regional electricity generation sustainability assessment. J. Cleaner Prod. 243 (118639), 1–13. doi:10.1016/j.jclepro.2019.118639.

Khan, I., 2021. Sustainability assessment of energy systems: Indicators, methods, and applications. In: Ren, J. (Ed.), Methods in Sustainability Science: Assessment, Prioritization, Improvement, Design and Optimization. Elsevier, Amsterdam, pp. 47–70. doi:10.1016/C2020-0-00430-5.

Khan, I., Kabir, Z., 2020. Waste-to-energy generation technologies and the developing economies: a multi-criteria analysis for sustainability assessment. Renewable Energy 150, 320–333. https://doi.org/10.1016/j.renene.2019.12.132.

Klein, S.J.W., Whalley, S., 2015. Comparing the sustainability of U.S. electricity options through multi-criteria decision analysis. Energy Policy 79, 127–149. doi:10.1016/j.enpol.2015.01.007.

Kuzemko, C., et al., 2016. Governing for sustainable energy system change: politics, contexts and contingency. Energy Res. Soc. Sci. 12, 96–105. doi:10.1016/j.erss.2015.12.022.

Larsen, M.A.D., Drews, M., 2019. Water use in electricity generation for water-energy nexus analyses: the European case. Sci. Total Environ. 651, 2044–2058. doi:10.1016/j.scitotenv.2018.10.045.

Lee, H.C., Chang, C.T, 2018. Comparative analysis of MCDM methods for ranking renewable energy sources in Taiwan. Renewable Sustainable Energy Rev. 92, 883–896. doi:10.1016/j.rser.2018.05.007.

Maxim, A., 2014. Sustainability assessment of electricity generation technologies using weighted multi-criteria decision analysis. Energy Policy 65, 284–297. doi:10.1016/j.enpol.2013.09.059.

May, J.R., Brennan, D.J., 2006. Sustainability assessment of Australian electricity generation. Process Saf. Environ. Prot. 84 (2), 131–142. doi:10.1205/psep.04265.

Mensah, J., 2019. Sustainable development: Meaning, history, principles, pillars, and implications for human action: literature review. Cogent Soc. Sci. 5 (1653531), 1–21. doi:10.1080/23311886.2019.1653531.

Munasinghe, M., 1993. Environmental economics and biodiversity management in developing countries. Ambio 22 (2/3), 126–135. Available at https://www.jstor.org/stable/4314057.

Nathan, H.S.K., Reddy, B.S., 2011. Criteria selection framework for sustainable development indicators. Int. J. Multicriteria Decis. Making 1 (3), 257–279. doi:10.1504/IJMCDM.2011.041189.

Neill, S.P., Hashemi, M.R., 2018. 'Introduction. In: Neill, S.P., Hashemi, M.R. (Eds.), Fundamentals of Ocean Renewable Energy. Academic Press, London, pp. 1–30. doi:10.1016/B978-0-12-810448-4.00001-X.

Pearce, D., Markandya, A., Barbier, E.B., 1989. Blueprint for a green economy. Blueprint for a Green Economy, 1st ed. Routledge, London. Available at: https://www.routledge.com/Blueprint-1-For-a-Green-Economy/Pearce-Markandya-Barbier/p/book/9781853830662#.

Edited by Pearce, D.W., 1993. Blueprint 3: measuring sustainable development. In: Pearce, D.W. (Ed.), Blueprint 3., 1st ed Earthscan. Edited by doi:10.4324/9781315070414.

Du Pisani, J.A., 2006. Sustainable development – historical roots of the concept. Environ. Sci. 3 (2), 83–96. doi:10.1080/15693430600688831.

Ren, J., Dong, L., 2018. Evaluation of electricity supply sustainability and security: multi-criteria decision analysis approach. J. Cleaner Prod. 172, 438–453. doi:10.1016/j.jclepro.2017.10.167.

Ribeiro, F., Ferreira, P., Araújo, M., 2013. Evaluating future scenarios for the power generation sector using a Multi-Criteria Decision Analysis (MCDA) tool : the Portuguese case. Energy 52, 126 136. doi:10.1016/j.energy.2012.12.036.

Richter, A., 2012. Capacity factors of geothermal plants, a global analysis by Bloomberg New Energy Finance, Think Geoenergy. Available at: https://www.thinkgeoenergy.com/capacity-factors-of-geothermal-plants-a-global-analysis-by-bloomberg-new-energy-finance/. (Accessed: 16 May 2021).

Richter, A., 2021. Geothermal energy is least land-use intense source of the renewable energy technologies, Think Geoenergy. Available at: https://www.thinkgeoenergy.com/geothermal-energy-is-least-land-use-intense-source-of-the-renewable-energy-technologies /#:~:text=Geothermal energy is by far the least land-use,we again and again stumble across interesting reports. (Accessed: 16 May 2021).

Roinioti, A., Koroneos, C., 2019. Integrated life cycle sustainability assessment of the Greek interconnected electricity system. Sustain. Energy Technol. Assessments 32, 29–46. doi:10.1016/j.seta.2019.01.003.

Sahabuddin, M., Khan, I., 2021. Multi-criteria decision analysis methods for energy sector's sustainability assessment: robustness analysis through criteria weight change. Sustain. Energy Technol. Assess. 47 (101380), 1–12. doi:10.1016/j.seta.2021.101380.

Santos, M.J., et al., 2017. Scenarios for the future Brazilian power sector based on a multi-criteria assessment. J. Cleaner Prod. 167, 938–950. doi:10.1016/j.jclepro.2017.03.145.

Schlömer, S., et al., 2014. Annex III: Technology-specific cost and performance parameters. IPCC Climate Change 2014: Mitigation of Climate Change, Fifth Asse(Contribution of Working Group III), pp. 1329–1356. Available at: https://www.ipcc.ch/site/assets/uploads/2018/02/ipcc_wg3_ar5_annex-iii.pdf.

Schönsleben, P., et al., 2010. The changing concept of sustainability and economic opportunities for energy-intensive industries. CIRP Annals – Manuf. Technol. 59, 477–480. doi:10.1016/j.cirp.2010.03.121.

Shaaban, M., et al., 2018. Sustainability assessment of electricity generation technologies in Egypt using multi-criteria decision analysis. Energies 11 (1117), 1–25. doi:10.3390/en11051117.

Shaaban, M., Scheffran, J., 2017. Selection of sustainable development indicators for the assessment of electricity production in Egypt. Sustain. Energy Technol. Assess. 22, 65–73. doi:10.1016/j.seta.2017.07.003.

Shortall, R., Davidsdottir, B., 2017. How to measure national energy sustainability performance: an Icelandic case-study. Energy Sustain. Dev. 39, 29–47. doi:10.1016/j.esd.2017.03.005.

Siksnelyte, I., et al., 2018. An overview of multi-criteria decision-making methods in dealing with sustainable energy development issues. Energies 11 (2754), 1–21. doi:10.3390/en11102754.

Stamford, L., Azapagic, A., 2014. Life cycle sustainability assessment of UK electricity scenarios to 2070. Energy Sustain. Dev. 23, 194–211. doi:10.1016/j.esd.2014.09.008.

Strantzali, E., Aravossis, K., Livanos, G.A., 2017. Sustainable evaluation of future electricity generation alternatives: the case of a Greek island. Renewable Sustainable Energy Rev. 76 (April 2016), 1–28. doi:10.1016/j.rser.2017.03.085.

Štreimikienė, D., Šliogerienė, J., Turskis, Z., 2016. Multi-criteria analysis of electricity generation technologies in Lithuania. Renewable Energy 85, 148–156. doi:10.1016/j.renene.2015.06.032.

UAlberta, 2012. What is Sustainability? University of Alberta, p. 4. Available at https://www.mcgill.ca/sustainability/files/sustainability/what-is-sustainability.pdf.

UNBC, 1987. Report of the World Commission on Environment and Development: Our Common Future, United Nations Brundtland Commission. Available at: http://www.un-documents.net/our-common-future.pdf. (Accessed: 10 November 2019).

Yilan, G., Kadirgan, M.A.N., Çiftçioğlu, G.A., 2020. Analysis of electricity generation options for sustainable energy decision making: the case of Turkey. Renewable Energy 146, 519–529. doi:10.1016/j.renene.2019.06.164.

CHAPTER 2

Prospects of bioelectricity in south Asian developing countries—A sustainable solution for future electricity

Pobitra Halder[a,b], Ibrahim Gbolahan Hakeem[a], Savankumar Patel[a], Shaheen Shah[c], Hafijur Khan[d] and Kalpit Shah[a]

[a]School of Engineering, RMIT University, VIC, Australia [b]Department of Industrial and Production Engineering, Jashore University of Science and Technology, Jashore, Bangladesh [c]Department of Oil and Gas Engineering, Memorial University of Newfoundland, Newfoundland, Canada [d]University of Chinese Academy of Sciences, Beijing, China

2.1 Introduction

The global primary energy consumption is currently estimated about 162.19 petawatt hour (PWh), of which approximately 84% (a decrease from 86% in 2000) came from fossil fuels and 16% from low-carbon sources (sum of nuclear and renewables) (Fig. 2.1A) (Ritchie, 2020a). Fossil energy is considered nonrenewable on any relevant timescale, and its reserve is believed to be finite and rapidly depleting (Moioli et al., 2018). Nevertheless, fossil fuels (i.e., oil, gas, coal) remain the dominant source of energy use despite the global annual increase in renewables energy production. Apart from the fact that most of the energy comes from fossil sources, there is a continuous increase in the production levels as the last ten years witnessed about a 20% increase in the total fossil energy production (Ritchie, 2020a). Fig. 2.1B shows the trend in the average energy consumption per capita by regions in the world. Energy consumption is divided into three components such as transport, heating and electricity (Ritchie, 2020b). Asia is the second-lowest per capita primary energy consuming region,

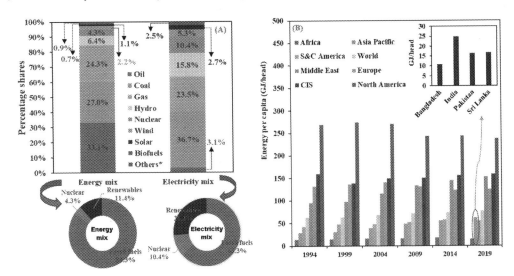

FIGURE 2.1 (A) Average primary energy consumption per capita by world regions. *Data source: BP Statistical Review of World Energy, 2020 (BP, 2020)* and (B) Global primary energy and electricity mixes by sources in 2019. *Data source: BP Statistical reviews of world energy, 2020 (BP, 2020) and Our World in Data (Smil, 2017; Ritchie, 2020a,b).* *Others include tidal, geothermal and biomass: NB: the estimation has not included traditional biomass, which may be an important energy source in rural settings.*

whereas north America consumes the highest amount of per capita primary energy in the world. Bangladesh is one of the lowest energy-consuming countries in the Asia Pacific region, with per capita primary energy consumption of around 10.8 GJ. India is the highest primary energy-consuming country in the south Asia with a per capita value of 24.9 GJ, followed by Pakistan and Sri Lanka. The inequities in per capita electricity consumption are even greater than that for primary energy consumption. Electricity consumption per capita varies more than 100-fold across the world against the 10-fold variation for per capita energy consumption with low-income developing countries at the end of the spectrum (Ritchie and Roser, 2020a).

In 2020, the global electricity production was about 25.87 PWh, which was slightly lower (25.9 PWh) than in 2019; however, higher by 24.5% than in 2010 (Ritchie, 2020b). Like primary energy, global electricity generation is also largely dominated by fossil energy sources (see Fig. 2.1A). In 2020, approximately 60.9% (nearly the same as 2019) of global electricity was produced from fossil fuels, 29.5% from renewable sources and the balance from nuclear source. However, the share of renewables in the electricity mix is higher than that for the primary energy mix (Fig. 2.1A). As of 2019, about 36.7% of the electricity mix is obtained from low-carbon sources, whereas only about 15.7% is the share of low-carbon sources for total energy. The other two energy components (transport and heating) rely more heavily on fossil fuels and will be difficult to decarbonize (Ritchie, 2020b). The availability of reliable electric power systems plays a foremost role in the functioning of modern and emerging economies in both developed and developing nations (Turdera and Garcia, 2018). Still, a large percentage of the world's population (ca. 2 billion) lack access to one or more form of basic energy services,

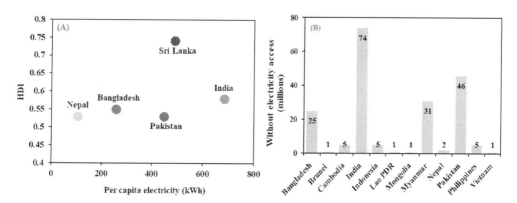

FIGURE 2.2 (A) Per capita electricity consumption vs human development index (HDI). *Data source: Center for Global Development (CGDEV, 2021)* and (B) electricity access in developing Asia. *Data source: Renewable Energy Policy Network (REN21, 2021a).*

including electricity, heating, cooking, and means of transportation (Ahuja and Tatsutani, 2009). As of 2016, about 13% of the world population (~950 million people) do not have access to electricity (Ritchie and Roser, 2020a). However, the global population without electricity access declined from about 1.4 billion in 2000 to 0.93 billion in 2019. Over 80% of this population are from least developed and developing countries, whereas the electricity required for basic living services is less than 1% of the overall global energy demand (Ahuja and Tatsutani, 2009). Therefore, developing countries face a two-level energy challenge in the current century; these are (i) meeting the inadequacy of access to basic and modern energy services, and (ii) participating in the global transition to clean, affordable and renewable energy systems (Ahuja and Tatsutani, 2009; Alvarez-Herranz et al., 2017). The human development index (HDI) is strongly linked to a country's electricity consumption. In 2013, Sri Lanka had the highest HDI (0.74) among the south Asian developing countries (see Fig. 2.2A); however, the value was below 0.8, benchmarked for most developed countries (CGDEV, 2021). The electricity access in developing Asia is still not reached to 100% (see Fig. 2.2B). Approximately, 74 million of the population in India and 25 million of the population in Bangladesh have no electricity access yet. However, the government of each country has set long-term plans to extend the electricity access. To this extent, the goal of tackling environmental pollution associated with fossil fuels with the pursuit of energy-related targets and the provision of other urgent human and societal needs has to be aligned to the development of indigenous and innovative renewable energy technologies (Vandaele and Porter, 2015).

Despite the slow transitioning from fossil energy to renewable energy, biomass remains the largest and the most versatile renewable source of energy in the world today (Roddy, 2012). Bioelectricity (electricity generated from biological sources) is one of the most significant areas in the renewable energy industry and can spark another revolution similar to biofuels (Altieri, 2012). Bioelectricity has great potential in carbon offsets and the economic utilization of enormous crop residues in many parts of the world, particularly in rural settings (Yang, 2011). The prospects of bioelectricity production in many developing countries, including south Asian countries, have been examined extensively in several studies. For instance, Bangladesh

TABLE 2.1 Installed capacity of renewable power technology in the world (IRENA, 2020; Ritchie and Roser, 2020b).

Technology	2010 Capacity (MW)	2019 Capacity (MW)	Relative change (%)
Bioenergy*	63,746	124,026	+94.6
Geothermal	9,992	13,909	+39.2
Hydro	880,827	1130,594	+28.4
Solar	40,279	578,553	+1336.4
Wind	177,790	594,253	+234.2

*Sum of solid biofuels, biogas, liquid biofuels and municipal waste.

has approximately 194 million tonnes of biomass resources, which can produce about 1345 petajoules (PJ) energy for electricity generation (Halder et al., 2015). India has about 131.5 million tonnes and 54.1 million tonnes of agricultural residues and forest surplus biomass, respectively, per year, which can be utilized for bioelectricity generation. Pakistan generates about 119.4 million tonnes of agricultural waste annually (Irfan et al., 2020). In China, available agricultural residues could achieve around 66.67% of the country's total bioelectricity target without compromising the current consumption pattern of the crops (Xiaoping et al., 2014). Therefore, green electricity generation from biomass has attracted significant attention to policymakers due to the huge availability and carbon neutrality of biomass resources.

This chapter discusses the current status and future prospects of bioelectricity generation in developing Asian countries. The chapter is structured into six sections; Section 2.1 discusses the energy challenges and reviews relevant previous literature, and Section 2.2 highlights the current and future energy mix in developing Asian countries. Section 2.3 presents the importance of bioelectricity generation in south Asian developing countries and its role in achieving a sustainable solution to future electricity security. The succeeding Section 2.4 provides insights into various bioelectricity generation technologies alongside their commercialization status and challenges. Section 2.5 evaluates the bioelectricity prospects of major bioenergy producing countries in south Asia, including Bangladesh, India and Pakistan, emphasizing their government bioelectricity mandates and roadmap. The final section provides conclusions and relevant recommendations for challenges identified.

2.2 Electricity generation mix in south Asian developing countries

Electricity consumption has been significantly increased over the last few years, particularly in developing countries. Electricity production in developing countries has been largely driven by coal combustion in traditional coal-powered plants; however, the beginning of the 21st century witnessed more gas usage in electricity production via steam turbines. Almost all developing countries have set their sustainable development goals and intending to shift their power consumption pattern more into renewable resources (see Table 2.1). Asia pacific consumed about 47% of global electricity in 2019. The total installed electricity generation capacity in Pakistan reached 35.6 GW in June 2020 (NEPRA, 2021). As of February 2021, the total

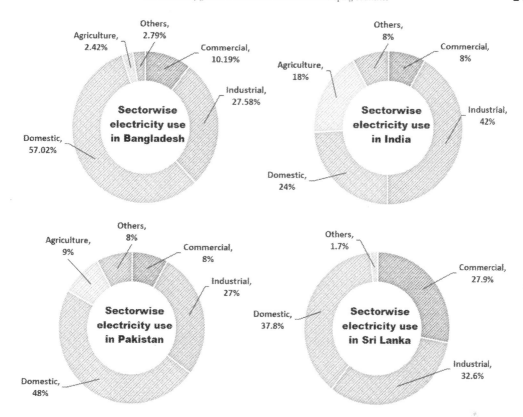

FIGURE 2.3 Sector-wise electricity consumption in some selected south Asian countries. *Data source: (Abbasi et al., 2020; BPDB, 2021; Statista, 2021; World Bank, 2021a).*

installed power generation capacity in India was 379.1 GW (MOP, 2021). Bangladesh increased the total installed power generation capacity to 25.2 GW, including captive power and off-grid renewable energy, as of April 2021 (BPDB, 2021). So far, Bangladesh served a maximum demand of approximately 13 GW. India is the largest electricity-producing and consuming country among the south Asian developing countries. The country produced approximately 1342 TWh of electricity in 2020, which was about 2.6% lower than in 2019 (Ritchie, 2020b). On the contrary, Bangladesh experienced a slight increase in electricity production from 2019 (~79.8 TWh) to 2020 (~80.4 TWh). Pakistan produced nearly the same amount of electricity (~145.4 TWh) in 2019 and 2020. Sri Lanka increased their electricity production by 6.6% in 2019 compared to the year 2018. Domestic utilization is the major sector for electricity consumption in Bangladesh, Pakistan and Sri Lanka, followed by the industrial sector (see Fig. 2.3). On the contrary, in India, the industrial sector consumed about 42% of total electricity, followed by the domestic sector (24%). In the fiscal year 2019, Bhutan had the highest per capita electricity consumption of 8664 kilowatt-hours (kWh), followed by Maldives (1173 kWh) and India (1009 kWh), whereas other south Asian countries are far lower than Bhutan (Ritchie, 2020b). In the fiscal year 2020, per capita electricity consumption in Bangladesh was 488 kWh, which

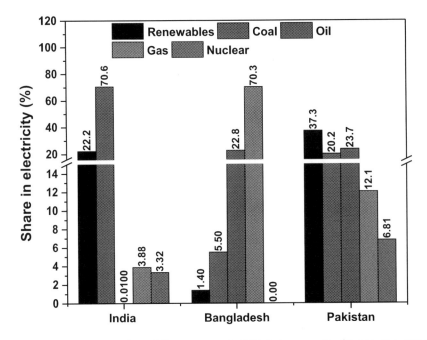

FIGURE 2.4 Electricity generation from different sources in 2020. *Data source: Our World in Data (Ritchie, 2020b).*

was the same as the previous year. However, per capita consumption for India and Pakistan decreased from 1009 to 972 kWh and from 671 to 658 kWh in 2020 compared to 2019 due to the COVID-19 outbreak, resulting in the slowdown in industrial consumption.

Electricity generation in south Asian developing countries, particularly India and Bangladesh, are still dependent on fossil fuels such as coal, gas and oil. In 2020, more than 70% of electricity in India was generated from coal, followed by 22.2% from renewables, including hydro, solar, wind, and bioelectricity (see Fig. 2.4). As of February 2021, the total installed renewable power capacity (including hydropower) in India was around 137.4 GW, with a per capita capacity of 0.15 kW, which was slightly higher than that of 2019. However, India has set a goal to achieve about 175 and 450 GW electricity generation from renewable resources by 2022 and 2030, respectively (REN21, 2021b). In Pakistan, renewables, including hydropower, shared the maximum percentage of electricity (37.3%), compared with individual fossil resources, which accounted for about 56% of the country's electricity generation. However, without including the hydropower generation, the share of renewables in power generation was only 4.7%. According to Alternative and Renewable Energies (ARE) policy 2019, Pakistan government has set a target to increase the share of renewables (excluding hydropower) to 20% and 30% of the country's national grid power by 2025 and 2030, respectively (NEPRA, 2021). The power generation in Bangladesh is dominated by gas (70.3%) and oil (22.8%); the share of renewables, including hydropower, was only about 1.4% of the country's total electricity generation in 2020. According to the renewable energy goal, Bangladesh had set a target to generate 10% electricity from renewables by 2021; however, the country is behind the target due to a number of factors. As of April 2021, Bangladesh produced about 723.97 MW of electricity from renewable resources, with another 543 MW under construction and 1416 MW in the planning stage.

2.3 Why bioelectricity?

Global primary energy and electricity demand are expected to increase considerably due to the rapid growth in population and the expansion of economic and industrial sectors in developing countries, including China, Bangladesh, and India. Energy production (i.e., primarily fossil forms of energy) and consumption (i.e., burning of fossil fuels) account for around three-quarters of the world greenhouse gas (GHG) emissions and is the largest driver of climate change (Ritchie and Roser, 2019). Also, other environmental issues such as pollution, resource depletion, flooding etc., can be linked to fossil energy production and use (GEA, 2012). Apart from the environmental issues, fossil fuels, including petroleum and gas, are associated with high cost and unstable supply. Despite the enormous attention devoted to increasing the shares of renewables in the global energy mix, there has not been a great shift towards low-carbon energy systems, implying that a great deal of concerted effort is needed to accelerate the transition. Although the traditional renewable power source is dominated by hydro, major development is now focusing on solar photovoltaic and wind power technologies; but these sources are generally nondispatchable due to their fluctuating nature (Kaygusuz, 2012; Ritchie, 2020b). Thus, integrating these renewable resources into a unified electricity grid can be challenging, particularly for large scale implementation (Ahuja and Tatsutani, 2009). Considering these issues, bioelectricity generation from waste biomass materials can play a significant role in obtaining a sustainable electricity mix.

Over the last few decades, the global CO_2 emission approaches 36.44 billion tonnes with huge negative impacts. Therefore, several nations are struggling to draw up effective policies to reduce CO_2 emission which accounts for more than 60% of the global greenhouse effect (Dogan and Seker, 2016; Turdera and Garcia, 2018). Asia Pacific region is the largest contributor to global CO_2 emission, accounting for about 49% (17.1 billion tonnes CO_2). However, the per capita CO_2 emission of this region is the second lowest (3.8 tonnes per person), followed by Africa (1 tonne per person). In 2019, India emitted the largest portion (~2.62 billion tonnes) of CO_2 among the south Asian countries, followed by Pakistan (0.249 billion tonnes) and Bangladesh (0.102 billion tonnes), (see Fig. 2.5). However, Maldives has the highest per capita emission (~3.14 tonnes), despite their lowest total CO_2 emission, followed by Bhutan (~2.237 tonnes). In India, coal combustion is the major cause of CO_2 emission, followed by oil and gas. In Bangladesh, the use of natural gas produces the highest amount of CO_2, whereas oil is the main source of CO_2 emission. Gas, coal and oil usage contributed equal proportion to CO_2 emission in Pakistan in 2019. In 2016, combined heat and power (CHP) production accounted for about 46.44% of India's total GHG emissions and remained the highest emission source. However, CHP generation was the second contributor to GHG emissions in Bangladesh, Pakistan, and Sri Lanka. Therefore, considering the rapid depletion of fossil reserves and the huge GHG emission from fossil fuels heighten by the continuous increase in electricity demand, harnessing low-cost and environmentally friendly bioresources for electricity generation is imminent.

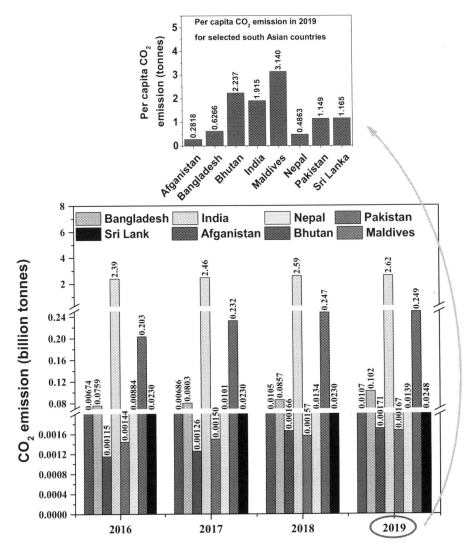

FIGURE 2.5 CO_2 emissions from selected south Asian developing countries. *Data source: Our World in Data (Ritchie and Roser, 2021).*

Bioelectricity is one of the most advanced routes of biomass utilization. Low-grade biomass-derived energy (bioenergy) has been used for thousands of years for cooking and heating; today, bioenergy is still considered one of the prime alternatives for fossil fuels in electricity generation. Several countries have taken initiatives for bioenergy expansion, which can significantly contribute to their electricity mix as bioelectricity (Souza et al., 2017). Bioelectricity is an attractive approach to meet any country's electricity demand, particularly in developing countries, due to several reasons highlighted in Fig. 2.6. For instance, (i) biomass is a renewable resource that can be sustainably developed, (ii) biomass have formidable

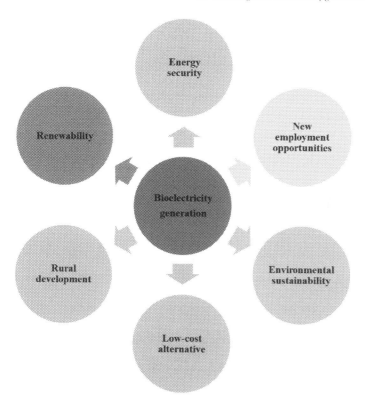

FIGURE 2.6 Various economic, environmental, and social benefits of bioelectricity generation.

positive environmental impacts, resulting in no net release of CO_2 as the released CO_2 from combustion of biomass is sequestered by the growing biomass, (iii) biomass appears to have significant economic potential, considering the spike in the fossil fuel prices. Apart from these, bioelectricity generation can create green job opportunities and food security under proper management practices and efficient conversion systems. The distributed renewable energy access sector has created approximately 95,000 formal jobs in India.

2.4 Technologies for bioelectricity generation

Several technologies have been explored for biomass to electricity conversion. Conventional techniques involve thermochemical and biochemical processes such as gasification, pyrolysis-combustion, hydrolysis-fermentation, and anaerobic digestion to produce biomass-derived fuels (syngas, biogas, biochar, biofuels), which are harnessed to drive turbines or gas engines for electricity generation (McKendry, 2002b,2002a; Diji, 2013; Ruiz et al., 2013; Poggi-Varaldo et al., 2014; Liu et al., 2020). These processes can be termed biomass-to-energy-to-electricity, and the generation of electricity via this route may compete with the rising bioenergy demands of developing countries despite the huge availability of biomass resources (Agbro and Ogie, 2012; Khatiwada et al., 2016). Moreover, these technologies are characterized by poor thermal efficiency, low biomass to energy conversion ratio, emissions and gas cleaning challenges and low

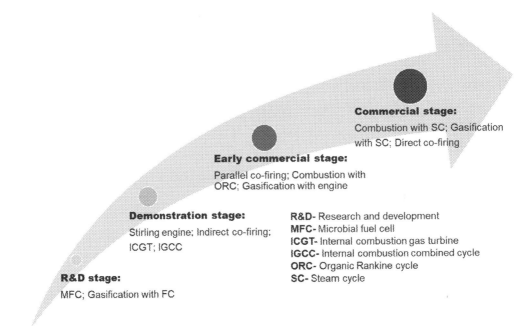

FIGURE 2.7 Technological advancement for biomass conversion into bioelectricity.

techno-commercial attractiveness (Asadullah, 2014; Sobamowo and Ojolo, 2018; Liu et al., 2020). Fig. 2.7 presents the advancements in bioelectricity generation technologies. The technologies capable of the direct conversion of biomass to electricity are attracting research interest in recent years. These include biomass-based flow fuel cells technology and electrolysis technology for co-producing hydrogen and electricity (Liu et al., 2020). Compared with thermal energy conversion into mechanical work, then to electricity, fuel cells (FC) convert the chemical energy of a fuel or biochemical energy stored in biomass into electrical energy with heat, water and CO_2 as by-products (Liu et al., 2020; Moradian et al., 2021). Earlier biomass fuel cells (such as solid oxides FC, direct carbon FC, molten oxides FC, etc.) types can convert biomass-derived fuels into electricity via series of electrochemical reactions occurring at high temperatures (>600°C) (Karl et al., 2009; Giddey et al., 2012; Zhao and Zhu, 2016; Jiang et al., 2017). The drawbacks of these high-temperature FCs technologies include the requirement of initial biomass conversion to liquid/solid/gaseous fuels for driving the FCs, special requirements of electrodes materials and high costs (Jiang and Chan, 2004; Gür, 2013). The latest efforts in bioelectricity generation focus on the development of more efficient, low temperatures microbial fuel cells, which is capable of direct conversion of biomass to electricity, avoiding the thermo-electric Carnot cycle (Venkata Mohan et al., 2008; Moqsud et al., 2015; Rahimnejad et al., 2015; Li et al., 2020; Liu et al., 2020; Moradian et al., 2021). Despite all the advances in bioelectricity generation technologies, biomass-fired power plant is the most matured and well-established technology that may meet the electricity need and expand access to remote communities in developing countries, at least in the near short and medium-term (Ahuja and Tatsutani, 2009).

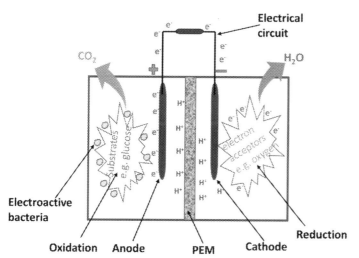

FIGURE 2.8 Dual-chamber microbial fuel cell with compartment components and redox reactions.

2.4.1 Microbial fuel cell

A microbial fuel cell (MFC) is a device that directly converts chemical energy into electrical energy via microbiological activities (Kim et al., 2007; Li et al., 2020). The process occurs through the action of special microbes (electroactive microorganisms) that degrade the biochemical energy stored in organic substrates, including biomass and release electrons for electricity generation (Zhao et al., 2017; Moradian et al., 2021). The interest in MFCs has increased tremendously due to the great potential in bioelectricity generation and wastewater treatment (Rahimnejad et al., 2015; Kumar et al., 2018). However, its performance in many of these applications is yet to reach maximum potential (Logan and Regan, 2006; Kim et al., 2007; Chandrasekhar et al., 2017). In general, there are two types of MFCs – single chamber and dual chamber MFCs (Kumar et al., 2018). Dual-chamber MFCs consist of anode and cathode compartments that are physically separated by a proton exchange membrane (PEM), an ion-exchange membrane, or other fitted membranes (such as reverse osmosis membranes, ultrafiltration membranes) (Kim et al., 2005). The anode chamber contains an electrode (anode), microorganism (exoelectrogens), growth media (anolyte), and a redox mediator (if desired), whereas the cathode chamber contains an electrode (cathode), an electron acceptor and a catalyst (if desired) (Pandit et al., 2017; Kumar et al., 2018). Earlier MFCs comprised single compartment housing both the anode and cathode with more compact size and comparative performance (Cheng et al., 2006; Di Lorenzo et al., 2009). The operation principle of MFCs involves the extraction of electrons from microbial cells and the transfer of electrons onto the anodic surface (Pandit et al., 2017). The anode is connected to the cathode through an external circuit from which electrons flow to generate electrical current (Kim et al., 2007). Generated electrons travel from the positive electrode (anode) surface to the negative electrode (cathode) due to the redox potential difference that exists between the two electrodes (Xu et al., 2016). Fig. 2.8 shows the schematic of dual-chamber MFCs set-up and components.

Electrochemically active microorganisms (anode-respiring bacteria) oxidizes organic substrates (e.g., cellulose, monomeric and oligomeric sugars, alcohols, organic acids) or inorganic matter (sulfide, ammonia) in the anode compartment to produce electrons and protons (Rahimnejad et al., 2015). Protons move through the proton exchange membrane (PEM) to reach the cathode, reacting with transmitted electrons from the external circuit and oxidizing substances (such as oxygen, hydrogen peroxide, potassium permanganate ferricyanide) in the cathode compartment (Kumar et al., 2018). Through this movement of ions, electrical current is generated, and organic contaminants are degraded (Logan and Rabaey, 2013). The anode chamber has to be operated in an anaerobic environment without active electron acceptors (like oxygen, nitrate), as oxygen in the anode chamber can inhibit electricity production (Najafpour et al., 2011). The bacteria in the anode chamber is shielded from oxygen by posing a membrane between the chambers, which allows charge to be transferred between the anode chamber, where the microbe grows (carbon oxidation) and the cathode chamber, where the electrons react the oxygen (oxygen reduction) (Rahimnejad et al., 2015). Several microorganisms can be used to generate electricity in MFCs, including the popular *Escherichia coli and Saccharomyces cerevisiae* (Rahimnejad et al., 2015; Kumar et al., 2018), and more robust exoelectrogens and mediator-less bacteria such as *Shewalla putrefaciens, Geobacter metallireducens, Geobacter sulferrenducens,* and *Rhodoferax ferrireducens* (Das and Mangwani, 2010; Li et al., 2019). The metabolic pathways and mechanisms used by these bacteria in MFCs are still largely being investigated. Direct electron transfer and mediated electron transfer are the two types of extracellular electron transfer mechanisms that have been proposed, but the metabolic pathways involved remain hypothetical (Kumar et al., 2018; Li et al., 2018). The opportunities and challenges associated with MFCs operation and commercialization are highlighted in Table 2.2 .

The potential of MFCs in bioelectricity generation and technological status is still at the research investigation level and requires several technical, economic, and performance improvements to reach the commercialization stage (Li et al., 2018). The first successful MFC prototype was developed by the University of Queensland, Australia (May 2007) in conjunction with Foster's Brewing Company. Gifu University, Japan (December 2014) was among the first to successfully recover phosphorus from wastewater and generate electricity using MFCs (Patel et al., 2019). Whilst many researchers have demonstrated the prospects of MFCs in bioremediation and electricity generation from various organic wastes in south Asian developing countries, including Bangladesh (Moqsud and Omine, 2010; Reza and Farzid Hasan, 2016), India (Khan et al., 2017; Patel et al., 2019), Sri Lanka (Gunathilake et al., 2008), Pakistan (Parkash, 2016; Raza et al., 2016); there are no full-scale reports on MFCs implementation in these countries. Globally, the electricity generated by MFCs is still at the demonstration stage; there is no industrial application yet (Obileke et al., 2021).

2.4.2 Anaerobic digestion-based technology

Anaerobic digestion (AD) is a process of harnessing methane-rich gas (50–70% methane) from biodegradable organic waste materials in an oxygen-free environment (Halder et al., 2016). Different anaerobic microbes are used to convert organic wastes into biogas fuel (Lim et al., 2020). The four successive steps involved in the AD process are depicted in Fig. 2.9A, and the application of AD products in a biorefinery context is shown in Fig. 2.9B. AD is a complex dynamic system comprising microbiological, biochemical and physical-chemical processes which can occur at either mesophilic (30–40°C) or thermophilic (50–60°C) temperatures, and varying pH (Náthia-Neves et al., 2018; Albihn, 2019). To achieve the optimum operating

TABLE 2.2 Opportunities and challenges of microbial fuel cells.

Indicator	Advantages	Disadvantages	Remarks	Reference
Power/energy density	Low power level generation may facilitate easy integration and deployment in small power devices operation.	MFCs power density is considerably lower than counterparts fuel cells such as chemical fuel cells and batteries fuel cell	The maximum power density obtainable in MFCs is impeded by high internal resistance from microbial consortia activities, organic substrates complexity, and reactor configurations.	(Kim et al., 2007; Ren et al., 2014)
Microbial inoculation	MFCs can utilize pure bacteria cultures, but mixed microbial cultures are suitable for complex substrates metabolism.	Microbes metabolic activities are very complex and highly impacted by their media and fuel substrate conditions. Poorly understood mechanisms of electrochemically active microbes.	The optimal microbial consortium needs to be enriched to enable efficient use of the substrate supplied under a given MFCs operating conditions.	(Phung et al., 2004; Ringeisen et al., 2006)
Costs	Mild operating conditions such as mesophilic temperature, electrode materials and fuel diversity make MFCs commercially attractive over the conventional DCFCs and SOFCs.	Dual-chamber MFCs require membranes, and the high cost of membranes may limit the wide range of large-scale application.	The use of low-cost and effective membrane material should be investigated. Chouler et al. (2017) demonstrated that naturally synthesized eggshell membrane performs similarly to other expensive membrane materials.	(Logan et al., 2015; Chouler et al., 2017)

(continued on next page)

TABLE 2.2 Opportunities and challenges of microbial fuel cells—cont'd

Indicator	Advantages	Disadvantages	Remarks	Reference
Fuel conversion efficiency	The single-step conversion of carbon-rich organic substrates in MFCs to directly generate electricity by the catabolic action of microorganisms gives better fuel conversion capacity as opposed to chemical fuel cells that requires two or multiple stage fuel consumption to generate electricity.	Fuel for MCFs must be readily biodegradable for higher conversion performance; therefore, complex carbohydrate matrix in biomass demands prior depolymerization to smaller units accessible for microbial degradation.	MCFs have been largely demonstrated with Bio Oxygen Dissolved (BOD) substrates conversion in wastewater, simple and pure organic feed material (acetate, glucose, propionate, etc.). More studies are required in the conversion of complex carbohydrate in biomass and how it affects the MCFs performance.	(Kim et al., 2007; Chandrasekhar et al., 2017)
Electrodes materials	Unlike high-temperature fuel cells requiring special electrode material to withstand harsh operating conditions, MFCs use various materials such as carbon cloth, paper and felt, graphite rods, plates, fiber brush, and reticulated vitreous carbon for the anode. Cathodes are made from similar materials doped with precious metals, such as Pt if oxygen is used as an electron acceptor or Ni, Ti for desalination application.	Some of the electrode materials used may not be suitable for scale-up due to the inherent lack of durability, electrical conductivity, and strength. The commercial production of MFCs is currently limited by the manufacturing capacity of the reactor cathodes.	Electrodes materials is a key factor for the cost and performance of MFCs. Anode materials must have large surface areas and biostructural stability for supporting biofilm growth, while cathode materials should have high redox potential and corrosion resistance.	(Logan and Regan, 2006; Zhou et al., 2011; Obileke et al., 2021)

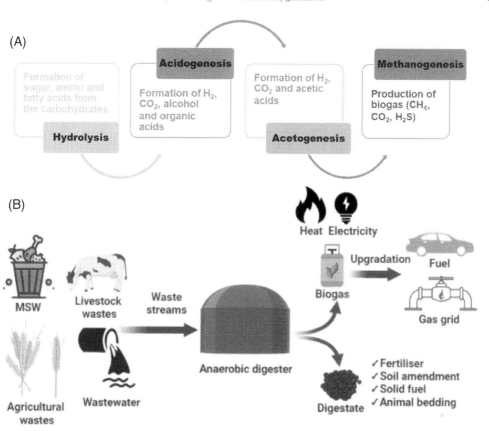

FIGURE 2.9 Process schematic of anaerobic digestion of biomass wastes to bioelectricity (A) process stages for biogas production (B) potential application routes of AD products.

temperature, both forms of digestion usually need additional heat sources. Lignocellulosic biomass from agricultural residues and forest wastes contains complex carbohydrates (i.e., cellulose, hemicellulose, and lignin), whereas organic wastes (such as food wastes, livestock litter, sewage sludge) contains complex proteins and lipids (Náthia-Neves et al., 2018). These feedstocks require pretreatment steps to break down into simple sugars, long-chain fatty acids, and amino acids via a hydrolysis mechanism. The simple molecules are then fermented to volatile fatty acids and alcohols (via acidogenic bacteria), which are then converted to organic acids (mainly acetic acid), H_2 and CO_2 through a process called acetogenesis. Methanogens bacteria catabolize the acetic acid, H_2, and CO_2 to CH_4 with CO_2 as a by-product (Meegoda et al., 2018). The biogas (gaseous mixture of CH_4 and CO_2) produced from AD can generate renewable power and heat (CHP) through biogas cogeneration technologies discussed below (Darrow et al., 2017). There are a number of technologies available for converting biogas energy into electrical power.

- *Steam engine/Steam turbine:* The biogas produced from the AD of organic biomass wastes is combusted in a boiler to generate high-pressure steam. The expansion of steam via a steam

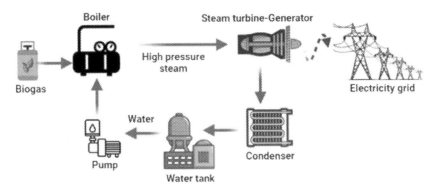

FIGURE 2.10 Schematic of steam turbine technology for bioelectricity production.

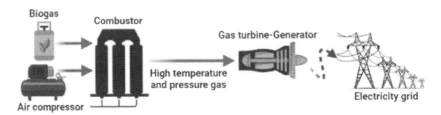

FIGURE 2.11 Schematic of gas turbine technology for biogas combustion to bioelectricity.

engine or turbine provides the heat or work required to drive an electric motor (generator) that converts the mechanical work into electricity. The generated electricity can be used to power nearby decentralized communities or transmitted through electrical systems for integration with the national grid. At the same time, the heat can be recycled as process heat for the digester, redirected for domestic use in surrounding buildings, or transferred to a central district heating system (Ong et al., 2014). A schematic of steam turbine technology for power generation is shown in Fig. 2.10. Steam turbines have been in use for about 100 years, and the capacity ranges from 50 kW to several hundred MW (EEAI and ERGI, 2007).

- *Gas engine/Gas turbine*: Biogas, as fuel is used in stationary engines, such as gas engines or gas turbines. A gas engine is an internal combustion engine, which converts the chemical energy in biogas to mechanical work and then to electrical power (Fig. 2.11). Approximately 30–40% of the energy in the fuel can be converted into electricity. The remaining energy is dissipated as heat. There are three main components in a gas turbine CHP system: the engine that combusts the biogas to mechanical energy and heat; a generator that converts the mechanical energy into electricity; and a heat recovery system, which converts the waste heat into useable energy. However, prior to using biogas in a gas engine, a purification step for removing certain impurities such as CO_2 and H_2S may be required to improve the engine efficiency (Halder et al., 2021). Reciprocating internal combustion engines are widely used gas engine for power generation from gaseous fuels. The engine sizes usually vary from 50 kW to several MW and can be used in small to large biogas plants. Although the

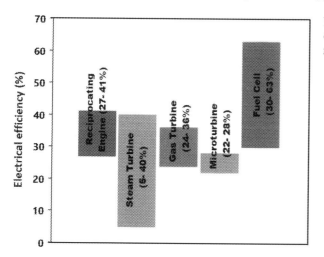

FIGURE 2.12 Comparison of electrical efficiency of various electricity generation technologies. *Data source: EPA 2017 (Darrow et al., 2017).*

reciprocating engines have the lowest initial costs, they have high operating costs due to the ongoing need for maintenance. However, they do not need biogas compression (Wellinger and Lindberg, 1999; Darrow et al., 2017). A gas combustion turbine or microturbine is another technology for electricity production from gaseous fuels, where the compressed air is mixed with the gaseous fuel in the combustion chamber, integrated with a turbine. The turbine then converts the energy of high temperature and high-pressure gas produced from the combustion chamber into mechanical energy and then to electrical power. Gas combustion turbines emit lesser pollutants than reciprocating engines and require less maintenance. Gas turbine with high capacity may not be suitable for community-based power generation in rural areas. Microturbines can be most effective for biogas-based small scale power plants of capacity ranging from 30 to 200 kW (Hosseini et al., 2016).

The various turbines/engines have different electrical efficiency, as shown in Fig. 2.12. Steam turbine has the lowest and least stable electrical efficiency (5–40%) due to reduced boiler thermal efficiency resulting from scaling and fouling. Microturbines have a more stable electrical efficiency.

2.4.3 Direct combustion/gasification-based technology

The gasification process involves the thermal conversion of solid fuels into a combustible gas composed primarily of H_2, CH_4, CO, and CO_2 (generally referred to as syngas) under partial oxidation condition. The produced syngas is used as fuel in the cogeneration of electricity and energy. Combustion is the oxidation of fuel materials to produce mainly heat and flue gas. When air is used as a gasifying agent and nitrogen is the main component of the gaseous product, the syngas is commonly referred to as "producer gas." The producer gas contains various contaminants, such as sulfur compounds and tars. After gas cleaning and conditioning, the gas can be used to generate heat and power in CHP systems (Raman and Ram, 2013; Bhaduri et al., 2015; Patuzzi et al., 2016). Like AD, biomass combustion/gasification

TABLE 2.3 Advantages and disadvantages of various CHP systems (Darrow et al., 2017).

CHP system	Advantages	Disadvantages
Spark ignition (SI) and compression ignition (CI) reciprocating engine	Provide high electrical efficiency with the flexibility of running at partial load condition	Maintenance cost is relatively high
		Not suitable for low-temperature cogeneration
	Easy to fast start-up	Produce relatively high air pollutive gases
	Low cost compared to other technology	
	Plant can run even in a low-pressure gas condition	Heat must be removed even if it is not used
Gas turbine	Higher plant reliability	High-pressure gas is required
	Associated emission is low	Low loading provides low efficiency
	Produce high-grade heat	Ambient temperature effect efficiency
	No requirement of cooling	
Steam turbine	Coupled to boilers, which can use a variety of gaseous, liquid or solid fuels	Plant start-up is slow
		Power to heat ratio is very low
	Can provide heat for more than one site	Requirement of boiler or other heat sources for steam generation
	High reliability and extended working life	
	Variable power to heat ratio	

technology is integrated with gas engine/turbine and steam engine/turbine for CHP generation. For small scale power (50 kW to 1-10 MW) generation, biomass gasification integrated with a gas engine is considered the most efficient, whereas, in large scale power (1-10 to 10-100 MW) generation, biomass combustion with a steam turbine is more efficient. However, a combination of both gas turbine and steam turbine with biomass gasification can significantly enhance the overall plant efficiency (Veringa 2009). Table 2.3 presents the advantages and disadvantages of available technologies for power and heat generation.

2.5 Bioelectricity prospects and technological status in south Asian developing countries

2.5.1 Biomass resources potential

Biomass is a huge renewable source that needs to be fully repositioned from low-grade heating fuel to produce modern biofuels and bioelectricity (Wang and Watanabe, 2020). Developed countries have recorded great success in the deployment and integration of most of these renewable energy technologies with their traditional power generation grid and off-grid systems (Madurai Elavarasan et al., 2020). Various biomass resources in several CHP technologies (see Fig. 2.13) are used to produce electricity in Bangladesh, India and Pakistan. In the fiscal year 2017-2018, Bangladesh had 100.9 million tonnes of agricultural residues, 16.94 million tonnes of forest residues, 93.04 million tonnes of animal residues and 14.74 million tonnes MSW, which could potentially produce about 383 TWh electricity

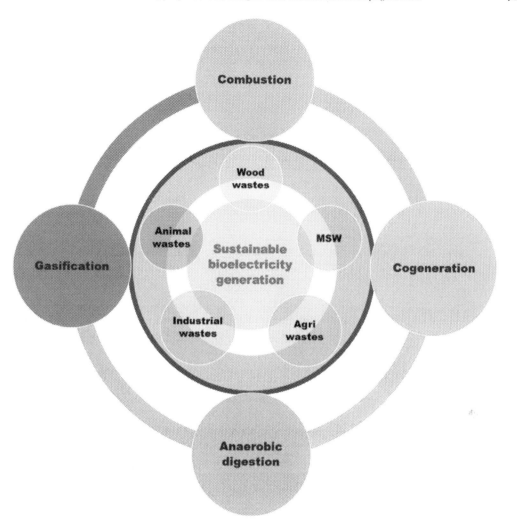

FIGURE 2.13 Major biomass resources and their conversion technologies for bioelectricity in south Asian developing countries.

(Himel et al., 2019). The bioelectricity generation potential in 2017-2018 was only 2.5% higher than the 2012-2013 value (Halder et al., 2014), indicating that Bangladesh has a consistent supply of biomass resources. India has an average biomass potential of about 500 million tonnes per annum, of which, about 120 to 150 million tonnes of surplus biomass (after the primary use of biomass for cooking, heating and other purposes) can produce about 18 GW of electrical power (MNRE, 2021a). The urban and industrial organic wastes, produced in the country, alone can generate about 5690 MW. Apart from these, sugarcane bagasse, produced from 500 sugar mills, has a potential of approximately 7,000 MW electrical power in India. Pakistan is another large bioenergy producing country with 4,800–5,600 MW electricity

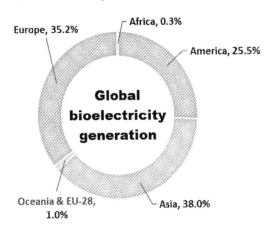

FIGURE 2.14 Share of bioelectricity generation by different continents in 2018. *Data source: World Bioenergy Authority (WBA, 2021).*

generation potential from animal dung only (Khatiwada et al., 2012). The per capita fuelwood consumption in Pakistan was 0.21 m³ and if half of the country's total fuelwood consumption is used for electricity production, it could generate about 45,000 GWh bioelectricity. Apart from these, the sugarcane bagasse produced in the country can generate about 5800 GWh of bioelectricity. In Nepal, the available sugarcane biomass can produce 209 to 313 GWh electricity (Irfan et al., 2020).

The developing countries have a massive potential for bioelectricity, given a viable bioelectricity mandate in place. In 2018, global bioelectricity (from all sources) production reached 637 TWh, with solid biomass taking up nearly 66.1% share, followed by biogas 14% and municipal waste 12% (WBA, 2021). As shown in Fig. 2.14, Asia was the major bioelectricity producing region (38% of global production), followed by Europe (35.2%). Among the south Asian developing countries, India, Pakistan and Bangladesh are major biomass and bioelectricity-producing countries. However, most of their biomass resources are underutilized. For instance, over 90% of the rural population in Bangladesh and 80% rural population in India directly depend on biomass fuel for their cooking activities (Billah et al., 2020). Nevertheless, diverting biomass for bioelectricity production via CHP technologies has received significant attention in recent times, and many initiatives are underway to increase bioelectricity production.

2.5.2 Bioelectricity generation: Current status and future mandates

2.5.2.1 *India scenario*

Combustion, cogeneration and anaerobic digestion are the major technologies used in India for bioelectricity generation from biomass. As of February 2021, India had installed about 10,532.7 MW biopower, including 10,314.6 MW grid-interactive and 218.1 MW off-grid power. Total biopower capacity in 2020 was approximately 3.8% higher than the previous year. Fig. 2.15 presents the state-wise installed biopower capacity in India. Maharashtra, Uttar Pradesh, Karnataka, and Tamil Nadu are major bioelectricity producing regions constituting about 74% of the country's total bioelectricity generation. Cogeneration of power and steam from biomass wastes, particularly sugarcane bagasse, is the foremost technology practiced in India for the generation of 8,319 MW bioelectricity (82% of total capacity). The power

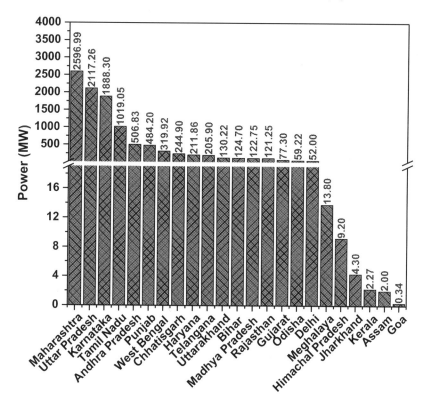

FIGURE 2.15 State-wise installed bioelectricity generation capacity in India, 2020. *Data source: Ministry of New and Renewable Energy, India (MNRE, 2021b).*

generated from the cogeneration-based technology is mainly used by industries and fed to the grid. Apart from the cogeneration, combustion-steam turbine and biogas to electricity have been used in India for converting the waste biomass and other organic wastes into green electricity. Under the Biogas-based Power Generation and Thermal Application Program (BPGTAP), India has started producing decentralized off-grid bioelectricity from biogas. So far, India has installed nearly 389 biogas power projects of a capacity of 87,990 m^3 biogas, producing about 8.96 MW bioelectricity (MNRE, 2021a). Tamil Nadu, Karnataka, Maharashtra, and Punjab are the major contributors to India's biogas-based off-grid biopower generation (see Fig. 2.16), accounting for about 75% of total biogas-based power. About 128 kW of bioelectricity is produced in Bihar using gasification technology. India has set a mandate to produce about 10 GW biopower by 2022. MFC technology has the potential to generate about 23.3 TW and 40 TW biopower by 2025 and 2050, respectively, from municipal wastewater generated in India (Khan et al., 2017)

2.5.2.2 Bangladesh scenario

Considering the geographical location and availability of biomass resources in Bangladesh, bioelectricity generation from biomass and biogas can be one of the predominant approaches

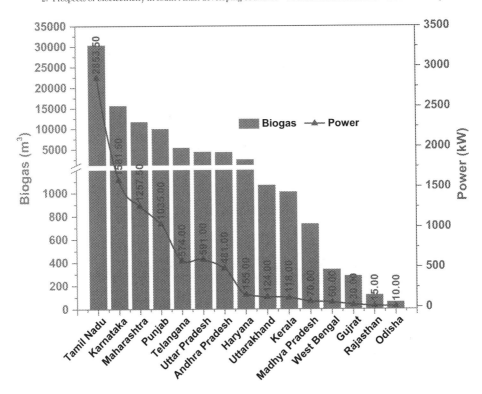

FIGURE 2.16 State-wise off-grid bioelectricity generation from biogas in India, 2019. *Data source: Ministry of New and Renewable Energy, India (MNRE, 2021a).*

to increasing the share of renewables in the country's future electricity mix. Currently, there is no significant share of bioelectricity in the country's electricity mix. Gasification and anaerobic digestion are being employed for biopower generation in Bangladesh. So far, Bangladesh has installed about 930 kW biogas-based off-grid electricity projects through funding by Infrastructure Development Company Limited (IDCOL). Nonetheless, the country has a huge potential for biogas generation from poultry manure, cow dung and other organic wastes (SREDA, 2021). Another two biogas-based power generation projects of 400 kW and 60 kW capacity are ongoing in Munshiganj and Gazipur, respectively, as presented in Table 2.4. Apart from these, IDCOL has installed two gasification-based power projects for electricity generation from rice husk; one is located at Kapasia, Gazipur (250 kW capacity), and the other is at Thakurgaon (400 kW capacity). The 250 kW plant at Kapasia is the first biomass gasification based power project in Bangladesh established in 2007 (IDCOL, 2021). About 5.1 tonnes/day rice husk is fed to the gasifier for electricity production, and 1.7 tonnes/day rice husk is used in the boiler for steam generation for processing approximately 36 tonnes/day paddy. The electricity generated from either biogas or biomass gasification technology goes to off-grid. However, one biogas-based power project of 1 MW capacity is underway to transmit electricity on-grid in Keraniganj by Bangladesh Power Development Board (BPDB). Bangladesh has aimed to achieve about 47 MW biopower, including biogas to power (7 MW)

TABLE 2.4 Bioelectricity generation capacity in Bangladesh (IDCOL, 2021; SREDA, 2021).

Project name, location (Agency)	Technology	Capacity (kW)	Present Status
Oasis Services (Agro) Ltd., Mymensingh (IDCOL)	Biogas to Electricity[a]	300	Running
Phenix Agro Ltd., Gazipur (IDCOL)	Biogas to Electricity[a]	400	Running
KKT Bio-Electricity Project, Panchagarh (IDCOL)	Biogas to Electricity[a]	100	Running
ZPL Bio-Electricity Project, Chuadanga (IDCOL)	Biogas to Electricity[a]	30	Running
UKAL Bio-Electricity Project, Tangail (IDCOL)	Biogas to Electricity[a]	30	Running
Seed Bangla Foundation Bio-Electricity Project, Gazipur (IDCOL)	Biogas to Electricity[a]	20	Running
RKKL Bio-Electricity Project, Mymensingh (IDCOL)	Biogas to Electricity[a]	50	Running
Dutch Dairy Ltd., Munshiganj (IDCOL)	Biogas to Electricity[a]	400	Ongoing[d]
UAL Bio-Electricity Project, Gazipur (IDCOL)	Biogas to Electricity[a]	60	Ongoing[d]
Pilot Project at Keraniganj on Turnkey Basis, Keraniganj (BPDB)	Biogas to Electricity[b]	1[c]	Under Planning
SEAL Biomass based Electricity Project, Thakurgaon (IDCOL)	Biomass to Electricity[a]	400	Running
Dreams Power Limited, Gazipur (IDCOL)	Biomass to Electricity[a]	250	Running

[a] Off-grid.
[b] On-grid.
[c] MW.
[d] Implementation ongoing.

and waste to energy (40 MW) by the end of 2021; however, the country is far away from the goal. IDCOL is the premier agency that is engaged in bioelectricity production programs. Various private companies are investing in bioelectricity production, but their progress has been limited mainly by a lack of funds. Therefore, the government should be more responsible and take necessary actions to overcome the technical and commercial bottlenecks, monitor and speed up the projects' implementation process, and increase research activities, funds, and mass awareness.

2.5.2.3 *Sri Lanka scenario*

Sri Lanka has installed ten biomass-based power projects with a total capacity of 26.1 MW (CEB, 2021). Half of the total installed capacity was based on fuelwood consumption under the "Dendro power project". Fig. 2.17 shows the installed bioelectricity at different locations in Sri Lanka. Dehiattakandiya was the first Dendro Power Project of 3.3 MW capacity, producing about 22 GWh of electricity per year. As of 2018, Sri Lanka has generated about 87 GWh bioelectricity, mostly from gasification technology.

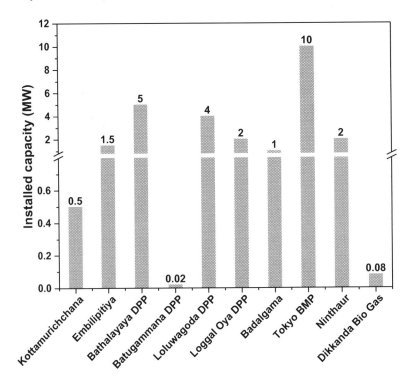

FIGURE 2.17 Installed bioelectricity capacity in Sri Lanka, 2017. *Data source: Ceylon Electricity Board, Sri Lanka (CEB, 2021).*

2.5.2.4 Pakistan scenario

Pakistan is another major biopower potential country in the developing south Asian region. Pakistan produces approximately 17.1 million tonnes of sugarcane bagasse per year from the 84 sugar mills, which could generate about 830 MW of electricity and steam for the mills using cogeneration technology (World Bank, 2021b). Gasification of rice husk (140,000 tonnes/year), generated from 54 rice mills, have a potential of 16 MW power. Apart from these, anaerobic digestion-based power generation from municipal solid wastes and cattle manure can produce about 360.4 MW electricity. Cogeneration of electricity and steam accounts for Pakistan's highest share of potential electricity capacity, as depicted in Fig. 2.18. Although Pakistan has a massive potential for biopower generation, up to date, only 259.1 MW electricity have been installed and running, mostly via cogeneration technology from bagasse (see Table 2.5). Currently, sixteen biopower plants are in the "Under Letter of Support" stage, and another nine plants are in the "Under Letter of Intent" stage (AEDB, 2021).

2.5.3 Biomass-fired plant for combined heat and power: Opportunities and challenges in developing countries

World bioenergy production reached about 84 million TOE in 2017, which is just about 1% of total global energy consumption (International Energy Outlook, 2017), of which, more

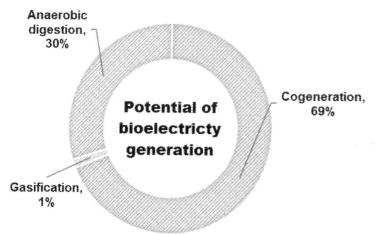

FIGURE 2.18 Technology-wise bioelectricity generation potential in Pakistan. *Data source: Alternative Energy Development Board, Pakistan (AEDB, 2021).*

TABLE 2.5 Bioelectricity generation capacity in Pakistan (AEDB. 2021).

Project	Location	Capacity (MW)
JDW Sugar Mills Ltd (Unit-II)	Rahim Yar Khan	26.35
JDW Sugar Mills Ltd (Unit-III)	Ghotki	26.35
Chiniot Power Ltd.	Faisalabad	62.4
RYK Mills Limited.	Rahim Yar Khan	30
Hamza Sugar Mill Limited.	Rahim Yar Khan	15
Layyah Sugar Mills	Layyah	41
Almoiz Industries Ltd.	Mianwali	36
Chanar Energy Limited	Faisalabad	22

than 90% is going to the transportation industry. Approximately, 50% of this biomass energy is consumed by developing countries as traditional fuel for heating and cooking. In contrast, developed and high-income developing countries use modern biomass for electricity production, which currently supplies 1.5% (about 280 TWh) of the world electricity demand (IRENA, 2015). Commercial biomass-fired plants for CHP production are located mainly in Europe (Padinger et al., 2019). Table 2.6 lists the major installed biomass-based plant in the world.

The developing countries from the south Asian region, particularly Bangladesh, India and Pakistan, have a massive potential for CHP generation from biomass for industrial application. These countries have already installed a number of coal-fired plants producing a significant amount of electricity. Apart from this, Sri Lanka produces about 300 kW power from Kiln waste gases using CHP technology. Therefore, the south Asian developing countries can use the existing coal-powered plant infrastructure with little modifications for biomass-driven power plant. However, some technological challenges and the socio-economic dilemma of biomass utilization remain the major hurdles. The growth of biomass-based CHP plants is still very low,

TABLE 2.6 Selected biomass-powered plants for combined heat and electricity generation (IRENA, 2015; Bester, 2021).

Location/Country (year)	Operator/Name	Biomass fuel	Capacity (MW)	Electricity (MW)	Heat (MW)	Other features
SevernGorge, United Kingdom (2013)	Ironbridge	Wood pellets	740			1000 MW coal-fired plant was converted
Jakobstad, Finland (2002)	Alholmenskraft				160	Used circulating fluidized bed boiler
Oulu, Finland	Toppila	Peat	210	220	340	
Staszów, Poland (2012)	Polaniec	Agricultural and wood residues	205			The facilities generate enough electricity to meet the needs of 600,000 households
Lathi, Finland (2012)	Kymijarvi	Plastic, paper, cardboard and wood	160	300[a]	600[a]	Transmitted electricity to the national grid using a 110 kV connection at Kymijärvi substation
Denmark (2009)	Dong	Straw		35	85	29.9% electric efficiency, used bubbling fluidized bed combustor
Germany (2004)	RWE	Wood and forest residues		20	65	19.4% electric efficiency, used CFBC[1]
Ireland (2004)	Barcas	Wood and forest residues		2.5	10	16.0% electric efficiency, used BFBC[+]
France (2010)	Solvay	Wood and forest residues		30		
Spain (2003)	EHN	Paper waste		25		32% electric efficiency used VG[#] boiler
Sweden (2009)	Soderenergi	Wood and forest residues		83	200	Used CFBC
Hungary (2010)	BHD	Paper waste		50		31.5% electric efficiency
Belgium (2010)	A&S	Wood and forest residues		24		
Finland (2020)	Fortum	Wood and forest residues		25	50	23.2% electric efficiency, used CFBC
United Kingdom (2007)	Semb	Wood and forest residues		30	10	29.5% electric efficiency, used BFBC

[1] CFBC – Circulating fluidized bed combustion; [+] BFBC – Bubbling fluidized bed combustion; [#] VG- Water-cooled vibrating grate; [a] GWh

even in developed countries. There is strong inertia in developing countries in commissioning biomass-fired plants due to perceived competition of biomass resources with animal feed and low-grade fuel and lack of viable bioenergy roadmap (Simonyan and Fasina, 2013; Jekayinfa et al., 2020).

2.6 Concluding remarks

Fossil fuels have primarily driven global economic growth and development; however, the strong dependency on this nonrenewable energy source is changing in recent days. Consequently, rapid advances in developing renewable energy technologies lead to the global transition in energy production patterns, including electricity production. Bioelectricity generation from waste biomass has received significant attention, and their share in the electricity mix is increasing remarkably in developing countries, particularly agriculture-intensive countries. So far, many decentralized bioelectricity technologies have been employed in rural areas of many south Asian developing countries. In order to achieve a sustainable electricity mix and increase the share of bioelectricity in future electricity mix for south Asian developing countries, the current study provides the following recommendations.

- Biomass supply chain logistics is one of the bottlenecks for large scale steady power generation. Therefore, comprehensive and up to date assessment of biomass resources (i.e., agricultural residues, forest residues, MSW, industrial wastes and other organic wastes), including data collection from ground level, geographic information system (GIS) analysis and geospatial planning, are necessary prior to installation of large-scale biomass-based power plants.
- In many developing countries, biomass is widely distributed in agricultural fields, poultry firms and cattle firms. The transportation of biomass to the power plant site may affect the techno-economic feasibility of the power generation. Additionally, the transmission and distribution will increase the overall economic footprint of the plant. Hence, community-based small-scale power generation using the wastes produced in nearby areas for their consumption could be an effective option for future bioelectricity generation.
- Biomass can be converted into electricity through a number of technologies; however, plants location and availability of biomass resources significantly affect the technology readiness and reliability. Therefore, location-specific comprehensive risk management, technological roadmap, life cycle and techno-economic assessment are required prior to the full-scale deployment of biomass-based combined heat and power plant.
- India, Bangladesh and Sri Lanka have a great potential of marine-based bioenergy (i.e., water hyacinth, seaweeds and microalgae) extraction from natural water bodies, including sea, rivers and lakes. However, currently south Asian developing countries do not have any marine-based bioenergy policy. Therefore, these countries require to develop a comprehensive policy for harnessing marine-based bioenergy.
- The lack of awareness in benefits of bioenergy, nonsystematic use of bioenergy funds and poor infrastructure are the major challenges of bioenergy dissemination. Therefore, the government should come forward to work with nongovernment organizations for enhancing the awareness program as well as for developing clear and structured policy for bioenergy project development and their dissemination.

Self-evaluation questions

1. What is your understanding of bioelectricity?
2. How can bioelectricity be a potential solution to sustainable electricity generation in developing countries?
3. Why is bioelectricity not getting adequate attention in developing economies? What are the barriers or challenges related to this?
4. What are the available bioelectricity generation technologies?
5. Explain the working principle of the microbial fuel cell.
6. What are the advantages and disadvantages of the microbial fuel cell?
7. Describe the bioelectricity generation process through anaerobic digestion.
8. Do you think bioelectricity generation processes are emission-free? Justify your answer.
9. What steps should the government of developing nations follow to make bioelectricity a promising sustainable electricity generation source?
10. Conduct a literature survey and present a comparative study of different available bioelectricity generation technologies and discuss their potential application in developing countries.

References

Abbasi, K.R., Hussain, K., Abbas, J., et al., 2020. Analyzing the role of industrial sector's electricity consumption, prices, and GDP: a modified empirical evidence from Pakistan. AIMS Energy 9, 29–49. https://doi.org/10.3934/ENERGY.2021003.

AEDB, 2021. Biomass Waste to Energy. Alternative Energy Development Board, Pakistan http://www.aedb.org/ae-technologies/biomass-waste-to-energy/current-status.

Agbro, E.B., Ogie, N.A., 2012. A comprehensive review of biomass resources and biofuel production potential in Nigeria. Res. J. Eng. Appl. Sci. 1, 149–155.

Ahuja, D., Tatsutani, M. (2009) Sustainable energy for developing Countries. Sapiens 2:

Albihn, A., 2019. Infectious waste management. In: Schmidt, TM (Ed.), Encyclopedia of Microbiology. Elsevier, London, United Kingdom, pp. 691–703.

Altieri, A., 2012. Bioethanol development in Brazil. Comprehensive Renewable Energy. Elsevier Ltd., Sao Paulo, Brazil, pp. 15–26.

Alvarez-Herranz, A., Balsalobre-Lorente, D., Shahbaz, M., Cantos, J.M., 2017. Energy innovation and renewable energy consumption in the correction of air pollution levels. Energy Policy 105, 386–397. https://doi.org/10.1016/j.enpol.2017.03.009.

Asadullah, M., 2014. Barriers of commercial power generation using biomass gasification gas: a review. Renew. Sustain. Energy Rev. 29, 201–215.

Bester (2021) The World's Largest Biomass Plants. In: Bester Newsl.

Bhaduri, S., Contino, F., Jeanmart, H., Breuer, E., 2015. The effects of biomass syngas composition, moisture, tar loading and operating conditions on the combustion of a tar-tolerant HCCI (Homogeneous Charge Compression Ignition) engine. Energy 87, 289–302. https://doi.org/10.1016/j.energy.2015.04.076.

Billah, S.M., Islam, S., Tasnim, F., et al., 2020. Self-adopted "natural users" of liquid petroleum gas for household cooking by pregnant women in rural Bangladesh: characteristics of high use and opportunities for intervention. Environ. Res. Lett. 15, 095008. https://doi.org/10.1088/1748-9326/ab7b25.

BP (2020) Statistical Review of World Energy 2020. https://www.bp.com/en/global/corporate/energy-economics/statistical-review-of-world-energy.html.

BPDB (2021) Bangladesh Power Development Board. https://www.bpdb.gov.bd/bpdb_new/. Accessed 6 May 2021

CEB, 2021. Sales and Generation Data Book 2018. Ceylon Electricity Board, Sri Lanka https://ceb.lk/publication_media/other-publications/41/en.

CGDEV (2021) Electricity Consumption and Development Indicators. https://www.cgdev.org/media/electricity-consumption-and-development-indicators. Accessed 6 May 2021

Chandrasekhar, K., Kadier, A., Kumar, G., et al., 2017. Challenges in microbial fuel cell and future scope. Microbial Fuel Cell: A Bioelectrochemical System that Converts Waste to Watts. Springer International Publishing, New Delhi, India, pp. 483–499.

Cheng, S., Liu, H., Logan, B.E., 2006. Increased performance of single-chamber microbial fuel cells using an improved cathode structure. Electrochem. Commun. 8, 489–494. https://doi.org/10.1016/j.elecom.2006.01.010.

Chouler, J., Bentley, I., Vaz, F., et al., 2017. Exploring the use of cost-effective membrane materials for microbial fuel cell based sensors. Electrochim. Acta 231, 319–326. https://doi.org/10.1016/j.electacta.2017.01.195.

Darrow, K., Tidball, R., Wang, J., Hampson, A., 2017. Catalog of CHP Technologies. US Environmental Protection Agency and the US Department of Energy, Washington D.C., United States.

Das, S., Mangwani, N., 2010. Recent developments in microbial fuel cells: a review. J. Sci. Ind. Res. 69, 727–731.

Di Lorenzo, M., Curtis, T.P., Head, I.M., Scott, K., 2009. A single-chamber microbial fuel cell as a biosensor for wastewaters. Water Res. 43, 3145–3154. https://doi.org/10.1016/j.watres.2009.01.005.

Diji, C.J., 2013. Electricity production from biomass in Nigeria: Options, prospects and challenges. Advanced Materials Research. Trans Tech Publications Ltd, Bach, Switzerland, pp. 444–450.

Dogan, E., Seker, F., 2016. The influence of real output, renewable and non-renewable energy, trade and financial development on carbon emissions in the top renewable energy countries. Renew. Sustain. Energy Rev. 60, 1074–1085.

EEAI, ERGI (2007) Biomass CHP Catalog of Technologies.

GEA, 2012. Global Energy Assessment-Towards a Sustainable Future. Cambridge UK and New York, NY, USA and the International Institute for Applied Systems Analysis, Laxenburg, Austria.

Giddey, S., Badwal, S.P.S., Kulkarni, A., Munnings, C., 2012. A comprehensive review of direct carbon fuel cell technology. Prog. Energy Combust. Sci. 38, 360–399.

Gunathilake, M.P., Gunawardane, S.H.P., De Alwis, A.A.P., 2008. Microbial Fuel Cell (MFC) Technology and Scale up Factors in a Sri Lankan Context. In: 14 th ERU Symposium. Faculty of Engineering.

Gür, T.M., 2013. Critical review of carbon conversion in "carbon fuel cells. Chem. Rev. 113, 6179–6206.

Halder, P., Patel, S., Kundu, S., et al., 2021. Potential of ionic liquid applications in natural gas/biogas sweetening and liquid fuel cleaning process. In: Azad, K, Khan, MMM (Eds.), Bioenergy Resources and Technologies. Academic Press, Elsevier Inc., London, United Kingdom, pp. 121–154.

Halder, P.K., Paul, N., Beg, M.R.A., 2014. Assessment of biomass energy resources and related technologies practice in Bangladesh. Renew. Sustain. Energy Rev 39, 444–460. https://doi.org/10.1016/j.rser.2014.07.071.

Halder, P.K., Paul, N., Joardder, M.U.H., et al., 2016. Feasibility analysis of implementing anaerobic digestion as a potential energy source in Bangladesh. Renew. Sustain. Energy Rev. 65, 124–134. https://doi.org/10.1016/j.rser.2016.06.094.

Halder, P.K., Paul, N., Joardder, M.U.H., Sarker, M., 2015. Energy scarcity and potential of renewable energy in Bangladesh. Renew. Sustain. Energy Rev 51, 1636–1649. https://doi.org/10.1016/j.rser.2015.07.069.

Himel, M., Khatun, S., Rahman, M., Nahin, A., 2019. A prospective assessment of biomass energy resources: potential, technologies and challenges in Bangladesh. J Energy Res. Rev. 3, 1–25.

Hosseini, S.E., Barzegaravval, H., Wahid, M.A., et al., 2016. Thermodynamic assessment of integrated biogas-based micro-power generation system. Energy Convers. Manag. 128, 104–119. https://doi.org/10.1016/j.enconman.2016.09.064.

IDCOL, 2021. Renewable Energy. Infrastructure Development Company Limited, Dhaka, Bangladesh https://idcol.org/home/other_re.

International Energy Outlook, 2017. International Energy Outlook 2017 Overview. US Department of Energy, Washington D.C., United States.

IRENA (2020) Renewable Energy Capacity by Energy Technology. In: Stat. time Ser.

IRENA (2015) Biomass for heat and power- Technology brief.

Irfan, M., Zhao, Z.Y., Panjwani, M.K., et al., 2020. Assessing the energy dynamics of Pakistan: prospects of biomass energy. Energy Rep. 6, 80–93. https://doi.org/10.1016/j.egyr.2019.11.161.

Jekayinfa, S.O., Orisaleye, J.I., Pecenka, R., 2020. An assessment of potential resources for biomass energy in Nigeria. Resources 9, 1–41. https://doi.org/10.3390/resources9080092.

Jiang, C., Ma, J., Corre, G., et al., 2017. Challenges in developing direct carbon fuel cells. Chem. Soc. Rev. 46, 2889–2912. https://doi.org/10.1039/c6cs00784h.

Jiang, S.P., Chan, S.H., 2004. A review of anode materials development in solid oxide fuel cells. J. Mater. Sci. 39, 4405–4439. https://doi.org/10.1023/b:jmsc.0000034135.52164.6b.

Karl, J., Frank, N., Karellas, S., et al., 2009. Conversion of syngas from biomass in solid oxide fuel cells. J. Fuel Cell Sci. Technol. 6, 0210051–0210056. https://doi.org/10.1115/1.2971172.

Kaygusuz, K., 2012. Energy for sustainable development: a case of developing countries. Renew. Sustain. Energy Rev. 16, 1116–1126.

Khan, M.D., Khan, N., Sultana, S., et al., 2017. Bioelectrochemical conversion of waste to energy using microbial fuel cell technology. Process Biochem. 57, 141–158.

Khatiwada, D., Leduc, S., Silveira, S., McCallum, I., 2016. Optimizing ethanol and bioelectricity production in sugarcane biorefineries in Brazil. Renew Energy 85, 371–386. https://doi.org/10.1016/j.renene.2015.06.009.

Khatiwada, D., Seabra, J., Silveira, S., Walter, A., 2012. Power generation from sugarcane biomass - a complementary option to hydroelectricity in Nepal and Brazil. Energy 48, 241–254. https://doi.org/10.1016/j.energy.2012.03.015.

Kim, B.H., Chang, I.S., Gadd, G.M., 2007. Challenges in microbial fuel cell development and operation. Appl. Microbiol. Biotechnol. 76, 485–494.

Kim, J.R., Min, B., Logan, B.E., 2005. Evaluation of procedures to acclimate a microbial fuel cell for electricity production. Appl. Microbiol. Biotechnol. 68, 23–30. https://doi.org/10.1007/s00253-004-1845-6.

Kumar, R., Singh, L., Zularisam, A.W., Hai, F.I., 2018. Microbial fuel cell is emerging as a versatile technology: a review on its possible applications, challenges and strategies to improve the performances. Int. J. Energy Res. 42, 369–394. https://doi.org/10.1002/er.3780.

Li, F., An, X., Wu, D., et al., 2019. Engineering microbial consortia for high-performance cellulosic hydrolyzates-fed microbial fuel cells. Front. Microbiol. 10, 409. https://doi.org/10.3389/fmicb.2019.00409.

Li, M., Zhou, M., Tian, X., et al., 2018. Microbial fuel cell (MFC) power performance improvement through enhanced microbial electrogenicity. Biotechnol. Adv. 36, 1316–1327.

Li, T., Cai, Y., Yang, X.-L., et al., 2020. Microbial fuel cell-membrane bioreactor integrated system for wastewater treatment and bioelectricity production: overview. J. Environ. Eng. 146. https://doi.org/10.1061/(ASCE)EE.1943-7870.0001608.

Lim, J.W., Park, T., Tong, Y.W., Yu, Z., 2020. The microbiome driving anaerobic digestion and microbial analysis. Advances in Bioenergy. Elsevier, London, United Kingdom, pp. 1–61.

Liu, W., Liu, C., Gogoi, P., Deng, Y., 2020. Overview of biomass conversion to electricity and hydrogen and recent developments in low-temperature electrochemical approaches. Engineering 6, 1351–1363.

Logan, B.E., Rabaey, K., 2013. Conversion of wastes into bioelectricity and chemicals by using microbial electrochemical technologies (Science (2012) (686)). Science 339 (80-), 906. https://doi.org/10.1126/science.339.6122.906-a.

Logan, B.E., Regan, J.M., 2006. Microbial challenges and fuel cells applications. Environ. Sci. Technol. 5172–5180.

Logan, B.E., Wallack, M.J., Kim, K.Y., et al., 2015. Assessment of microbial fuel cell configurations and power densities. Environ Sci Technol Lett 2, 206–214. https://doi.org/10.1021/acs.estlett.5b00180.

Madurai Elavarasan, R., Selvamanohar, L., Raju, K., et al., 2020. A holistic review of the present and future drivers of the renewable energy mix in Maharashtra, State of India. Sustainability 12, 6596. https://doi.org/10.3390/su12166596.

McKendry, P., 2002a. Energy production from biomass (part 3): gasification technologies. Bioresour. Technol. 83, 55–63. https://doi.org/10.1016/S0960-8524(01)00120-1.

McKendry, P., 2002b. Energy production from biomass (part 2): conversion technologies. Bioresour. Technol. 83, 47–54. https://doi.org/10.1016/S0960-8524(01)00119-5.

Meegoda, J.N., Li, B., Patel, K., Wang, L.B., 2018. A review of the processes, parameters, and optimization of anaerobic digestion. Int. J. Environ. Res. Public Health 15. https://doi.org/10.3390/ijerph15102224.

MNRE (2021a) Bioenergy. Ministry of New and Renewable Energy, Government of India. https://mnre.gov.in/bio-energy/current-status. Accessed 6 May 2021.

MNRE (2021b) Physical Progress. Ministry of New and Renewable Energy, Government of India. https://mnre.gov.in/the-ministry/physical-progress. Accessed 6 May 2021.

Moioli, E., Salvati, F., Chiesa, M., et al., 2018. Analysis of the current world biofuel production under a water – food – energy nexus perspective. Adv. Water Resour. 121, 22–31. https://doi.org/10.1016/j.advwatres.2018.07.007.

MOP (2021) Power Sector at a Glance ALL INDIA . https://powermin.gov.in/en/content/power-sector-glance-all-india. Accessed 6 May 2021.

Moqsud, M.A., Omine, K., 2010. Bio-electricity generation by using organic waste in Bangladesh. In: Proc. of International Conference on Environmental Aspects of Bangladesh (ICEAB10), pp. 122–124.

Moqsud, M.A., Yoshitake, J., Bushra, Q.S., et al., 2015. Compost in plant microbial fuel cell for bioelectricity generation. Waste Manag. 36, 63–69. https://doi.org/10.1016/j.wasman.2014.11.004.

Moradian, J.M., Fang, Z., Yong, Y.-.C., 2021. Recent advances on biomass-fueled microbial fuel cell. Bioresour. Bioprocess 8, 1–13. https://doi.org/10.1186/s40643-021-00365-7.

Najafpour, G., Rahimnejad, M., Ghoreshi, A., 2011. The enhancement of a microbial fuel cell for electrical output using mediators and oxidizing agents. Energy Sources A Recover. Util. Environ. Eff. 33, 2239–2248. https://doi.org/10.1080/15567036.2010.518223.

Náthia-Neves, G., Berni, M., Dragone, G., et al., 2018. Anaerobic digestion process: technological aspects and recent developments. Int. J. Environ. Sci. Technol. 15, 2033–2046.

NEPRA (2021) National Electric Power Regulatory Authority. https://www.nepra.org.pk/. Accessed 6 May 2021.

Obileke, K.C., Onyeaka, H., Meyer, E.L., Nwokolo, N., 2021. Microbial fuel cells, a renewable energy technology for bio-electricity generation: a mini-review. Electrochem. Commun. 125, 107003.

Ong, M.D., Williams, R.B., Kaffka, S.R. (2014) Comparative Assessment of Technology Options for Biogas Clean-up

Padinger, R., Aigenbauer, S., Schmidl, C., Bentzen, J.D., 2019. Best practise report on decentralized biomass fired CHP plants and status of biomass fired small-and micro scale CHP technologies. IEA Bioenergy 32, 1–83.

Pandit, S., Chandrasekhar, K., Kakarla, R., et al., 2017. Basic principles of microbial fuel cell: Technical challenges and economic feasibility. Microbial Applications. Springer International Publishing, New York City, United States, pp. 165–188.

Parkash, A., 2016. Bio-electricity generation from different biomass using microbial fuel cell. Artic. J. Proteomics Bioinforma. 3, 1–5. https://doi.org/10.15406/mojpb.2016.03.00098.

Patel, R., Zaveri, P., Munshi, N.S., 2019. Microbial fuel cell, the Indian scenario: developments and scopes. Biofuels 10, 101–108. https://doi.org/10.1080/17597269.2017.1398953.

Patuzzi, F., Prando, D., Vakalis, S., et al., 2016. Small-scale biomass gasification CHP systems: comparative performance assessment and monitoring experiences in South Tyrol (Italy). Energy 112, 285–293. https://doi.org/10.1016/j.energy.2016.06.077.

Phung, N.T., Lee, J., Kang, K.H., et al., 2004. Analysis of microbial diversity in oligotrophic microbial fuel cells using 16S rDNA sequences. FEMS Microbiol. Lett. 233, 77–82. https://doi.org/10.1016/j.femsle.2004.01.041.

Poggi-Varaldo, H.M., Munoz-Paez, K.M., Escamilla-Alvarado, C., et al., 2014. Biohydrogen, biomethane and bioelectricity as crucial components of biorefinery of organic wastes: a review. Waste Manag. Res. 32, 353–365. https://doi.org/10.1177/0734242X14529178.

Rahimnejad, M., Adhami, A., Darvari, S., et al., 2015. Microbial fuel cell as new technol ogy for bioelectricity generation: a review. Alexandria Eng. J. 54, 745–756. https://doi.org/10.1016/j.aej.2015.03.031.

Raman, P., Ram, N.K., 2013. Performance analysis of an internal combustion engine operated on producer gas, in comparison with the performance of the natural gas and diesel engines. Energy 63, 317–333. https://doi.org/10.1016/j.energy.2013.10.033.

Raza, R., Akram, N., Javed, M.S., et al., 2016. Fuel cell technology for sustainable development in Pakistan - an overview. Renew. Sustain. Energy Rev. 53, 450–461.

Ren, H., Torres, C.I., Parameswaran, P., et al., 2014. Improved current and power density with a micro-scale microbial fuel cell due to a small characteristic length. Biosens. Bioelectron. 61, 587–592. https://doi.org/10.1016/j.bios.2014.05.037.

REN21 (2021a) Renewables 2020. Global status report. https://www.ren21.net/wp-content/uploads/2019/05/gsr_2020_full_report_en.pdf. Accessed 6 May 2021.

REN21 (2021b) Renewables in Cities 2021 Global Status Report. https://www.ren21.net/reports/cities-global-status-report/. Accessed 6 May 2021.

Reza, F., Farzid Hasan, M., 2016. Potential for recovery of resources from food market refuse: a case study in Dhaka Bangladesh. Int. J. Environ. Ecol. Fam. Urban Stud 6, 59–70.

Ringeisen, B.R., Henderson, E., Wu, P.K., et al., 2006. High power density from a miniature microbial fuel cell using Shewanella oneidensis DSP10. Environ. Sci. Technol. 40, 2629–2634. https://doi.org/10.1021/es052254w.

Ritchie, H. (2020a) Energy mix- Our World in Data. https://ourworldindata.org/energy-mix. Accessed 26 Apr 2021.

Ritchie, H. (2020b) Electricity Mix - Our World in Data. https://ourworldindata.org/electricity-mix. Accessed 26 Apr 2021.

Ritchie, H., Roser, M. (2020a) Access to Energy- Our World in Data. https://ourworldindata.org/energy-access. Accessed 26 Apr 2021.
Ritchie, H., Roser, M. (2020b) Renewable Energy- Our World in Data. https://ourworldindata.org/renewable-energy. Accessed 6 May 2021.
Ritchie, H., Roser, M. (2019) Fossil fuels- Our World in Data. https://ourworldindata.org/fossil-fuels. Accessed 26 Apr 2021.
Ritchie, H., Roser, M. (2021) CO2 emissions - Our World in Data. https://ourworldindata.org/co2-emissions. Accessed 6 May 2021.
Roddy, D.J., 2012. Biomass and biofuels - introduction. Comprehensive Renewable Energy. Elsevier Ltd., Hoboken, United States, pp. 1–9.
Ruiz, J.A., Juárez, M.C., Morales, M.P., et al., 2013. Biomass gasification for electricity generation: review of current technology barriers. Renew. Sustain. Energy Rev. 18, 174–183. https://doi.org/10.1016/j.rser.2012.10.021.
Simonyan, K.J., Fasina, O., 2013. Biomass resources and bioenergy potentials in Nigeria. Afr. J. Agric. Res. 8, 4975–4989. https://doi.org/10.5897/AJAR2013.6726.
Smil, V. (2017) Energy Transitions: Global and National Perspectives, Second. Praeger
Sobamowo, G.M., Ojolo, S.J., 2018. Techno-economic analysis of biomass energy utilization through gasification technology for sustainable energy production and economic development in Nigeria. J. Energy 2018, 1–16. https://doi.org/10.1155/2018/4860252.
Souza, G.M., Ballester, M.V.R., de Brito Cruz, C.H., et al., 2017. The role of bioenergy in a climate-changing world. Environ. Dev. 23, 57–64.
SREDA (2021) Suatainable and Renewable Energy Development Authority, Bangladesh. https://ndre.sreda.gov.bd/index.php. Accessed 6 May 2021.
Statista (2021) Distribution of electricity consumption across India in financial year 2020, by sector. https://www.statista.com/statistics/1130112/india-electricity-consumption-share-by-sector/. Accessed 6 May 2021.
Turdera, MV., Garcia M da, S., 2018. Bioelectricity's Potential Availability from Last Brazilian Sugarcane Harvest. In: Silva, V, Hall, M, Azevedo, I (Eds.), Low Carbon Transition - Technical, Economic and Policy Assessment. InTech, London, United Kingdom, pp. 89–105.
Vandaele, N., Porter, W., 2015. Renewable energy in developing and developed nations: outlooks to 2040. J. Undergrad. Res. 15, 1.
Venkata Mohan, S., Mohanakrishna, G., Reddy, B.P., et al., 2008. Bioelectricity generation from chemical wastewater treatment in mediatorless (anode) microbial fuel cell (MFC) using selectively enriched hydrogen producing mixed culture under acidophilic microenvironment. Biochem. Eng. J. 39, 121–130. https://doi.org/10.1016/j.bej.2007.08.023.
Veringa, H.J. (2009) Advanced techniques for generation of energy from biomass and waste
Wang, L., Watanabe, T., 2020. The development of straw-based biomass power generation in rural area in northeast China-An institutional analysis grounded in a risk management perspective. Sustain 12, 1973–1995. https://doi.org/10.3390/su12051973.
WBA (2021) Global bioenergy statistics 2020. http://www.worldbioenergy.org/. Accessed 6 May 2021.
Wellinger, A., Lindberg, A. (1999) Biogas upgrading and utilisation. IEA Bioenergy. Task 24: Energy from biological conversion of organic waste.
World Bank (2021a) Sri Lanka - Energy Infrastructure Sector Assessment Program: Executive Summary. https://documents.worldbank.org/en/publication/documents-reports/documentdetail/843901561438840086/sri-lanka-energy-infrastructure-sector-assessment-program-executive-summary. Accessed 6 May 2021.
World Bank (2021b) Biomass resource mapping in Pakistan: final report on biomass atlas. https://documents.worldbank.org/en/publication/documents-reports/documentdetail/104071469432331115/biomass-resource-mapping-in-pakistan-final-report-on-biomass-atlas. Accessed 6 May 2021.
Xiaoping, J., Gan, X., Fang, W., Zhiwei, L., 2014. Opportunities and challenges for bioelectricity in rural China. Energy Procedia 50, 171–177. https://doi.org/10.1016/j.egypro.2014.06.021.
Xu, L., Zhao, Y., Doherty, L., et al., 2016. The integrated processes for wastewater treatment based on the principle of microbial fuel cells: a review. Crit. Rev. Environ. Sci. Technol. 46, 60–91. https://doi.org/10.1080/10643389.2015.1061884.
Yang, J., 2011. Crop residue based bioelectricity production prospect in China. In: ICMREE2011 - Proceedings 2011 International Conference on Materials for Renewable Energy and Environment, pp. 267–270.

Zhao, X., Liu, W., Deng, Y., Zhu, J.Y., 2017. Low-temperature microbial and direct conversion of lignocellulosic biomass to electricity: advances and challenges. Renew. Sustain. Energy Rev. 71, 268–282.

Zhao, X., Zhu, J.Y., 2016. Efficient conversion of lignin to electricity using a novel direct biomass fuel cell mediated by polyoxometalates at low temperatures. ChemSusChem 9, 197–207. https://doi.org/10.1002/cssc.201501446.

Zhou, M., Chi, M., Luo, J., et al., 2011. An overview of electrode materials in microbial fuel cells. J. Power Sources 196, 4427–4435.

CHAPTER 3

Environmental, social, and economic impacts of renewable energy sources

Zobaidul Kabir[a], Nahid Sultana[b] and Imran Khan[c]

[a]School of Environmental and Life Sciences, University of Newcastle, Ourimbah, NSW, Australia [b]University of Southern Queensland, School of Business, Toowoomba, QLD, Australia [c]Department of Electrical and Electronic Engineering, Jashore University of Science and Technology, Jashore, Bangladesh

3.1 Introduction

Renewable energy sources have the potential to facilitate universal access to electricity. The ratification of the Paris Agreement is an obligation for all countries to tackle climate change by reducing greenhouse gas (GHG) emissions. It is widely recognized that the technologies for generation of renewable energy are an indispensable part of addressing the threat of climate change impacts by reducing the emission of GHGs. It has been estimated that the generation of every 1 GW of extra renewable energy has the potential to reduce CO_2 emissions by an average 3.3 million tons (IRENA, 2016a). About 1000 million people are now living without electricity globally and 2700 million people lack access to fresh fuels and technologies for cooking (Shahsevari and Akbari, 2018) in developing countries. With this situation, it will be one of the key challenges to ensure the access of all people globally to both electricity and clean cooking by 2030, one of the seventeen Sustainable Development Goals (SDGs) of United Nations.

Human well-being and economic development significantly depend on energy and the demand for energy is rapidly increasing globally. One of the key reasons is that fossil fuel-based energy is not environment friendly. Overall, there are two main drawbacks for the use of fossil fuel for energy development: on the one hand, the stock of fossil-fuel is limited; on the other hand, fossil-fuel based energy generation affects the environment negatively. Climate change continues to be an increasing problem mainly due to the emission of GHGs from the generation of fossil-fuel based energy. In addition to GHGs, fossil-fuel based energy generation

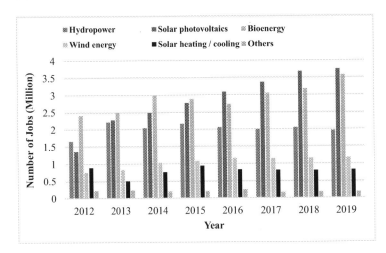

FIGURE 3.1 Trend of employment generation in renewable energy sector. *(Source: IRENA, 2020a).*

produces many other pollutants including Suspended Particular Matter (SPM) (Kabir and Khan, 2020). It should be noted that conventional coal is used for the generation of almost 40% of total electricity (IRENA, 2016b). The use of diesel and petrol also accounts for 3.28% of total electricity generation. Burning of these fossil-fuels releases large amount of pollutants including GHGs, each year.

Renewable energy sources are becoming attractive because of their relatively low emission of GHGs. The key potential of renewable energy sources is economic development with reduced emission of GHGs. Currently, the share of renewable energy generation is about 25% of total energy production globally (IRENA, 2018). It has been estimated that the share of renewable energy by 2030 would increase global GDP by up to 1.1% or USD 1.3 trillion. Doubling the share of renewable energy by 2030 would increase global welfare up to 3.7% against 1.1% of GDP. Increasing investment in renewable energy will trigger ripple effects throughout the economy at both the global and also the regional scale (IRENA, 2017).

It is assumed that the demand for primary energy globally will be increased by 1.5 to 3 times by 2050 (Dinc and Zamfirescu, 2011). The demand for energy is rapidly increasing in developing countries due to rapid urbanization, rural-urban migration, the rise of living standards and population growth (BP, 2016; Omer, 2008; Kalogirou, 2004), for example, Bangladesh, China, India and countries in Southeast Asia. The benefits of renewable energy are manifolds. In hard-to-reach areas where on-grid supply of electricity is not possible, renewable energy, for example, solar energy or wind energy can play a vital role in social and economic development. Table 3.1 shows the potential benefits of renewable energy with some examples of different countries.

From Table 3.1, it is apparent that one of the key benefits from renewable energy production is employment generation. Another is the development of the green economy. Importantly, the generation of employment opportunities applies to both developed and developing countries. These countries have already taken policy initiatives and targets for renewable energy generation. Fig. 3.1 shows the trend of global employment generation from renewable energy production activities. Although this is a global scenario, employment generation is likely to be increased in developing countries more rapidly than that of in developed

TABLE 3.1 Potential benefits from renewable energy.

Country/Region	Forecast year	Analysed policy interventions	Impact on GDP	Impact on employment
Chile	2028	20% renewables in electricity generation.	+0.63 (USD 2.24 billion)	7800 direct and indirect jobs
European Union	2030	-40% greenhouse gas emissions in 2030	+0.46%	+125 million economy-wide jobs
Germany	2030	Different targets for renewable energy development	Up to +3%	+1% of net employment
Ireland	2020	Meeting the target for wind by 2020	+0.2% to +1.3 %	+1150 to 7450 net employment
Japan	2030	Adding 23.3% (capacity in GW) of solar PV	+0.9% (USD 47.5 billion)	NA
Mexico	2030	21 GW of renewable capacity	+0.2%	+134,000 in the sector
Saudi Arabia	2030	54 (GW of renewable capacity	+4% (USD 51 Billion)	+137,000 in the sector
United Kingdom	2032	Large role of off-street wind instead of natural gas	+0.8%	+70,000 net employment
USA	2030	Decentralisation driven by renewable energy	+0.6%	+0.5 to +1 million

(Source: IRENA, 2017).

countries because many people in developing countries are unemployed or underemployed. Developing countries have taken initiatives for renewable energy production and training to avoid installation of coal-based power plants following the example of developed countries.

While there are benefits of renewable energy, there are also negative impacts of renewable energy (Vezmar et al., 2014; Kumar, 2020). There are negative social, environmental, and economic impacts of renewable energy sources (Khan, 2019a). In some cases, the negative impacts of renewable energy may become complex and contested. For example, hydroelectricity although provide cheaper electricity, may have destructive negative impacts. The overflooding of a dam or reservoir of hydropower plant may severely affect agricultural land, forest, wildlife, and nearby residents (Vezmar et al., 2014). The adverse impacts of solar system may include the use of large areas of land and thereby may cause the loss of wildlife habitat and vegetation (Khan, 2020). Similarly, biomass used for bioenergy generation may affect agricultural land and forest land (Kabir, 2021). The installation of renewable energy plants may cause conflicts between communities and proponents or companies if the local communities are not well informed and involved in the installation process.

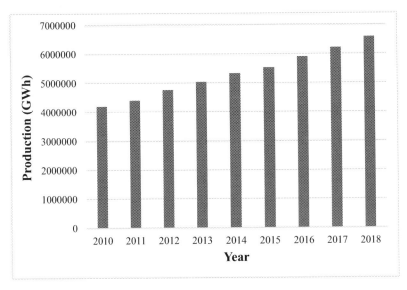

FIGURE 3.2 Global production of renewable energy. *((Source: IRENA, 2020b).*

The aim of this chapter is to provide an overview of the impacts of electricity generation from renewable energy sources. This chapter is divided into four sections. The second section outlines the global scenario of production of renewable energy following an introduction. The third section elaborates the social, environmental, and economic impacts of renewable energy. The fourth section provides a discussion based on the impacts identified in the third section. This will be followed by conclusions. It is to be noted that this chapter identifies the impacts are limited to solar, hydro, wind, biomass and solid waste, the key sources of renewable energy with available technology. There are other sources of renewable energy such as geothermal, tidal, and wave. However, there are technological limitations (e.g., tidal and wave energy) to generate energy from these sources though these sources also have potential to generate energy.

3.2 Global scenario of renewable energy production

Developed countries are already utilizing potential sources of renewable energy to produce electricity. Developing countries, although they depend mostly on fossil fuel-based electricity generation, are also taking the initiative to replace them by renewable energy. Encouragingly, the total global production of renewable energy in 2018 was approximately 7,000,000 GWh, compared with 4,000,000 GWh in 2010 as shown in Fig. 3.2.

Fig. 3.2 shows that the renewable energy production is increasing over the years. The sources of this renewable energy production include solar, hydro, wind, geothermal, biomass, in addition to solid waste, wave, and tidal. Technologies are already available for converting these renewable sources into energy. Table 3.2 shows a pen-picture of global renewable energy production (in million tons oil equivalent) by sources with projection of production by 2040.

3.2 Global scenario of renewable energy production

TABLE 3.2 Global renewable energy scenario by 2040 (Source: Kralova and Sjoblom, 2010).

Sources of renewable energy	Years				
	2001	2010	2020	2030	2040
Total production (Million tons oil equivalent)	10,038	10,549	11,425	12,352	13,310
Biomass	1080	1313	1791	2483	3271
Hydro	32.2	285	358	447	547
Geothermal	43.2	86	186	333	493
Wind	4.7	44	266	542	688
Solar	4.3	17.4	93	481	1332
Marine (tidal/wave/ocean)	.05	0.1	0.4	3	20
Total Renewable Energy	1365.5	1745.5	2964.4	4289	6351
Contribution of RES sources (%)	13.6	16.6	23.6	34.7	47.7

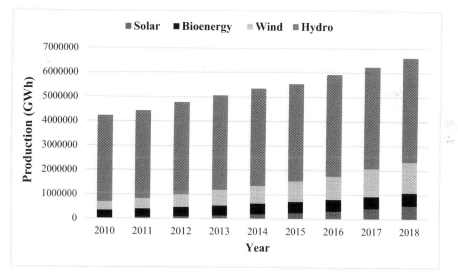

FIGURE 3.3 Trend of renewable energy production by sources (IRENA, 2020b).

In general, the potential of generating renewable energy from solar, bioenergy, wind, and hydro is noticeable as shown in Fig. 3.3. Given the available technology, solar energy production is rapidly increasing. The production of solar energy has increased from approximately 50,000 GWh in 2010 to 600,000 GWh in 2018 (REN21, 2019). Developing countries have high potential to receive adequate solar heat due to their geographic location (mostly in tropical zone) compared to developed countries (Teske et al., 2015). Bioenergy is receiving increasing attention (Poppa et al., 2014; Overend, 2019) where the production of bioenergy has reached from about 300,000 GWh in 2009 to 600,000 GWh in 2018, almost double in one decade (World Bioenergy Association, 2019). It is to be noted that the availability of feedstocks in the forest and agricultural sectors contributing to the increase of bioenergy (Wu et al., 2018a; Ashwath and Kabir, 2019).

Wind energy is commercially viable and therefore maintaining a double-digit growth since 2000 complying with a relatively mature technology (Jaber, 2013). Global production of wind energy has increased from 342,831 GWh in 2010 to 1,262,914 GWh in 2018. The application of wind energy technology may improve livelihood in developing countries through its contribution to agri-food chain (Leary et al., 2019; Tesfahunegny et al., 2020). It is estimated that there are more than 45,000 hydropower dams globally used for hydro electricity generation and about one-fifth of the global energy is supplied from these dams (Zarfl, 2014). The generation of hydropower increased by 2.5% in 2019 and the production of hydroelectricity have increased from 3,500,000 GWh in 2010 to about 45,000,000 GWh in 2018 (International Hydropower Association, 2020). There are thousands of nonpowered dams worldwide and it is possible to produce more hydroelectricity from these dams.

3.3 Impacts of renewable energy

The key renewable sources include solar, hydro, wind, and bioenergy (including waste). The impacts of each of the sources depends on the nature of sources, technologies used to produce energy, socioeconomic and biophysical context of a country and so the extent of social, environmental, and economic impacts (Sayed et al., 2021). The following sections outline the impacts of renewable energy generation from solar, hydro, wind, and bioenergy.

3.3.1 Solar energy

3.3.1.1 *Environmental impacts*

Among the renewable energy sources, solar energy is an infinite and almost pollution-free. Due to the release of insignificant GHGs and environmental pollutants, solar energy has already been popularized globally as a clean and alternative source. It has been estimated that the use of PV systems can reduce 69 to 100 million tons of CO_2, 126,000 to 184,000 tons of SO_2, and 68,000 to 99,000 tons of NOx by 2030. In countries where concentrating solar power (CSP) systems are applied, each square meter of concentrator surface is enough to save about 200 to 300 kg of CO_2 emissions annually (Shahsavari and Akbari, 2018).

One of the key negative impacts of solar energy is that for large-scale energy generation, huge amounts of land are required for solar energy plants. It is estimated that for small and large photovoltaic power plants, the area of land required ranges from 2.2 to 12.2 acres per megawatt (MW). In some countries where population density is very high, for example, in Bangladesh, development of solar plants is not feasible due to the scarcity of land. In this case, roof-top solar panel installation for households could be an alternative. Moreover, development of large-scale solar energy plants may have negative visual impacts at a place where ecologically critical areas or areas with natural beauty and heritage exist (Hohmeyer and Ottinger, 1992). The intensity of land-use may possible be reduced, given the increasing efficiency of solar panel (Sean et al., 2013). Furthermore, consumption of water in the production of photovoltaic modules is another negative environmental impact. It is estimated that to produce 1 kWh electricity generation from photovoltaics, 10 kg water is consumed (Evans et al., 2009). It is to be noted that the cost of solar power continues to be reduced (Evans et al., 2009).

3.3.1.2 Social impacts

For households in remote rural areas or islands where extension of the grid is costly and the supply of fossil-fuel based energy is technically difficult, there is an opportunity to overcome this difficulty by installing solar panels which is affordable and reliable (Saim and Khan, 2021). In rural areas of developing countries, social impacts of solar energy have been considered to a significant degree. The social impacts include the rise and creation of alternative livelihoods for poor people, increased educational facilities for students, improvement of health services, demographic impacts such as reduction of fertility rate, and improvement of women conditions (Bhandari et al., 2017; Grimm and Peters, 2016; Matungwa, 2014; Pramanik, 2012).

Access to indoor lighting of children enables them to study more effectively. In developing countries, solar energy plays a key role in promoting education or the level of education of communities especially in remote rural areas (Saim and Khan, 2021). The electricity supplied by solar panel not only provide facilities to the students to stay at school during daytime but gives them the opportunity to remain at night if required (Saim and Khan, 2021; Bhandari et al., 2017). In the absence of electricity, schools lack audio-visual teaching facilities. The solar energy system enables schools to use audio-visual facilities in addition to the use of information technology. It has been reported by the World Bank that the absence of electricity, both at school and home may affect students' performance significantly (Bahandari et al., 2017). For example, in Nepal, the performance of students was found poor in a village prior to electricity. Access to electricity using solar system, the academic performance of students was significantly improved. The electricity provided flexibility of study for students for extended hours at night (Bahandari et al., 2017).

Solar energy provided not only education facilities for students but also recreation for community members (Grimm and Peters, 2016; Pramanik, 2012). Television is now one of the recreation sources where electricity is available. Furthermore, because of solar electricity, tea stalls are available for social gathering where people gossip as part of their culture and thereby can spend their leisure time more effectively. Experiences shows that the diversification of recreation facilities due to electricity has direct impact on the reduction of population growth where population growth is high. The increased exposure of rural people to television increases the use of contraception and thereby leads to a reduction in the fertility rate. In the absence of electricity, recreation facilities can be limited and therefore couples may not have alternative choices other than intercourse (Grimm and Peters, 2016) and this may lead to a high birth rate.

Also, it has been possible to provide health services in rural areas for twenty-four hours and this can improve the quality of human health. In African countries, of 58% institutions, those providing health services have no access to electricity. Solar energy has the potential to solve this problem (Lee, 2013). Indeed, electricity from solar energy has opportunity to offer modern health services utilizing the necessary medical equipment and other facilities such as cooling of pharmaceutical products, and sterilization of medical equipment used in operations (UNESCO, 2014).

3.3.1.3 Economic impacts

The efficiency of PV systems is increasing and the cost for PV system continues to decrease (IRENA, 2019; Cengiz et al., 2015); since 2010 the cost of energy from solar systems has diminished significantly by 82% (IRENA, 2019). Increased efficiency and decreased costs will continue to play a significant role in providing low cost and affordable electricity (Cengiz et al., 2015) particularly unprivileged people living in rural areas. The economic impacts of solar energy include creation of employment, increase of income, women's empowerment, promotion of small entrepreneurships, increased quality of life and overall development (Wassie and Adaramola, 2021; Khan, 2020; Cengiz et al., 2015; Matungwa, 2014) at the household, community, and regional scale.

One of the key benefits from the solar PV system is the avoidance of cost resulting from the access to electricity. In Ethiopia, for example, the use of PV has helped people to charge their mobile phones without incurring any expense. It has been estimated that on average each household could save ETB 480-720 (US$ 18–27) per year due to the availability of electricity generated from solar PV systems. Furthermore, the consumption of kerosene was reduced with an associated reduction in overall household expenditure. It was estimated that on an average each of the household could save ETB1278 (US$48) per year (Wassie and Adaramola, 2021). The avoidance of these expenses is significant, particularly for the rural poor.

The availability of electricity from the PV system also promotes new entrepreneurships. Households who use solar energy and micro-business owners can lift their productivity (Wassie and Adaramola, 2021). This is made possible because electricity from PVs can considerably increase their working hours and so the income. Experiences elsewhere show that opening hours of business shops can be longer where solar PV is available. The opening of business shops relatively earlier (e.g., at 6 AM instead of 9 AM) may increase the productivity in business and services (Matungwa, 2014). Access to solar-electricity can create jobs for local communities by enabling the opening of small-business for example, rural barbershops and kiosks. The young people who are unemployed in addition to men and women can start small-businesses (Wassie and Adaramola, 2021) and create innovative ideas for different income generating activities although the opportunities for small-scale businesses may vary from one country to another given the socio-economic and geographic context.

The electricity from solar energy has enabled local communities in Nepal to take on more income generating activities in rural areas. The availability of solar electricity has significantly increased income generating training activities such as beekeeping, agriculture, computer use not only during the day but also at night. One of the key characteristics of solar energy is its reliability; that is, there is no problem with load shedding. The electricity from solar energy has allowed the villagers to start small income generation activities such as knitting, weaving, and other handicraft skills to earn extra money by utilizing their leisure time. Thus, small and cottage industries have been increasing in the village and this has played a vital role in income generation. More than 75% of villagers have stated that they could increase their income after the introduction of electricity. Some of the previously unemployed youths opened workshops for repairing mobile phones in addition to the development of entrepreneurships for income generation based on renewable energy (Bhandari et al., 2017).

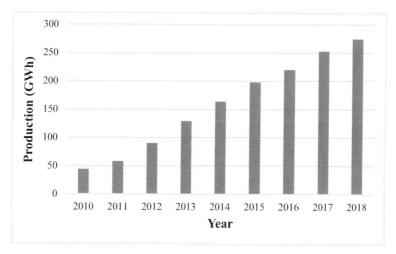

FIGURE 3.4 Solar energy production in Bangladesh (IRENA, 2020b).

Another example is Bangladesh, a densely populated country, with 3.8 million roof-top solar panels (solar home systems) which helped in the creation of 115,000 jobs by March 2015 (IRENA, 2017) and the number of jobs continues to increase. Fig. 3.4 shows the increasing production of electricity from Solar Home System (SHS). The availability of solar electricity in rural areas and islands of Bangladesh has created opportunities for irrigation and many other economic activities (Saim and Khan, 2021).

3.3.2 Hydro energy

3.3.2.1 Environmental impacts

Hydro energy is another renewable energy source that does not directly release any gaseous pollutants. It is also one of the cheapest sources of electricity. However, hydropower may have negative environmental and social impacts particularly where large hydropower plants are built (Elliott, 2015). The dam built for hydropower generation may change river flow characteristics including river ecosystems. The aquatic ecosystem of the river may cause the loss of habitats of fish species, affect the migration of aquatic species and disrupt normal aquatic biodiversity due to sedimentation and deterioration of water quality (Opperman et al., 2017; von Sperling, 2012). The construction of dams and the reservoir for water behind a dam may affect agricultural land, forest, human settlements, wildlife habitats and natural heritage. The consumption[1] of water by hydropower plants is another negative environmental aspect. It is estimated that 36 kg water is consumed to produce 1kWh hydroelectricity (Evans et al., 2009). Furthermore, a dam for storage of water is required for large hydroelectricity plants. These dams reserve huge volumes of water and may cause the loss of large amount

---

[1] For renewable energy technologies, it is difficult to quantify water consumption accurately during electricity generation. According to Inhaber (2004), it is not easy to distinguish between water withdrawal (water that is taken, then returned to circulation) and water consumption that is water removed from circulation outside the plant.

of water due to surface evaporation although this may vary according to the size of dams and ambient temperature (Inhaber, 2004). For large hydropower plants, the reservoirs may become vast and therefore may inundate large areas of land (Elliott, 2015) and affect both terrestrial and aquatic ecosystems.

Reservoirs of large hydropower plants located in tropical regions usually generate from 8 gCO_2 eq./kWh up to 6,647 gCO_2 eq./kWh due to the decomposition of biomass and organic matter (Song et al., 2018). However, run of river (RoR) hydropower plants do not need flooding reservoirs and therefore decay of biomass is negligible. Therefore, the emission of GHGs from decayed biomass may be low (Aung et al., 2020; Fearnside, 2016). Overall, hydropower plants may generate relatively lower impacts per unit of electricity generation (Aung et al., 2020; Siddiqui and Dincer, 2017) when compared with other electricity generation technologies.

3.3.2.2 Social impacts

The reservoirs can be overflooded and the surrounding forest, wildlife, agricultural land and therefore local communities may be affected adversely. Due to the loss of forests, local communities may suffer from livelihood where they were dependent on the forest or other resources. Because of inundation of agricultural land, the affected people may suffer from food shortage or even become poor. Importantly, there is a risk of failure of big dam and people who live nearby the hydropower plant may even be killed from over flooding. For instance, the Banqiao or Simantan dam failure in China in 1975 took 30,000 lives (Xu et al., 2008). Proper planning for protection of reservoirs is necessary where large hydroelectric plants are located.

One of the major impacts of large hydropower plants is the displacement of people (Boyé and deVivo, 2016). The construction of dam for water reservoirs force people to leave places where they used to live for many generations. The affected people are displaced involuntarily, or they move in another unknown place for their livelihood (Vezmar et al., 2014). Although resettlement plans are in many cases available and include socioeconomic support for the affected people, these supports often are not sufficient to reconcile the community cohesion and the emotional attachment to the place where they used to live. Due to the loss of relatives, friends and community, the loss of community cohesion is often irreparable (Yu et al., 2020; VanCleef, 2016). The indigenous communities may experience life-altering consequences including the loss of their own land, traditional lifestyle, livelihood, homes, and cultural heritage. The place, where the displaced people are resettled, may not be welcomed by the local communities at the beginning due to the "fear of the unknown." The displaced people may suffer from difficulties finding sources of income as they already have lost their networks. Thus, displaced people, particularly the most vulnerable such as pregnant women, children and older people may suffer from impoverishment (Cernea, 2004).

3.3.2.3 Economic impacts

During the construction and operation of hydropower plants, local people have opportunity to be employed and thereby improve the quality of life by earning more income. During the construction the economic activity may grow and reach a peak before the plant operation. Because of the significant investment in the project (millions or even billions of dollars) and employment of the workers for a significant period (e.g., 3–10 years) who spend their money on local goods and services economic activities continues to grow. In addition, because much of the money is spent by the workers locally, indirect jobs for local business are created to supply goods and services (de Faria et al., 2017). Local communities can benefit from irrigation of their cultivated land and fish farming. For example, in California the utilization of water from reservoirs made the surrounding land arable that would otherwise be desert (Boye and deVivo, 2016). However, the influx of foreign people during construction may affect the employment opportunities of local communities.

Furthermore, local communities benefit from the supply of electricity generated from hydropower (Vezmar et al., 2014). Local infrastructure can be improved including road construction. Using electricity facilities, local communities may improve their overall quality of life though various income generating activities and better education facilities and social recreational activities. However, local communities can be deprived of these benefits if the purpose of the construction of the plant is to transmit thousands of kilometers away from the generating plant (Boye and de Vivo, 2016; Severnini, 2014). Careful planning for the distribution of benefits of electricity is therefore necessary.

3.3.3 Wind energy

3.3.3.1 *Environmental impacts*

The wind farms can be located on open land, coastal land, on mountain ranges or also in water bodies, typically in lakes or in the oceans where wind speed is available. The establishment of wind power plant or wind turbines may affect biodiversity or affect habitat structure and function (Strickland et al., 2011). The disturbance of wildlife habitat may occur due to wind energy plant where it is located. The rotor blades of windmills create noise during the operation of windmills (Tesfahunegny, 2020) that may affect wildlife of the nearby forest and their quality of life (for example, by disturbing sleeping animals at night due to noise pollution) of local habitats. Another environmental aspect of windmills is landscape change or visual impacts (Tesfahunegny, 2020) although land use impacts vary based on where the windmills are located. It is to be noted that for a windmill with one MW capacity the use of land ranges from 30 to 141 acres.

Often, wind turbines rotor blades kill or injure birds both residents and migratory bats that fly into the rotors in the area (KPMG, 2019). One of the key concerns is the increased mortality of bats caused by wind turbines. This is a concern because bats are long-lived and have a low productivity rate and therefore population growth is relatively slow. Therefore, the mortality rate of bats caused by wind farms may limit their ability to recover from this decline and maintain a sustainable population. In North American countries such as the USA and Canada, the average annual fatality rate of bats caused by wind energy facilities located in the forested ridges was found to be high, that is 6 bats/MW of energy production (Arnett and Baerwald, 2013). Similarly, in the Black Forest region in Germany, the average annual fatality rate was also found to be high (10 bats killed/MW) respectively (Rydell et al., 2010). The EUROBATS reported that 6,429 bats were killed by wind energy plants in Europe between 2003 and 2014 (EUROBATS, 2014). Although, the fatality rates vary across the world given the locations of wind energy facilities and size of farms, the issue remains problematic.

Wind farms may also be the case that the loss of bird habitats may occur due to the clearance of forestland (Arnett et al., 2016). Feeding and nesting of birds and bats are also affected. The noise of turbines may prevent birds from using the trees located in the surrounding areas from feeding, breeding, and nesting (Arnett et al., 2013). The interference of the wind farms may force the movement routes of migratory birds and bats to change. Similarly, the offshore windmills may affect the aquatic ecosystem including habitats of marine animals and fish species. The habitats of marine animals and fish species may be disrupted during the construction and operation of wind farms. The noise of offshore wind farms may affect the movement of fishes and their breeding activities. Overall, the reduction, fragmentation or loss of available habitats may occur due the establishment of wind farms (KPMG, 2019).

Another environmental aspect of wind energy, in this case a positive aspect, is that it saves water since the generation of electricity from wind power does not require water. The generation of wind energy requires negligible amount of water. Given the lowest water consumption footprints, wind energy eventually helps in the reduction of consumption. This means, that the increase of wind energy instead of fossil fuel-based energy generation will save more water in future.

3.3.3.2 Social impacts

The esthetics of landscape and visual disturbances are key disadvantages of wind farms. A wind energy plant may cover, for example, 15 kilometers of the surrounding areas (Fernández-Bellon et al., 2019), disturbing the scenic natural resources. Often this esthetic issue is not adequately assessed involving local communities. The perception of communities in relation to wind energy plants varies and so does the perception of visual impacts and esthetic disturbance (Stigka et al., 2014). Though some in the community see the wind farms as beautiful, other community members may not like the project if the windmill with large turbines is located close to their homes and recreational areas. The community close to the windmills can be more affected than others (Jami and Walsh, 2014). Therefore, it is imperative to identify the visual character and scenic resources where the wind energy plant will be installed, in addition to local issues such as geographic location, site-selection and the distance of the windmill to be installed from residents (Fernández-Bellon et al., 2019; Stigka et al., 2014). Scenic landscape features of the site, and the surrounding context including community perception is an important way to understand the esthetic impacts.

One of the major social impacts of wind farms is noise generated from turbines. Wind energy systems create tonal noise (due to nonaerodynamic instabilities and unstable airflows) and broadband noise (is nonperiodic noise with a frequency of more than 100 Hz generated by interaction of wind turbine blades and atmospheric turbulence). Wind farms may disturb residents living close to the mills by making irritating sound. When wind farms create noise with high frequency, they may affect residents' quality of life in many ways, including sleep and hearing loss, headaches, irritability, fatigue, and annoyance and mental dissatisfaction (Dai et al., 2015). There can be other undesired noise created due to mechanical reasons. This type of noise can be generated from turbine's internal gearbox, generator, cooling fans and often they can be irritating and noticeable more than aerodynamic noise (Møller and Pedersen, 2011). It is to be noted that wind turbines generate about 105 dB of sound and therefore selection of sites for windmills should be far away from residents so that residents are not significantly affected by noise (Nazir et al., 2020). In general, with faster wind speed, windmills may cause louder noises with frequencies of 100 Hz or more (Dhar et al., 2020). Technological advances such as improved blade design may reduce wind turbine noise and enhance productivity of energy (Feder et al., 2015). Similarly, noise generated from mechanical reason can be reduced through insulation during manufacturing operations.

Another aspect of the social impact of wind farms is potential conflict between communities and companies. Local communities, the ultimate stakeholders of the project, if not included during the planning of the project, including the site-selection, and if they do not provide proper information about the impacts of the wind farms, both positive and negative conflicts may occur (Gonyo et al., 2021). Therefore, transparency is required from the company during planning and construction of windmills in addition to fair distribution of income and other benefits and including the community in the active decision-making process rather than being passive (Hall et al., 2013). The inclusion of the community from the very beginning of the project in the decision-making process helps to form mutual trust and respect between the companies and the communities. The local community cooperates with the companies if they are included and treated with trust and fairness. If it is not possible to remove the fear and mistrust from the communities, then social dynamics may be altered, discrimination may be felt and there may be a tendency to resistance. The community may assume that companies will violate their land rights, and their culture if they are not consulted (Momtaz and Kabir, 2018).

3.3.3.3 Economic impacts

From the economic viewpoints, the potential of wind energy to achieve energy transition is enormous. It is estimated that the role of wind energy in the supply of electricity could be nine times larger and could supply around 34% of global electric power demand in 2040 (up from 4% today). This is equivalent to the total power generation (14,000 TWh) of China, Europe, and the USA today. It could support a reduction of 23% of the carbon emissions needed in 2050 which is equivalent to 5.6 billion tons of CO_2. The reduction of carbon emission by 23% will provide huge economic benefits (equivalent to 386,400 million $ of social cost of carbon) globally and especially for developing countries who still lag in their adoption of renewable energy technologies and production of energy. Wind farms also contribute to the creation of jobs. For example, in European Union countries, the wind sector has created more than 300,000 jobs till 2019 and the number is increasing. The creation of jobs in this sector is also contributing to local employment in rural or remote and disadvantaged areas (KPMG, 2019). The economic impact will be substantial as well as positive.

3.3.4 Bioenergy

3.3.4.1 Environmental impacts

One of the key negative environmental aspects of bioenergy is the deterioration of water in terms of quantity and quality. It is to be noted that the conversion of land use to produce bioenergy crops could reduce the sources of water as well as the amount of water resources. For example, the cultivation of sugarcane in North Queensland in Australia to produce ethanol (biofuel) has potential impacts on the water consumption (Ashwath and Kabir, 2019). Most of the sugarcane fields in this region are located along the Great Barrier Reef (GBR) catchment area. The use of land to produce sugarcane in this region requires more water compared to other crops (e.g., wheat or soybean) (Jones et al., 2015).

The demand for bioenergy feedstock is increasing and this may lead to the clearance of forestlandsxsxss to grow more grains for biofuels. The clearance of forestland affects biodiversity including wildlife habitats, availability of food and may cause the imbalance of ecosystem. The conversion of land for bioenergy corps production may affect biodiversity including the loss of natural habitat (Kabir, 2021). To produce bioenergy, the extensive practice of energy crops monocultures may negatively affect the local biodiversity. This may include the loss of local vegetation and habitats for animal species and the expansion of invasive plants. The demand for biofuels is rising and this may motivate government or companies to clear natural habitats such as tropical forests. For example, the increasing expansion of soy cultivation for biofuels in the Amazon forests may deteriorate the quality of biodiversity (Kabir, 2021; UNEP, 2008).

Production of crops for bioenergy may affect the quality of water as there is an association between water quality and farming of bioenergy crops. The nutrient pollution resulting from surface runoff may cause degradation of water quality or aquatic ecosystems of nearby waterbodies including rivers, lakes, or coastal areas. It has been found that sugarcane production is one of the highest fertilizer users among the bioenergy crops and there is low nutrient use efficiency. The use of fertilizer for sugarcane production may release nitrate, for example, to water ways and decrease the content of soil nitrogen as well (Ashwath and Kabir, 2019).

However, the production of bioenergy also has environmental benefits/positive impacts. One of the key benefits of bioenergy is the reduced emission of GHGs. Biofuel, for example, emits the amount of GHGs far fewer than fossil fuel (Fu et al., 2014; Dunn et al., 2013; Wang et al., 2012). It is evident that ethanol used for transportation may reduce 40 to 85% of GHG emissions compared with fossil fuel (e.g., gasoline or diesel) per megajoule (MJ) (Li and Khanal, 2016; World Energy Council, 2013). Furthermore, the use of marginal land to produce bioenergy may have a positive impact on GHG emission (Liu et al., 2017) . To produce bioenergy crops, the transition of land use from arable or perennials, for example, growing grasses may reduce 30 to 40% of total nitrogen loss in comparison with traditional system of cotton production (Chen et al., 2017). The production of bioenergy crops using marginal land may protect the loss of soil-nutrient (Guo et al., 2018). In addition, the use of marginal land for recurrent production of grasses may be beneficial to the improvement of water quality as the production of grasses requires almost no pesticides (Hoekman et al., 2018). Another source of bioenergy is biogas which can play a significant role in reducing GHGs and thereby improve the air quality (Paolini et al., 2018). Bangladesh is an example where thousands of biogas plants are functioning using biowaste from poultry farms and thereby playing a key role in both reducing GHGs emission and improving the quality of life by reducing the risk of indoor air pollution and illness.

3.3.4.2 Social impact

The traditional use of bioenergy or biofuel may affect human health (Khellaf, 2018). Although the generation of heat or energy in developed countries mostly in place using biofuel efficiently, the use of biomass is still traditional in developing countries. The use of primary biomass such as firewood or cow dung is used for cooking in many developing countries. The stove that uses these primary biomasses is with low efficiency and cause indoor air pollution by releasing smoke, carbon monoxide and other toxic matters. As a result, people, particularly women and infants who spent most of the time at home, may suffer from respiratory illness (Chum et al., 2011).

One of the key issues relating to social impacts is the involvement of women in planning bioenergy projects for the establishment of bioenergy plants at local level. This is important because, women depend on biomass for example, firewood and residues from agricultural land for cooking. Also, agricultural activities done by women are well recognized in some countries. The biofuel plants may make the women and female-headed households marginalized and they may suffer from food security as well as energy security (Beall and Rossi, 2011). Therefore, when a bioenergy plant is installed, women need to be involved in the project planning. Furthermore, active involvement of women in project planning for bioenergy is important because they may contribute to the development of the project so that it is suitable to the local environment. Without the participation of women, some key issues such as food security, conservation of nature and energy security may be omitted (Edrisi and Abhilash, 2016).

Due to the increasing demands of renewable energy the expansion of cultivation of bioenergy crops in agricultural land may threaten food security. Countries such as in Canada, USA, Australia and even Brazil are expanding cultivation of bioenergy crops to produce biodiesel. On the other hand, many countries suffer from scarcity of land for food production. The production of bioenergy crops therefore competes not only with food but also for the same resources for example, labor force, land, and water (Hoekman et al., 2018; Dale et al., 2013). It is well recognized that food security is one of the key development goals. Therefore, a potential conflict between food security and energy security (where land is used for bioenergy) may occur at the global, national, regional, and local scale (UNEP, 2013). The competition of bioenergy and food security may affect national or even global policy. For example, in the USA, to produce biofuels, about 40% of US corn is now required. For this reason, it has been estimated that world food prices will rise by about 32% by 2022 (OECD and FAO, 2013). This means the increased use of bioenergy will affect the food prices when food or feed crops are used for energy production (Poppa et al., 2014) and this (biofuel-induced) price may have significant impacts on developing nations where food shortage is a common issue (Gheewala et al., 2013).

3.3.4.3 Economic impacts

It has been recognized that bioenergy production requires a bigger labor force than many other renewable energy sources. Studies show that five times more manpower is required for the production of bioenergy than is needed for the production of fossil fuels (Remedio and Domac, 2003). Another study shows that the production of per unit of energy from biofuel may employ about 100 times more workers than the fossil fuel industry (Gheewala et al., 2013). Furthermore, the level of direct jobs required for the operational activities of bioenergy production is about four times higher than the level that is needed for the operation of fossil fuel-based energy production (Paredes-Sanchez et al., 2019; Jackson et al., 2018).

Bioenergy offers employment and development opportunities in rural areas, especially in developing countries. It is to be noted that the potential for employment in bioenergy is very high among all renewable energy resources. The production and preparation of biomass, construction of power plant and operation and management of plants require massive labor forces (Vezmar et al., 2014). Overall, 11 million people are employed in the renewable energy sector (IRENA, 2019) where bioenergy is the second largest employer after solar energy among the renewable energy sources. It is to be noted that a significant part of the bioenergy industry is not considered, largely because traditional biomass is not included in the renewable energy sources. This would indicate a much higher employment figure for the sector (World Bioenergy Association, 2019). With the aim of job creation through bioenergy production, some developing countries, for example the Philippines, have adopted bioenergy policies with the purpose of job creation. The policy of Philippines government states that the country will become one of the leading exporting countries of biofuels by 2030. By exporting biofuels, the country will be able to create both employment opportunities and economic growth (Gheewala et al., 2013). Similarly, Indonesia hopes to create substantial number of employment opportunities by exporting 1225 million liters of biofuel produced from oil palm in 2011 (USDA, 2012). However, there is a risk of the creation of employment opportunities depending on the policy options of a country. If the country allows foreign investors and the foreign investors can bring their own workers, the domestic employment opportunities will be limited (Domac and Segon, 2020). A careful policy development particularly in developing countries is therefore necessary.

3.3.5 Solid waste

The increasing amount of solid waste, particularly in municipal areas is going to be a huge source of renewable energy globally. Given the rapid urbanization, population growth and economic growth and change of lifestyle, the amount of Municipal Solid Waste (MSW) is increasing, and technologies have developed for generation of energy from it, that is, Waste to Energy (WtE) plants. These WtE plants have environmental, social, and economic impacts.

3.3.5.1 Environmental impacts

The environmental impacts of solid waste management or the generation of energy from solid waste are discrete issues, but both are important. In developing countries, solid waste is still poorly managed. MSW is dumped at open land and therefore there is a possibility of deterioration of water quality both of ground and of nearby water bodies such as lakes or rivers. The toxic materials including heavy metals such as cadmium, mercury and in addition to nutrients can be transported from the dumped sites into the nearby water due to heavy rain and water quality can deteriorate. The aquatic habitats for various species including fish live in water can be heavily affected due to additional nutrients. It is to be noted that additional nutrients released from water often cause eutrophication and therefore Bio Oxygen Dissolved (BOD) is reduced. All these issues could be minimized through WtE plants. However, the emission of CO_2, CO, SO_2, NOx, N_2O, HCl, NH_3, and HF can be released from WtE plants and affect the air quality. The fly ash and bottom ash and suspended particulate matters such as PM_{10} and $PM_{2.5}$, are also released from plants. The bottom ash may spread down over the land nearby the WtE plant and the crops and soil quality can deteriorate.

The technologies used for generation of energy from MSW require land. The selection of sites for the establishment of plants for energy production requires clearance of land and this may cause habitat loss and removal of species such as birds and wildlife (Kabir and Khan, 2020) in addition to fragmentation of habitats. The fragmentation of habitats may isolate the wildlife species, reduce the availability of food through damaging the food chain and they might be forced to migrate to other places.

During the construction of plants, the noise may affect wildlife living nearby. In addition, frequent transportation of MSW from municipalities to the plant site for energy generation may cause disturbance to the local communities. The noise from operation of plants may affect the local communities and wildlife species those live nearby. Another issue could be the odor, which could occur during MSW transportation and even during plant operation if not managed properly.

3.3.5.2 Social impacts

Although WtE plants release relatively less GHGs than fossil-fuel based power plants (Roberts and Chen, 2006) there are public health risks. The release of smoke, dust, noise from plant and frequent traffic congestion can be the causes of mental and physical illness for local communities particularly for vulnerable people such as children, aged people, and pregnant women. Dioxins may negatively affect the reproductive function, sexual development, and immune system of human bodies (Robert and Chen, 2006). If the MSW is dumped/stored at a site for longer time than usual, this may cause health risks for local communities and workers because of odour released from the waste (Wu et al., 2018a). This is obvious where many people live close to the WtE plants. The exposure to odours for a long period of time may negatively affect local people by putting physical and mental stress including 'including anxiety, headaches, vomiting, and respiratory problems' (Hayes et al., 2017; Wu et al., 2018b).

There are esthetic impacts that may be caused by WtE plants. The local residences, recreational places, educational and religious institutions, heritage can be affected due to visual impacts. Overall, the sites for WtE generation plants, if not selected properly, may decrease the "esthetic value of the surrounding community places" (Weinstein, 2006). The transformation of landscape due to plant installation or dumping sites may create adverse visual impact and the value of land and other assets can be reduced in addition to the value of natural and cultural heritage (Tudor, 2014; Tolli et al., 2016).

3.3.5.3 Economic impacts

Energy generation from MSW can create new jobs and business opportunities, not only during the construction of a plant, but also during the operation of a WtE plant. Some backward and forward linkage businesses can be established, and this can open up opportunities for local people to be employed (Weinstein, 2006). From collection of waste to deliver it to the plant through transportation, a good number of workers are required. In addition, operation of the plant requires skilled workers, usually hired from local communities. The workers hired from local communities are provided with training for the development of skills. It is evident that a WtE generation plant with medium capacity may employ about 100 employees in developing countries (GIZ, 2017). The electricity generated from MSW may open new opportunities for employment and self-employment through raising the small entrepreneurships.

3.4 Discussion

The impacts of renewable energy vary according to the type of renewable energy sources and technology for renewable energy production. The extent of one type of impacts may vary according to the type of energy sources (e.g., solar or wind) in addition to size of the plant. For example, while the environmental impacts of solar energy may be lower than other sources of renewable energy, the environmental impacts of other sources such as hydro or bioenergy can be significant if proper precautions are not taken. The noise pollution for example, is not significant for solar energy sources. However, noise pollution for wind energy sources can be significant if there are residents close to the wind energy plant. In general, the environmental impacts of all renewable energy sources may be significant depending on the situation[2]. Table 3.3 shows the summary of impacts of renewable energy sources (solar, hydro, wind, bioenergy, and MSW) irrespective of size and socioeconomic context of renewable energy plants.

Similarly, the social impacts of renewable energy from all relevant sources are not the same. The analysis of social impacts of renewable energy sources may indicates that there are both positive and negative social impacts for all renewable energy sources. However, the extent of the impact varies according to the source type, size of plant and technology used. The scope and extent of social impacts of a large renewable energy plant will not be as same as those of a small or medium-size renewable energy plant. In the case of a big dam for hydroelectricity, the extent of social impacts such as involuntary displacement of people is much higher than that of any other renewable energy plant. In addition, experience shows that the social impacts of renewable energy sources also depend on the socioeconomic and geographic contexts of countries. In general, developing countries' experience of social impacts of renewable energy sources are not always same given it socioeconomic context.

The economic impacts such as employment generation, income generating activities (e.g., development of entrepreneurship, avoidance of the cost of GHGs) also depend on the types of renewable energy sources and size of the plants to generate renewable energy. Geographically, not all countries have same opportunities for generation of renewable energy. This depends on resources availability and capacity. For example, countries located in coastal areas have opportunities to establish wind farms in addition to tidal and wave energy. Countries located in tropical areas have more opportunities to use solar energy than others. So, the overall economic impacts will vary from one country to another given the available renewable energy sources and their utilization. Where countries have same opportunities to generate renewable energy, the technologies to be adopted may not be the same. For example, Bangladesh is suitable for solar energy production, but this is a very densely populated country. Therefore, due to scarcity of land, it is not possible to use land for solar panel. Only roof top solar system is suitable in the context of Bangladesh. Similarly, for the generation of energy from MSW, there are four key technologies widely used such as anaerobic, gasification, pyrolysis, and incineration. Studies show that in the case of Bangladesh, anaerobic digestion is the most suitable technology (Khan and Kabir, 2020) because of the characteristics of MSW.

Furthermore, the impact of renewable energy may vary given the local, national, regional, and global scales. The impact of renewable energy at the global scale is undoubtedly positive due to the release of reduced GHGs and thereby curbing the impacts of negative climate change. At national level, the overall impacts are also found to be positive. At the local level, communities' experience about the impacts of renewable energy can be both positive and

---

[2] The key focus of this chapter is to identify possible impacts of renewable energy rather than assessment of the significance of the impacts.

TABLE 3.3 Major impacts of renewable energy sources.

	Environmental	Social	Economic
Solar energy	The release of GHGs is negligible.	Education for kids is facilitated by solar energy in remote rural areas.	More income due to extended time for work.
	Vast land is required for large-scale solar energy plant.	Provide more recreational activities among community people.	Opportunities for businesses or entrepreneurships.
	Solar Home System (SHS) does not require any land.	Improved quality of life.	Employment generation.
Hydro-energy	Release of small scale GHGs for large reservoirs.	Involuntary displacement of people due to construction of dam.	Employment generation.
	Damage of aquatic ecosystems of rivers due to change of river water flow.	Resettlement of displaced people but loss of community cohesion, network for finding jobs and even suffer from impoverishment.	More income generating activities.
	Loss of natural heritage and habitats for wildlife and residents due to construction of dam.	Local communities are benefited (e.g., education, improved quality of life, recreation) if they are provided with electricity.	Industrialization due to the production of high amount of electricity.
Wind energy	No GHGs is released during operation and no water is required.	Noise may affect residents physically and mentally who live close to the plant.	Employment generation is an opportunity.
	Noise generated by turbines affects wildlife including nesting and breeding.	Price of land may be reduced due the location of the plant.	People have more income generating activities.
	Vast amount of land is required for high-capacity wind power plant that may affect agriculture or forest in addition to affecting aquatic ecosystem by offshore wind energy plants.	Disturbance or affect esthetic view or that has visual impact.	Comparatively less local economy development.

(continued on next page)

TABLE 3.3 Major impacts of renewable energy sources—cont'd

	Environmental	Social	Economic
Bioenergy	Release some GHGs due to burning of biomass.	Negative impact on health during cooking and heating.	Burning of biodiesel releases negligible GHGs in comparison with fossil fuel.
	Production of sugarcane as a source of bioenergy (e.g., ethanol or biodiesel) consumes more water and negatively affects soil quality.	Social conflict may occur if the local communities are not considered during project planning.	Offer employment opportunities for rural people.
	Biomass from forest-wood may affect soil quality and illegal logging.	The project proponents or companies need to include local women in decision making as they are to collect firewood for cooking.	More income generation due to opening of new business or entrepreneurship.
Solid waste	Air pollution due to emission of GHGs from WtE plants.	There are public health risks due to the release of smoke, dust, toxic materials, odor from WtE plants.	Creation of new jobs for local communities during the construction and operation of WtE plants. Development of skills of local people through training to operate the plants.
	WtE plants release fly ash and bottom ash those may affect the quality of soil and may decrease the soil fertility.	Frequent traffic congestion may occur while transporting MSW and this may cause mental and physical illness for local communities.	Construction of WtE plants may provide with business opportunities for local communities through the creation of new entrepreneurship.
	The noise of WtE plant may disturb animal species living nearby. Animal species may be forced to leave their living places and affect breeding activities.	WtE plants may affect residents, recreational places, and heritage. WtE generation plants may decline the "esthetic value of the surrounding community places."	The transformation of landscape due to installation of WtE plants may generate adverse visual impact and this may negatively affect the value of land and other assets of local communities.

negative. At this level, the installment of a renewable energy may make the community people both looser (for example, due to the loss of land) and winners (e.g., a business opportunity). Proper planning and policies may address the negative impacts and maximize the positive impacts (Kabir et al., 2021; Khan et al., 2020).

From sustainability points of view renewable energy is more sustainable than fossil fuels-based energy generation. The environmental cost of the generation of energy from fossil fuels, although cheaper than renewable energy sources, is very high and it is obvious that fossil fuels are responsible for global warming. However, it is apparent that the generation of renewable energy is not without negative environmental, social, and economic impacts. Therefore, maximizing the positive social, environmental, and economic impacts and minimizing the negative impacts of renewable energy technology generation are essential (Khan, 2020a, 2020b; Khan, 2019b). The key issue is how to minimize these negative impacts and maximize the benefits.

While developed countries are in well positioned to generate renewable energy, developing countries are not even though most of the developing countries have renewable energy policies in places and are taking initiative to generate renewable energy (Khan et al., 2020). In developing countries, many people not yet have access to electricity which is essential for sustainable development. In developed countries, all people have access to electricity long ago either renewable or nonrenewable electricity. To avoid GHGs emissions, the developed countries are now replacing fossil fuel-based energy with more renewable options in addition to ensure consistency of energy supply and energy security. Similarly, developing countries are also adopting renewable technologies toward sustainable development. In general, impacts of renewable energy adoption in developing countries will be positive given its socio-economic conditions.

3.4.1 Future impacts in developing countries

3.4.1.1 Social impacts

The renewable energy, for example, solar system will provide electricity in the remote areas where grid electricity supply is not possible. Many people in the developing countries particularly in Asia and Africa are still lack of access to electricity. Solar energy in this case, may support them not only to get access to electricity but also to create income generating activities (Khan, 2020). Most of the developing countries are agriculture based and therefore agricultural residue is a viable source of energy. The renewable energy will also provide employment opportunities in developing countries and will play a vital role in poverty reduction. Also, the renewable energy will play a crucial role in women empowerment through self-employment and income generation as mentioned earlier.

Overall, renewable energy will contribute to more income generation, improved health system, education quality, and other social wellbeing. The contribution of solar power in Nepal in socioeconomic development is a good example. Furthermore, the decentralization of renewable energy technology, for example solar energy may contribute to rural development (Khan, 2020) including employment generation and promotion of cultural activities. According to Khan (2020), the obvious positive impacts of solar energy (i.e., SHS) are quality education for children in rural area, access to information and communication devices, and health benefits.

Furthermore, access to energy for people particularly in rural areas in developing countries will contribute to social development. Developing countries have already taken initiatives to improve the access to electricity and facilities for clean cooking and heating (Farabi-Asl et al., 2019) through the installation of solar panel, bioenergy, and wind energy plants. For example, the biomass can support to produce biogas and biogas can be used by improved stove for cooking and heating at the household level in local rural areas. The use of bioenergy means a shift from the use of cow dung, agricultural residues to renewable energy-based cooking system (Shaibur et al., 2021).

3.4.1.2 Economic impacts

It is apparent that the increasing demand for energy is being influenced by economic growth. The developing countries, for example Southeast Asian countries and other countries including Brazil, India, China, and Bangladesh have increasing economic growth and therefore increasing demand for energy. This requires a shift from conventional fossil fuel-based energy to renewable energy. Developing countries will be benefited from this shift to sustain continuous economic growth. The economic benefit will increase given the increase of renewable energy sources. The costs of renewable energy technologies are gradually reducing (Evans et al., 2009). Therefore, it is expected that developing countries will enjoy more benefits in future. For example, the costs of generating solar and wind power have reduced in recent years (IRENA, 2020; Cengiz et al., 2015) and renewable energy will outprice fossil fuel in future.

Moreover, in the energy sector, uncertainty of energy supply may disrupt the continuation of economic growth both at local level and national level in developing countries. Renewable energy may ensure uninterrupted supply of energy and thereby can promote economic development. Renewable energy produced locally in developing countries may reduce the dependency on the import of fossil fuel such as coal and petroleum (Sen and Ganguly, 2017) and thereby save many. The adoption of solar technology at households in rural areas will contribute to economic development through additional and variety of economic activities. The additional economic activities will generate more household income than before.

3.4.1.3 Environmental impacts

While developed countries are emitting more GHGs than those of developing countries, developing countries are also affected badly. For example, Bangladesh is one of the worst victims of negative climate change impacts even though the emission of GHGs by Bangladesh is negligible. With increasing adoption of renewable energy technology given the available renewable energy sources, developing countries will contribute to the reduction of environmental pollution. It will be possible to avoid the emission of GHGs through renewable energy use that would be emitted if fossil-fuel based energy were used. This avoided emission may come from the production of electricity from solar, hydropower or bioenergy.

In addition to reduction of GHGs due to renewable energy, there are other environmental benefits such as reduced air pollution with suspended matters and improved health conditions (less respiratory infection and other relevant diseases). It is to be noted that seven million people are killed annually due to air pollution largely from burning fossil fuels. Developing countries carry the highest burden of this consequence (mortality). The increasing use of renewable energy will reduce this burden in developing countries. Also, in developing countries, most of the households in rural areas use fuel wood from nearby forest, coal, or dung primarily for cooking and heating. Often collection of fuelwood and dung may affect biodiversity including wildlife habitat and soil fertility respectively. The use of renewable energy will address these environmental issues significantly.

In summary, with increasing renewable energy production, developing countries will be benefited financially, socially, and environmentally. Although there are some barriers such as shortage of skilled manpower, lack of awareness among people about the benefit of renewable energy, insufficient technology and technical knowhow, and institutional set up (Khan, 2020), it is expected that developing countries will be able to overcome all of these barriers with proper policies and action plans (Kabir et al., 2021; Khan et al., 2020).

3.5 Conclusion

The aim of this chapter was to identify the social, environmental, and economic impacts of renewable energy. In this chapter four key sources of renewable energy, including solar, hydro, bioenergy, and wind energy were considered in addition to waste energy. This chapter indicates that there are both positive and negative impacts of renewable energy generation. The positive impacts in general are reduced emission of GHGs, improved quality of life due to solar energy, increased income, and employment opportunities. The negative impacts include the use of arable land due to establishment of plants for energy generation, noise generation that may affect social life and human health, and the clearance of land that affects wildlife or biodiversity. The impacts identified indicate that the extent of common types of impacts varies according to the sources of renewable energy, technology used for the generation of renewable energy and socioeconomic context of a country. Furthermore, the feasibility of renewable energy generation depends on the availability of renewable energy resources because the sources of renewable energy vary from one country to another. Moreover, it is apparent that the future impacts of renewable energy in developing countries are encouraging. While there are some negative impacts, these can be addressed by adopting proper policies and plans and thereby developing countries can maximize the benefits of renewable energy. Additionally, this chapter did not focus on the comparison of positive and negative impacts and identification of the best renewable energy sources that is more sustainable than any other sources. A comprehensive research can solve this issue.

Self-evaluation questions

1. Why renewable energy is important for developing countries?
2. Renewable energy generation have both positive and negative social, environmental, and economic impacts. Explain.
3. Compare between positive and negative impacts of renewable energy generation.
4. Hydropower is one of the cheapest renewable energy sources but has significant social impact such as involuntary displacement. How to minimize this impact?
5. Outline in brief the environmental impacts of key renewable energy sources.

6. What are the negative social, environmental, and economic impacts of renewable energy and how to minimize these impacts?

References

Ashwath, N., Kabir, Z., 2019. Environmental, economic, and social impacts of biofuel production from sugarcane in Australia. In: Khan, M., Khan, I. (Eds.), Sugarcane Biofuels. Springer, USA available at https://doi.org/10.1007/978-3-030-18597-8_12.

Arnett, E.B., Baerwald, E.F., 2013. Impacts of wind energy development on bats: implications for conservation. In: Adams, RA, Peterson, SC (Eds.), Bat Evolution, Ecology, and Conservation. Springer, New York, pp. 435–456.

Aung, T.S., Fischer, T.B., Azmi, A.S., 2020. Are large-scale dams environmentally detrimental? Life-cycle environmental consequences of mega-hydropower plants in Myanmar. Int. J. Life Cycle Assess. 25, 1749–1766.

Arnett, E.B., Baerwald, E.F., Mathews, F., Rodrigues, L., Rodríguez-Durán, A., Rydell, J., 2016. Chapter-11: Impacts of wind energy development on bats: a global perspective. In: Voigt, C.C., Kingston, T. (Eds.), Bats in the Anthropocene: Conservation of Bats in a Changing World. Springer, USA, pp. 232–295.

Boyé, H., and de Vivo, M., 2016. The environmental and social acceptability of dams, Field Actions Science Reports, Special Issue 14, available at http://journals.openedition.org/factsreports/4055.

Beall, E., Rossi, A., 2011. Good Socio-Economic Practices in Modern Bioenergy Production: Minimizing Risks and Increasing Opportunities for Food Security. FAO, Rome, Italy.

BP, 2016. Statistical Review of World Energy. Energy Outlook to 2035.

Bhandari, B., Lee, K., Chu, W., Lee, C.S., Song, C., Bhandari, P., Ahn, S., 2017. Socio-Economic Impact of Renewable Energy-Based Power System in Mountainous Villages of Nepal. Int. J. Precis. Eng. Manuf. -Green Technol. 4 (1), 37–44.

Cengiz, M.S., Mami, M.S., 2015. Price-efficiency relationship for photovoltaic systems on a global basis. Int. J. Photoenergy 256101, 12 pages, available at http://dx.doi.org, accessed on 25/3/2021.

Chen, Y., Ale, S., Rajan, N., Srinivasan, R., 2017. Modeling the effects of land use change from cotton (Gossypium hirsutum L.) to perennial bioenergy grasses on watershed hydrology and water quality under changing climate. Agric. Water Manag. 192, 198–208.

Cernea, M., 2004. Social Impacts and Social Risks in Hydropower Programs: Pre-emptive Planning and Counter-risk Measures. In: United Nations Symposium on Hydropower and Sustainable Development. Beijing, China 27–29 October, 2004.

Chum, H., Faaij, A., Moreira, J., Berndes, G., Dhamija, P., Dong, H., et al., 2011. Bioenergy. In: Edenhofer, O., et al. (Eds.), IPCC Special Report on Renewable Energy Sources and Climate Change Mitigation. Cambridge University Press, New York, USA, pp. 209–331 ISBN: 978-1-107-60710-1.

Dinc, I., Zamfirescu, C., 2011. Sustainable Energy Systems and Applications. Springer, New York, Dordrecht Heidelberg London, p. 823 http://dx.doi.org/10.1007/978-0-387-95861-3.

Dhar, A., Naeth, M.A., Jennings, P.D., Gamal El-Din, M., 2020. Perspectives on environmental impacts and a land reclamation strategy for solar and wind energy systems. Sci.Total Environ 718, 134602.

Elliott, D., 2015. Green Energy Futures: A Big Change for the Good. Palgrave Macmillan, Basingstoke doi:10.1057/9781137584434.0005.

Dunn, J.B., Mueller, S., Kwon, H., Wang, M.Q., 2013. Land-use change and greenhouse gas emissions from corn and cellulosic ethanol. Biotechnol. Biofuels 6, 51.

Dale, V.H., Efroymson, R.A., Keith, L., Kline, K.L., 2013. Indicators for assessing socioeconomic sustainability of bioenergy systems: a short list of practical measures. Ecol. Indic. 26, 87–102.

Dai, K., Bergot, A., Liang, C., Xiang, W.N., Huang, Z., 2015. Environmental issues associated with wind energy — a review. Renew. Energy 75, 911–921.

Domac, J., and Segon, V., 2020. IEA Bioenergy Update-31, International Energy Agency, http://www.ieabioenergy.com/wp-content/uploads/2013/10/IEA-Bioenergy-Update-31-Task-29-Technology-Report.pdf.

de Faria, F.A.M., Davis, A., Severnini, E., Jaramillo, P., 2017. The local socio-economic impacts of large hydropower plant development in a developing country. Energy Econ. 67, 533–544.

Evans, A., Strezov, V., Evans, T.J., 2009. Assessment of sustainability indicators for renewable energy technologies. Renewable Sustainable Energy Rev. 13, 1082–1088.

References

Edrisi, S.A., Abhilash, P.C., 2016. Exploring marginal and degraded lands for biomass and bioenergy production: an Indian scenario. Renewable Sustainable Energy Rev. 54, 1537–1551.

Farabi-Asl, H., Taghizadeh-Hesary, F., Chapman, A., Bina, S.M., Itaoka, K., 2019. Energy challenges for clean cooking in Asia, the background, and possible policy solutions. ADBI Working Paper 1007. Asian Development Bank Institute, Tokyo available at: https://www.adb.org/publications/energy-challenges-clean-cooking-asia .

Feder, K., Michaud, D.S., Keith, S.E., Voicescu, S.A., Marro, L., Than, J., et al., 2015. An assessment of quality of life using the WHOQOL-BREF among participants living in the vicinity of wind turbines. Environ. Res. 142, 227–238.

Fu, J., Jiang, D., Huang, Y., Zhuang, D., Ji, W., 2014. Evaluating the marginal land resources suitable for developing bioenergy in Asia. Adv. Meteorol. 4, 1–9.

Fernández-Bellon, D., Wilson, M.W., Irwin, S., O'Halloran, J., 2019. Effects of development of wind energy and associated changes in land use on bird densities in upland areas. Conserv. Biol. 33 (2), 413–422. doi:10.1111/cobi.13239.

Fearnside, P.M., 2016. Greenhouse gas emissions from Brazil's Amazonian hydroelectric dams. Environ. Res. Lett. 11 (1), 1–3. https://doi.org/10.1088/1748-9326/11/1/011002.

Gonyo, S.B., Fleming, C.S., Freitag, A., Goedeke, T.L., 2021. Resident perceptions of local offshore wind energy development: modelling efforts to improve participatory processes. Energy Policy 149, 112068.

IRENA, 2019. Renewable Energy: Market Analysis, Dubai, available at http:irena.org, accessed on 22/2/2021

Grimm, M., Peters, J., 2016. Solar off-grid markets in Africa. Recent dynamics and the role of branded products. Field Actions Sci. Rep. Special Issue 15, 161–163 available at http://journals.openedition.org/factsreports/4222. accessed on 22/2/2021.

Gheewala, S., Damen, B., Shi, X., 2013. Biofuels: economic, environmental and social benefits and costs for developing countries in Asia. Wiley Interdiscip. Rev. Clim. Change 4 (6), 497–511.

Guo, T., Cibin, R., Chaubey, I., Gitau, M., Arnold, J.G., Srinivasan, R., Kiniry, J.R., Engel, B.A., 2018. Evaluation of bioenergy crop growth and the impacts of bioenergy crops on streamflow, tile drain flow and nutrient losses in an extensively tile-drained watershed using SWAT. Sci. Total Environ. 613–614, 724–735.

GIZ, 2017. Waste-to-Energy Options in Municipal Solid Waste Management. Deutsche Gesellschaft für Internationale Zusammenarbeit (GIZ) GmbH, Bonn, Germany.

Hayes, J.E., Stevenson, R.J., Stuetza, R.M., 2017. Survey of the effect of odour impact on communities. J. Environ. Manage. 204 (1), 349–354.

Hall, N., Ashworth, P., Devine-Wright, P., 2013. Societal acceptance of wind farms: analysis of four common themes across Australian case studies. Energy Policy 58, 200–208. https://doi.org/10.1016/j.enpol.2013.03.009.

Jaber, S., 2013. Environmental impacts of wind energy. J. Clean Energy Technol. 1 (3), 251–254.

Hoekman, S.K., Broch, A., Liu, X., 2018. Environmental implications of higher ethanol production and use in the US: a literature review. Part I—impacts on water, soil, and air quality. Renew. Sustain Energy Rev. 81, 3140–3158.

Hohmeyer, O., Ottinger, R.L., 1992. Social Costs of Energy: Present Status and Future Trends. Springer Verlag, USA, p. 432.

IRENA, 2017. Prospective for the Energy Transition, Investment Need for a Low-Cost Carbon Energy System. Dubai available at https://www.irena.org/publications.

IRENA, 2016a. REmap: Roadmap for a Renewable Energy Future, 2016 Edition International Renewable Energy Agency (IRENA), Abu Dhabi available at www.irena.org/publications.

IRENA, 2018. Global Energy Transformation: A Roadmap to 2050. International Renewable Energy Agency, Abu Dhabi available at www.irena.org/publications.

IRENA, 2019. Renewable Energy: Market Analysis, Dubai, available at http:irena.org, accessed on 22/2/2021

IRENA, 2016. Investment Opportunities in West Africa: Suitability maps for grid-connected and of-grid solar and wind projects. International Renewable Energy Agency (IRENA), Abu Dhabi.

IRENA, 2020a. Renewable Energy and Jobs: Annual Review 2020. International Renewable Energy Association, Dubai.

IRENA, 2020b. Renewable Energy Statistics. International Renewable Energy Association, Dubai.

International Hydropower Association (IHA), 2020. Hydropower Status Report, https://www.hydropower.org/resources/status-report.

Inhaber, H., 2004. Water use in renewable and conventional electricity production. Energy Sources 26, 309–322.

Jackson, R.W., Amir, B., Neto, A.B.F., Erfanian, E., 2018. Woody biomass processing: potential economic impacts on rural regions. Energy Policy 115, 66–77.

Jami, A.A.N., Walsh, P.R., 2014. The role of public participation in identifying stakeholder synergies in wind power project development: the case study of Ontario. Can. Renew. Energy 68, 194–202. https://doi.org/10.1016/j.renene.2014.02.004.

Jones, M.R., Singles, A., Ruane, A., 2015. Simulated impacts of climate change on water use and yield of sugarcane in South Africa. Agricul. Syst. 139, 10.

Kalogirou, S.A., 2004. Environmental benefits of domestic solar energy systems. Energy Convers Manag. 45, 3075–3092.

Kabir, Z., Khan, I., 2020. Environmental impact assessment of waste to energy projects in developing countries: general guidelines in the context of Bangladesh. Sustain. Energy Technol. Assess. 21, 1–13.

Kabir, Z., 2021. Chapter-13: Social, economic, and environmental aspects of bioenergy resources. In: Kalam, AK., Khan, MMK. (Eds.), Bioenergy Resources and Technologies. Academic Press, Elsevier, USA, pp. 391–423.

Kabir, Z., Yusuf, M.A., Khan, I., 2021. Chapter-14: An overview of policy framework and measures promoting bioenergy usage in the EU, the United States, and Canada. In: Kalam, AK., Khan, MMK. (Eds.), Bioenergy Resources and Technologies. Academic Press, Elsevier, USA, pp. 425–463.

Khan, I., 2020. Impacts of energy decentralization viewed through the lens of the energy cultures framework: Solar home systems in the developing economies. Renewable Sustainable Energy Rev. 119, 109576.

Khan, I., 2020a. Critiquing social impact assessments: Ornamentation or reality in the Bangladeshi electricity infrastructure sector? Energy Res. Soc. Sci., 60:101339

Khan, I., 2020b. Sustainability challenges for the south Asia growth quadrangle: a regional electricity generation sustainability assessment. J. Cleaner Prod. 243, 118639.

Khan, I., 2019a. Drivers, enablers, and barriers to prosumerism in Bangladesh: a sustainable solution to energy poverty? Energy Res. Soc. Sci. 55, 82–92.

Khan, I., 2019b. Power generation expansion plan and sustainability in a developing country: a multi-criteria decision analysis. J. Cleaner Prod. 220, 707–720.

Khan, I., Kabir, Z., 2020. Waste-to-energy generation technologies and the developing economies: a multi-criteria analysis for sustainability assessment. Renewable Energy 150, 320–333.

Khan, I., Chowdhury, S., Kabir, Z., 2020. Chapter-3: An overview of energy scenario in Bangladesh: current status, potentials, challenges and future directions. In: Asif, M. (Ed.), Energy and Environmental Outlook for South Asia. CRC Press, Tailor and Francis, USA, pp. 39–62.

Khellaf, A., et al., 2018. Overview of Economic Viability and Social Impact of Renewable Energy Deployment in Africa, Chapter 5. In: Mpholo, M., et al. (Eds.), Africa-EU Renewable Energy Research and Innovation Symposium (RERIS 2018), Springer Proceedings in Energy, pp. 59–70.

KPMG, 2019. The socioeconomic impacts of wind energy in the context of the energy transition, Report for European Commissioner for Climate Action and Energy, European Commission.

Kralova, I., Sjöblom, J., 2010. Biofuels-renewable energy sources: a review. J. Dispersion Sci. Technol. 31 (3), 409–425.

Kumar, M., 2020. Social, economic, and environmental impacts of renewable energy resources. In: Okedu, KE., Tahour, A., Aissaou, AG. (Eds.), Wind Solar Hybrid Renewable Energy System (edt). IntechOpen, London, UK.

Leary, J., Schaube, P., Clementi, L., 2019. Rural electrification with household wind systems in remote high wind regions. Energy Sustain. Dev. 52, 154–175.

Li., Y., Khanal, S.K., 2016. Bioenergy: Principles and Applications. John Wiley Sons, USA available at http://ebookcentral.proquest.com.

Lee, j., 2013. 5 reasons to care about access to electricity. United Nations Foundation. http://www.unfoundation.org/blog/5-reasons-electricity.html.

Liu, T., Huffman, T., Kulshreshtha, S., McConkey, B., Du, Y., Green, M., Liu, J., Shang, J., Geng, X., 2017. Bioenergy production on marginal land in Canada: Potential, economic feasibility and greenhouse gas emissions impacts. Appl. Energy 205, 477–485.

Momtaz and Kabir, 2018. Evaluating Environmental and Social Impact Assessment in developing countries, 2nd Ed Elsevier, USA.

Matungwa, B., 2014. An analysis of PV solar electrification on rural livelihood transformation: a case of Kisiju-Pwani in Mkuranga District, Tanzania, Master's Thesis, Centre for Development and the Environment, University of Oslo, Norway.

Møller, H., Pedersen, C.S., 2011. Low-frequency noise from large wind turbines. J. Acoust. Soc. Am. 129, 3727–3744.

Nazir, M.S., Ali, N., Bilal, M., Iqbal, H.M.N., 2020. Potential environmental impacts of wind energy development: a global perspective. Curr. Opin. Environ. Sci. Health 13, 85–90. https://doi.org/10.1016/j.coesh.2020.01.002.

Omer, A.M., 2008. Energy, environment and sustainable development. Renew. Sustain. Energy Rev. 12, 2265–2300.

Overend, R.P., 2019. Biomass energy heat provision for cooking and heating in developing countries. In: Kaltschmitt, M. (Ed.), Energy from Organic Materials (Biomass). Encyclopedia of Sustainability Science and Technology Series. Springer, New York, NY.

Opperman, J., Hartmann, J., Raepple, J., Angarita, H., Beames, P., Chapin, E., et al., 2017. The Power of Rivers: A Business Case. The Nature Conservancy, USA.

OECD and FAO, 2013. OECD – FAO Agricultural Outlook 2013–2022, available at http://www.agri-outlook.org, accessed on 25/3/2021

Poppa, J., Laknerb, Z., M. Harangi-Rákosa, M., et al., 2014. The effect of bioenergy expansion: food, energy, and environment. Renewable Sustainable Energy Rev. 32, 559–578.

Paredes-Sanchez, J.P., Lopez-Ochoa, L.M., Lopez-Gonzalez, L.M., 2019. Evolution and perspectives of bioenergy applications in Spain. J. Cleaner Prod. 213, 553–568.

Paolini, V., Petracchini, F., Segreto, M., Tomassetti, L., et al., 2018. Environmental impact of biogas: a short review of current knowledge. J. Environ. Sci. Health A 53 (10), 899–906. doi:10.1080/10934529.2018.1459076.

Pramanik, M.A., 2012. Impact of Solar Electricity on Rural Development: a study of some villages in Dinajpur and Thakurgaon of Bangladesh, Conference Paper, Technology in Sustainable Energy, 19-20 June 2012, Montréal, Canada, pp. 55–59.

Roberts, R.J., Chen, M., 2006. Waste incineration - how big is the health risk? A quantitative method to allow comparison with other health risks. J. Public Health 28, 261–266.

Remedio and Domac, 2003. Socio-Economic Analysis of Bioenergy Systems: A Focus on Employment. FAO, Italy.

Rydell, J., Bach, L., Dubourg-Savage, M., Green, M., Rodrigues, L., Hedenstrom, A., 2010. Bat mortality at wind turbines in north-western Europe. Acta Chirop. 12, 261–274.

REN21., 2019. Renewables 2019 Global Status Report. REN21 Secretariat, Paris.

Sayed, E.T., Wilberforce, T., Elsaid, K., Rabaia, M.K.H., Abdelkareema, M.A., Chae, K., et al., 2021. A critical review on environmental impacts of renewable energy systems and mitigation strategies: wind, hydro, biomass and geothermal. Sci. Total Environ. 766, 144505.

Siddiqui, O., Dincer, I., 2017. Comparative assessment of the environmental impacts of nuclear, wind and hydro-electric power plants in Ontario: a life cycle assessment. J. Cleaner Prod. 164, 848–860. doi:10.1016/j.jclepro.2017.06.237.

Sean, O., Clinton, C., Paul, D., Robert, M., Heath, G., 2013. Land-Use Requirement for Solar Power Plants in the United States. National Renewable Energy Laboratory, USA available at http://www.nrel.gov.

Shaibur, R.M., Husain, H., Arpona, S.H., 2021. Utilization of cow dung residues of biogas plant for sustainable development of a rural community. Curr. Res. Environ. Sustain. 3, 100026.

Sen, S., Ganguly, S., 2017. Opportunities, barriers and issues with renewable energy development-a discussion. Renewable Sustainable Energy Rev. 69, 1170–1181.

Shahsavari, A., Akbari, M., 2018. Potential of solar energy in developing countries for reducing energy-related emissions. Renewable Sustainable Energy Rev. 90, 275–291.

Strickland, M.D., Arnett, E.B., Erickson, W.P., Johnson, D.H., Johnson, G.D., Morrison, M.L., et al., 2011. Comprehensive Guide to Studying Wind Energy/Wildlife Interactions. National Wind Coordinating Collaborative, Washington. http://www.batcon.org accessed on 23/3/2021

Stigka, E.K., Paravantis, J.A., Mihalakakou, G.K., 2014. Social acceptance of renewable energy sources: a review of contingent valuation applications. Renewable Sustainable Energy Rev. 32, 100–106.

Severnini, E.R., 2014. The Power of Hydroelectric Dams: Agglomeration Spill overs. Institute for the Stufy of Labor (IZA), Bonn, Germany, pp. 1–70.

Song, C., Gardner, K.H., Klein, S.J.W., Souza, S.P., Mo, W., 2018. Cradle-to grave greenhouse gas emissions from dams in the United States of America. Renewable and Sustainable Energy Review 90, 945–956.

Saim, M.A., Khan, I., 2021. Problematizing solar energy in Bangladesh: benefits, burdens, and electricity access through solar home systems in remote islands. Energy Res. Soc. Sci. 74, 101969.

Tesfahunegny, W., Datiko, D., Wale, M., Hailay, G.E., Hunduma, T., 2020. Impact of wind energy development on birds and bats: the case of Adama wind farm, Central Ethiopia. J. Basic Appl. Zool. 81, 41.

Teske, S., Sawyer, S., Schäfer, O., 2015, Green-peace international, global wind energy Council, solar power Europe energy revolution a sustainable world energy outlook.

Tudor, C., 2014. An Approach to Landscape Character Assessment. Natural, England.

Tolli, M., Recanatesi, F., Piccinno, M., Leone, A., 2016. The assessment of aesthetic and perceptual aspects within environmental impact assessment of renewable energy projects in Italy. EIA Rev 57, 110–117.

EUROBATS, 2014. Report of the intersessional working group on wind turbines and bat populations, http://www.eurobats.org, accessed on 20/2/2021.

UNEP, 2013. Impacts of Biofuel Production Case Studies: Mozambique, Argentina and Ukraine, Final Report, available at: http://www.unido.org/en/resources, accessed on 15/2/2021

UNEP, 2008. The emerging biofuels market: regulatory, trade and development implications. In: United Nations Conference on Trade and Development. New York and Geneva.

USDA, 2012. Foreign Agriculture Service, Global Agriculture Information Network Report, Washington, US

UNESCO Institute for Statistics, 2014. A view inside schools in Africa: regional education survey.

Vezmar, S., Spajic, A., Topic, D., Sljivac, D., Jozsa, L., 2014. Positive and negative impacts of renewable energy sources, https://www.researchgate.net, accessed on 23/3/2021

VanCleef, A., 2016. Hydropower development and involuntary displacement: T y displacement: toward a global solution. Indiana J. Global Stud. 23 (1), 349–374 available at: https://www.repository.law.indiana.edu.

von Sperling, E., 2012. Hydropower in Brazil: overview of positive and negative environmental aspects, Energy Procedia, 18:110–118

Wu, C., Liu, J., Liu, S., Li, W., Yan, L., Shu, M., Zhao, P., Zhou, P., Cao, W., 2018a. Assessment of the health risks and odour concentration of volatile compounds from a municipal solid waste landfill in China. Chemosphere 202, 1–8.

Weinstein, P.E., 2006. Waste-To-Energy as a Key Component of Integrated Solid Waste Management for Santiago, Chile: A Cost-Benefit Analysis. Columbia University, Columbia.

Wassie, Y.T., Adaramola, M.S., 2021. Socio-economic and environmental impacts of rural electrification with solar photovoltaic systems: evidence from southern Ethiopia. Energy Sustain. Dev. 60, 52–66.

World Bioenergy Association, 2019. Global Bioenergy Statistics, available at https://worldbioenergy.org/uploads/191129%20WBA%20GBS%202019_HQ.pdf, accessed on 22/2/2021.

Wu, M., Demissie, E., Yan, E., 2018b. Simulation impact of future biofuel production on water quality and water cycle dynamics in the upper Mississippi river basin. Biomass Bioenergy 41, 44–56.

Wang, M., Han, J., Dunn, J.B., Cai, H., Elgowainy, A., 2012. Well-to-wheels energy use and greenhouse gas emissions of ethanol from corn, sugarcane and cellulosic biomass for US use. Environ. Res. Lett. 7 (4), 045905.

World Energy Council, 2013. World Energy Resources: Bioenergy, Chapter-7: Bioenergy, available at https://www.worldenergy.org, accessed on 20/2/2021

Xu, Y., Zhang, L., Jia, J., 2008. Lessons from catastrophic dam failures in August 1975 in Zhumadian, China. GeoCongress March 9–12, 2008 available at https://doi.org, accessed on 15/3/2021.

Yu, X., He, D., Phousavanh, P., et al., 2020. Chapter-8: Changes in women's livelihood in areas affected by hydropower projects. In: Yu, X., et al. (Eds.), Balancing River Health and Hydropower Requirements in the Lancang River Basin. Springer Nature, Singapore, pp. 217–257 available at https://doi.org, accessed on 25/2/2021.

Zarfl, C., 2014. Global Boom in Hydropower Expected This Decade. University of Copenhagen, Denmark available at http://news.ku.dkallnews.

CHAPTER 4

Application of solar photovoltaic for enhanced electricity access and sustainable development in developing countries

Majbaul Alam

Sustainable Energy Research Group, Energy and Climate Change Division, University of Southampton, United Kingdom

4.1 Introduction

Solar energy is one of the mainstream renewable resources which can serve a great proportion of present and future global energy needs if sustainably harnessed. Solar photovoltaic (PV) is a well-developed technology as well as the most widely accepted energy-service solution at different scale across the world, especially in the regions with poor electricity access. Despite, abundance of solar resources in the global south, impressive development in PV technologies and customer acceptance of PV-based electrification, access to electricity in many developing countries are still far behind to catalyze balanced socio-economic development. There are many socio-political, economic and technological factors and issues which differ in their own extents and characteristics in different countries are yet to be addressed. It is therefore very important to understand different PV technologies and their applications, especially for the people in the off-grid and poor-grid areas in various countries.

4.1.1 Solar energy: Global and developing world context

A balanced interplay among the ongoing global and regional development activities, climate change and human response to future environmental challenges is crucial than ever before to achieve an equitable and consistent low carbon pathway limiting global warming to 1.5°C. Reducing greenhouse gas (GHG) emissions requires phasing out of existing fossil fuel-based energy infrastructures and integration of renewable resources into future energy generation at scale. While the growing number of 'net-zero emission by 2050 and 2060' pledges are made by several governments and organizations across the developed regions of the world, many developing nations are struggling to provide basic energy services for their own people. Globally almost 770 million people in 2019 had no access to electricity, of which 579 million are in Africa and 155 million in developing Asia (IEA, 2020a). United Nation's Sustainable Development Goal-7 (SDG 7) which is the backbone for other SDGs emphasizes on access to affordable, reliable and sustainable electricity for all (IEA, IRENA, UNSD, World Bank, WHO, 2020). Most of the countries with poor electrification are endowed with rich solar resources, which can be sustainably utilized for electricity generation to reduce the country specific access deficits, and support national and regional growth.

While renewable sources and technologies are set to lead the global energy sector, growth in solar and wind energy capacity additions have been unprecedented in the recent years. Despite the uncertainties due to COVID-19 pandemic, global energy sector witnessed a record growth of 23% solar Photovoltaic (PV) capacity expansion in 2020 accounting about 135 GW (IEA, 2021). China (aggregated capacity of 254 GW) leads the total global installed PV capacity followed by USA, Japan, Germany and India respectively. Almost all developing nations have embraced applications of solar resources and technologies into their national electrification strategies. With the rapid declining cost and technology leapfrogging, PV has become more and more applicable to reduce energy access gaps in many developing countries.

4.2 Solar technology

History of harnessing solar power for the development of mankind can be backtracked to many centuries. However, refinement and commercial development of solar technologies begun during the middle of the nineteenth century. There are two key matured technologies to convert solar resource to electrical energy. These are:

i) Solar photovoltaic (PV), and
ii) Concentrating solar power (CSP)

Solar thermal technology is also used in many applications where different types of solar collectors are used to generate thermal energy.

4.2.1 Solar photovoltaic

Photovoltaic modules use solar cells made of semiconducting materials, mainly silicon which absorbs sunlight to produce electricity. PV modules may consist different number of solar cells based on their designed capacity. A typical solar cell can produce about 5 Watts of electricity at 15% - 18% efficiency under ideal sunlight and ambient temperature conditions. Three types of PV modules based on their electro-chemical characteristics are commercially available in the market. These are:

i) **Monocrystalline:** *Uses single silicon crystals and presents around 18% efficiency.*
ii) **Polycrystalline:** *Uses multiple crystalline fragments of silicon and presents around 15–17% efficiency.*

iii) Thin-film: *Typically uses amorphous silicon. Also uses Cadmium Telluride and Copper Indium Gallium Selenide. Efficiency may vary between 11% and 15% or higher in some cases.*

PV technologies have dominated the globally installed solar capacity because of their ease of installation and wider applicability at scale. To date, installed PV capacity is around 585 GW (IEA, 2020b) with more capacity expected to add up. Applications of such technologies related to their market segments can be categorized as below:

1. Grid connected systems
    1.1 Utility scale solar farm
    2.2 Rooftop solar (domestic and commercial)
2. Off-grid systems
    1.2 Large scale decentralized solar system
    2.2 Mini-grid and micro-grid
    3.2 Solar home system (SHS)
    4.2 Pico-solar system

While, large scale PV installations (1 MWp and larger) are mainly designed to support the utility grids, smaller rooftop systems (usually several kilowatts, and less than 1 MWp) are designed for self-consumptions as well as exporting any excess generation to the grid (where applicable). Some solar farms are combined with energy storage facilities to support utility grids.

Solar mini-grids and micro-grids ranging from 4 kWp to several hundred kilowatts which have their own independent distribution networks can serve remote and rural areas beyond the reach of national grid. Such electrification approach is considered as cost competitive compared to national grid extension in the developing country context.

Off-grid small PV systems of different sizes have been playing a vital role in enhancing energy access in many developing countries. Solar home systems (SHS) are the most widely accepted and used PV technology in many off-grid areas. Commonly used SHSs by the rural communities in many African and South Asian countries range between 20 Wp and 50 Wp. However, larger systems (50Wp and above) are also available. The pico-solar systems are much smaller (around 5 Wp–10 Wp), and have the target market at the bottom of the economic pyramid. Most of these systems are integrated with LED lighting, mobile phone charging and digital audio facilities (i.e., radio), and sold as complete service packages rather than individual products.

4.2.2 Concentrating solar power (CSP)

This solar technology converts concentrated sunlight into thermal energy. It uses concentrating collectors to generate heat to produce electricity through a turbine or heat engine. Thermal energy produced through CSP during the day time can be stored using convenient "thermal energy storage" (TES) and electricity can be generated when there is no sunlight. Thus, CSP presents great dispatchability, and this technology can minimize the intermittency of solar resource availability, and avoid expensive electro-chemical (i.e., battery) storage. Globally installed CSP capacity is about 6 GW (IEA, 2020c), which is very small compared to the total installed PV capacity. Spain has the highest installed CSP capacity followed by the USA. Unlike the PV, CSP projects so far have only been deployed at utility scale.

Different types of CSP technologies are in use. These are, (a) Parabolic through, (b) Solar tower, (c) Dish Stirling, and (d) Fresnel reflectors. Among these, parabolic trough (PT) is the most developed and commonly used technology. The PT uses parabolic reflectors to concentrate sunlight into linier receivers to heat up the fluid to run the electric generator.

4.3 Solar energy for developing countries

Energy is a key prerequisite for development. "It is evident from the recent global history that countries lacking the access to basic energy services and their applications are backtracked in the race of social and economic development" (Alam, 2017). With the poor energy access in many developing countries, governments are ramping up national electrification programs largely based on fossil fuels, which are related to their own socio-economic and political agendas. However, the emerging trend of exponential investment growth in renewable energy projects supported by innovative financing models acted as catalyst for many developing nations to get along. As such, poor nations may not get the full benefit of this trend; a holistic approach is required through integrated global policy framework, financial support packages and multilateral cooperation.

Plummeted costs of PV technologies, suitable business models and financing options that integrate potential stakeholders including the private sectors can enhance its dissemination process in developing countries. Global development activities around SDG-7 have revitalized this sector in the recent years, which is evident from the extended initiatives in financing PV-based electrification by the development partners and governments across the electricity deficit regions in Africa and Asia.

Although some developing countries have made good progress to meet some of the "17 Sustainable Development Goals"[1] outlined by the United Nations, many are yet to progress a lot. While all the defined goals are interlinked, access to energy remains at the core of all goals (Fig. 4.1). For example, some of the basic human needs embedded into several SDGs, that is, no poverty (SDG 1), zero hunger (SDG2), good health (SDG3), quality education (SDG 4), clean water and sanitation (SDG 6) are vital to deliver on sustainable growth in industry innovation (SDG9), gender equality (SDG5), reduced inequalities (SDG10), decent work and economic growth (SDG8), none of which can be achieved without having access to affordable and clean energy provisions. Similarly, SDG 11 to SDG 16 are dependent on combined outcomes of SDG 1 to SDG 10. Overall achievement of all such goals are somewhat related to SDG 17, which emphasizes on global partnership around capacity development, trade, finance, technology, science, and sustainable strategies.

As many of the stipulated SDGs are dependent on access to affordable and clean energy to kick start or fast track the sustainable development journey for a developing nation, *supply of sustainable energy for all* is a prerequisite. Here, sustainable energy supply means: (i) energy should be available for all, (ii) should be from renewable sources or at least should have

---

[1] Sustainable Development Goals: 17 Goals to Transform our World (available at: https://www.un.org/sustainabledevelopment/).

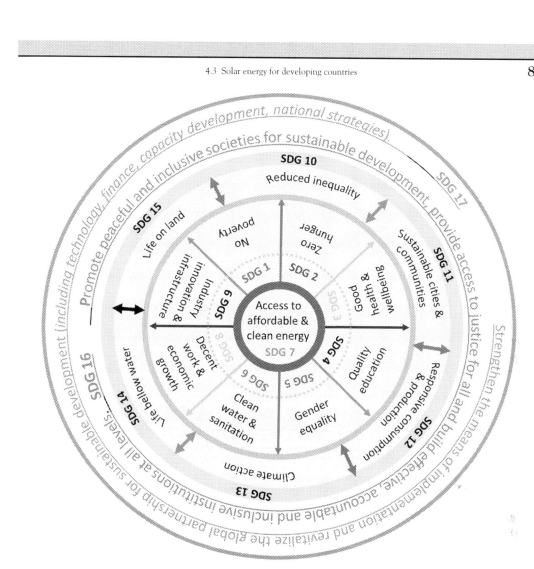

FIGURE 4.1 Interlinked sustainable development goals outlined by the United Nations. Access to affordable and clean energy is shown at the core for attaining other goals (Author's own illustration).

low carbon footprint, (iii) should be affordable to all (individual and commercial uses), (iv) generation, distribution and consumption infrastructures should be built around resilient supply chain, and (v) should not compromise any long-term sustainability of the society and the environment. However, access to energy only can initiate or enhance the development process, but it would need other inputs to complete the process. Some of these inputs are policy support, good governance, individual and institutional capacity building, access to finance and support, and broader awareness. For developing countries, making all these elements and inputs available remain as the key challenge. Therefore, regional and global partnership in win-win fashion is very important.

Infrastructure development for making energy available for all is a key priority for developing countries. Thus, attaining affordable and sustainable energy access for all remains a secondary target for many governments. Lack of needed investment is the main obstacle

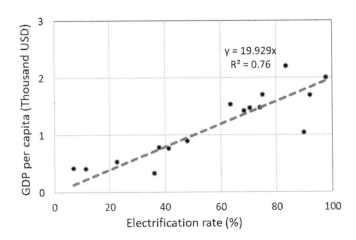

FIGURE 4.2 Relationship between national electrification rate and per capita GDP of some selected countries (n = 16) in Table 4.1.

in this essence, where most of the conventional energy resources are import dependent. Furthermore, historically unplanned urbanization and urban-centric industrialization have forced development of centralized energy infrastructures. This results in serious disparity between urban and rural energy access in most developing countries. Having the nature of very poor per capita energy demand among the marginalized rural and remote areas, grid extension is not often financially viable. This phenomenon makes the vicious cycle of energy poverty everlasting at the bottom of the economic pyramid. To alleviate poverty in such areas, limited access to electricity has been provided through various sizes of diesel generators (from few kilowatts to megawatt capacity) in decentralized mode. It is notable that electricity from diesel generators is too expensive to be used for cooking, and access to cooking fuel in such communities is still coal and biomass based. While electricity tariff form the centralized utility grid is subsidized, poor communities in rural and remote areas have to pay the premium price because of high cost of unsubsidized fuel in the most cases and other supply chain challenges.

Considering the aforementioned issues and challenges of access to energy, especially access to electricity, solar resource availability, and application of PV technologies present great solutions for most of the developing nations. At present, the development of PV technologies and various PV products are aligned at different customer segments and their requirements. For example, large scale (utility type) PV installations can support heavy industrial loads, small solar home systems (SHS) can provide basic residential electrical needs and pico-solar can supply lighting requirements. Thus, solar energy through different PV technologies can reduce the electricity access gap to enhance economic growth. Table 4.1 presents some countries with very poor and moderate rate of electrification related to their "per capita gross domestic product" (GDP). Fig. 4.2 indicates a strong positive relationship ($R^2 = 0.76$) between the electrification rate and per capita GDP of these countries. However, as the "per capita GDP growth" also depends on many other factors (i.e., access to finance, appropriate policy support, good governance), higher electrification rate may not always deliver relatively higher per capita GDP in every case and vice versa. For example, South Africa with a lower electrification rate of 85% compared to 95.6% in the Philippines presents a higher per capita GDP of USD 6374

TABLE 4.1 List of some selected countries with their national electrification rate and per capita GDP during 2018 and 2019.

Country	Electrification rate[a] (%)	GDP per capita[b] (USD)
South Sudan	6.7	424
Malawi	11.2	411
Sierra Leone	22.7	534
Somalia	36	332
Kenya	75	1707
Mali	48	894
Cameroon	63.5	1534
Rwanda	37.7	783
Ghana	83.5	2202
Senegal	70.4	1466
Uganda	41.3	770
Pakistan	74	1482
Bangladesh	92	1698
India	97.8	2005
Myanmar	68.4	1418
Nepal	89.9	1039

[a] World Bank, SE4ALL Global Tracking Framework https://data.worldbank.org/indicator/EG.ELC.ACCS.ZS.
[b] World Bank national accounts data and, OECD National Accounts data https://data.worldbank.org/indicator/NY.GDP.PCAP.CD.

compared to USD 3252 in the later. On the other hand, Djibouti with only 61.3% electrification rate has almost similar per capita GDP value (USD 3142) as the Philippines.

For maintaining a sustainable growth, every developing country has to determine their own future development pathways underpinning the crucial agenda of providing "access to energy for all" according to the SDG-7. Six common areas of development needed by the most of such countries are, (i) food security, (ii) health services, (iii) education, (iv) communication, (v) transport, and (vi) industrial development. All of these six areas need resilient energy supply provisions. As it is mentioned earlier that grid extension to many dispersed areas is not economically viable and therefore, may not happen in the foreseeable future for many of those 770 million people globally at present living beyond the grid, applications of PV-based electrification at different scales will open the opportunity for such communities to get on board the sustainable development journey.

The most challenging part of deploying solar power as the single source of energy is its intermittency and seasonality. To have energy supply readily available, PV systems must be coupled with energy storage. Most common type of energy storage used with PV is battery bank. Lead acid and Li-ion batteries are commonly used with different types and sizes of PV systems. Although battery technology is quite matured, its cost is yet to reduce by many folds to be suitable for large scale applications. In the case of large scale PV systems, to complement the intermittent nature of solar resources, other renewables (i.e., wind, hydro, geothermal) based on their availability can be coupled in the energy system. However, such coupling of different renewable energy generation is a complex process. For, comparatively smaller PV systems, diesel hybridization is a common practice.

4.3.1 Present status

With a 22% increase in generation capacity in 2019, photovoltaic (PV) has contributed to over 2.7% of global electricity generation (IEA, 2020d). Sharp declining trend in PV module cost leading to a staggering reduction of about 80% since 2000 has pushed the global installed capacity of 135GW in 2020 (IEA, 2021). Major contributors of such PV uptake are both developed and developing countries. China, USA, Japan, Germany, India, Italy, Australia and Vietnam installed the most of the capacities in 2020. The lowest winning bid of USD 0.01567/kWh for PV power supply under a 25-year power purchase agreement (PPA) was witnessed in Qatar for an 800MWp capacity solar park at the first quarter of 2020 (PV Magazine 2020). Although, the utility scale installations are dominating global capacity installation, off-grid solar market is growing year on year as well. While the cost reduction of PV modules and components is driven by the large scale installations, off-grid solar sectors in developing countries are taking the benefit of this economy of scale.

Photovoltaic (PV) based electrification is well accepted by now in the off-grid areas along with its growing demand in areas under unreliable or week grid. Table 4.2 presents accumulated installed capacity comparisons of different off-grid solar photovoltaic technologies in Africa and Asia indicating the growth trend of the sector between 2012 and 2018. In this essence, Table 4.3 depicts the total number of people served as of 2018 through different off-grid photovoltaic technologies in Africa, Asia, and globally. These two tables (Table 4.2 and 4.3) summarize the disseminations of six key PV technology segments, which are briefly described below with focus on their off-grid applications.

i. *Solar light systems (<11 Wp):* Represents pico-solar systems with integrated Li-ion batteries. Most of the items come with USB (Universal Serial Bus) charging options for mobile phones, and sometimes with radio.
ii. *Solar home system (11–50 Wp):* Mainly DC (direct current) coupled systems, suitable for one or multiple LED bulbs, USB charging, radio, other audio systems and small cooling fans. Often sold with Pay-as-you-go (PYG) or installment payment options. PYG systems are remotely lockable by the vendors in the case of nonpayment incidents by the customers.
iii. *Solar home system (>50 Wp):* Comparatively bigger systems. Also sold as PYG. Sometimes customers choose their components (PV, charge controller, battery) and appliances. Systems less than 100Wp are normally DC coupled (12V, 24V, 48V), and systems greater than 100Wp are AC (Alternate Current) coupled which require inverters (DC to AC). Besides powering small appliances (LED, mobile charger, audio, TV), such systems are also used for other purposes (i.e., small refrigerator, small electric motors). Traditionally used with

TABLE 4.2 Installed capacity comparison of different off-grid solar photovoltaic technologies between Africa and Asia in 2012, 2015 and 2018 indicating the growth trend (Adapted from IRENA 2020, 2020a).

Continent	Solar light <11 W			Solar home system 11–50 W		
	Total installed capacity (MW)					
	2012	2015	2018	2012	2015	2018
Africa	4.8	28.8	27.2	0.54	6.0	23.4
Asia	6.8	25.5	47.4	86	155	146.2
	Solar home system >50 W			Solar mini-grids		
	Total installed capacity (MW)					
	2012	2015	2018	2012	2015	2018
Africa	6.5	26.7	84.4	4.6	74.2	280.4
Asia	12.8	13	17	25	50	60
	Solar water pumps			Other off-grid solar PV		
	2012	2015	2018	2012	2013	2018
	Total installed capacity (MW)					
Africa	3.5	7.2	14.2	3.1	63	135.3
Asia	15.7	110.4	557.2	131	370	627.8

TABLE 4.3 Number of people served through different off-grid photovoltaic technologies in Africa, Asia, and globally in 2018 (Adapted from IRENA 2020, 2020a).

	Solar light <11 W	Solar home system 11–50 W	Solar home system >50 W	Solar mini-grids (Tier 1*)	Solar mini-grid (Tier 2+*)
	Total number of people connected as of 2018 (Thousands)				
Africa	42,097	4982	3570	563	1052
Asia	86,129	15,685	646	298	940
Global total	130,837	20,798	4357	1082	2147

*Level of energy access is measured against a Multi-Tier Framework (see: https://mtfenergyaccess.esmap.org/). According to this framework while Tier 1 should have available power of ≥3 W for >4 hours a day, Tier 2 should have ≥50 W for >4 hours a day, Tier 3 should have ≥200 W for >8 hours a day, and Tier 4 should have ≥800 W for >16 hours a day.

Lead acid battery storage. However, new market trend is pushing the integration of Li-ion batteries.

iv. *Inter connected clusters of solar home systems (IC-SHSs):* Research and innovation in power electronics and control mechanisms have enabled connecting multiple SHSs in a small footprint for power sharing among them. This approach is termed as swarm electrification (i.e., prosumerism), which enables individual SHS owners to earn revenue for exporting the excess power generated by their PV systems to other SHS owner or electricity user. Thus clustered IC-SHSs can enhance access to electricity and offer a more resilient business model.

- **v.** *Solar mini-grids*: Decentralized independent grids, generally range between 4 kWp and 100 kWp. Solar mini-grids can be of megawatt scale. However, most of the existing and planned PV mini-grids in Africa and Asia are within 100 kWp range. Mini-grids can be fully renewable energy based (PV-battery only) or hybridized with suitable size of diesel generators. Smaller mini-grids normally have single phase distribution, while bigger ones may have three phase supply to support heavy loads (i.e., mills, motors and other industrial uses).
- **vi.** *Other off-grid solar PV systems*: This segment covers range of PV systems installed in any public or private organizations. Such PV systems are used for power supply for self consumption, power backup and communication towers. Number of these systems have been installed in many refugee camps by the support organizations (i.e., UNHCR, UNICEF) in off-grid locations to deal with sudden influx of refugee situations.
- **vii.** *Solar water pump*: This is a very important segment of the PV-based energy systems. Solar pumps are mainly used for irrigation and in some cases for other water uses. Such systems do not require a battery bank as water can be stored as an alternative to energy storage. Thus cost of operations can be reduced. Both the AC and DC pumps are used in different type of applications.

While Africa historically showed a slow pickup of PV-based off-grid electrification approach, developing Asia indicated a comparatively faster growth. India and Bangladesh have been the early responders followed by Sri Lanka, Vietnam and Indonesia in deploying smaller off-grid PV technologies, that is, pico-solar, small SHS and medium SHS. However, during the recent years' technology leapfrogging and innovative leasing models to serve the marginalized off-grid communities in Africa have sped up the sales of such small PV systems. Nevertheless, with the faster dissemination of solar mini-grids, Africa has outpaced Asia. By the end of 2018, Africa had a total installed capacity of 280MWp PV mini-grids compared to 60MWp in Asia (Table 4.2). It is notable that both continents have thousands of planned mini-grids to be installed in the near future. Many Asian countries have adapted solar irrigation at bigger scale and the continent has an aggregated installed capacity of 557 MW, where India alone has contributed 512 MW.

Globally around 131 million people had access to lighting through pico-solar systems as of 2018 (Table 4.3). At the same time approximately 21 million people had access to basic access beyond lighting through SHSs ranging between 11 Wp and 50 Wp, while only 4.4 million had comparatively higher tiers of electricity access through +50 Wp PV systems. Solar mini-grids have been serving almost 32.3 million people globally during the same period.

4.3.2 Future prospects and challenges

Global energy sector decarbonization and low carbon future pathway include two parallel activities, which are, (i) new renewable capacity additions, and (ii) transitioning of existing fossil fuel-based generation to renewables. Solar energy being the most available resource and suitable option for most of the developing countries to deliver on "energy access for all" presents a future low carbon energy mix prospect. Major emerging economies (i.e., China, India, Brazil, Mexico, Indonesia, and South Africa) with considerable GDP growth through rapid industrialization which contribute almost 40% of global carbon dioxide emission are also increasing their solar energy portfolio using both the PV and CSP technologies (IRENA, 2020b). Growing number of large scale solar projects for centralized utility grid integrations are evident in many developing countries. For example, government of Senegal in collaboration with the "Scaling Solar" project of the World Bank has finalized delivering 60MW solar power at a unit cost of USD 0.04/kWh[2]. Many other developing countries have followed similar path and planned more solar power generation in the future. However, despite an expected 8% post-COVID-19 decline in distributed PV dissemination across many poor nations due to shrinking investment flow, overall PV capacity growth will remain stable. The volume shortfall in this sector is expected to be recovered by the utility scale PV installations in many countries.

Many of the developing countries in Asia and Africa already have planned to support their own national renewable energy generation targets with a "dual-path strategy" (DPS). The term DPS here refers to a parallel growth of both the centralized and decentralized generation capacities. As industrialization in these regions is mainly urban centric, increasing the capacity of centralized energy generation to maintain the GDP growth is very important while reducing energy access gaps in off-grid communities through expanding related decentralized portfolio. The "duel-path strategy" of the developing countries integrates application of renewable energy resources in line with the United Nation's Sustainable Development Goals (SDGs). Although many countries have abundance of different renewable energy resources (i.e., geothermal, hydro, on- and off-shore wind), use of solar energy especially PV technologies remains at the center of the focus by almost all countries. Table 4.4 indicates the solar energy generation targets by selected developing countries. Many developing countries in Asia have ambitious targets of PV power generation compared to African countries. However, the recent trend of off-grid solar sector growth in Africa is the lifeline for achieving SDG-7.

Keeping the off-grid PV sector's growth in pace with the required extent and speed to hit the "electricity access for all by 2030" target, needs integrated support at different areas. Major areas of focus are listed below followed by brief explanations in each area.

i) Quality of off-grid PV systems (pico-solar, SHS, mini-grid)
ii) Sustainable business models for wider dissemination
iii) Supportive policy frameworks
iv) Local capacity building
v) Integration of "energy access" into other development agendas
vi) Regional and global collaboration

Quality of PV systems in the pico-solar, SHS and mini-grid categories which may include some or all the components (i) PV module, (ii) charge controller, (iii) battery, (iv) inverter, and (v) other related switch gears and system integrated appliances has been a concern over the years. Markets in many developing countries have been flooded with poor quality pico-solar and SHSs. This has happened due to inadequate quality control measures both

---

[2] Scaling Solar: Senegal (https://www.scalingsolar.org/active-engagements/senegal/).

TABLE 4.4 Future solar energy generation targets (installed capacity) by different countries in Asia and Africa.

Country	Solar energy target	Details
Bangladesh	40 GW (ambitious plan) by 2041. Conservative estimates indicate 8 GW by this time[a].	Utility scale PV farms, floating solar in Kaptai lake and Meghna river estuary, countrywide rooftop solar. Also, increased penetration of solar home systems.
India	100 GW by 2022[b]. This target might be missed.	Large scale PV farms, utility scale CSP, rooftop solar. Decentralized solar projects. Different states emphasize their own targets besides the central government.
Vietnam	20 GW by 2030[c]	The country had record growth in PV in 2020. Utility scale PV projects and rooftop PV expected to grow fast.
Pakistan	27 GW by 2047[d]	Countrywide application of large scale projects like Shind Solar project as well as rooftop solar. This PV target is a part of its "30% renewable" in energy mix by 2030.
Indonesia	9 GW by 2030. However, it has a potential of 47GW[e]	Installed capacity may go beyond 9 GW if millions of those who currently lack access in remote areas, especially in Java and Bali are connected through decentralised PV.
Ghana	750 MW by 2030[f]	Ghana's Renewable Energy Masterplan-2019 outlines conservative targets for PV installation by 2030. 450 MW will come from utility scale PV. Distributed PV will contribute 200 MW, 100 MW from standalone solar including solar irrigation.
South Africa	5.2 GW Solar by 2030[g]	4.6 GW of 2030 solar energy target to come from photovoltaic, and 600 MW from CSP.

[a] Renewable Energy in Bangladesh (https://energytracker.asia/renewable-energy-in-bangladesh-current-trends-and-future-opportunities/).
[b] India plans to produce 175 GW of renewable energy by 2022 (https://sustainabledevelopment.un.org/partnership/?p=34566).
[c] Renewable Energy: Vietnam, Climate Action Tracker (https://climateactiontracker.org/countries/vietnam/policies-action/).
[d] REN21, Renewables 2017: Global Status Report (https://www.ren21.net/wp-content/uploads/2019/05/GSR2017_Full-Report_English.pdf).
[e] IRENA 2017: Renewable Energy Prospects Indonesia, Remap, March 2017.
[f] Ghana Renewable Energy Masterplan 2019 (http://www.energycom.gov.gh/files/Renewable-Energy-Masterplan-February-2019.pdf).
[g] IRENA 2020: Renewable Energy Transition in Africa.

in overseas import and local production of whole system or selected components. Common quality issues are: (i) poor quality or inefficient PV modules, (ii) low quality power electronics in charge controllers and inverters, (iii) sub-standard battery chemistry, and (iv) poor quality and inefficient appliances. Any combination of one or more of these issues shortens the PV-system's life leading to user dissatisfactions. Thus, persistent quality issues affecting user confidence may hinder the expected growth of pico-solar and SHS sector. On the other hand, quality issues PV mini-grids are both technological and nontechnological. These two categories need more clarifications for clear understanding about the issues.

Technological quality issues of PV mini-grids are dominated by the poor quality battery bank and maintenance of the battery bank. While poor quality of lead acid cells is common

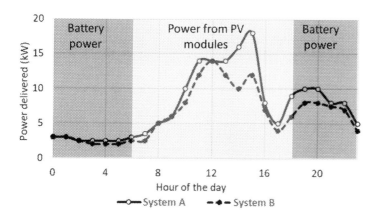

FIGURE 4.3 Comparison of two different load scenarios of a 100 kWh lead acid battery bank of a 35 kWp PV-battery mini-grid. System A (black solid line, daily load 182.5 kWh) battery bank usually delivers 69 kWh/d energy, and System B (black dotted line, daily load 150kWh) battery bank delivers 57.5 kWh/d energy.

in many Asian and African markets, inappropriate sizing, design and maintenance of battery banks also lead to early energy storage failure. Poor quality batteries cannot deliver rated power for the designated periods and lead to system failures which require expensive replacements risking projects' financial sustainability. Besides the battery quality, poor battery bank management also causes energy storage failure. The most common causes of battery bank failure in African mini-grids are, very high ambient temperature and unregulated discharge (i.e., very fast discharge, discharge below the standard threshold). Combination of these two factors results in high internal resistance development inside the cells, thus incapacitating the battery bank. Fig. 4.3 presents an example of two different load scenarios for a 100kWh lead acid battery bank of a 35kWp PV-battery mini-grid system in Africa. Mini-grid "System A" has a daily combined load of 182.5 kWh and its battery bank usually delivers 69 kWh/d energy, whereas the mini-grid "System B" has a daily load of 150 kWh and the same size battery bank (100 kWh) delivers 57.5 kWh/d energy. Where, manufacturers' recommended depth of discharge (DoD) for the lead acid batteries is 50%, the battery bank of mini-grid "A" usually delivers 69 kWh/d energy leading to a DoD well beyond the recommended level compared to the Battery bank "B" (delivers 57.5kWh/d, DoD is ~57%). In these two scenarios, battery bank of mini-grid "A" may suffer other consequences besides deep discharge, which are: (i) in the case of poor irradiance for longer periods in a day or over multiple days, the battery bank may not be fully charged while meeting daytime required loads, and will either stress itself to deliver needed power in the evening or will result in blackouts, (ii) in the case of an increase in ambient temperature beyond the standard limit (i.e., +45°C) it would not be able to deliver needed power.

Battery bank failure of mini-grid can be reduced in many ways, these are (i) carefully designing the energy storage size in response to the required loads to serve, (ii) hybridizing the mini-grid with a diesel generator, (iii) automated load control and demand side management to support battery bank discharge, and (iv) battery bank ambient temperature control. However, considerable development in Li-ion battery (LiB) technology has been witnessed in the recent years, which is driving this technology to the mini-grid mainstream with its increased economy of scale. LiBs can be fast and deep discharged without any negative impact on its longevity and performance. This technology will give a boost to the global mini-grid sector. However, quality aspects throughout the supply chain should be maintained.

To date 19,000 mini-grids in distributed mode have been serving almost 47 million people (Day and Kurdziel, 2020), most of which are fossil fuel powered. It is estimated that approximately 210,000 mini-grids are needed at a cost of USD 220 billion to provide electricity access to almost 500 million people by 2030 (ESMAP, 2019). Almost 80% of such mini-grids can potentially be of PV-battery and PV-battery-diesel hybrid architecture. Besides the key technological failure of mini-grids mentioned above, the nontechnological issues for mini-grid projects' failure are crucial to understand, which are mainly: (i) unsustainable business model, (ii) policy framework not been aligned with the real life needs, and (iii) lack of operational knowledge and poor techno-economic performance.

For the commercial success of any PV mini-grid project, income generation is essential. Income generation cannot be ensured without a suitable business model in place. Different business models have been adapted by different developing countries with various outcomes at initial stages. Examples of Kenya mini-grid business models are highlighted in the following section as a case study. Whether the business model is dominated by government backed subsidies (GBS) or guaranteed by a long-term power purchase agreement (PPA), private sector involvement is crucial to attract more finance into this sector. Asian Development (ADB), African Development Bank (AfDB), World Bank (WB), Islamic Development Bank (IDB) and many global development partners, that is, UK Aid, USAID, German Corporation for International Cooperation (GIZ) are some of the main funders in mini-grids and other off-grid electrification programs across the developing countries. All of these, and many other likeminded organizations have been supporting countries to deploy the most suitable business models taking public and private sector stakeholders on board. Innovative financing with pay-as-you-go (PYG), rent-to-own or lease-to-own and usages-based-pay options for pico-solar and solar home systems of different sizes facilitated through mobile payment and smart prepay meters for mini-grids has revolutionized growth of this sector.

As mentioned above, the growth in PV-based off-grid electrification (pico-solar, solar home systems of different sizes, PV mini-grids in "AC and DC"[3]) fuelled by the declining cost of PV and components, innovative financing and business models and appropriate policy supports has been supplementing grid extension for millions of people. With all these recent developments and advancements, it is estimated that conventional PV mini-grids (AC mini-grids) can attain acceptable level of grid price parity. Analysis based on Kenyan utility grid tariff (average)[4] and estimated levelized cost of electricity (LCOE) form different off-grid PV electrification technologies in Kenya indicates various grid price parity (GPP) as presented in Fig. 4.4. While pico-solar inherently being expensive presents poor grid price parity (<20% of KPLC tariff), AC mini-grids serving Energy Access Tier 2-4 (EA Tier 2-4) and higher tiers (EA Tier 5 and above) present much better grid price parity (GPP) ranging between 44% and 68%. DC mini-grid can offer the best GPP compared to other off-grid PV electrification approaches. However, scope and opportunities for application of DC mini-grids at a larger scale are limited.

---

[3] AC mini-grids: mini-grids using inverters to convert direct current (DC) generated by the PV modules to alternate current (AC) to be used in many conventional devices. DC mini-grids: mini-grids using direct current (DC) without any conversion to power up appliances. DC mini-grids are normally applicable in small footprints.

[4] Average KPLC, the national utility tariff as of 2018 considered as USD 0.10/kWh for residential users consuming up to 100 kWh/month, and for commercial customers consuming up to 15,000 kWh/month tariff was considered as USD 0.14/kWh.

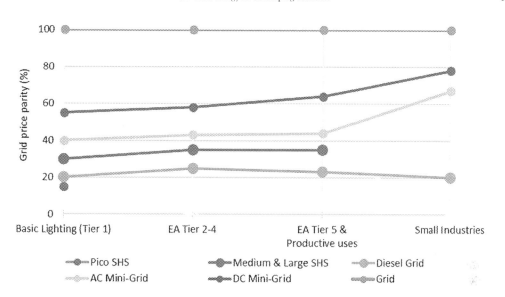

FIGURE 4.4 Grid price parity (tariff of Kenya Power and Lighting Company) comparisons for estimated levelized cost of electricity (LCOE) form different off-grid PV electrification technologies in Kenya at different tiers[8] of electricity access and consumption. Also includes comparison for diesel grid. "EA Tier 2-4" and "EA Tier 5 and Productive uses" in the figure refer to Energy Access Tiers of different levels.

Diesel grids always remain expensive with related poor GPP compared to Solar home systems, AC and DC mini-grids at all levels of energy access. It is clear from Fig. 4.4 that with increasing consumptions of electricity from different PV-based off-grid solutions grid price parity gap reduces, which means levelized cost of electricity (LCOE) or the tariff becomes cheaper.

Off-grid customer segments in the developing countries are characterized by very low per capita electrical demand, which is linked to their financial inabilities. Unfortunately, such customers historically have been paying very high price for the small amount of electricity they consume. Cost of electricity only can be cheaper with the increase in consumptions (see Fig. 4.4, where, pico-solar users with basic lighting at Tier 1 pay the highest tariff). This "vicious cycle of energy poverty" (VCEP) hinders the future mass dissemination of off-grid PV electrification in a balanced manner to attain the SDG-7. Planned exit from the VCEP requires tailored multi agency and multi stakeholder based holistic approach. For further understanding, an example of energy access and productive uses of electricity for women empowerment is included in the next section.

Future of solar energy uses in the developing countries to increase their renewable energy mix portfolio while assuring energy securities, sector friendly policy frameworks are the key prerequisites. Existing policies related to this sector should continuously be revised and updated to address any new issues and to support expected future growth. Policy frameworks for different countries may vary based on their requirements, current status and strategic pathways around national renewable energy targets, integration of 'energy access' into other development agendas, and regional and global collaboration initiatives.

Despite all the challenges mentioned earlier alongside the opportunities for future growth, solar energy may open up tremendous economic benefits at scale for many African countries.

"Solar resource can be the clean energy alternative to replace fossil fuel." As African continent has the abundance of solar resources along with potential access to vast land areas to deploy mass-scale solar power generation projects (i.e., PV, CSP), especially the northern and north-eastern Africa closer to Europe can be used as the inter-continental solar power house. Electricity generated from such power plants can be transmitted to Europe through subsea cables. Thus, many African countries have the potential of being the hub for continental and inter-continental solar power generation. The existing and proposed subsea interconnectors between Morocco and Spain, and the proposed longest subsea interconnector between UK and Morocco delivering clean electricity through high voltage DC (HVDC) transmission lines are the confidence building mega projects to support intercontinental super grids. "Future African Solar Power Hubs" have the potential of creating enormous financial boost in the continent's economy. Similar, energy hubs can be developed in Asia as well.

Although this may be true, intermittent nature of solar resources is the biggest challenge of using PV, CSP or other solar technologies as a single energy supply source while conventional energy storage (battery) are expensive. Fortunately, technological developments in generating hydrogen using electricity form solar and using hydrogen as energy carrier has created opportunity for solar energy to be used more conveniently. Decarbonization of transport sector globally requires more renewable energy integrations into this sector. Electric vehicles and hydrogen powered transports will accommodate massive scale solar power generation projects by many developing countries in the near future.

Another major energy access issue yet to solve is clean cooking fuel. Recent World Bank study reported that globally 2.8 billion people do not have access to clean cooking facilities (The World Bank, 2020). Serious health issues and environmental concerns related to cooking based on solid biomass, kerosene and coal are well documented. Access to sustainable electrification at the lower tiers through pico-solar and solar home systems does not create an immediate path to access to clean cooking for millions of people. However, current effort of making solar energy cheaper and accessible at scale will create the techno-economic potential of using it for cooking. For example, electricity from photovoltaic technologies, that is, large solar home systems, centralized or decentralized large solar projects, and mini-grids can be used for cooking if (i) cost of electricity can be reduced to an affordable level supported by efficient electric cookers, and (ii) the energy generating system have enough resilience to support such loads. On the other hand, hydrogen produced form solar energy can be used for cooking as a clean fuel. However, such approach needs more research and development related to cost-compatibility, technological adaptation and to a major extent safety concern.

4.4 Contribution to sustainability

Solar energy and its contributions to sustainability to a broader scale need multiple inputs in the process of "inclusive socio-economic growth"[5] of a society in a developing country. Providing access to clean energy to a community may not deliver sustainable growth in that community as a lone catalyzer. Energy needs to be used for more productive purposes to enhance income generation, so that users can climb up the energy ladder. To use energy in

---

[5] Inclusive growth refers to a fair growth process and opportunity that includes all in the society regardless of their gender, age, race, ethnicity, and physical or mental limitations (please see: https://www.oecd.org/inclusive-growth/#introduction).

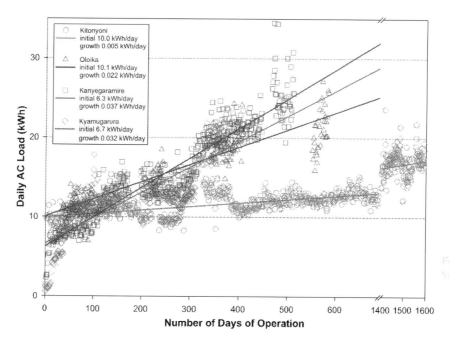

FIGURE 4.5 Electricity access with different tariffs in four off-grid communities in Kenya and Uganda through community run PV mini-grid intervention under the "Energy for Development program (e4D)"[9] at the University of Southampton, UK. Figure shows the linear correlation of the measured daily demand growth with a number of days of operation (For Kitonyoni y = 0.0045x + 10.0, R^2 = 0.71, Oloika y = 0.022x + 10.1, R^2 = 0.81, Kanyegaramire y = 0.037x + 6.3, R^2 = 0.88, and Kyamugarura y = 0.032x + 6.7, R^2 = 0.68). *(Source: Bahaj and James, 2019).*

productive applications, users should be provided with (i) affordable tariff, and (ii) access to support with training, finance and suitable policy. Here an example of four off-grid communities with PV mini-grid intervention for electricity access in Kenya and Uganda to understand the underlying factors contributing to sustainability is presented (Fig. 4.5). The Kitonyoni mini-grid was installed in 2012, and a higher tariff of USD 1.85/kWh applied in the first year. Tariff was reduced to USD 0.93/kWh and USD 0.70/kWh in the second and fourth year respectively. The second mini-grid in Kenya was installed in 2015 with a tariff of USD 0.74/kWh. Two Ugandan mini-grids, Kanyegaramire and Kyamugarura were installed in 2015 with a cheaper tariff of USD 0.20/kWh. All four projects installed by the 'Energy for Development' team (www.energyfordevelopment.net) of the University of Southampton, UK were identical in capacity (13.5 kWp PV, 38.4 kWh battery bank, 28 kWh/d designed capacity, single phase distribution) and operated by elected local cooperatives. It is clear from the study (Fig. 4.5) that higher tariff in Kitonyoni for first three years resulted in very slow consumption growth compared to the Oloika mini-grid with cheaper tariff. In the case of two mini-grids in Uganda with much lower tariff compared to Kenya showed faster consumption growths. Therefore, it is clear that the affordable tariff helps the communities to grow and consume more electricity towards better lifestyle. With declining cost and innovative business models in PV mini-grid sector, off-grid communities can be served with more affordable tariff to have sustainable growth in the society.

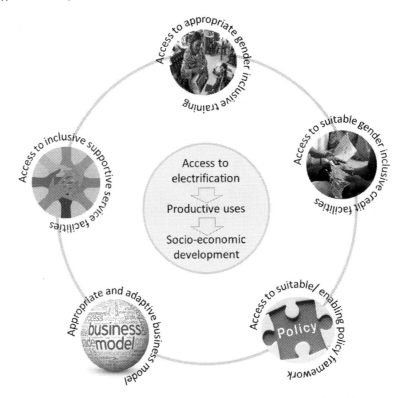

FIGURE 4.6 Access to PV- based electrification, productive uses and inclusive sustainable socio-economic growth.

Access to electricity with affordable tariff can enhance sustainable inclusive growth. Here the model of an integrated approach is proposed for an off-grid PV mini-grid project in Daulatpur, Manikganj district in Bangladesh. The 242kWp project[6] with 14 kilometers distribution line was financed by the infrastructure development company limited (IDCOL) of Bangladesh government at a cost of USD 1.88 million to serve 1000 households, 96 businesses, 4 rice mills and 16 social institutes. A local cooperative with the focus on inclusive growth has been using the benefit of access to electricity with a tariff almost similar to the national utility tariff in Bangladesh. It has trained local women to produce children dresses through the support of available microcredit. The proposed inclusive socio-economic growth model is presented in Fig. 4.6, where the marginalized local women attained the maximum benefit of PV-based off-grid electrification being supported through the access to (i) gender inclusive training and credit facilities, (ii) suitable policy and business model, and (iii) inclusive support.

4.5 Case study: Kenya

Kenya is progressing fast with its "electricity access for all" plan. As of 2018, the country has achieved 75% national electrification (IEA, 2019), which is phenomenal across the

---

[6] Super Star Solar PV Mini-grid, Bangladesh. (Information provided by SSG Solar).

sub-Saharan Africa. Although renewable energy (mainly geothermal and hydro) accounts for 85% of Kenya's installed capacity, almost 55% and 87% households still use kerosene for lighting in urban and rural settings, respectively. Per-capita electricity consumption in Kenya remains as low as 162.5kWh. The planned national target of electricity access for all by 2022 and attaining 'middle-income country' status target by 2030 has been hit hard by COVID-19. To revive country's economy while maintaining National Climate Change Action Plan (NCCAP) 2018-2022 and being on track with its Nationally Determined Contributions (NDC) to reduce greenhouse gas emissions by 30% (110 $MtCO_2e$) by 2030 energy sector will face a bumpy road ahead. Kenya's clean energy pathway (KCEP) is planned around the NCCAP which focuses on expansion of renewable energy production, enhancing energy and resource efficiency, and adaptation of low-carbon technology both in off-grid and on-grid sectors. Government assessed an additional investment requirement of USD 2.5 billion under the National Electrification Strategy (NES) towards its plan for the universal electrification.

Kenya Off-Grid Solar Access Project (KOSAP) is the flagship off-grid electrification program (KOSAP component-1, 2018–2023) which targets 14 underserved counties and planned to install 120 mini-grids (ESMAP, 2017). Through KOSAP, government will deploy mini-grids for community facilities, enterprises and households deploying a public-private partnership (PPP) model using power purchase agreement (PPA). This model is designed to be implemented by three key actors, Rural Electrification and Renewable Energy Corporation (REREC), Kenya Power and Lighting Company (KPLC), and private service provider (PSP), and be overseen by the Ministry of Energy (The World Bank 2017).

REREC and KPLC support county governments in finalizing the potential sites for PV mini-grids based on two key criteria. These are, (i) number of potential customers between 100 and 700, and (ii) electrical demand of 20-300 kW. Once, sites are identified, KPLC will determine mini-grid system size and architecture (combination of PV, battery and diesel generator) based on number of customers and level of required services at each site. KOSAP's public-private partnership (PPP) model for PV mini-grids is laid out around specified roles of different actors as indicated below:[7]

(i) REREC to deliver off-grid/rural electrification through the KOSAP program.
(ii) PSP to build electricity generating systems and distribution networks, and to operate and maintain these for an agreed period (usually between 7 and 10 years).
(iii) KPLC to deal with the customer connections and revenue collection.
(iv) PSP will receive regular payments for their services from KPLC according to the agreed PPA.

KOSAP program is designed to ensure recovery of the equity of the private investor (part cost of generation and distribution infrastructure) within the agreed service contact period. After this period all generation and distribution assets will be owned by the government. Success of KOSAP project will strengthen PV-based electrification potential of hundreds of off-grid health centers and clinics under the fourteen focused counties (Fig. 4.7) in Kenya to deliver better healthcare services.

---

[7] The Kenya Off-Grid Solar Access Project (K-OSAP) is a flagship Project financed by the World Bank and jointly implemented by the Ministry of Energy (MoE), Kenya Power and Lighting Company (KPLC) and Rural Electrification and Renewable Energy Corporation (REREC). https://kosap-fm.org/about.

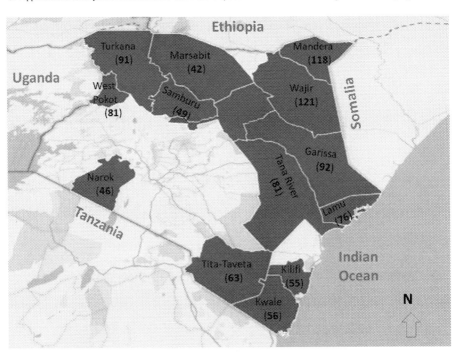

FIGURE 4.7 Potential number of health centers in fourteen counties focused under KOSAP project. Number of health centers is indicated in each selected counties which can be benefited through PV-based electrification using similar business model as KOSAP (Author's own illustration).

The KOSAP project and Kenya's collaborative partnership with regional and international partners seek to support the PV-based on-grid and off-grid electrification through eliminating existing national capacity gaps in the areas of:

(i) Renewable resource mapping for optimal applications in urban and rural settings,
(ii) Addressing energy efficiency for end users, managing demand through micro generation in both urban and rural settings,
(iii) Making mini-grids more resilient to demand growth, providing utility grid quality power, supporting performance through automation, monitoring and operation,
(iv) Quality assurance of solar home systems, and
(v) Integrating energy uses for inclusive socio-economic development.

The KOSAP model of Kenya can be applied with necessary adjustments in many other developing countries for sustainable development using solar energy.

4.6 Conclusions

Developing countries should use solar technologies both in on- and off-grid applications. However, present grid infrastructures in many countries may not be capable enough to cope with sudden surge of solar power integration. Provisions of developing regional independent grid with new infrastructures should be considered. Besides the established concept of feed-in tariff, building integrated PV microgenerations for self-consumptions in the urban areas can be promoted through financial rewarding schemes to support reducing peak loads from utility grids. In the case of off-grid PV electrification, making power available for small productive use and medium industrial applications is a big challenge. Techno-economic feasibility of multiple interconnected mini-grids is worthwhile to study which can resolve the issue of seasonal or occasional heavy power requirements.

Large scale on- and off-grid projects in developing countries should be deployed through innovative public-private partnership models. Replication of a successful models and their application at a scale require minimum bureaucracy and maximum transparency at the government level. Early adaptation of the emerging technologies like "hydrogen as energy carrier and fuel for cooking" can fast-track solar energy integration in developing countries. This needs research and development, collaboration, and long-term strategic planning.

Self-evaluation questions

1. What are the two key matured technologies to convert solar resource to electrical energy? – Explain in brief.
2. What are the six key PV technology segments that have off-grid applications?
3. What are the common technical problems associated with solar PV technologies?
4. Do you think solar photovoltaic could be a potentially sustainable solution to energy poverty in the developing world? – Explain.
5. What are the challenges of access to energy, in particular, electricity in developing countries? How could these challenges be overcome?
6. Conduct a literature survey and compare between different off-grid energy generation technologies.

References

Alam, M., 2017. Decentralized renewable hybrid mini-grid based electrification of rural and remote off-grid areas of Bangladesh, PhD thesis, De Montfort University, UK. (e-print). Available at: https://dora.dmu.ac.uk/handle/2086/14535?show=full (accessed: 23.04.2021).

Bahaj, A.S., James, P.A.B., 2019. Electrical minigrids for development: lessons from the field. Proc. IEEE 107 (9), 1967–1980.

Day, T., Kurdziel, M., 2020. The Role of Renewable Energy Mini-Grids in Kenya's Electricity Sector: Evidence of a Cost-Competitive Option for Rural Electrification and Sustainable Development. New Climate Institute. https://newclimate.org/wp-content/uploads/2019/11/The-role-of-renewable-energy-mini-grids-in-Kenya%E2%80%99s-electricity-sector.pdf.

ESMAP, 2017. Mini-Grids in Kenya: A Case Study of a Market at a Turning Point, The World Bank Group, Washington DC.

ESMAP, 2019. Mini-grids for half a billion people: Market outlook and handbook for decision makers, ESMAP Technical Report: 014/19. 2019.

IEA, 2019. Kenya Energy Outlook: Key Indicators and Policy Initiatives. International Energy Agency, Paris. Available at: https://www.iea.org/articles/kenya-energy-outlook (accessed 08.02.2021).

IEA, 2020a. Data and Projections. International Energy Agency, Paris. Available at: https://www.iea.org/reports/sdg7-data-and-projections (accessed: 12.02.2021).

IEA, 2020b. Renewables 2020: Analysis Forecast 2025. International Energy Agency, Paris. https://www.iea.org/reports/renewables-2020 (accessed: 12.02.2021).

IEA, 2020c. CSP Capacity Additions in Selected Countries, 2018–2021. International Energy Agency, Paris. https://www.iea.org/data-and-statistics/charts/csp-capacity-additions-in-selected-countries-2018-2021 (accessed: 07.03.2021).

IEA, 2020d. Solar PV: Tracking Report, 2020. International Energy Agency, Paris. Available at https://www.iea.org/reports/solar-pv (accessed: 21.03.2021).

IEA, 2021. Renewable Energy Market Update 2021: Outlook 2021–2022. International Energy Agency, Paris. Available at: https://www.iea.org/reports/renewable-energy-market-update-2021 (accessed: 14.05.2021).

IEA, IRENA, UNSD, World Bank, WHO, 2020. Tracking SDG 7: The Energy Progress Report. The World Bank, Washington DC.

IRENA, 2020a. Off-grid renewable energy statistics 2020. International Energy Agency, Abu Dhabi ISBN 978-92-9260-333-5.

IRENA, 2020b. Renewable power generation costs in 2019. International Renewable Energy Agency, Abu Dhabi ISBN 978-92-9260-244-4.

Magazine, P.V., 2020. Qatar's 800 MW tender draws world record solar power price. Available at: https://www.pv-magazine.com/2020/01/23/qatars-800-mw-pv-tender-saw-world-record-final-price-0-01567-kwh/ (accessed: 28.01.2021).

The World Bank, 2020. Access to Clean Cooking: The efficient, clean cooking and heating programme and the clean cooking fund. Available at: https://www.worldbank.org/en/results/2020/11/10/accelerating-access-to-clean-cooking-the-efficient-clean-cooking-and-heating-program-and-the-clean-cooking-fund (accessed: 12.02.2021).

The World Bank, 2017. Kenya Off-Grid Solar Access Project for Underserved Counties', Washington, D.C. Available at: http://documents.worldbank.org/curated/en/212451501293669530/Kenya-Off-grid-Solar-Access-Project-for-Underserved-Counties (accessed: 22.03.2021).

CHAPTER 5

Hydropower–Basics and its role in achieving energy sustainability for the developing economies

Arun Kumar

Department of Hydro and Renewable Energy, Indian Institute of Technology Roorkee, Uttarakhand, India

5.1 Introduction

Hydropower is the energy generated from water flowing from higher to lower levels. Hydropower technology has the best conversion efficiencies (about 90% efficiency) and the highest energy payback ratio of all known energy sources. Though a relatively high initial investment is required for hydropower projects, such projects have a long life with quite low operation and maintenance costs.

Hydropower plants are not water consumptive for running the turbine. The water, after generating power, is available for various other uses. Hydropower serves both large and small, centralized, or isolated electric grids. It supports both solar and wind energy by providing the backup that variable technologies need. In addition, it provides grid flexibility services, such as frequency response, black start capability, and spinning reserves. Many hydropower projects (HPP) are multipurpose providing energy and water supply services such as irrigation, drinking water supply, drought and flood management, navigation, and recreation along with socio-economic benefits. For constructing a hydropower project, in addition to the environmental, social, physical, and economic considerations, legal considerations are equally important as in most of the countries the necessary permissions are required from the respective government (Kumar et al., 2011).

Hydropower is one of the cleanest sources of electricity, emitting lesser greenhouse gases than other kinds of energy sources. Globally, hydropower is the largest source of renewable electricity (15.9%) out of a total of 27.3%, generating more electricity than all other renewables combined (11.4%). However, in the era of the energy transition with the aim of limiting the global temperature rise to no more than 2°C, the development of sustainable hydropower construction is needed to increase significantly. The International Energy Agency's (IEA) flagship Net Zero by 2050 report (IEA, 2021) suggests that about 2600 GW of hydropower capacity is required by 2050 for keeping global temperature rises below 1.5°C.

The energy in today's context needs to be sustainable. Coal-based power generation is now getting the least priority and most of them after completing their life are being abandoned. Globally, a determination toward emission-free energy is placed and most of the developed, as well as developing countries have started working on this. The hydropower being the most important renewable energy has been around for over a century and aims to provide energy security by complementing wind and solar energy in sustainable energy storage (Kaunda, 2012).

Hydropower constitutes nearly all of the total electricity generation in several developing countries (Mozambique, Tajikistan, Zambia, and Nepal) and developed countries such as Norway, Canada, and Switzerland.

5.2 Historical background of hydropower development

Hydro energy has been used for irrigation and the operation of various end-use equipment, like watermills, textile machines, lumber, and weaving machines, in numerous nations since ancient times. The mechanical power of flowing water is an old natural resource used for services and productive applications. Over 2000 years ago, Greeks harnessed power from flowing water to use watermills to crush wheat into flour. In the 1700s, mechanical hydropower was used extensively for milling and pumping. There are a number of places in the world where such technology is still being utilized. During the 1700s and 1800s, modern hydraulic turbine development continued. On September 30, 1882, the world's first hydropower plant, with a capacity of 12.5 kW, was commissioned at the Vulcan Street Plant on Fox River, in Appleton, Wisconsin, USA, lighting two paper mills and a residence.[1] During the initial stages of hydropower development, small hydropower generation was vital to the socio-economic growth of various areas. During the year 1897, the first hydropower plant in India was commissioned in Sidrapong of Darjeeling district.

After the development of a high voltage transmission network in the early 1920s, a shift occurred from small hydropower plants serving local electricity requirements to large hydropower plants, through extensive electricity grids supported with other sources of electricity, serving industrial, agricultural, and domestic needs.

5.3 Present status of hydropower development

The International Journal on Hydropower & Dams *World Atlas & Industry Guide* (IJHD, 2020) provides the most comprehensive inventory of current hydropower installed capacity and annual generation, and hydropower resource potential. Country-wise hydropower technical potential in terms of annual generation and installed capacity (GW); and current generation, installed capacity, average capacity factors in percent, and undeveloped potential as of 2019 are listed in Table 5.1.

---

[1] United States Bureau of Reclamation: www.usbr.gov/power/edu/history.html.

TABLE 5.1 Region-wise technical potential of hydropower as of 2019 (Source: (IJHD, 2020)).

Region	Technical potential, annual generation (TWh/yr)	Technical potential, installed capacity (MW)	Total generation as of 2019 (TWh/yr)	Installed capacity as of 2019 (MW)	Un-developed potential (%)	Average regional capacity factor (%)
Africa	1627	401	139	34	91	46
Asia	7981	2298	2165	624	73	40
Australasia/Oceania	186	63	41	14	78	34
Europe	1196	435	550	200	54	31
North America	1909	482	717	181	62	45
Latin America	2850	724	700	178	75	45
World	15,749	4403	4312	1231	72	40

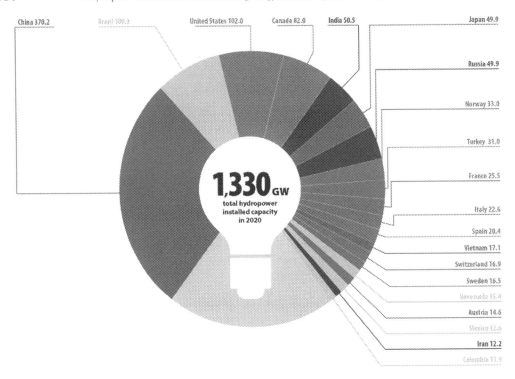

FIGURE 5.1 Hydropower installed capacity in 2020 (IHA, 2021a).

Hydropower installed capacity reached 1231 GW in the year 2019 and generation reached 4312 TWh (IJHD, 2020). China, Brazil, the USA, Canada, and India are the largest hydropower producers by installed capacity. Hydropower installed capacity in 2020 is shown in Fig. 5.1 (IHA, 2021a).

Hydropower is the leading renewable energy technology toward low-carbon electricity generation. Hydropower development needs better visibility on its direct and indirect revenues on the grid, society, and environment and is key to attract investment at a large scale (DFID, 2009). According to the latest analysis, state-owned utilities are anticipated to add over 75% of new hydropower capacity globally through large hydropower projects in Asia and Africa. The flexibility of hydropower is essential for incorporating increasing solar PV and wind penetrations into the electricity grid (IEA, 2021). The world's present hydropower capacity will need to grow by around 60% by the year 2050 to reach 2150 GW for assisting in limiting the rise in global temperature to below two degrees Celsius (IRENA, 2021).

Even with stringent environmental and social compliances as needed for sustainable hydropower, the potential for an additional 850 GW of installed capacity is well within our reach by 2050. As per the International Hydropower Association (IHA) study, at least 500 GW of hydropower projects are at different stages of development. About 156 GW of hydropower is under construction, 165 GW is looking for awaiting construction, 138 GW is awaiting clearances (IHA, 2021b).

The world's total estimated potential and installed capacity of small hydropower (SHP) up to 10 MW are 229,111 MW and 78,046 MW respectively. The largest potential of SHP lies in Asia with a total capacity of 138,275 MW, out of which only 37% is harnessed. Following this, the second-largest potential of SHP lies in America with a total capacity of 41,861 MW, out of which only 14.91% is harnessed. In Europe, Africa, and Oceania, the total SHP potential is 37,526 MW, 10,241 MW, and 1208 MW respectively, out of which only 52.30%, 5.80%, and 36.50% are harnessed, respectively (UNIDO, 2019).

It may be noted that a large number of existing water resource facilities, where hydropower is not developed (nonpowered), are available worldwide. These facilities may be utilized for hydropower generation by installing new generating equipment. Based on a study in the United States, installing generating equipment in 2500 nonpowered reservoirs may result in the installation of about 20,000 MW capacity. For example, as a cost-effective option, small hydropower plants of about 1000 MW capacity were built on existing water resources facilities in India between 1997 and 2016.

The majority of reservoirs are designed to hold seasonal storage, but some can also perform the multi-year regulation, in which water is kept during high flow seasons for more than one year and released during low flow seasons. Other water uses, such as irrigation, water supply, and navigation, and flood control, need water storage. Reservoirs can also be used for a variety of purposes, including recreation and aquaculture. The bulk of the world's 45,000 big dams were built for irrigation, flood control, navigation, and urban water supply programs, rather than electricity (WCD, 2000). Only around 25% of big reservoirs are presently used for hydropower generation in addition to other applications, thus, retrofitting these existing facilities with hydraulic turbines represents a significant decentralized hydro potential.

International Hydropower Association[2] (IHA) has published details showing the number of dams with single or multipurpose use constructed globally as presented in Table 5.2.

5.4 Hydropower basics

Hydropower resource potential is estimated as the product of head, available discharge, and a conversion factor. The major portion of precipitation usually occurs in the mountainous region with the largest possible elevation differences, these locations have the greatest potential for hydropower generation. The total potential of hydropower may be estimated using Eq. (5.1).

$$P = 9.81 \times Q \times H \times \eta \tag{5.1}$$

where P represents the power generated in kW, Q represents discharge in m^3/s, H represents the head in meters and η represents the overall efficiency of generating equipment.

In run of river (discussed later in the chapter) hydropower projects, the variation in the head is less except for some losses which occur when discharge varies in the water conductor system. In dam-based or canal falls schemes, the variations in head occur during variations in the release of water. In order to work out the rated head of the turbine, the weighted average

---

[2] www.hydropower.org

TABLE 5.2 Number of dams constructed globally with single or multipurpose use (Source: IHA website).

S. No.	Purpose	No. of dams		Total
		Single purpose	Multipurpose	
1.	Irrigation	6002	14,011	20,013
2.	Hydropower	3913	5751	9664
3.	Water supply	4342	3276	7618
4.	Flood control	4861	2480	7341
5.	Recreation	2933	1349	4282
6.	Fish breeding	1399	41	1440
7.	Navigation	580	96	676
8.	Others	1322	1594	2916
	Total	25,352	28,598	53,950

of the head is calculated. The design head is chosen in such a way that the turbine operates for the maximum amount of time, resulting in maximum energy output.

For the run of river scheme, it is always desirable to study the temporal variation of water discharge, for a year, to select a suitable configuration of generating unit and estimate the power generation. Flow duration curves represent temporal variation in river discharge or release of water from the dam to carry out the water power studies. These curves are the basis of determining firm power and secondary power for a particular project.

Head is defined as gross head, net head, and rated head. The *gross head* is the water level elevation difference between the forebay/reservoir and the tail water at the outlet. *Net head*, also known as the effective head, is the head available at the inlet of the turbine for power generation. As illustrated in Fig. 5.2, this head is the gross head minus the hydraulic losses of water passage. The hydraulic losses in a closed conduit may be estimated using standard hydraulic textbook concepts. The hydraulic losses through the turbine and draft tube are accounted for in the turbine efficiency. The net head at which the full-gate opening of the turbine generates the rated output is known as the *rated head*. At this head, the turbine nameplate rating is generally stated. The selection of the rated head requires planning and deliberation.

Primarily, the hydropower projects are classified on the basis of their installed capacity and availability of head. The main parameters for selecting the type and capacity of hydraulic turbines are available head and water discharge. Hydropower is categorized differently in different nations based on the installed capacity of hydropower plants.

Based on the installed capacity, the hydropower projects in India are classified as given in Table 5.3.

In other countries, such as China, the small hydropower schemes are categorized up to 50 MW whereas United Nations Industrial Development Organization (UNIDO) classified small hydropower schemes as capacity up to 10 MW. Irrespective of this classification hydropower

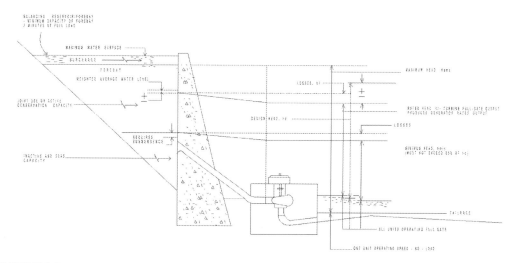

FIGURE 5.2 Definition of head in reaction turbines (Adapted from USBR (USBR, 1976) – Selecting Hydraulic Reaction Turbine, Engineering Monograph No. 20).

TABLE 5.3 Classification of hydropower projects in India based on the installed capacity.

Type of project	Installed capacity (MW)
Pico hydro	≤0.005
Micro hydro	>0.005 to ≤0.1
Mini hydro	>0.1 to ≤2
Small hydro	>2 to ≤25
Medium hydro	>25 to ≤100
Large hydro	>100

TABLE 5.4 Classification of hydro projects in India based on the head.

Type of project	Head (m)
Ultra low head	<3
Low head	3 to ≤40
Medium head	>40 to ≤75
High head	>75

covers a continuum in project scale. Further, the hydropower projects can be developed as grid-connected or standalone, single or in cascade on a river, with or without pondage.

As per Indian Standard 12800 – Part III 1991, the classification of hydropower projects based on the availability of head is given in Table 5.4.

The system may utilize all of the electricity generated by a grid-connected hydropower plant, depending on the power demand. It is necessary to forecast the power demand for isolated plants. The power potential from a hydropower plant is evaluated using several parameters such as available water discharge at the site which depends upon the catchment characteristics, head available at the site which depends upon the topography, and overall efficiency of generating equipment.

5.5 Services from hydropower development

Hydropower exists in a range of forms and sizes, and it can fulfill both large centralized and small decentralized demands. Hydropower being a tradable commodity, there are many hydropower plants in the world constructed or under construction with the focus of selling electricity internationally having the higher demands and better price. Hence, hydropower is being utilized to fulfill the electricity requirement of various nations.

Storage hydropower and/or pumped storage power (PSP) plants enable grid operators to manage their networks in a secure and flexible manner by balancing the intermittent renewable energy sources (e.g., wind and solar PV) and increasing protection for various ancillary services. Hydropower is the most efficient energy source to fulfill peak demand, making it ideal for auxiliary services and balancing unreliable transmission networks. Transmission, security, and qualitative electrical operations are all ensured by storage-based hydropower schemes and pumped storage schemes. As a result, hydropower development is a part of both water and energy management systems, each of them is more influenced by climate change and hence, offers water and energy security, respectively. Hydropower services for electricity, environment and society have been well presented in the literature such as Egré and Milewski (2002), IEA (2000), and Kumar et al. (2011) and these are summarized in Table 5.5.

5.6 Required surveys and investigations

The surveys and investigations required for hydropower development are well understood and being developed by using national and other international standards available. The required survey and investigations are listed as follows:

- Hydrological survey
- Topographical survey
- Geological survey
- Socioeconomic and environmental survey
- Assessment of power requirement
- Power evacuation network
- Meteorological survey
- Construction material survey
- Muck disposal survey

The primary purpose of the hydrological survey is to fix the design discharge and number of generating units of a hydropower plant. In addition, the flood estimation at the project

TABLE 5.5 Energy, environmental, and social impacts of different types of hydropower projects (adapted from Kumar et al., 2011; IEA, 2000; Egré and Milewski, 2002; Killingtveit, 2019).

Type	Energy and water management services	Main environmental and social impacts
All types	-Renewable electricity generation. -Increased water management options.	-Barrier for fish migration and navigation, and sediment transport. -Physical modification of riverbed and shorelines.
Run-of-river	-Limited flexibility and increased variability in electricity generation output profile. -Water quality management.	-Unchanged river flow when powerhouse in dam toe; when localized further downstream reduced flow between intake and powerhouse.
Reservoir (Storage)	-Storage capacity for energy and water. -Flexible electricity generation output. -Water quantity and quality management; groundwater stabilization; water supply and flood management.	-Alteration of natural and human environment by impoundment, resulting in impacts on ecosystems and biodiversity and communities. -Modification of volume and seasonal patterns of river flow, changes in water temperature and quality, land use change-related GHG emissions.
Multipurpose	-Water release from reservoir of hydropower plants. -Dependent on water consumption of other uses downstream.	-Possible water use conflicts. -Driver for regional development.
Pumped storage	-Storage capacity for energy and water; net consumer of electricity due to pumping. -Can be used for peak time electricity generation.	-Impacts confined to a small area; often operated off the river as a separate close system. -Land use change-related GHG emissions.

location is used to design the diversion structure/dam and maintain the power plant safe from flooding.

The topographical survey provides availability of head for power generation, project layout as well as alignment of various works, and length of the transmission network. The geological survey involves the systematic investigation of rocks and geological formations available at the project location for the stability of different components. Social and environmental impacts of the proposed hydropower plant are assessed through environmental and social impact assessment[3] (ESIA). On the other hand, the economic impacts of the proposed project are assessed as a part of the feasibility study.

A load demand survey (i.e., assessment of power requirement) is carried out to determine whether the power produced by the hydropower plant is being used in a stand-alone mode or supplied to the grid. Following this, the power evacuation survey is conducted to decide the optimal route of power evacuation and connection to the grid connection. The meteorological survey gives the estimation of rainfall, snowfall, and ambient temperature conditions at project locations during different seasons. Construction material surveys are usually carried out to get an understanding of material availability in terms of quality and quantity in relation

---

[3] https://www.iucn.org/sites/dev/files/iucn_esms_esia_guidance_note.pdf

to the project requirements, as well as to evaluate the techno-economic feasibility of the project. During the construction of the project, muck is generated from soil material as well as excavation of rock. This muck should be disposed off at the designated areas keeping in view the prevailing environmental and land regulations.

5.7 Hydrology

"Hydrology is the science that encompasses the occurrence, distribution, movement and properties of the waters of the earth and their relationship with the environment within each phase of the hydrologic cycle"[4]. The development of hydropower projects requires the estimation of available discharge at the proposed site. Long-term information on river discharge would be ideal for estimating the design discharge for a hydropower scheme.

When discharge records are unavailable, as frequently in hilly and underdeveloped areas, the discharge data is generally synthesized by combining the records of rainfall and catchment area characteristics. In some cases, the discharge data are obtained by correlating records from adjacent catchments having hydrological similarities, however, such data is less precise and has a greater level of uncertainty. In order to extend the discharge data using long-term run-off data from adjacent sites, several methods can be used for hydrological analysis. These methods are double mass curve, index-station, Langbein's log-deviation, regression analysis, flood frequency, and method of correlation with catchment areas (Jain et al., 2007).

Various flood estimation methods are utilized to design the diversion structure and maintain the power plant secure from flooding. Based on the statistical relationships of recorded peak floods and catchment area characteristics, empirical equations are developed, however, these equations are site-specific and considered to be inaccurate.

The rational formula for flood estimation is based on an assumption of isosceles triangle hydrograph and is applicable only for small catchments. The deterministic method based on unit hydrograph is frequently used to estimate the design flood hydrograph. The choice of a particular approach is influenced by various factors like data availability, catchment characteristics, and relevance of the project determined by criteria such as the available data, the catchment area, and the project's relevance.

River discharge is routinely controlled for societal benefit. *Environmental flows* are referred to as the requirement of water discharge in the river/stream to sustain aquatic life and human livelihoods and well-being having competing water uses. For ecological sustainability, most nations have rules assuring environmental flow requirements in the dry river stretch by releasing a certain amount of water downstream of the diversion structure.

5.8 Cumulative impact assessment of hydropower plants

In addition to project-specific environmental impact assessment studies, the process of getting cumulative environmental impact assessment studies has been started for hydropower projects for river basins. The subject of cumulative impact assessment and thereafter deciding

---

[4] https://www.usgs.gov/special-topic/water-science-school/science/what-hydrology?qt-science_center_objects=0#qt-science_center_objects

the carrying capacity (decision on what should be maximum number of projects and river stretches for hydropower development) of the river is new and there is no single broadly agreed method for carrying out such studies. Further such studies are also handicapped due to a deficit of required data and expertise. State governments and developers are still gearing up to understand and appreciate the concept of cumulative environmental impact assessment (CEIA) and implement the outcome of the studies. There are complexities at the government level and the process of CEIA carried out by the different consulting groups.

5.9 Climate change impact on hydropower potential

Climate change is likely to increase uncertainty regarding water supply for hydropower generation, affecting the long-term prospects of hydropower infrastructure. Climate change may also increase the conflict over water sharing due to decreasing availability and increasing variability of water for several sensitive areas between areas and countries like South Asia.

Hydropower potential is estimated using historical discharge time series. However, if the environmental conditions change, the hydropower potential may vary as well (Berga, 2016) due to the following:

- Variations in river discharge, and volume, as well as seasonal variation of discharge.
- Variations in major events like floods and droughts may reduce the economic feasibility.
- Increased sediment in river flows due to higher sediment erosion on account of extreme events resulting in increased turbine abrasive erosion, decreased generating efficiency, higher maintenance cost, breakdown, and loss during repairs.

Reliability of energy systems having a high portion of hydropower in the total electricity supply may be reduced due to change in hydrology due to climate change unless supported with storage. Intermittent rainfall patterns may become a risk to developing countries like Ethiopia which is constructing Africa's largest 6450 MW Grand Ethiopian Renaissance Dam-based hydroelectric project and may require a diversified energy portfolio.

Hydropower being low-carbon emission energy contributes to mitigating climate change. As per IHA (2018), based on data of 178 and 320 single- and multi-purpose hydropower reservoirs respectively, the world's average greenhouse gas emission intensity from the reservoirs was found to be 18.5 g CO_2 eq/kWh -over a life-cycle against the median lifecycle carbon equivalent intensity of onshore wind energy as 11 g CO_2 eq/kWh, solar PV as 48 g CO_2 eq/kWh, gas as 490 g CO_2 eq/kWh, and for coal as 820 g CO_2 eq/kWh (IHA, 2018 and Bruckner et al., 2014).

5.10 Technology

Hydropower plants are divided into three categories based on their operation and available discharge at the project site. Based on the available discharge and head, there are run-of-river (RoR) hydropower plants (HPP) with or without pondage, reservoir based HPPs, and pumped storage-based HPPs. The in-stream hydrokinetic turbine is a recent technology. Efforts are being made throughout the world to make it commercially feasible.

Run of river hydropower plants involve the diversion of water from the river for power generation. These schemes are built with or without small pondage and operated as base

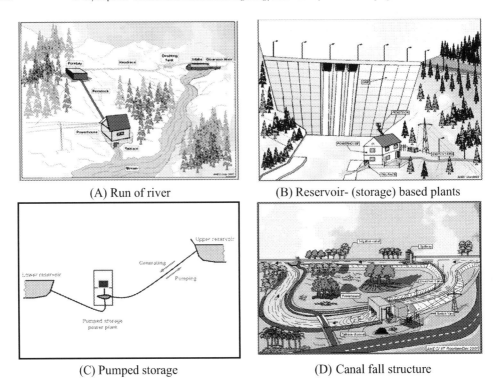

FIGURE 5.3 Types of hydropower plants: (A) run of river, (B) reservoir-based plants, (C) pumped storage, (D) canal fall structure.

load plants. *Reservoir-based hydropower plants* are normally placed at the toe of the dam or farther downstream via tunnel or penstock depending upon the available site conditions. In addition, these reservoirs may serve as drinking water, irrigation and flood management schemes. The *pumped storage plants* have a unique ability to recirculate the water between the two reservoirs. During off-peak period, water is pumped from the lower reservoir to the upper reservoir using low-cost off-peak power, whereas, during the peak load period, the electricity is produced using water from the upper reservoir. In addition, the power can be extracted from the falls available in the canals, by constructing a hydropower plant across the main canal or bypass canal depending upon the site conditions. The typical illustrations of these types of hydropower schemes are shown in Fig. 5.3.

5.11 Components of hydropower plant

A hydropower plant comprises of the following components:

- Civil works components
- Hydro electrical components comprising of hydro-mechanical and hydro-electrical equipment

5.11.1 Civil works components

5.11.1.1 Diversion structure or reservoir

The diversion structures are built to divert the discharge needed for power generation to the settling basin through a feeder channel, whereas, the reservoir is built to store the water during the period of excess flows to low flows. These structures are built either across the river or maybe in the form of weirs, barrages, dams, or inflatable rubber dams. The structures should withstand piping, overturning, and sliding.

5.11.1.2 Desilting structures

In hilly regions, the rivers carry high sediment that can harm the turbine components, penstock, and other underwater components of the run of river hydropower plants. Desilting structures are provided as a part of the water conductor system to remove the sediment particles carried by the diverted water. The approach velocity of water in the desilting structure is reduced by increasing the cross-sectional area of the structure, thereby enabling particles to settle. In vortex type sediment removal structure, the sediment is thrown on the outer surface and travels through a common outlet and such structures are cost-effective in specific instances for smaller discharges.

5.11.1.3 Power channel

The power channel conveys water to the forebay tank through the desilting structure. The primary purpose of a power channel is to convey water and, sometimes to absorb water surges. To decrease water loss due to seepage and increase the carrying capacity, the power channel might be coated with concrete or stones. The power channel may have a rectangular or trapezoidal shape in form. The power channel may be constructed as an uncovered or cut & cover section, depending upon the site conditions. The power channel may be covered or uncovered. Design of a channel is a trade-off between excavation and structural cost and the saving of energy (additional head) leading to more revenue. Manning's equation is normally used to find out the channel head losses (Eq. 5.2).

$$Q = 1/n A R^{2/3} S^{1/2} \tag{5.2}$$

where Q is discharge in the channel (m^3/s), n is Manning coefficient, A is channel cross-section area (m^2), R is channel hydraulic radius (A/P), where P is wetted parameter (m), S is slope of energy gradient, that is, head loss per unit length.

5.11.1.4 Tunnels

In hydropower projects, tunnels are often used for carrying water from the water source to the turbines installed in the powerhouse which may be on the surface or underground. The tunnels are also being used as penstock and known as pressure shafts for a specific site. When the power station is underground, tunnels are also used for access, power cables, surge shafts and ventilation, and tailrace, etc. Tunnels are increasingly preferred for HPPs against surface structures like open channels and exposed penstocks to avoid land acquisition and environmental landscape issues as well as to reduce maintenance costs.

Due to the development of efficient systems, the tunneling technique has been considerably enhanced. The drill and blast method (DBM) and tunnel boring machines (TBM) are currently the two prominent tunneling techniques. TBMs use the explosives and can excavate the whole tunnel cross-section of diameter from less than 1 m up to 15 m but they should not be utilized in soft strata. Flow in tunnels flowing full is pipe flow whereas tunnels flowing partially are open channel flow and are designed accordingly.

5.11.1.5 Forebay tank or surge tank

A storage tank that connects the channel or tunnel from one end and the penstock from another end located at the end of the power channel/tunnel and from here the penstocks intake start. It supplies instantaneous water demand when the turbine units are turned on, as well as the necessary water seal over the penstock entrance to prevent air entrainment. The forebay is provided with a spillway on either side to spill excessive discharge during load rejection. A surge tank is a type of storage that allows the surge to go up and down without causing any damage to the penstock.

5.11.1.6 Penstock

Penstock is a closed conduit that conveys water to the turbine through a forebay or surge tank. Penstocks are made to resist maximum water pressure, especially water hammer. They are also quite expensive and play a vital role in the water conductor system. The penstock is laid on the surface or embedded in the concrete or buried in the ground as per site-specific conditions. The penstocks are fitted with specials like bends, expansion joints, manholes, matching pieces, reducer or expander, and bell mouth intake. Flow in the penstocks is pipe flow and the losses are calculated using the Darcy Weisbach equation (Eq. 5.3).

$$\text{Headloss} = fLV^2(2gD)^{-1} \tag{5.3}$$

where f is friction coefficient of the inner surface of penstock and depends upon the material, L is the length of penstock (m), V is the velocity in the penstock (m/s), g is gravitational coefficient (m/s^2) and D is diameter (m).

5.11.1.7 Powerhouse

The purpose of the powerhouse is to support and house the generating units (turbine and its auxiliary units) and their accessories against various climatic conditions. The type of powerhouse selected is based on the economic analysis carried out on the construction cost and operation cost. The powerhouse may be located on the surface or underground, depending upon the prevailing site conditions. The layout of the powerhouse depends on the type of generator selected for the project. Most of the projects accommodate one or more units of similar or different capacity depending on potential and variation in water availability. The arrangements of handling the generating equipment for installation and maintenance are carried out by indoor cranes for which capacity is decided as per the largest piece to be handled. An erection or service bay is constructed within the powerhouse building for loading and unloading the equipment from transport and subsequently shifted to the respective place.

The capacity, number, and type of turbine unit, as well as the proportion of turbine runner diameter, determine the design and size of the powerhouse building. Depending upon the site conditions, the layout of the powerhouse building may vary with the type of project. The powerhouse building includes a machine hall having main hydro generating equipment

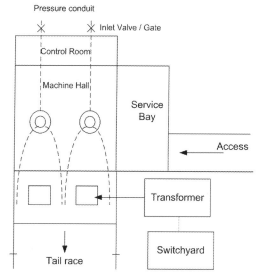

FIGURE 5.4 Typical powerhouse layout for hydropower plant.

(turbine, governor, and generator), a service bay for erection and maintenance purposes, a control room having control panels, relays, etc. and a tailrace for the outlet of water, as can be seen in Fig. 5.4.

Following equipment are placed in the powerhouse building:

Mechanical systems	Electrical systems
• Inlet gate or valve • Turbine • Speed increaser (if needed) • Water for cooling and service • Heating and ventilation • Compress air • Fire protection • Dewatering and sanitary drainage • Draft tube gates	• Generator • Excitation equipment • Control system • Condenser, switchgear • Protection systems • DC emergency supply • Power and current transformers • Neutral grounding

The powerhouse is designed keeping in view three parts of its structure i.e. substructure, intermediate, and superstructure. The substructure comprises of draft tube and sump well along with a ground mat. The turbine manufacturer provides the details for the draft tube and relatively loading. The powerhouses design for their stability in terms of water uplift, sliding, and overturning. The typical joints water stops and waterproofing are used in powerhouse buildings. The powerhouse structure is designed taking into account the maximum dead and live load, hydrostatic load, wind, and earthquake loads. The intensity of these loads is selected as per the equipment and site-specific conditions. In the powerhouse, the location of the turbine setting is an important feature which is decided as per altitude and head available at the site to avoid any cavitation and vibration. Most of the powerhouses are located below the normal tail water level, thus, requires special attention and care for avoiding any accidental flooding of the powerhouse. The interior of the powerhouse is well illuminated and ventilated.

Underground powerhouses are constructed as these have higher flexibility for location and layout, if no suitable site for surface powerhouse is available but have favorable geological conditions. Underground powerhouses have the potential for cost-saving, large capacity plants, uninterrupted working under adverse climatic conditions, better safety for penstock, better turbine governing, safer from defense and security consideration, and have the least environmental impacts since they cause the least disturbance to biodiversity and landscape.

5.11.1.8 Tail race channel

Tail race may be a channel or tunnel, through which water is released after passing to the hydraulic turbines. It should be constructed such that a minimal tail water level is maintained to avoid the issue of cavitation in hydraulic turbines.

5.11.2 Hydroelectrical equipment

5.11.2.1 Hydraulic turbine

The type of hydraulic turbine appropriate for operation at a particular location is decided by the net head available to the turbine. The capacity of hydraulic turbine is determined by the availability of design discharge. Hydraulic turbine manufacturers are responsible for the mechanical design and hydraulic efficiency of the turbine. The turbine and generator unit should have the maximum feasible speed at the least cost to generate supplied power at the specified head. However, the speed may be restricted due to mechanical design, cavitation and vibration consideration, and reduction in peak efficiency, as well as the reduction in overall efficiency because the best efficiency range of the power efficiency curve is narrowed.

The selection of a suitable type of hydraulic turbine is completed on the basis of specific speed (Ns) which classifies different types of turbines and their characteristics. *Specific speed* is defined as the speed in revolutions per minute at which a particular turbine would operate if reduced homologically in size to produce one metric horsepower at full gate opening under a one-meter head. Low-specific speeds are associated with high heads and high-specific speeds are associated with low heads. There are a variety of specific speeds that may be appropriate for a certain head, as shown in Fig. 5.5.

High-specific speed for a given head means a smaller turbine and generator, which reduces capital cost. The reaction turbine, on the other hand, will have to be placed lower, which may offset the cost savings. The specific speed of a turbine is mathematically expressed as Eq. (5.4).

$$N_s = \frac{N\sqrt{P}}{H_r^{1.25}} \quad (5.4)$$

where N is the rotational speed in revolutions per minute, P is power generated in metric horsepower at full gate opening, and H_r is rated head in m.

In Europe, the specific speed (Nq) is based on discharge and mathematically expressed as Eq. (5.5).

$$N_q = \frac{N\sqrt{Q}}{H^{0.75}} \quad (5.5)$$

where Q is discharge in m^3/s and H is net head in meters.

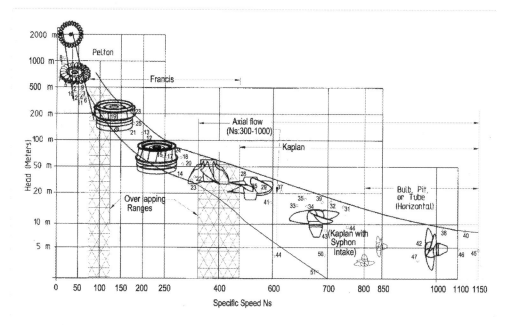

FIGURE 5.5 Head versus specific speed. It shows different types of turbines as a function of head and specific speed (Ns). (Adapted from HCI-ASME, 1996).

TABLE 5.6 Ranges of specific speed for hydraulic turbines.

Type of turbine	Range of specific speed (metric horse power units)
Propeller with fixed blades	300–100
Kaplan with adjustable blades	400–1400
Francis	55–500
Pelton	16–20 per jet
Cross flow	12–80

The specific speed for different types of hydraulic turbines is listed in Table 5.6.

5.11.2.2 Generators

The electric generator uses an electromechanical energy conversion mechanism to transform the turbine's mechanical energy into electrical energy. In HPP schemes, two types of generators are typically used: synchronous and asynchronous. An electric generator comprises the rotor (moving part) and stator (stationary part). The rotor is the rotating assembly to which the mechanical torque from the hydraulic turbine shaft is connected. A voltage is induced in the stator, the stationary component, by magnetizing or "exciting" the rotor. The exciter-regulator in the generator is used to fix and stabilize the output voltage. The generator speed matches the turbine speed except when coupled to a speed gear box. The provision of a gear box reduces the cost and size of a generator for a given power value.

The rotational speed of the synchronous generator connected to the grid is kept constant and is in synchronization with grid frequency. The rotation speed of the generator (revolution per minute) has a relation equal to the frequency (cycles per second) divided by twice the number of poles. The number of poles indicates the number of sets of coils in the generator stator producing electricity. Two poles make one pair of poles and hence generator will always have an even number of poles. The speed of generators for large units normally is lower than 600 rpm whereas for small units this may be up to 1500 for economic consideration. The turbine and generator are connected through a shaft.

The type and orientation of the turbine govern the location and orientation of the generator. For instance, there is an in-house arrangement of generators in the bulb turbine. For tube turbines, a horizontal generator is generally required, whereas a vertical turbine requires a vertical shaft generator with a thrust bearing. *Synchronous* generators are provided with a permanent magnet excitation system or DC electric system, as well as an automated voltage regulator to regulate the output voltage. These generators can be used for both isolated and grid-connected projects, ranging from a few kW to MW capacity. *Asynchronous* generators are squirrel-cage induction motors that do not have a voltage regulator and run at a speed that is proportional to the grid frequency. These generators with capacitors are useful for isolated plants (below 50 kW) when the required power supply is not particularly high. Asynchronous generators up to 2 MW may be utilized in HPP's that are connected to a stable grid.

5.11.2.3 *Governing system*

A governor is a feedback controller that senses the speed and power of the hydraulic turbine and adjusts the guide vanes or wicket gates to regulate the discharge and load depending upon the deviation of the actual setpoint from the reference point. The governors can be purely mechanical, mechanical-hydraulic, electrical-hydraulic, and mechanical-electrical.

5.11.2.4 *Transformers*

A transformer is a static device that step-up or step-down the voltage to transmit electrical energy from the primary to the secondary end. Step-up transformer is used to increase the produced voltage for transmission. Transmitting higher voltage over a long distance has several benefits, including low power transmission loss, reduced cable and insulator size, and cheap.

Step-down transformer is used at the distribution end to reduce the voltage to meet the needs of the consumers. Large hydropower projects include a station transformer for the auxiliary consumption of a powerhouse.

5.11.2.5 *Circuit breakers and relays*

Circuit breaker is a device that protects electrical networks from system disturbances. Its primary function is to separate defective parts from networks, allowing the issue to be localized. Previously, fuses were utilized; currently, circuit breakers are used.

It is also a protection device that detects abnormalities in the electrical system. When an abnormal situation occurs, the relay closes its connections to send a signal to the circuit breaker, which opens the circuit breaker contacts and isolates the problem.

5.12 Present day operation strategy

In due course of time, the old hydropower plants, which form the part of base load and generate a very high efficiency, are now forced to generate not according to the load, but, according to the variations of the solar and wind energy sources (Gupta et al., 2019). The system operator/regulator decides how these hydropower plants should schedule their operation considering the technical aspects of the grid and keeping the must-run plants of solar, wind and coal. It does not take into consideration the interest of the hydropower plants. In some instances, the hydropower plants are just made to run as instructed by the system operator. The hydropower plants, which used to serve the base and peak load in the past are now made to run as stand-by units to take up the generation whenever the generation from renewable energy sources such as wind and solar energy goes down, under policy support. This decreases both the generation and income of these plants, resulting in a conflict of interest between various power generation firms. This becomes further complicated as the tariff for different sources are different, even higher, for solar and wind. This situation affects the financial conditions of the existing hydropower generating companies, as they are not able to generate at their rated capacities, consequently, losing revenue. Therefore, it is necessary to review the rules and regulations for hydropower plant operation considering the existing and future capacities of wind and solar.

The World Bank studied the operation and maintenance of few hydropower utilities across the globe and brought out the handbook for practitioners and decision-makers on strategies for operation and maintenance of hydropower (WB, 2020a) and recommended strategies (WB, 2020b) as follows:

- Improve hydropower plant efficiency and reliability by considering the full life cycle of the plant covering planning, design, construction, testing, commissioning, operation, and refurbishment and decommissioning
- Safeguard the ecological landscape, workers, and the surrounding communities
- Optimize stakeholder advantages, such as providing low-cost, dependable renewable energy.

The handbook also observed different practices for operation and maintenance viz., (i) the owner is solely responsible for O&M, (ii) the owner outsources some part of O&M to consultants and contractors, and (iii) the owner outsources all O&M to a contractor.

5.13 Renovation, modernization, and upgrading

Hydropower plants typically have a lifespan of 30 to 80 years. After 30 to 40 years, electromechanical equipment may need to be renovated or replaced, but civil components such as diversion structures, channels, and powerhouse buildings have a longer lifespan. When compared to developing a new hydropower plant, renovation, modernization, and uprating (RM&U) of an existing hydropower plant are less expensive. In addition, RM&U of an existing hydropower plant takes lesser time to set up, is less ecological, and is easier to license. During the next three decades, the majority of existing hydropower plants will need to be upgraded for better efficiency, increased power output, and better environmental measures.

The maintenance or replacement of identified components whether electro-mechanical or hydromechanical, is based on thorough inspection and evaluation including residual life assessment. Furthermore, RM&U focuses on the restoration and improvement of environmental regulation compliances.

For run-of-river schemes in Himalayan and Sub-Himalayan range where silt contained in the inflows causes enormous erosion to the hydraulic structures and turbines, the rehabilitation of turbine parts in contact with water is required to be conducted almost every year. Such situations call for technological innovations and modernization of rehabilitating techniques for optimizing the renovating down-time.

Thus, it is recommended that renovation and modernization may be initiated after 15 years for adopting modern technology and arresting the dropping efficiency especially in the Himalayan region (Kumar, 2016). For enhancing reliability and generation quality, online instruments for suspended sediment in the water, governing system, machine vibrations and SCADA, etc. should be used. Performance of the hydropower plant included not only turbine efficiency and generator efficiency but also dependability, availability, minimal maintenance, minimal capital cost, and minimal per unit electricity cost as well as minimal environmental impacts.

5.14 Hydropower industry

The hydropower industry is focusing on relicensing and modernization as well as capacity addition to existing hydropower plants in developed markets like the United States, Canada, Europe, Japan, and Norway, where hydropower plants were constructed 30 to 60 years ago. In emerging markets such as China, Brazil, Ethiopia, India, Malaysia, Iran, Laos, Turkey, Venezuela, Ecuador, and Vietnam, utilities and private developers are pursuing large-scale new hydropower construction with increasing policy support of governments for new hydropower construction, hydropower industrial activity is expected to be higher in the coming years. As hydropower and its industry are mature, it is expected that the industry will be able to meet the demand that materializes.

IEA (2021) studied the case of the hydropower market and brought out several findings. Long-term power purchase contracts shall ensure economic viability and shall make the hydropower business sustainable. It is recommended that noneconomic barriers should be removed to enable greenfield hydropower development and refurbishment of aging hydropower. It is recommended that hydropower's multiple benefits should be properly evaluated so that the hydropower development is tenable.

5.15 Cost of hydropower

Hydropower is a capital-intensive technology that often necessitates significant lead periods, particularly for large hydropower projects. The lead period involves the time required for the different stages of project development. These stages include survey & investigations,

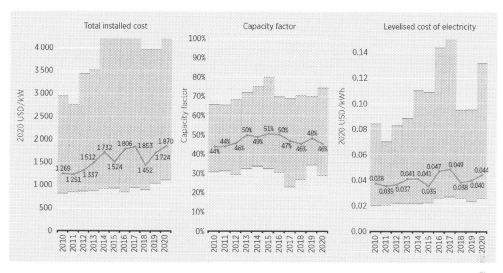

FIGURE 5.6 Global weighted-average total installed costs, capacity factors, and levelized cost of electricity for hydropower, 2010-2020 (IRENA, 2021). *Source: IRENA Renewable Cost Database.*

clearances, construction, and commissioning. Among the various cost components, the construction of civil structures often accounts for the majority of the overall expenditure. The cost of setting up the powerhouse is the second largest capital investment, accounting for almost 30% of the overall expenditure.

IRENA (2021) published the installed cost, capacity factor and levelized cost of electricity (LCOE) of global hydropower projects for the year 2010-2020 as presented in Fig. 5.6.

5.16 Achieving energy sustainability through transboundary hydropower cooperation

Transboundary cooperation in river basin water sharing and hydropower development and management increases the benefits to all riparian populations in the respective country though there may also be the possibility of negative impacts. Several instances have demonstrated that different approaches and considerations may be applied to mitigate social and environmental impacts, sharing of costs and benefits. These instances like Columbia River Project (Canada and USA), Itaipu Dam (Paraguay and Brazil), Manantali Dam (Senegal, Mali, and Mauritania), Kariba Dam (Zambia and Zimbabwe), and Kosi barrage (Nepal and India) demonstrate river basin planning in terms of the hydropower developments. One needs to understand that cost and benefits sharing is to be fair, transparent and flexible, active and reliable cooperation at regional and local level exist. Social and environmental mitigation measures are adequately envisaged, planned, and implemented (GIZ, 2014).

As per World Commission on Dams report (WCD, 2000), an approach based on the riparian and common rights and the assessment of risks taking into consideration the core values of participatory decision-making, efficiency, equity, ownership, and sustainability, and accountability, needs to be adopted. Based on the experience gained from the past water resource development projects constructed in emerging and developing economies during the last few decades in the 20th century, there is a strong need for sustainable development and water resource's structure particularly dam-based reservoirs but there are several developmental challenges.

With the increasing number of interconnections between neighboring countries and different national and regional electricity systems, and more than 260 global rivers passing at least one national boundary, upcoming hydropower development will require a high level of cooperation and strategic partnership with surrounding nations. The possibilities and advantages related to local development of hydropower in a sustainable manner need harmonized governance for river basins having transboundaries. Regional power pools for the electricity technology may enhance the viability of hydropower development through regional energy markets. This shall also make the case for large-scale hydropower development and shall reduce the dependence on fossil fuels which are often imported. For example, hydropower generation from Canada is traded with the United States and establishes a mutually beneficial relationship between them (Bin, 2021). Bhutan's excess hydropower is transported to India and benefits by expanding its electricity market. However, institutional expertise and governmental involvement are required to navigate the complexities and costs connected with local hydropower schemes and related linkages.

Thus, for developing countries, creating transboundary grids connection shall be an excellent opportunity for economic empowerment similar on lines of the large number of hydropower producing countries where, in many instances, finance is provided by the neighboring country. However, the inherent complexity of transboundary power transport through hydropower projects requires strong and decisive governments and political will. There is immense opportunity for cross-border electricity trade in many developing countries, for example, complementary nature of load power demand profile as well as power supply capacity in Bhutan, Nepal, India, and Bangladesh transboundary electricity trade augurs very well in view of the rapid expansion of solar photovoltaic and meeting the climate change-related targets (Vaidya et al., 2021).

5.17 Contribution of hydropower to sustainability in developing countries

Historically, hydropower has been a driver for social and economic growth in numerous developed and developing nations by offering electricity and water management facilities (Savocool et al., 2018). The storage schemes can help to alleviate water shortage by ensuring drinking water supply, agriculture, navigation, and flood management (World Bank, 2005). The multipurpose hydropower schemes play an important role outside the energy sector in the face of climate change as a funding mechanism for reservoirs that serve to guarantee freshwater supplies.

Hydropower has been developed as an indigenous source in line with social and economic conditions. Tennessee Valley Authority of the USA developed its over 90% potential by having the social benefits such as power, flood control, navigation, irrigation by providing local inputs to develop its economy (Bin, 2021). Ethiopia, a country striving hard for achieving rapid economic growth is developing hydropower by using its rich water resources. The tapping of hydro resources is not only to provide much-needed sustainable energy but to provide water for agriculture and bring revenue by selling hydropower to its neighboring countries as well (Degefu et al., 2015).

Despite frequent discussions and concerns shown in the literature, media, meetings regarding socio-economic consequences of hydropower on affected communities, quantitative studies that look at long-term impacts are sparse. Moreover, only economic parameters are typically employed to assess the effects. Faria et al., (2017) found in their study for the hydropower projects built during 1990-2010 in Brazil, that, states having hydropower plants constructed had a greater gross domestic product (GDP) and tax revenues during their first few years of development compared to others. However, those favorable economic impacts were temporary (>15 years) and social indices were not much different compared to those which do not have hydropower projects. Through industrialization powered with hydropower, several developed nations (UK, Switzerland, and the USA, etc.) and developing economies (China, Brazil, etc.) have achieved poverty alleviation leading to prosperity. Hydropower in addition to conventional power, also provides water, food, and energy security.

Hydropower development is synonymous with regional development in many instances. An overview of all the benefits associated with perspectives of the river basin, regional connections, and bringing markets to the resource is presented by IHA through its publication (IHA, 2013). Strategic development, flood management, water security, multifunctional usage, and regional governance are all provided by river basin viewpoints (Mekong river) (WWF, 2021). It may be noted that the regional interconnections result in cost savings, increased system dependability and affordability, and environmental advantages through increased efficiency and enhanced cross-national interactions. By sharing the advantages, it also allows effective utilization of resources. Connecting markets to the resource creates economic possibilities and serves as a stimulus for the development of energy resources for the local people (IHA, 2013).

For large hydropower plants globally and specifically in countries like China, Brazil, and India, strong debates are happening about the positivity (income, infrastructure and health, etc.) or negativity (traffic congestion, altered flows, loss of forest land, and landslides, etc.). While, dam-based hydropower increases agricultural production due to the regulated availability of water but may decrease stress upstream of the dam. Due to the high voltage transmission line, the electricity is easily transmitted to remote areas to other areas of the high population, thus, giving them the benefits of electricity but may not directly benefit the local population around the dam.

Under these circumstances, small hydropower projects (SHP) can be a cost-effective source of power, as these projects are capable of providing a decentralized power supply in rural regions where exists significant hydropower technical potential. One of the most effective instances is China's small hydropower based rural electrification. SHP is generally found in isolated grids, off-grid, and central-grid settings. Since 60–70% of expenditures are site-specific, appropriate site selection and capacity fixing are major issues. In isolated grid systems, the variation in natural seasonal flow variations may require the integration of hydropower plants with other sources of power generation to maintain uninterrupted supply during dry periods.

To attract climate finance, hydropower projects may be part of the transformation of energy and water systems by offering transition services such as energy storage and grid stabilization which is very much required for other carbon-friendly energy technologies such as wind and solar. Climate funds may incentivize hydropower projects to include characteristics to support low carbon transformation (Patel et al., 2020).

5.18 Possible multiplier effects of hydropower projects

Dam projects have a wide range of effects on the region in which they are built, as well as at inter-regional, national, and even global levels. The effects may be in terms of socio-economic, environmental, cultural, health, and institutional. The relevance and difficulty of analyzing a number of these consequences have been explored by the World Commission on Dams (WCD) and several other studies. One of the concerns addressed by these studies is the necessity to take into account, the indirect benefits and costs of dam projects (Bhatia et al., 2003). It may be noted that the social advantages linked with direct benefits such as agriculture, electricity, municipal and industrial water supply, and flood management from major dams are often not analyzed and taken into consideration (WCD, 2000). Multiplier impacts also known as indirect impacts are the consequence of both inter-industry linkage impacts and consumption-induced impacts. Multipliers are summary measures expressed as a ratio of the total effects (direct and indirect) of a project to its direct effects (Bhatia et al., 2007). It is estimated that the multipliers values for the large hydropower projects range from 1.4 to 2.0, implying that for every dollar of value generated by the sectors directly involved in dam-related activities, another 40 to 100 cents could be generated indirectly in the region. Though these multiplier effects are not specific to hydropower projects and may be seen in every energy project to various degrees, they, nonetheless represent benefits that might be considered by communities considering hydropower development.

5.19 Conclusion

Hydropower is the energy generated from water flowing and is the largest source of energy globally and emits lesser greenhouse gas than other kinds of energy. Hydropower constitutes nearly all of the total electricity generation in several developing countries and developed countries. A large number of existing water resource facilities, where hydropower is not developed (nonpowered), are available worldwide. These facilities may be utilized for hydropower generation by installing new generating equipment. Hydropower plants typically have a lifespan of 30 to 80 years. Renovation and modernization work normally required after 30 to 40 years is less expensive than building a new hydropower plant. The re-licensing and modernization as well as capacity addition to existing hydropower plants are being focused in developed markets. It is necessary to review the rules and regulations for hydropower plant operation considering the existing and future capacities of wind and solar.

Being a capital-intensive technology, it is often necessitating significant lead periods, particularly for large hydropower projects. Transboundary cooperation in river basin water sharing and hydropower development and management increases the benefits to all riparian populations in the respective country. There is immense opportunity for cross-border electricity trade in many developing countries. Complementary nature of load power demand profile as well as power supply capacity in several countries offer transboundary electricity trade in view of the rapid expansion of solar photovoltaic and meeting the climate change-related targets. Therefore, hydropower could play a vital role in achieving regional as well as local sustainable electricity sector development in the developing world.

Historically, hydropower has been a driver for social and economic growth in numerous developed and developing nations by offering electricity and water management facilities. Hydropower development is synonymous with local and regional development in many instances. By sharing the advantages, it also allows effective utilization of resources. Connecting markets to the resource creates economic possibilities and serves as a stimulus for the development of energy resources for the local people.

The major barriers in hydropower development are environmental flow requirements to sustain aquatic life and land acquisition. Thus, hydropower projects need to be developed following the international sustainability protocol to attract foreign investment and meet environmental and social sustainability. The protocol as a tool needs to be applied to check the sustainability of existing and future projects.

Self-evaluation questions

1. How hydropower works? - Explain.
2. Classify and explain different types of hydropower projects based on capacity, head, and technologies.
3. Explain the components of a hydropower plant.
4. How hydropower could contribute to sustainable electricity sector development in developing countries? - Explain.
5. Is there any impact of negative climate change on hydropower projects? – Justify your answer.
6. With the help of recent literature, assess (using one of the recognized methods/techniques) the sustainability of hydropower generation.
7. Conduct a literature survey and prepare a report on - how energy sustainability can be achieved through transboundary hydropower cooperation. Justify your answer with a case study.

References

Berga, L., 2016. The role of hydropower in climate change mitigation and adaptation: a review. Engineering 2, 313–318.
Bhatia, R., Scatasta, M., Cestti, R., 2003. Study on the multiplier effects of dams: methodology issues and preliminary results. In: Third World Water Forum, Kyoto, Japan, 16–23 March 2003.
Bhatia, R., Malik, R.P.S., Bhatia, Meera, 2007. Direct and indirect economic impacts of the Bhakra multipurpose dam, India. Irrigation Drainage 56, 195–206. doi:10.1002/ird.315, 04/2007. Wiley InterScience.
Bin, D., 2021. Discussion on the development direction of hydropower in China. Clean Energy 2021, 10–18.
Bruckner, T., Bashmakov, I.A., Mulugetta, Y., Chum, H., de la Vega Navarro, A., Edmonds, J., Faaij, A., Fungtammasan, B., Garg, A., Hertwich, E., Honnery, D., Infield, D., Kainuma, M., Khennas, S., Kim, S., Nimir, H.B., Riahi, K., Strachan, N., Wiser, R., Zhang, X., 2014. Energy Systems. In: Edenhofer, O., Pichs-Madruga, R., Sokona, Y., Farahani, E., Kadner, S., Seyboth, K., Adler, A., Baum, I., Brunner, S., Eickemeier, P., Kriemann, B., Savolainen, J., Schlömer, S., vonStechow, C., Zwickel, T., Minx, J.C. (Eds.), Climate Change 2014: Mitigation of Climate Change. Contribution of Working Group III to the Fifth Assessment Report of the Intergovernmental Panel on Climate Change. Cambridge University Press, Cambridge, United Kingdom and New York, NY, USA.

Degefu, D.M., 2015. Hydropower for sustainable water and energy development in Ethiopia. Sustain. Water Resour. Manag. I (2015), 305–314.

DFID, 2009. Water storage and hydropower: supporting growth, resilience and low carbon development, A evidence-into-action paper, p. 21.

Egré, D., Milewski, J.C., 2002. The diversity of hydropower projects. Energy Policy 30 (14), 1225–1230.

Faria, F.A.M., Davis, A., Severnini, E., Jaramillo, P., 2017. The local socio-economic impacts of large hydropower plant development in a developing country. Energy Econ. 67, 533–544. http://dx.doi.org/10.1016/j.eneco.2017.08.025.

GIZ, 2014. Training Manual Hydropower and Economic Development, Network for Sustainable Hydropower Development in the Mekong Countries (NSHD-M). P, p. 182.

Gupta, A., Kumar, A., Khatod, DK., 2019. Optimized scheduling of hydropower with increase in solar and wind installations. Energy 183 (2019), 716–732.

HCI-ASME, 1996. Guide to Hydropower Mechanical Design. American Society of Mechanical Engineers, USA, p. 370.

IEA, 2000. Hydropower and the environment: present context and guidelines for future action. Volume II: main report. Implementing agreement for hydropower technologies and programmes, Annex III, International Energy Agency, Paris, France, 172 pp. Available at: www.ieahydro.org/reports/HyA3S5V2.pdf.

IEA, 2021. Hydropower Special Market Report – Analysis and Forecast to 2030, 2021. International Energy Agency, Paris, France, p. 122.

IHA, 2013. Hydropower and Regional Development: Case studies, 2013. International Hydropower Association, London, United Kingdom, p. 14.

IHA, 2018. Hydropower Status Report, Sector trends and insights. International Hydropower Association, London, United Kingdom, P, p. 101.

IHA, 2021a. Hydropower Status Report, Sector trends and insights. International Hydropower Association, London, United Kingdom, pp. 1–48.

IHA, 2021b. Hydropower 2050: Indentifying the Next 80+ GW Towards Net Zero. International Hydropower Association, London, United Kingdom, pp. 1–20.

IJHD, 2020. World Atlas & Industry Guide. International Journal on Hydropower and Dams, London, United Kingdom.

IRENA, 2021. Renewable Power Generation Costs in 2020. International Renewable Energy Agency, Abu Dhabi, p. 178.

Jain, S.K., Agarwal, PK., Singh, V.P., 2007. Hydrology and Water Resources of India, 2007. Springer Science and Business Media LLC, Dordrecht, The Netherlands.

Killingtveit, Å., 2019. Chapter 8 -Hydropower, Managing Global Warming. Elsevier BV, Academic Press, Amsterdam, Netherlands, pp. 265–315.

Kumar, A., Schei, T., Ahenkorah, A., Caceres Rodriguez, R., Devernay, J.-M., Freitas, M., Hall, D., Killingtveit, Å., Liu, Z., 2011. Hydropower. In IPCC Special Report on Renewable Energy Sources and Climate Change Mitigation. O. Edenhofer, R. Pichs-Madruga, Y. Sokona, K. Seyboth, P. Matschoss, S. Kadner, T. Zwickel, P. Eickemeier, G. Hansen, S. Schlömer, C. von Stechow (Eds.), Cambridge University Press, Cambridge, United Kingdom and New York, NY, USA.

Kaunda, C.S., 2012. Hydropower in the context of sustainable energy supply: A Review of Technologies and Challenges, Department for International Development, International Scholarly Research Network. ISRN Renewable Energy 2012, 15.

Kumar, Arun, 2016. Why and how: Renovation, modernization and uprating of hydro power plants. Workshop on Renovation, Modernization, Uprating & Life Extension of Hydro Power Plant - Diverse Issues & Handling Strategies. Central Electricity Authority, New Delhi, India, pp. 23–30.

Patel, S., Shakya, C., Rai, N., 2020. Climate Finance for Hydropower: Incentivising the Low-Carbon Transition. International Institute for Environment and Development, London, p. 41 January 2020.

Sovacool, B.K., Walter, W., 2018. Internationalizing the political economy of hydroelectricity: security, development and sustainability in hydropower states. Rev. Int. Polit. Econ. 26 (1), 49–79. doi:https://doi.org/10.1080/09692290.2018.1511449.

UNIDO, 2019. The World Small Hydropower Development Report 2019. United Nations Industrial Development OrganizationInternational Center on Small Hydro Power, ViennaHangzhou.

USBR, 1976. Selecting Hydraulic Reaction Turbine, Engineering Monograph No 20, 49.

Vaidya Ramesh, A., 2021. The role of hydropower in South Asia's energy future. Int. J. Water Resour.; Dev., 2021 37 (3), 367–391.

WCD, 2000. Dams and Development – A New Framework for Decision-Making. World Commission on Dams, Earthscan, London, UK, p. 356.

World Bank, 2005. Shaping the Future of Water for Agriculture: A Sourcebook for Investment in Agricultural Water Management. World Bank, Washington, DC, USA.

World Bank, 2020a. Operation and Maintenance Strategies for Hydropower—Handbook for Practitioners and Decision Makers. World Bank, p. 155.

World Bank, 2020b. Operation & Maintenance Strategies for Hydropower – Six Case studies. World Bank, p. 48.

WWF, 2021. 10 Rivers at Risk – Hydropower Dams Threaten Diverse Benefits of Free Flowing Rivers, 2021. World Wide Fund for nature, Gland, Switzerland, p. 33.

CHAPTER 6

Wind energy and its link to sustainability in developing countries

Mahfuz Kabir[a], Navya Sree BN[b], Krishna J. Khatod[b], Vikrant P. Katekar[c] and Sandip S. Deshmukh[b]

[a]Bangladesh Institute of International and Strategic Studies (BIISS), Dhaka, Bangladesh [b]Department of Mechanical Engineering, Birla Institute of Technology & Science, Pilani, Hyderabad, India [c]Department of Mechanical Engineering, S. B. Jain Institute of Technology, Management and Research, Nagpur, Maharashtra, India

6.1 Introduction

Wind power is becoming an increasingly important source of renewable electricity, which notably decreases CO_2 emissions to the atmosphere and helps mitigate climate change (Peri and Tal, 2020). Wind energy conversion systems (WECS) convert the wind energy into mechanical power using wind turbines. The electricity generated through turbine is utilized for pumping water, mill grains, drive machinery, or supply electricity to households. Rapid development of wind electricity technologies can be observed in over the last three decades, which has been accompanied by policy incentives. The design of blades and turbine generator have been undergoing significant transformation–from smaller system with many parts to smaller number of parts and direct magnetic drives. Most of the countries that generate wind power have single, clustered and organized farms while power storage and metered hook-ins are introduced in order to supply in local and national grids (Andersen et al., 2017).

In general, the wind is the movement of air caused by the uneven heating of the earth by the sun. The two main origins of wind are the unequal temperature distribution caused by solar irradiation and the earth's rotation. Winds have different speed levels, such as breeze and gale, depending on how fast they blow (National Geographic Society, 2021). Wind power, often known as wind energy, generates electricity by using wind turbines to create mechanical power. Wind energy is nonpolluting, nontoxic, and environment friendly. Wind energy is a

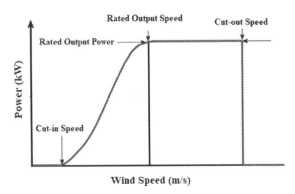

FIGURE 6.1 Typical wind speed vs. power curve.

renewable energy source that has a lower environmental impact than fossil fuels. Wind energy has tremendous promise as a form of energy (Sharma et al., 2012).

The rest of the chapter has been organized as follows. Section 6.2 provides a detailed account of wind turbines and their classification, and wind energy conversion. Section 6.3 describes the basic and modern wind power technologies along with energy storage system. Section 6.4 elaborates the present status on wind energy from the perspective of global and developing countries. It covers installed capacity, costs, job creation and future outlook. Section 6.5 explains the performance, progress and challenges faced by developing countries, such as China, India, Brazil, Philippines, and Vietnam. Section 6.6 analyzes the implications of wind energy for sustainability. Finally, Section 6.7 concludes the chapter.

6.2 Foundational content of wind energy

The longitudinal airflow from steep locations to down regions is known as wind (Cappucci, 2018). Wind energy is the kinetic energy related to the movement of large volumes of air. The kinetic energy of air is converted to mechanical energy with the help of a wind turbine. Power generation from wind depends on its speed. A typical wind speed vs. power curve is shown in Fig. 6.1.

The minimum wind speed at which the turbine blades start to rotate by overcoming the friction is known as cut-in speed. The maximum wind speed at which the generator is able to produce usable output power and beyond this point the blades could be damaged due to over speed is known as cut-out speed. Power curves for different wind turbines capacities can be found in Gul et al. (2019).

6.2.1 Historical development of wind energy

Wind energy system was found along the Nile River even in 5000 BC. The first wind machines can be traced back to 200 BC, which perhaps vertical axis windmills with several arms to mount sails, which were utilized to grind grain in Persia. Horizontal axis-mounted windmills were observed in the Mediterranean in the tenth century, which were permanent installations to generate mechanical power using coastal winds. In the 11th century, wind

FIGURE 6.2 (A) Persian, (B) American, and (C) traditional Dutch windmill. *Source: Salameh (2014).*

pumps, and wind turbines were commonly used in the Middle East for food production. Subsequently, horizontal windmills were constructed in Europe, which had been operated to generate energy through capturing available wind by rotating the entire windmill manually. The energy generated through the windmills was utilized to grind grains and fetch/pump water. Since earliest recorded history, wind power has been used to move ships, grind grain, and pump water. Simple windmills were utilized to fetch/pump water in China several centuries BC. Electricity was first produced using wind turbine generator by Denmark in 1890, while many European countries started to produce electricity by the next two decades. Conversely, millions of windmills were constructed in the United States in the late nineteenth century mainly to pump water for crop and cattle farms. Small electric wind systems were installed in the country by 1900, while the first modern 1.25 MW wind-turbine generator was installed in 1941 (Sebestyén, 2021). Fig. 6.2 shows Persian, American, and Dutch windmill. Table 6.1 presents the trend in the development of wind turbine technologies (Shahan, 2014).

TABLE 6.1 The historical development of wind turbines technologies.

Year	Development
1700 B.C.	The earliest known reference to a windmill was discovered in Mesopotamia, now part of Iran and Iraq.
50 A.D.	The Hero of Alexandria, an Egyptian Greek mathematician, was the first to depict a wind turbine. It's been debated whether the windmill ever existed or if it was just a drawing.
700 A.D.	The first practical windmill, known as "The Persian Windmill," can be seen in Siestan, Iran.
1185	The first historical mention of a windmill in Europe was in Yorkshire, Great Britain, in 1185.
1219	The Chinese legislator Yehlu Chhu-Tshai was the first to be documented in 1219. These vertically axed windmills were very similar to Persian windmills.
1870	The mass of the wings was lowered after the introduction of steel blades in 1870.
1887	The vertical-axis wind turbine developed by Scottish academic James Blyth in Glasgow, Scotland, in 1887 was the very first known electric power wind turbine.
1888	Charles Brush was recognised for his first large-scale wind energy generation in 1887 in Ohio, USA.
1931	Georges Jean Marie Darrieus, a French aeronautical engineer, built and trademarked "The Darrieus wind turbine" in 1931.
1941	The Smith-Putan turbine, the world's first-megawatt wind turbine, was constructed and linked to the local electrical power system in a mountain near Castleton, Vermont, USA, in 1941.
1978	Horizontal axis wind turbines had been turning counter-clockwise until 1978 when a transition occurred, and now all significant horizontal axis turbines rotate clockwise to offer a consistent picture.
1980	The world's first wind farm, comprising 20 wind turbines with 30KW rated power, was constructed near Southern New Hampshire, USA.
2008	Enercon developed the E-126, the world's most robust onshore wind turbine, near Ernden, Germany. It had a capacity of 7 MW. It had a rotor diameter of 126 metres and stands 131 metres tall.
2009	Statoil installed the world's first offshore wind turbine at Karmoy, Norway. Siemens designed the 2.3 MW wind turbine, and additional of these turbines were scheduled to be deployed in the North Sea.

Source: Mathew and Philip (2012), Mathew (2006).

6.2.2 Classification of wind turbines

The transformation of wind's kinetic energy to electrical energy by turbines is known as wind power (US Department of Energy, 2021). Wind energy is harvested by letting it pass through turbine blades, which generate the force on a rotor. The rotor size, swept area, and wind speed all determine the amount of power delivered. Wind turbines range from small four-hundred-watt generators for household use to multi-megawatt machines for wind farms and offshore use. Unlike their larger counterparts, tiny turbines use a mass flywheel with DC (direct current) output, aero-stiff blades, and permanent bearings to aim toward the wind. The larger ones often have geared trains, but they actively alternate current output, flaps, and winds (Kaldellis, 2012). The capacity factor determines the annual power generation of the wind turbines as wind velocity is not constant. A well-located wind generator has a capacity

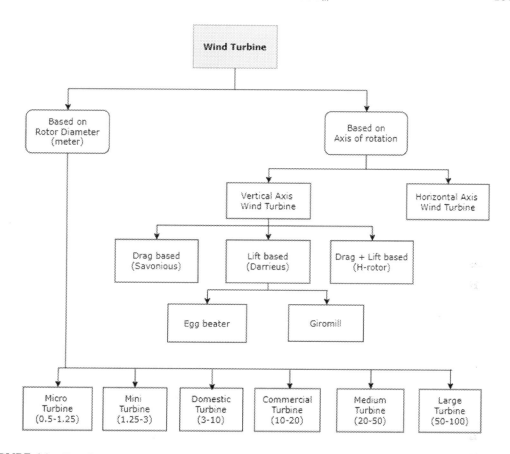

FIGURE 6.3 Classification of wind turbines (Tasneem et al., 2020).

factor of roughly 35% (Center for Sustainable Systems, 2020). In most cases, wind turbines will not function if the mean wind speed is less than 4.5 m/s. Wind resources are often preselected depending on an on-site wind study and validated using a prevailing winds map (Renewables, 2021).

Small wind turbines provide modern electrical services to families that previously did not have access to them, lowering the cost of electricity on islands and other remote regions that rely on fuel generation and allowing people and small companies to produce their power. The main rotor shaft of a vertical axis wind turbine is positioned vertically. This increases the structural stability of wind turbines typically located near heavily populated areas, lowering the chance of accidents. VAWTs (vertical axis wind turbines) are becoming increasingly popular, particularly in the residential market. Although VAWTs are less prevalent than horizontal axis wind turbines (HAWTs) due to their lower efficiency, they are well suited for residential use (see, Figs. 6.3–6.4 for the type of turbines). Apart from domestic applications, VAWTs are often utilized to power street lighting because they are small and attractive. In

FIGURE 6.4 Horizontal axis wind turbine located near Anantapur, Andhra Pradesh, India. *Source: Authors.*

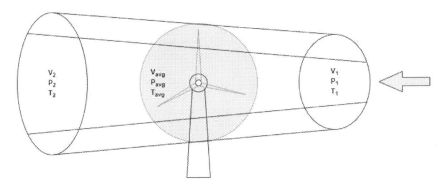

FIGURE 6.5 Thermodynamic analysis of wind energy systems (Hu et al., 2020).

addition, when compared to horizontal positioning systems, vertical axis wind turbines do not require a directed positioning system. With recent technology developments and considerable reductions in wind turbine manufacturing costs, vertical axis wind turbine use is likely to rise in the future years (Mordor Intelligence, 2020).

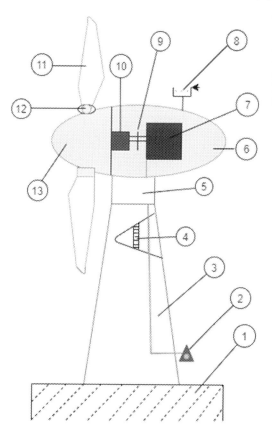

FIGURE 6.6 Wind turbine components: (1) Foundation, (2) Connection to the electric grid, (3) Tower, (4) Access ladder, (5) Wind orientation control-Yaw control, (6) Nacelle, (7) Generator, (8) Anemometer, (9) Electric or mechanical brake, (10) Gearbox, (11) Rotor blade, (12) Blade pitch control, (13) Rotor hub. *Source: (Arne Nordmann, 2021)[1], under CC BY-SA.*

6.2.3 Thermodynamics of wind energy conversion systems

Wind turbines use the kinetic energy generated by rotor blades, which act similarly to an aeroplane wing or a helicopter rotor blade, to convert wind power into electricity. Fig. 6.6 depicts the fundamental terminology for a wind turbine. When the air flows across the blade, the pressure of the air on one side reduces. The change in atmospheric pressure around the two sides of the blades creates lift and drag. The rotor spins because the force of lift is greater than that of the drag force. The rotor is directly connected to a generator or through a shaft and several gears that allow the turbine to be significantly lighter by increasing the spin speed. The whirling of a generator generates aerodynamic force, which is converted into energy (Hu et al., 2020).

The flow of air spins the propeller of the windmill with the maximum efficiency of the "Betz's constant." This constant is calculated from the equation of kinetic energy (KE) that is stored in the area of the wind stream in the location of the wind turbine as follows:

$$\text{KE} = \frac{1}{2}(\text{Mass})V^2 \tag{6.1}$$

---

[1] https://upload.wikimedia.org/wikipedia/commons/a/ac/Wind_turbine_int.svg (accessed 25-Nov-2021).

The power in the wind is given by the rate of change of energy:

$$P = d(KE)/dt$$
$$= 1/2 v^2 dm/dt \tag{6.2}$$

As mass flow rate is given by:

$$dm/dt = \rho * A * dx/dt \tag{6.3}$$

and the rate of change of distance is given by:

$$dx/dt = v \tag{6.4}$$

Hence, the power of wind turbine can be defined as

$$P = 1/2 * \rho * A * V^3 \tag{6.4}$$

where P indicates the power available in the wind, ρ implies air density that is equivalent to 1.2 kg/m^3, A stands for the area of the wind turbine blades in m^2 and $A = \pi\left(D/2\right)^2$ where D is the rotor diameter, and V is velocity or wind speed in m/s.

The blades swept area can be calculated from the empirical relation:

$$A = \pi[(l+r)^2 - r^2] = \pi l(l+2r) \tag{6.5}$$

where l is the length of rotor blades and r is the radius of the hub.

The wind retardation before the rotor ($v_1 - v_{avg}$) is the same as wind retardation after rotor ($v_{avg} - v_2$). The rotor power may be computed using the formula:

$$P = \frac{1}{4}\rho A (v_1 + v_2)^2 (v_1 - v_2) \tag{6.6}$$

where v_1, P_1, T_1 and v_2, P_2, T_2 are wind speeds, wind pressure and temperatures at the inlet and outlet, respectively, of the stream tube as shown in Fig. 6.5.

$$\text{where } v_{avg} = \frac{v_1 + v_2}{2}, P_{avg} = \frac{p_1 + p_2}{2}, T_{avg} = \frac{T_1 + T_2}{2} \tag{6.7}$$

The tower supports the turbine's structure. Taller buildings allow turbines to catch additional energy to produce more electricity since wind velocity increases with height.

$$\text{Tower height} = \text{hub height} - r \tag{6.8}$$

where r is the radius of the hub.

The mechanical power output of a wind turbine (P_m) can be expressed as-

$$P_m = \frac{1}{2}\rho V^3 A C_p \tag{6.9}$$

where C_p stands for the coefficient of aerodynamic power that correspond to the efficiency of the turbine. The maximum value of C_p or the ceiling performance for turbine is given by Betz as $16/27 = 0.593$. Thus, P_{mx}, the maximum extractable mechanical power from the wind through the wind turbine (generator) is-

$$P_{mx} = 0.593 P_a \tag{6.10}$$

where P_a stands for the available power in the wind. It implies that wind turbines cannot convert more than 59.3% of the KE available in the wind. However, C_p hinges on the type of the windmill, and it is a function of λ–the ratio of blade tip speed to wind speed, and

$$\lambda = \frac{\pi D n_r}{60V} \qquad (6.11)$$

where n_r stands for the rotor speed in revolutions per minute (rpm).

The immediate power output of the generator, P_0, can be written as-

$$P_0 = \frac{1}{2}\rho V^3 A C_p \eta \qquad (6.12)$$

where η stands for the efficiency of the wind power generator. The average output power of the generator can be expressed as-

$$P_{av} = \int_0^\infty P_0 f(V) dV \qquad (6.13)$$

where $f(V)$ indicates the probability density function (PDF) (Salameh, 2014).

The precise calculation of output of the wind power generator is hinges heavily on the pattern and distribution of wind speed. Weibull distribution is the most commonly referred distribution to explain the wind system (Aririguzo and Ekwe, 2019; Deep et al., 2020) for altitude of up to 100 meter. The PDF of the Weibull distribution can be written as-

$$f(V) = \frac{k}{c}\left(\frac{V}{c}\right)^{k-1} e\left[-\left(\frac{V}{c}\right)^k\right] \qquad (6.14)$$

where c and k are scale and shape factors, respectively. Now, $f(V)$ of a specific wind speed ranges between two different wind speed, viz. V_1 and V_2:

$$f(V_1 \leq V \leq V_2) = \int_{V_1}^{V_2} f(V) dV \qquad (6.15)$$

Here, c is related to the mean wind speed $\bar{V}$ and k inversely proportional to the standard deviation σ^2 as follows:

$$\bar{V} = c\gamma\left(1 + \frac{1}{k}\right) \qquad (6.16)$$

and

$$\sigma^2 = c^2\left[\gamma\left(1 + \frac{2}{k}\right) - \gamma^2\left(1 + \frac{1}{k}\right)\right] \qquad (6.17)$$

where γ stands for the Gamma function.

Conversely, Rayleigh distribution, a rather simpler than Weibull, is a popular single-parameter distribution to calculate the wind power potential (Paraschiv et al., 2019; Serban et al., 2020). It produces nearly similar results based only on $\bar{V}$ as follows:

$$f(V) = \frac{\pi V}{2\bar{V}^2} e\left(\frac{-\pi V^2}{4\bar{V}^2}\right) \qquad (6.18)$$

And the mean wind speed is calculated as-

$$\bar{V} = \sum_{i=1}^{n} V_i f(V_i) \qquad (6.19)$$

where n is the number of bins considered and V_i is the average score of each bin (Zafirakis, 2021).

In addition to the above two, some other distributions include γ-II, log-normal, inverse Gaussian, normal truncated, Maxwell, beta, square-root normal, maximum entropy principle, Weibull mixture, Gamma Weibull mixture, Kappa, and Burr distributions (Wang et al., 2021; Li and Miao, 2021; Zafirakis, 2021).

Conversion efficiency of wind power technologies has been increasing over time through development of rotor aerodynamics, modes of operating turbines and increase in turbine size among others. Good wind speed helps achieve high efficiency of the modern wind power system, which can be made available through installing the turbines at the locations of highest wind speed, such as hilltops, offshore or coastal areas (Salameh, 2014).

The height of wind turbines determines the power output of the wind plant. The tip of the blade needs to be minimum 15m higher than the ground to maintain safety. The usual hub height is 10-50m, while 80m height is observed in some cases. However, the world's tallest wind turbine had been declared in Maasvlakte Rotterdam in 2019 with a height of 260m and a rotor diameter of 220m (Sebestyén, 2021).

6.2.4 Design of a wind turbine

6.2.4.1 Wind speed

The wind speed is defined as the average incidence speed on the blade's swept area and is critical in calculating the value of the Reynold's number. It is stated by equation as-

$$R_e = \frac{\rho v c}{\mu} \qquad (6.20)$$

where, ρ is the density of air (1.225 kg/m³), v is the rated output wind velocity (e.g., 18 m/s), c is the chord length of the blade (e.g., 0.1 m), and μ is the dynamic viscosity of the air-fluid (1.983×10^{-5} NS/m²).

Thus,

$$R_e = \frac{\rho v c}{\mu} = \frac{1.225 \times 18 \times 0.1}{1.983 \times 10^{-5}} = 111195.15$$

As the Reynolds number is more than 4000 in this case, the flow is considered turbulent.

6.2.4.2 Swept area

The rotor radius is computed as the hub radius and blade span sum, say, 0.5 m. The swept area A is computed as follows using a 0.50 m rotor radius(r):

$$A = \pi \times r^2 = 0.785 \, m^2$$

6.2.4.3 Wind turbine power

Wind power can be computed using Eq. (6.4). Say, a wind speed V of 18 m/s, an air density of 1.225 kg/m³, and a rotor radius r of 0.50 m.

$$P = \frac{1}{2}\rho A V^3 = \frac{1}{2} \times 1.225 \times 0.785 \times 18^3 = 2.804 \text{ kW}$$

However, when the Betz limit and machine efficiencies are considered, the power to be generated may be calculated using the following equation.

$$P = 0.593 \times C_p N A V^3 \tag{6.21}$$

where, C_p denotes the power coefficient, which is typically 0.4; N denotes the efficiency of driven machinery, which says 0.7; and A denotes the swept rotor area, which is calculated as 0.785 m².

As a result, the turbine's power output is-

$$P = 0.593 \times C_p N A V^3 = 0.593 \times 0.4 \times 0.7 \times 0.785 \times 18^3 = 0.760 \text{ kW}$$

6.3 Available technologies

To catch the greatest energy, wind turbines, like windmills, are positioned on a tower. They can take full advantage of the quicker but less turbulent wind at 100 feet or higher above the earth. Turbines use propeller-like blades to capture wind energy. A rotor is usually composed of 2-3 blades positioned on a shaft. A blade functions similarly to an aeroplane wing. Low pressure forms on the blade's underside when the wind blows. The blade is then pulled toward it by the low-pressure air, forcing the rotor to rotate. This is referred to as lift force (perpendicular to airflow). The lift force is substantially more significant than the drag force exerted by the wind on the blade's front side (parallel to airflow). The rotor spins like a propeller due to the drag and lift, and the rotating shaft revolves around the generator to generate power (Twidell and Weir, 2021).

6.3.1 Conventional wind energy conversion system

A wind energy conversion system (WECS) is shown in Fig. 6.7, a mechanical system that converts the energy of the wind into mechanical energy that may be used to power machines and drive the electrical generator. Wind energy is a stable and cost-effective source of electricity. Modern wind turbines have a life expectancy of 20 to 25 years, similar to several other power generation systems (Jacobson, 2016). Professional wind power plants are currently available in roughly 98% of cases. Wind power costs have continued to fall due to technological advancements, greater output levels, and the use of massive turbines.

A wind turbine, a generator, connecting apparatus, and control systems are the main components of a conventional wind energy conversion system (see Fig. 6.7). Because of their dependability and cost-effectiveness, permanent magnet synchronous generators and squirrel-cage induction generators are commonly used in small to medium-power wind turbines (Katekar et al. 2021). Various high-power wind turbines use induction generators, synchronous generators, and wound synchronized field generators.

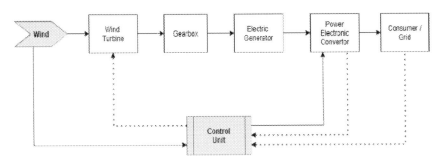

FIGURE 6.7 Basic terminology of wind energy conversion system (Taylor, 1984).

The devices that perform power regulation, soft start, and connectivity operations are known as interconnection apparatuses. Electric power converters are frequently employed as such devices. Most current turbine generators employ forced commutated pulse width modulation (PWM) inverters to produce constant voltage and frequency output with maximum power. In wind turbines, dual voltage source voltage-controlled converters and voltage source current-controlled converters are being used. Two-fold PWM converters, which allow a bidirectional power flow between the wind turbine and the utility grid, can provide effective power control for specialized high-power wind turbines (Vasar et al., 2018).

6.3.2 Advanced wind energy conversion system

Modern wind power conversion systems use advanced technologies and variable-speed generation converters to generate the most power possible. Many wind power conversion system proposals have been proposed in recent years. Fixed-speed wind turbines were the first type of wind power conversion device. The converter system should be designed to run at a constant rotor speed dictated by the grid frequency, regardless of wind speed. At high wind speeds, this constraint limits the ability to extract maximum power. The classification of improved WECS according to speed control is shown in Fig. 6.8. To address this constraint, the wind power conversion system typically uses power converters between the stator and the grid. However, because the total power passes through the converters, this construction

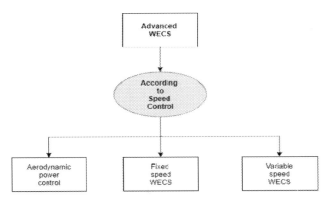

FIGURE 6.8 Classification of advanced WECS according to speed control (Nouh and Mohamed, 2014).

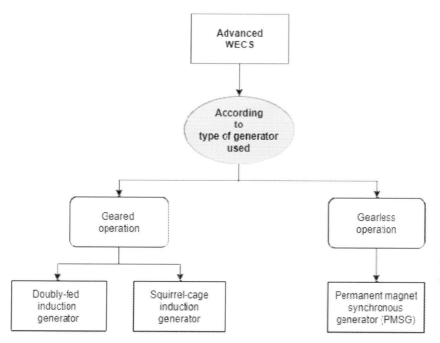

FIGURE 6.9 Classification of advanced WECS according to the type of generator used (Nouh and Mohamed, 2014).

necessitates a cooling system (Nadour et al., 2018). The advanced wind energy conversion systems are classified as (Nouh and Mohamed, 2014):

i) According to speed control
ii) According to the type of generator used

Variable speed generators are now found in the majority of current wind energy conversion systems. Fig. 6.9 depicts the categorization of advanced WECS based on the kind of wind turbine generator. The terms Doubly-Fed Induction Generator (DFIG) and Squirrel Cage Induction Generator (SCIG) refer to variable speed generators. The DFIG is the most popular application due to its numerous benefits over other machines in the same class. It is an induction generator with an armature winding with direct connection to the grid and an armature winding connected to a back-to-back inverter which consists of two converter topologies, that is, Rotor Side Converter (RSC) and Grid Side Converter (GSC), isolated by a DC link capacitor as well as an energy storage device to keep voltage variations for the inverter operation to a minimum. The DFIG converts available wind power with excellent efficiency, lower mechanical stresses, and more electronic complexity over a broader range of wind speeds (Simonetti et al., 2018).

6.4 Current status: Global and in developing countries

According to the Global Wind Energy Council, 2020 will be the best year in the history of the global wind industry, with 93 GW of new installed wind capacity, a 53% increase year

TABLE 6.2 Top ten countries with cumulative installed capacity in 2020.

	Cumulative installed capacity (GW)	Percentage of the world wind energy
China	282.85	38.62
United States	117.74	16.08
Germany	62.18	8.49
India	38.56	5.26
Spain	27.09	3.70
United Kingdom	24.48	3.34
France	17.38	2.37
Brazil	17.20	2.35
Canada	13.58	1.85
Italy	10.84	1.48
Top ten countries	611.90	83.55
Rest of the World	120.51	16.45

Data source: IRENA (2021a).

over year. However, this growth will not be enough to achieve net-zero by 2050 (Global Wind Energy Council, 2021). To avoid or reduce the worst effects of climate change, the globe has to construct three times the current wind capacity in the next decade. The worldwide wind energy market has grown over the last decade due to revolutionary technology, establishing wind energy as the most cost-competitive and reliable power source on the planet. Of 75% of all new installations in the United States and China, the world's two largest wind power markets, its record growth in 2020 has increased half of the total global wind power capacity (Paul, 2021).

6.4.1 Global status of wind energy

According to IRENA database (2021), the installed capacity of the global wind power has reached 732.41 GW in 2020 from merely 183.91 GW in 2010. Despite Covid-19 pandemic, as high as 110.76 GW within only one year, that is, 2020, which indicate a significant development of global wind power sector. Ten countries, viz. China, United States, India, Germany, United Kingdom, Spain, Brazil, Canada, France, and Italy occupied 83.55% share of the global cumulative capacity of the wind power (see Table 6.2).

China holds the highest share (38.62%) of the world's total wind energy. Its cumulative installed capacity is 282.85 GW which is even higher than the capacity of other countries except the top ten. The rest of the world except the top ten countries has the capacity of installing 120.51 GW of wind power which holds only 16.45% of the total capacity. While the top ten countries alone hold an 83.55% share which means their total installed capacity is 611.90 GW.

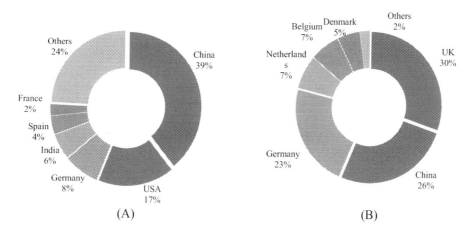

FIGURE 6.10 Global share of installed capacity: (A) onshore, and (B) offshore wind power by country in 2020. *Source: Based on IRENA (2021a) data.*

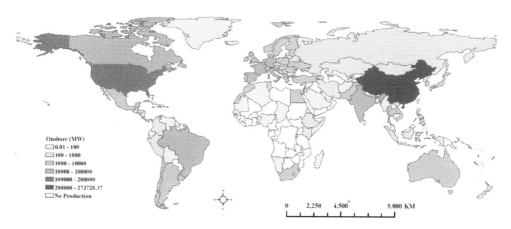

MAP 6.1 Installed capacity (in MW) of onshore wind power in 2020. *Source: Authors' presentation based on IRENA (2021a) data.*

In the list of the top ten, India and Brazil secured the fourth and eighth positions, respectively, which are also developing countries. India has 38.56 GW cumulative installed capacity, which is 5.26% of the world's total installed capacity. Brazil's cumulative installed capacity is 17.2 GW which is 2.35% of the world's total installed capacity (see Table 6.2) (IRENA, 2021a). Fig. 6.10 shows the percentage of global on- and off-shore wind power installed capacity. Map 6.1 and 6.2 illustrate global map of actual installed wind capacity (in MW) for on- and off-shore, respectively.

Installed capacity for both onshore and offshore wind power increased from the starting of the last decade to the starting of the present decade. From 177.8 GW in 2010 to 698.04 GW in 2020, the onshore wind power experienced more than three-fold increase in its installed

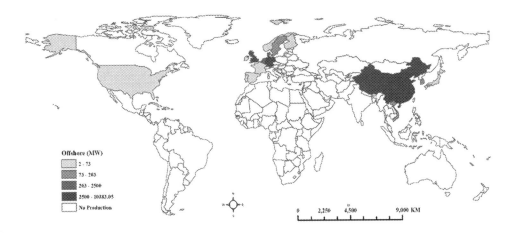

MAP 6.2 Installed capacity (in MW) of offshore wind power in 2020. *Source: Authors' presentation based on IRENA (2021a) data.*

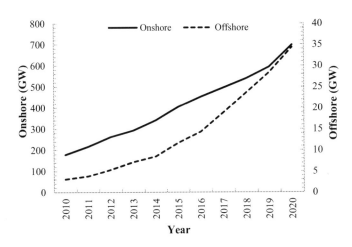

FIGURE 6.11 Cumulative installed capacity (GW) for on- and off-shore wind. *Source: Based on IRENA (2021a).*

capacity. Though offshore wind power also experienced more than 11-fold increase in its install capacity at the same time, the amount is insignificant compared to the onshore wind power. It was 3.06 GW in 2010 and reached to 34.37 GW in 2020 (see Fig. 6.11).

6.4.2 Power generation

The percentage of onshore wind energy was increasing every year of the last decade. The decade started with 8% share in 2010, it almost doubled in 2016 (15.4%) and ended with 19.1% share in 2019. The similar trend is also seen in the amount of wind energy production. In 2010, the total produced onshore wind energy was around 336 TWh. Notably, it tripled at the end of the decade with a production of around 1328 TWh (Fig. 6.12A).

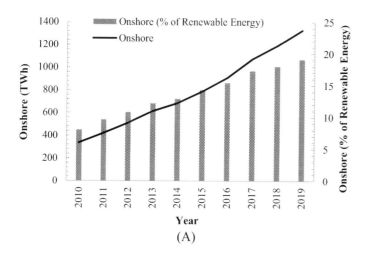

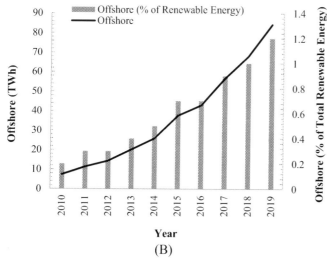

FIGURE 6.12 Year-wise wind energy generation from (A) onshore, and (B) offshore sites (in TWh). *Source: Based on IRENA (2021a).*

The offshore wind power plants do not hold a significant amount of share like onshore ones. In 2010, offshore plants produced only 0.2% of total renewable energy. The share increased year by year, but the highest amount was 1.2% in 2019 which is still very low compared to the share of onshore wind. The percentage of share also reflects the condition of the amount of production. In 2010, it produced around 7.4 TWh power only and in 2019, the amount was around 84.33 TWh (Fig. 6.12B).

6.4.3 Technology manufacturing and supply chain

Strong local supply chain, investment in manufacturing of equipment and grids, and port infrastructure and specialized vessels for offshore projects determine the jobs in the wind

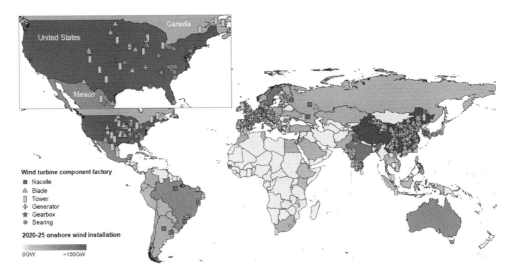

MAP 6.3 Global location of factories of wind turbine components. *Source: BNEF (2021b) Bloomberg NEF.*

energy sector. China leads the global production of the full range and some major parts of wind turbines. Even though around 40 countries produce various components, the full range of parts that include nacelles, blades, towers, generators, gearboxes and bearings are produced only in five countries, viz. China, India, the United States, Spain and Germany (in 300, 48, 39, 31 and 24 companies, respectively) (see Map 6.3) (BNEF, 2021b).

Among the component-producing countries, China had the highest capacity in the manufacturing of wind turbine nacelle 2020 (about 58%) followed by the United States and India (about 10% each). The largest proportion of global commissioned turbine blade-manufacturing plants is also located in China, which is followed by India and the United States (59, 12 and 5%, respectively). Brazil and Turkey are the developing countries which have 3 and 2% share of plants. However, even though China is the largest blade-manufacturer, it mainly oriented toward domestic market. Since blades production is a labor-intensive industry, India enjoys a cost advantage for exporting and low-cost installation in the domestic market. Conversely, nearly half the global commissioned wind tower manufacturing plants are located in China. The country has the highest 50 plants, which is followed by Spain (13) and the United States (9). India and Brazil are the other developing countries have 4 and 2 commissioned tower manufacturing factories, respectively. Similarly, China occupies the highest 38% of global commissioned turbine generator manufacturing plants, which is followed by India, Spain and Brazil (with 10, 8 and 4% share, respectively). The highest number of gearbox producing plants is located in China (18 plants), which is followed by Germany, Spain and India (6, 5 and 2 plants, respectively). Finally, China has the largest number of active bearing factories in the world (one-third of the total), which is followed by the United States and Japan (20 and 9%, respectively). Among developing countries, India and Brazil have 6 and 5% operating plants, respectively (BNEF, 2021b).

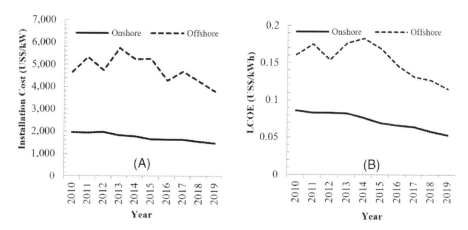

FIGURE 6.13 Global weighted average (A) total installation cost of wind farms (US$ per kW), (B) levelized cost of electricity in 2019 (US$ per kWh). *Source: Based on IRENA data (2021).*

6.4.4 Cost of wind energy

The cost of electricity generation from wind power projects has been decreasing rapidly over time. The capital costs of wind power installation projects decreased from about US$ 2,500/kW in early-1980s to nearly US$ 1000/kW in large projects in mid-1990s. In addition, the cost of irregular and precautionary maintenance decreased, from about US$ 0.05/kWh to nearly US$ 0.01/kWh over the same period. Thus, levelized cost of electricity (LCOE) from wind projects has reduced from US$ 0.15 to less than US$ 0.05 per kWh because of the significant reduction of capital and maintenance costs. If the current pace of research and development (R&D) continues and the volume of production increases through economies of scale, the cost would be further reduced to US$ 0.035-0.035 per kWh over the next 10 years (Salameh, 2014).

Though onshore wind power plant installation cost experienced a smooth and steady decline, the installation cost of offshore plants experienced rise and fall in every other year (Fig. 6.13A). As of 2019, it still took almost three times more installation cost per kW in offshore plants than the onshore ones.

For global weighted average LCOE of onshore wind power projects, it decreases from US$0.086 per kWh to US$0.053 per kWh in the course of ten years. Offshore plants' average LCOE experienced the same up and down in every other year of 2010s as installation cost. It was highest in 2014 (US$0.183 per kWh). It started at US$0.161 per kWh in 2010 and ended at US$0.115 per kWh (Fig. 6.13B).

6.4.5 Employment generation

Wind power has been contributing significantly toward creating employment opportunities all over the world. Jobs in this energy sector have increased to 1.17 million in 2019 from around 0.75 million in 2012. Despite Covid-19 pandemic, onshore and offshore wind power sector has increased the number of jobs in 2020, which reached nearly 1.25 million. It was about 10.4%

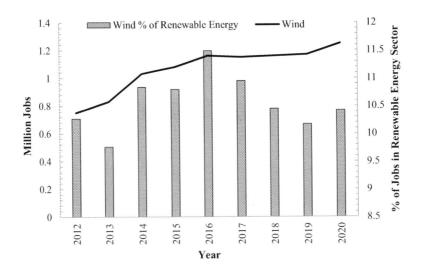

FIGURE 6.14 Job creation by wind power sector. *Source: Based on IRENA data (2021).*

of the total jobs in renewable energy sector. Offshore wind energy sector is emerging as a potential job provider as it requires more workers than the onshore projects.

Wind power sector has introduced a wide range of jobs in the job market and the number increased from 75 thousand to 1.16 million in the last decade. In 2019, this sector offered 1.16 million jobs which are the highest in the 2010s. A steady increasing rate in the number of jobs is noticeable at the same decade. However, if the percentage of jobs in renewable energy is considered, it shows a mixed trend, a steady increase from 2012 to 2016 and then a steady decline from 2016 to 2019 (see Fig. 6.14). In 2016, wind power offered 11.49% jobs of jobs offered by renewable energy sources, which is the highest percentage of jobs offered by wind energy sector in 2010s. The 2020 also marked a great start of the decade with 1.17 million jobs and 10.42% share of total renewable energy.

Top ten wind energy producing countries provide about 80% of the world's job. About 44% global job in this sector was created by China in 2020, followed by the United States and Germany (about 9 and 7%, respectively). The other developing countries except China in the top job providers are India and Brazil with the fifth and six positions in employment, respectively (about 4 and 3% of global jobs, respectively) (see Table 6.3) (IRENA, 2021; IRENA and ILO, 2021). The expansion of wind energy sector was marked by significant addition of installed capacity, which nearly doubled to 111 GW in 2020 from 58 GW added in 2019. China led this expansion with additional capacity of 72 GW (about 65% of new installations) followed by 14 GW of the United States, while ten other countries individually installed higher than 1 GW (IRENA, 2021a).

Wind energy would be the third largest job provider (5.5 million) in 2050 after solar and bioenergy (19.9 and 13.7 million, respectively) (IRENA, 2021a). Women's share was 21% of total job of wind energy sector. The offshore wind market is emerging as a promising sector of power generation as well as creating jobs because of declining costs and massive installation

TABLE 6.3 Employment by country in wind energy sector in 2020.

Country	Jobs (*1000)	Jobs per GW of installed capacity (*1000)
China	523.57	1.85
USA	120.00	1.02
Germany	121.70	1.96
India	62.80	1.63
Spain	22.46	0.83
UK	44.14	1.80
France	15.70	0.91
Brazil	18.75	1.09
Canada	7.60	0.56
Italy	8.10	0.75

Source: Based on IRENA data (2021).

plans. Offshore wind farms are more labor-intensive than onshore projects because of complex construction and installation processes with special towers, blades, turbines, foundations, installation vessels, substations and undersea cables to supply electricity to the mainland and nearby islands (GWEC, 2020a). As per the outlook of IRENA's (2021b) 1.5°C pathway for renewable energy, wind power would create 5.6 million jobs in 2030, while by 2050 it would create 5.5 million jobs.

6.4.6 Future outlook

Based on IRENA (2021) database, the calculated compounded annual growth rates (CAGR) indicate that the growth rate of cumulative installed capacity of global wind power was 15.01% over the period 2010-2020 per annum. However, offshore capacity has grown at a much faster rate than onshore projects. For onshore projects the growth rate was 14.66% while for offshore plants it was 27.38% during the same period. Onshore capacity grew at 17.87% over 2010-15 while it was 11.53% over 2015-20. On the other hand, the growth of offshore capacity was 30.84% over 2010-15 while it was 24.01% over 2015-20. Thus both onshore and offshore projects witnessed slower average growth over the period 2015-2020. Conversely, new installations grew at 12.21% per annum for new installation. For onshore and offshore, the annual average growth rates were 11.75 and 26.59%, respectively.

With the current progress in wind power, it is possible to achieve only half of the total amount of wind power needed to reach cumulative target for well below 2°C pathway. The wind sector needs to have installed 224 GW by 2026 to reach the cumulative target. Then it should be continuously increasing and reach to 280 GW by 2030 (Fig. 6.15).

As per GWEC's (2021) forecasts, global wind power market would experience an annual average growth of 4%. It would experience decline till 2022 and then would start to revive

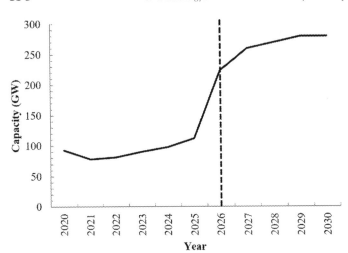

FIGURE 6.15 Estimated installation to reach cumulative target for well below 2°C pathway (IRENA TES) 224 GW in 2026. *Source: Based on IRENA (2021a).*

from 81.1 GW. In 2022 the new wind projects would have 112.2 GW capacity. A total of 469 GW of installed capacity will be added over the period 2021-2025. New installation of onshore wind plan would increase to 88.3 GW in 2025 from 73.4 GW in 2022. China would be the top wind power adopter in the world. The country's new onshore installation would decline to 30 GW in 2021, which would start increasing in the following year. It would be 45 GW in 2025. India's new installation of onshore plants would mostly increase compared to the previous years. It would be 3.1 GW in 2021, which would be at the peak in 2023 and then would finally be 4.2 GW in 2025. Beside India, Vietnam would be a major driver in Asia keeping in mind its ongoing projects. In North America, the new installation of onshore wind plants would experience a fluctuation in the coming years with a declining trend till 2023 (from 14.7 GW in 2021 to only 6.5 GW in 2023) and subsequently move upward (10.6 GW in 2025). In Latin America, the new installation would be a record 5.3 GW in 2021, then it would continue to decline in subsequent years because of uncertainty in government support, possible economic instability and lack of fostering grid capability at country level–in 2025 it would be 4 GW. However, Brazil, Chile, Mexico, Argentina and Colombia would be the highest contributors to regional wind adopters till 2025. Considerable fluctuations would be observed in Europe in new onshore installations. It would range from 14.1 GW in 2022 to 16 GW in 2025. Germany, France and Spain, Sweden, Norway, Turkey and Russia would be the top contributors in the new onshore projects. However, Africa would remain the exception in the coming years. The new onshore installations would continue to increase from 2 GW in 2021 to 4.3 GW in 2025. South Africa, Egypt and Morocco would be the driving countries in such an impressive performance of the region.

On the offshore front, the new installation would increase to 23.9 GW in 2025 from merely 7.7 GW in 2022. New installations would continue to increase in North America, and it would reach 3.6 GW in 2025. European region experience rise till 2023 (6.5 GW) and fall in 2024 and then sudden upward movement in 2025 (10.3 GW). Conversely, new installations in Asian offshore projects would steadily increase from 4.5 GW in 2022 to 10 GW in 2025, which would be led by South Korea, Japan and Vietnam. Thus, there would be a mixed trend of new installations in both onshore and offshore wind power projects across the world (GWEC, 2021).

Annual average investment in wind power was US$ 80 and 18 billion for onshore and offshore projects, respectively over the period 2017-19. However, US$ 212 and 177 billion would be required for onshore and offshore projects, respectively, per year over the period 2021-50 to meet 1.5°C scenario (IRENA, 2021b).

6.5 Wind power in developing countries

6.5.1 Wind power in selected developing countries

China holds the highest share (38.62%) of the world's total wind energy. Its cumulative installed capacity is 282.85 GW which is even higher than the capacity of other countries except the top ten. The rest of the world except the top ten countries has the capacity of installing 120.51 GW of wind power which holds only 16.45% of the total capacity. While the top ten countries alone hold an 83.55% share which means their total installed capacity is 611.90 GW. In the list of the top ten, India and Brazil took the fourth and eighth positions, which are also developing countries. India has 38.56 GW cumulative installed capacity, which is 5.26% of the world's total installed capacity. Brazil's cumulative installed capacity is 17.20 GW which is 2.35% of the world's total installed capacity (IRENA, 2021a).

6.5.1.1 China

With the world's largest wind capacity installation and substantial, sustained expansion in new wind facilities, China is the global leader in wind generation. China has tremendous wind power resources due to its enormous landmass and extensive coastline. On land, China is estimated to have 2380 GW of exploitable potential. In 2020, wind power was China's third-largest source of electricity, accounting for 12.8% of total power generation capacity (G.W.E. Council, 2021). China had cumulative installed capacity of 282 GW in 2020, of which 9 GW was in offshore and 273 GW was in onshore projects. The country's installed and expanded capacity was the world's largest in 2020 (see Fig. 6.16) (IRENA, 2021a).

According to GWEC (2021), the installations of onshore wind plants will decrease considerably in China in 2021 after an exponential growth in 2020. It is mainly because most of the wind power projects in pipelines which were approved before the end of 2019 were already installed. All onshore projects which are being undertaken from 2021 are not receiving subsidy. However, installation of onshore wind plants would again increase gradually to promote the country's target of carbon neutrality in the near future.

An estimated 0.55 million jobs were created in wind power sector in China in 2020, which was higher from 0.518 million in 2019. More than 90% of total onshore installations were supported by domestic companies over the preceding decade. Domestic wind power projects are the main sources of jobs in the country. The country's expiring feed-in tariff for offshore wind at the end of 2021 caused rapid completion of the new and ongoing projects. Joint collaboration between state-owned companies and provincial governments helped in developing offshore wind power in the country. In addition, coastal provinces, viz. Guangdong, Jiangsu, Zhejiang and Fujian set ambitious targets in offshore wind, which stimulated employment growth. Local supply-chain clusters have developed rapidly because of the country's expertise in onshore wind (World-Energy, 2020). As many as 18 industrial

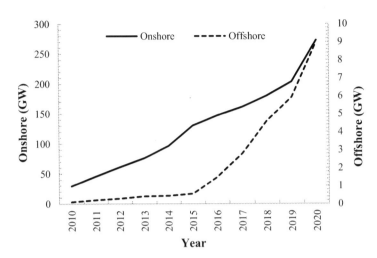

FIGURE 6.16 Installed wind power capacity (GW) of China. *Data source: (IRENA, 2021a).*

parks are being developed along the coast–half of which are located in Jiangsu and Guangdong provinces.

6.5.1.2 India

India is the world's fourth-largest country in terms of cumulative installed capacity of wind power. However, new installations have significantly reduced from 2.2 GW in 2019 to merely 1 GW in 2020 because of the Covid-19 pandemic (IRENA, 2021b). Employment in the country's wind sector declined correspondingly, from nearly 62,800 in 2019 to about 44,000 in 2020. However, India is getting prepared to as an alternative manufacturing hub of blade and gearbox for exporting to the US and other markets in order to reap benefits of the US-China trade war. The presence of foreign manufacturers and supplies is notable in India–about 100% in bearings, gearboxes and converters; 72% in blades; 53% in nacelles and 29% in towers in 2019 (Barla, 2020).

India's progress in the wind energy sector is praiseworthy. The wind energy sector experienced notable growth, from 13.18 GW in 2010 to 38.56 GW in 2020 (see Fig. 6.17). It happened due to the government's huge investment in wind energy R&D. Their aim is to produce 60 GW exploiting wind energy technology by 2022. Technical innovations have made it possible to increase the range of turbines (from 250 kW to 2100 kW) and rotor diameters (from 28 m to 80 m). From 2000 to 2013, installed wind turbines' average electricity generating capacity reached from 400 kW to 1.5 MW and installation cost significantly decreased at the same time.

R&D also resulted in less maintenance and operation costs of larger turbines. Suzlon is the best example of it. This company is India's largest and the world's fifth largest turbine maker. It collaborated with a Belgian company "Hansen" in 2006 and a German company "Repower" in 2007. Therefore, now Suzlon can produce wind turbines of 6 MW range. To meet the demand for larger turbines (Irfan et al., 2020), Bharat Heavy Electricals Ltd (BHEL) started manufacturing turbine blades and towers in 2010. TATA power also started increasing

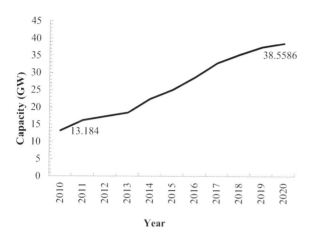

FIGURE 6.17 Installed wind power capacity (GW) of India. *Data source: (IRENA, 2021).*

its capacity in 2013 and reached 2000 MW in 2017. Karnataka has a wind power plant of 3000 MW capacity which was established by Airvoice. Some government organizations constructed wind power plants in different places, such as a 100 MW wind power plant at Madhya Pradesh by NHPC Ltd, a 50 MW wind farm in Tuticorin city by Neyveli Lignite Corporation Ltd, and a 2000 MW wind capacity in Gujarat by Oil and Natural Gas Corporation Ltd (Arora et al., 2010). Due to several initiatives and robust policies by the Indian government, private enterprises and foreign organizations are now investing in the Indian wind energy market. India's 24% of total electricity demand will come from wind energy by 2030 and this sector will attract US$ 10.637 billion of investment annually (GWEC, 2011). However, unsatisfactory performance and failure of imported wind turbines raised suspicions in investors' minds about the effectiveness of the technology at the very beginning. Investors were also discouraged when there was a reduction in tax concessions (Jagadeesh, 2000). Danish enterprises are manufacturing wind turbines and towers, they consult with their local partners in India for transferring necessary skills, technical capacity and experience and India manufactures key wind energy equipment and parts (Rennkamp and Perrot, 2016). As per the Statistic Review of World Energy 2021 report, carbon emissions have decreased by 6.3%, and coal use has reduced by 4.2%, while renewable energy has grown by 9.7% in India (Statistical Review of World Energy, 2021).

6.5.1.3 Brazil

Brazil's installed capacity in wind power has become about 2.3 GW in 2020, which was significantly greater than that of 2019. The country's employment in this sector is nearly 40,200 who are engaged mostly in construction, operations and maintenance (IRENA and ILO, 2021).

Brazil's first wind farm was established in 1992 in a small archipelago named Fernando de Noronha. But that does not indicate the beginning of wind power exploitation in Brazil (Gonçalves et al., 2020). Even in 2000, wind power contributed only 0.1% of Brazil's total installed capacity (Soares et al., 2021). In 2001, Brazil experienced an energy crisis which led

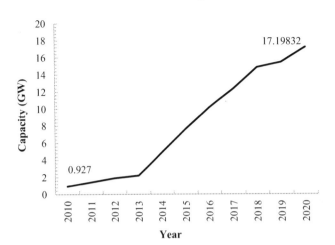

FIGURE 6.18 Installed wind power capacity (GW) of Brazil. *Data source: (IRENA, 2021).*

them to explore new sources and they found wind power as one of the most developed sources (González et al., 2020). To explore the commercial potential of wind power, Brazil launched Program for Incentive to Alternative Energy Sources (PROINFA). However, PROINFA focused on other renewable energy sources too (Gonçalves et al., 2020). In 2008, wind power competed with other energy sources in auction for the first time (Gonçalves et al., 2020). Wind power was one of the most developed sources, accounting for more than one-fifth of the capacity expansion of the Brazilian electricity matrix between 2012 and 2016 (WWF, 2015; Bell et al., 2018). In 2018, Brazil's wind power installed capacity was 9% of its total electricity and wind power projects held 54% of the total registered projects which was less than 10% in 2007 (Gonçalves et al., 2020). This is the result of the increasing installation of wind farms since 2006 which skyrocketed from 2014 onwards (Gonçalves et al., 2020). The installed capacity increased from 0.93 GW in 2010 to 17.2 GW in 2020 (see Fig. 6.18) (IRENA, 2021). Brazil committed to reducing its carbon emission by 37% in 2025 and 43% in 2030 (EPE, 2016) and wind power is going to underpin these goals.

In the first half of 2021, Brazil installed wind farms with a total capacity of 1429.9 MW. The country added 284.46 MW of new installed capacity in June alone. Its total wind capacity installation was 1787.4 MW in the first half of 2021, accounting for 10.47% of the country's total installed capacity. Brazil today has an allowed power capacity of 176,157.6 MW installed (Renewables, 2021b).

6.5.1.4 Vietnam

Vietnam has the potential to expand its wind energy market considering its increasing energy demand. The expansion of the wind energy market will also ensure energy security and deliver socio-economic benefits. Every year the electricity demand increases 10% than the previous year and so giant turbines in windy areas have become a common sight in the last five years. A positive factor is that there is a 30% decline in the capital cost of wind turbines. However, wind energy supplies only 1% of the total electricity production of Vietnam which is 600 MW in 2020, of which 500.8 MW was in onshore and 99.2 MW in offshore projects (see

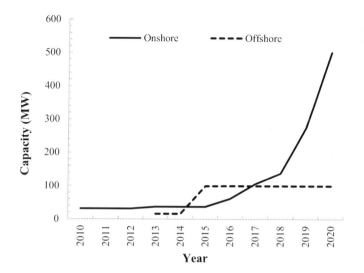

FIGURE 6.19 Installed wind power capacity (GW) of Vietnam. *Data source: (IRENA, 2021).*

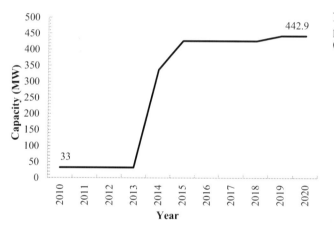

FIGURE 6.20 Installed wind power capacity (GW) of Philippines. *Data source: (IRENA, 2021).*

Fig. 6.19) (IRENA, 2021). This amount cannot even complete their unambitious target of 800 MW wind power back in 2018(67) In June 2020, the prime minister approved 7 GW of wind projects. Later, ministry of industry and trade proposed to include an additional 6.4 GW of wind projects. These announcements also offered the private sector to participate in energy development which also marks Vietnam's ambition to become the leading market for wind energy development in Asia (GWEC, 2021).

6.5.1.5 Philippines

The wind power sector in Philippines has increased rapidly from merely 33 MW in 2013 to 443 MW in 2020 as illustrated in Fig. 6.20 (IRENA, 2021a). The Philippines published its energy plan in 2020 which covers the period of 2018-2040. However, their plan does not include a solid

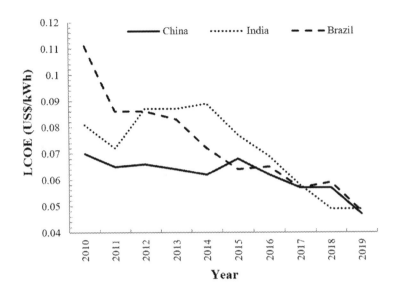

FIGURE 6.21 Weighted average LCOE of newly commissioned onshore wind projects (US$ per kWh) in selected developing countries. *Source: Based on IRENA data (2021).*

plan for achieving the 2030 renewable energy target. Also, the last four years did not show any progress in the wind energy sector. However, the green energy auction in 2021 along with the moratorium on Greenfield coal plants and the development of the Green Energy Option Program for large power consumers give a ray of hope for the growth of renewable energy. A 132 MW wind power plant in 2021 inaugurates almost 5 GW of proposed wind projects. However, the major challenges the Philippines are facing in terms of establishing wind energy as a renewable energy source are lack of incentivization, legal obstacles and limitations on transmission capacity. The problems can be addressed by improving the economics of wind development and green energy auctions (GWEC, 2021).

All the economies presented experienced very unstable rise and fall of average LCOE over the decade. Even the year-wise comparison among them varied greatly as depicted in Fig. 6.21. For instance, in 2010, Brazil had the highest average LCOE and China had the lowest. In 2017, both Brazil and China had lower average LCOE than the other mentioned economies.

Wind energy technology adaptation and refining will become more efficient and cost-effective. High-altitude devices, for example, make use of more powerful and steady winds at higher altitudes. Such slightly elevated devices are modern wind-harvesting technologies, also known as Air-borne Wind Energy Systems (AWES), which are equipped with wind turbines. These components of air-borne wind generators float on any gas, such as helium, or rely on their aerodynamics to remain at higher elevations in the air, where the wind is more incredible. Many engineers are currently working on developing air-borne wind turbines. Because installing traditional turbines on tall towers is difficult and expensive, AEWS can also be deployed offshore (Daniel, 2020).

6.6 Contribution to sustainability

Wind power has been one of the world's fastest-growing and most widely accessible renewable energy sources today. As a result, it plays important role in meeting energy demand. The quantity of carbon released during the generation of power by wind turbines is minimal, showing that wind energy is a clean energy source. Production and efficient utilization of wind energy could contribute to long-term development with positive impact on environment and society as it reduces the usage of fossil fuels and contributes significantly to economic growth. It is the least polluting and inexhaustible source of renewable energy with a negligible environmental impact. In addition, it is an infinite energy source that has low cost and also creates local jobs (NCTCE, 2019).

6.6.1 Wind energy and its link to sustainability

Sustainable development satisfies current needs while safeguarding future generations' ability to meet their own. According to the UN report of Sustainable Development Goals (SDGs), at least 759 million people of the world's population are without electricity (UN Sustainable Development Goals, 2021).

As a result, developing countries or areas have a dual-energy challenge in the twenty-first century: addressing the needs of billions of humans who need clean, modern, low-cost energy including renewable energy (SDG-7) while increasing productive and decent employment (SDG-8), reducing material footprint through lessening the use of fossil fuel (SDG-12), climate change mitigation (SDG-13), and harnessing lives below water (SDG-14) and on land (SDG-15). To do so, historical rates of progress toward higher efficiency, decarbonization, increased fuel diversity, and lower pollutant emissions must be significantly accelerated. Lowering greenhouse gas emissions has been achieved by developing indigenous alternative energy sources and minimizing local pollution.

Globally, substantial wind power potential exists, which can be observed from Map 6.4. Specifically, high wind power density potential is evident in most parts of the Northern Hemisphere, while some countries of Southern Hemisphere, such as Brazil, Argentina, Australia, New Zealand, etc. have high wind power potential. However, the potential can be realized in a sustainable manner through minimizing the adverse impacts of wind power projects on environment and society.

The share of cumulative installed capacity of wind power in renewable energy has increased significantly in recent years. It is mainly because of its environmental benefits as it is a clean and sustainable source of electricity as well as economic benefits, such as significantly declining per unit cost and large job opportunities (Yazdani et al., 2019).

However, nonrenewable energy sources emit significantly higher CO_2 than renewable ones. Lignite emits 1069 tonne CO_2-e/GWh which is the highest carbon an energy source can emit. Natural gas emits the lowest carbon among traditionally used nonrenewable energy sources which is 500 tonne CO_2-e/GWh. Among the renewable energy sources, solar PV is the highest emitter. It emits 85 tonne CO_2-e/GWh which is a lot less than the lowest emitter of the traditional nonrenewable sources. Interestingly, despite being a nonrenewable energy source, nuclear energy emits very low carbon. Its emission rate (26 tonnes CO_2-e/GWh) is less than solar PV and biomass (45 tonnes CO_2-e/GWh). The lowest emitters among all

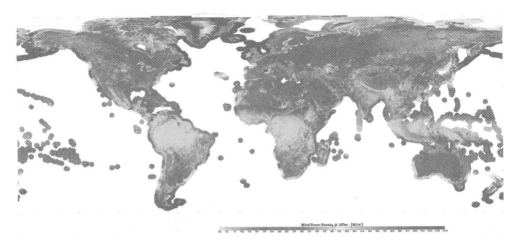

MAP 6.4 Wind power density potential of the world. *Source: Global Wind Atlas, https://globalwindatlas.info.*

renewable sources are hydroelectric and wind energy sources. Both energy sources emit only 26 tonnes CO_2-e/GWh (World Nuclear Association, 2011). A comparison has been illustrated in Table 6.4.

Wind power is regarded as the cleanest and eco-friendly source of renewable energy (Nazir et al., 2020). Wind technologies are constantly being improved for complying with safety and convenience of humans and animals. Wind energy will become the second-highest source of renewable energy after solar PV over the next couple of decades (IEA, 2021). It is mainly because of its significantly lower adverse impacts on environment and global climate, while it generates a number of benefits to the land and ocean environment, physical atmosphere, economy and society (Dai et al., 2015).

Wind plant does not produce greenhouses gases (GHGs) such as COx, SOx, NOx, particulate matter (PM) or any other air pollutants during the production of power like fossil fuel-based power plants (Zhu et al., 2020; Khan, 2020). Instead, it helps reduce GHG emission by supplying clean electricity. Wind power has helped to evade 189 and about 160 million metric tonnes of CO_2 in 2019 and 2018 which could otherwise be generated in the electricity sector by using fossil fuels.

Beside climate change mitigation and protecting environment, reduction of the pollutants has considerably positive impact on human health. For example, reduction in COx, SOx, NOx and PM helped save US$9.4 billion in public health in 2018. Moreover, wind power does not require massive amount of water like thermal electricity projects. Thus, it releases water for better alternative use. For example, (AWEA, 2020) found that in 2019, the active wind power plants decreased water consumption by nearly 103 billion gallons.

As one of the fastest growing energy sources, the contribution of wind power to global electricity generation would increase from 5 to 30% in 2050. In order to support sustainability of this energy, environmental safeguards must be ensured while sourcing materials to support the expected growth of global installed capacity. In 2050, the carbon footprint from globally materials requirement for wind turbines would be 9.3 times lower than the

TABLE 6.4 Emission of pollutants per kWh of electricity produced by wind vs. other sources.

Pollutants	Onshore wind	Offshore wind	Average wind	Solar PV	Solar thermal	Biomass	Hard coal	Lignite	Natural gas combined cycle (NGCC)	Nuclear	Combined heat and power (CHP)
Carbon dioxide, fossil (g)	8	8	8	53	9	83	836	1,060	400	8	0
Methane, fossil (mg)	8	8	8	100	18	119	2554	244	993	20	12
Nitrogen oxides (mg)	31	31	31	112	37	814	1309	1,041	353	32	1
Nonmethane volatile organic compounds (NMVOC) (mg)	6	5	6	20	6	66	71	8	129	6	0
Particulates (mg)	13	18	15	107	27	144	147	711	12	17	2
Sulfur dioxide (mg)	32	31	32	0	31	250	1548	3,808	149	46	14

Source: (Kondili, 2021).

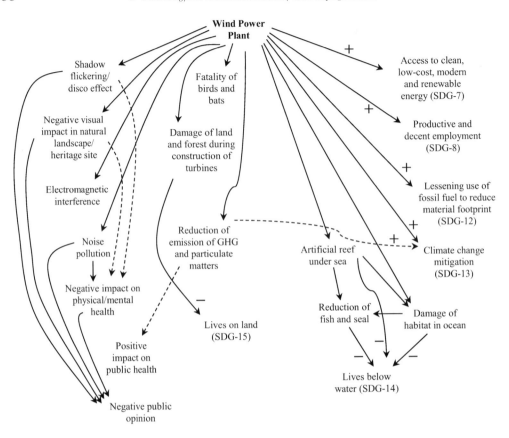

FIGURE 6.22 Wind energy and its link to sustainable development. *Source: Authors.*

present level of CO_2 emission from coal-fired power plants of the USA (Farina and Anctil, 2022).

As a cheap source of power, such a positive impact of wind turbines on environment and climate change has attracted attention of the planners and policymakers toward gaining public acceptance. However, Peri and Tal (2020) found significant anthropocentric environmental impacts from wind power projects if the turbines sites were located less than 750 m from the settlement.

The negative environmental effects of wind power plants are negligible compared to traditional sources of energy and even to other renewable energy. However, it can affect human life and health, land and ocean ecosystems, and wildlife (Dhar et al., 2020). The major impacts of wind energy are noise pollution (Koschinski and Lüdemann, 2020), visual effects (Kowalczyk, 2020), local climate change, electromagnetic obstructions, mortality and injury of birds and bats (Marques et al., 2020; Conkling et al., 2021), soil erosion and deforestation (Nazir et al., 2020), and impacts on marine wildlife (Sanders and Beecham, 2017). All these links are illustrated in Fig. 6.22.

6.6.2 Impacts of wind energy

The most noted visual impact of wind power plants is the shadow flickering and the disco effect or periodic flashes of light that can occur simultaneously during sunlight. Rotating blades of the turbine generate shadow flickers by disrupting the sunlight. Even though the flickers do not pose direct threat to human health, it creates a considerable annoyance among residents (Voicescu et al., 2016; Smedley et al., 2010). Turbines create significant visual impact since it can be viewed from long distance. It can distort the view of the natural landscape. Such an involuntary exposure to the turbines can produce negative psychological and physiological impacts among bystander residents (Henningsson et al., 2013; Voicescu et al., 2016). However, these effects can be minimized by selecting onshore wind power sites far away from residential areas and using coating materials in the blades that prevent flickering. These effects can also be minimized by optimizing the smoothness of the rotor blade surface and by coating the blades with less reflecting material (Hartmann and Apaolaza-Ibáñez, 2012).

Wind turbines generate a continuous and repetitive sound that can affect sleep quality of the nearby residents, especially in the rural areas (Schmidt and Klokker, 2014). Two types of noise are generated from wind turbines: aerodynamic, which is produced by flow of air above the blades, and mechanical. The level of aerodynamic noise depends on turbine size, wind speed, and rotation speed of the blade (Dai et al., 2015). Large turbines that are exposed to strong winds generate greater noise. However, mechanical noise can be decreased through optimizing the design of the side gears or by sound insulation inside the turbine housing, using sound-proofing curtain and anti-vibration support feet during electricity generation (Nazir et al., 2019).

The wind turbine impacts the wildlife indirectly through avoidance, habitat disruption, and displacement. Avian casualties give rise to serious concerns against the growth of wind energy and their rapid installations in developing countries. Wind turbines can cause injury and mortality to birds and bats through their collisions during rotation of blades (McNew et al., 2014; Spellman, 2014; Dai et al., 2015). It has been found that mean annual mortality of birds and bats due to a single turbine are 2.3 and 2.9 (Rydell et al., 2012) with a considerable deviation between turbines (0–60 for birds and 0–70 for bats) depending on the location of the turbines vis-à-vis the level of avian activity. A 1 GW wind power project located in wildlife-caring site kills a total of 20 fauna a year, while hunters kill 1500, and vehicle movement and electricity lines lead to 2000 deaths in the same area (Nazir et al., 2020). Turbines can indirectly impact the habitats of migrating and nesting species through primarily affecting their aviation routes and feeding sources, which depend on the number, height and locations.

Bird mortality was 20,000 to 234,000 in the USA with a range of 0 to 40 per turbine per year (Loss et al., 2013). However, it is considered to be much lower than that caused by human activities and other sources of power. Sovacool (2013) found that mean bird mortality from wind plants was approximately 20,000, while bird deaths from nuclear plants and fossil fuel based power plants had 330,000 and greater than 14 million in the USA in 2009. Similarly, Marques et al. (2020) found that wind turbine is source of death of only 1 per 10,000 fatalities of bird, while communication tower, pesticides and vehicles, high-voltage electricity lines, cats and buildings/windows cause 250, 700, 800, 1000, and 5500 per 10,000 fatalities. Thus, it causes significantly less fatalities to birds than other sources of casualty of birds.

Turbines can damage environment if they are installed in the center of open areas as many ecosystems require large, continuous and dedicated open spaces. Therefore, wind turbines need to be placed in ecologically distressed locations and far away from the human settlements, built environments, and wildlife habitats and movement routes (Peri and Tal, 2020). The risk is lower if turbine is located in spatial plains and higher if it is installed on the edge of a hill or adjacent to the shoreline (Loss et al., 2013).

Offshore wind power plants are installed on different water bodies, which can cause dolphin and seals fatalities during the construction unless they are planned and implemented carefully (Lee et al., 2019). Shadow flickering can also affect marine wildlife located nearby the water surface adjacent to a wind farm, which can distort their normal marine life.

The electromagnetic interferences emanating from wind turbines can distort some radio or television transmissions of adjacent stations. Thus, wind farms can be located far away from broadcast stations to avoid high electromagnetic interferences by the turbines as reported in (Katsaprakakis, 2012). Moreover, blades can be manufactured by using synthetic materials to lessen electromagnetic interferences.

Construction of wind power plants can affect nature and the local climate first by taking out plants and vegetations of the project area. In addition, excavation of foundation, usage of large machinery, carrying massive parts of turbine, and establishing transmission/feeder lines to the local and national electricity grids can again damage soil and land (Spellman, 2014; Bijian et al., 2017). At the same time, construction of wind power plants can destruct habitat, create barriers of movement to wildlife, and impact breeding and feeding behavior of the birds (Dai et al., 2015). Tang et al. (2017) found that wind power plants suppressed the content of soil water and increased water stress, which caused to a reduction of 8.9% summer gross primary output and 4% annual net primary output in Northern China.

Offshore wind towers can damage the benthic community and block the sunlight in the ocean surface water by introducing objects into the seabed. Ocean biodiversity is also affected by the installation of wind power plant that produces an artificial reef. In addition, the turbine's electromagnetic field and noise pollution generated can exert negative effect on fish, dolphins and seals (Lintott et al., 2016). In addition, seawater can be polluted through operation and maintenance that include lubrication or parts replacement and insertion of debris into the water. Therefore, offshore wind farms can be installed, operated and maintained carefully in order to minimize the negative impacts on the ocean ecosystem (Lee et al., 2019).

The adverse environmental impacts of wind power are negligible compared to traditional sources of energy (Saidur et al., 2011). However, the environmental effects of wind power that are related to sustainability include noise, visual impact, safety of wildlife and ecology, distortion of bio-system and physical environment, electromagnetic intrusion, and change of local and micro climate (Spellman, 2014; Dai et al., 2015). Some of these adverse effects may influence public support and endorsement of wind energy projects, which would pose threat to sustainability.

In addition, simpler design fewer numbers of turbines can produce a visually balancing and consistent image of the pant (Dai et al., 2015). Since visual effects of wind turbines cannot be evaded, decreased or masked, opposition of the local people can be minimized by educating the environmental effects of the plants compared to fossil fuels. These measures would help increase public support for wind power plants especially in developing economies, thereby fostering sustainability.

During the production of electrical energy using wind turbines through a fraction of the wind's momentum, a similar amount of momentum is lost from the existing wind that distorts the natural exchanges of energy between the land/sea surface and the adjacent atmospheric layers. This distortion may modify local hydrometeorology and create a cascading effect on atmospheric dynamics (Tabassum-Abbasi et al., 2014; Dai et al., 2015). Tang et al. (2017) found that a downwind warming effect in Northern China because of a wind power plant, while (Wang and Prinn, 2010) projected that global temperature could increase by 1°C in 2100 if wind power meets 10% of global demand for energy. On the other hand, Shang (2010) found that wind power plants mitigated sandstorms by reducing wind speed in Gansu Province of China. Thus, the wind power has both positive and negative impacts on local climate and climatic hazards.

Installation of wind power plants requires lesser amount of land compared to other energy technologies. Nearly 1600 m^2 of land is required to install a typical 2 MW wind turbine (Katsaprakakis, 2012). Therefore, there is ample scope to use the remaining land in other productive activities, such as cropping, grazing, aquaculture, besides operating the windmills in land-scarce developing countries (Weiss et al., 2018). In addition, wind turbines can be collocated with other renewable energy plants to lessen the land use footprint. Thus, the wind plants can be collocated with solar, biomass, geothermal, power plants by optimizing combination of technologies. For instance, India has installed combined wind and solar energy as a hybrid production system of electricity (Sharma et al., 2012).

Wind energy can generate 5.4 times greater jobs than coal-fired, geothermal and nuclear power plants (Scheer, 1995). The gross socioeconomic benefits during the life of a plant would be US$3.12 billion, with multiplier effect of 2.2 and indirect per direct employment generation ratio of 1.21 (Rodríguez et al., 2017). Public education and awareness should be raised on socioeconomic and environmental benefits of wind power, which include lowest GHG emission among renewable energy sources per unit of electricity production, public health benefits, increased job opportunities compared to other dominant power technologies, lower social cost, and power supply in rural, remote, and isolated areas.

Public acceptance is an important precondition for achieving sustainability of the energy projects. Wind power plants are visible in rural areas and natural settings, which usually exert visual impact of the nearby population. In addition to negative visual effect and noise pollution, the presence of wind turbines can be visually incompatible to some of the residents of adjacent locality (Spellman, 2014). Generally, people have negative perception regarding wind power projects mainly because of noise pollution and visual impacts (Olson-Hazboun et al., 2016). Moreover, wind plants have strong negative impact on people's perception if these are installed in the locations adjacent to natural beauty and cultural heritage sites (Dhar et al., 2020). Therefore, a fraction of local population may oppose installation of a new wind plant closed to their localities albeit of knowing its environmental benefits compared to traditional energy, which is called Not-In-My-Back-Yard (NIMBY) effect (Swofford and Slattery, 2010; Tabassum-Abbasi et al., 2014).

Wind power projects should be planned and installed considering the negative impacts on land and ocean environment. Safety and preventive measures should be adopted to minimize the casualty of birds and bats, and effects on marine fishery and benthic community.

6.7 Conclusion

Wind energy is environment friendly compared to fossil fuel-driven electricity generation. Even though solar power generates electricity, wind energy stands as a prominent and excellent source for developing countries, which provides a steady supply of electricity because of the enormous wind potential. The wind turbines last for almost 20 to 25 years, and as long as the wind blows, it can generate electricity. Hence, manufacturers can increase the structure's material qualities and design capability to sustain more considerable stresses and maintain the same stress level for a more extended period. This chapter has discussed the current situation of wind energy worldwide, mainly in developing countries. It also includes case studies of China, India, Brazil, Vietnam, and the Philippines.

Developing countries, such as China, India, Brazil, Vietnam, and Philippines which are leading manufacturers and markets of wind energy may make a significant contribution to energy demand to some amount if they place a greater emphasis on capacity expansion. Devising sophisticated technology has been needed to optimize the characteristics of wind turbines and make them better suited for grid integration. Some of these countries are the world's fastest producers and consumers of energy. It, however, requires additional attention for further growth, and technological development of systems in recent years has been encouraging.

Developing countries need to encourage value-added commerce by concentrating their efforts on cooperative energy development in these countries to achieve long-term growth. As the world's biggest consumers of oil, natural gas, and coal, China and India should take the necessary steps to reduce their reliance on coal consumption. Inadequate technology and infrastructure necessitate the development of modern technologies that can be overcome through R&D, allowing for widespread application of technologies for resource utilization in offshore regions and hill areas. Governments should take necessary steps to attract investors and allow more funds to support research activities in this field. Focusing on a particular technical development system and its manufacturing system is critical for future progress. As a result, emerging countries will make a rapid and global transition to wind energy technology soon, resulting in sustainable growth, and climate change mitigation.

6.8 Self-evaluation questions and numerical problems

6.8.1 Short questions

1. Differentiate between air and wind.
2. Why is the use of wind energy known as a reflection of solar energy?
3. What causes the wind to form in the atmosphere?
4. Identify the variables that influence the output of a wind turbine.
5. How does a vertical axis wind turbine differ from a horizontal axis wind turbine in terms of benefits and drawbacks?
6. Define the term "power coefficient."

7. What is an advanced wind energy conversion system? How is it classified?
8. What are the different factors which influence the amount of power produced by the wind turbine?
9. What are the different efforts taken and policies implemented to enhance wind energy utilization in developing countries?

6.8.2 Long questions

1. Explain the mechanism of conversion of wind energy into electrical power through wind turbines.
2. Derive steady-flow energy equation for wind energy conversion.
3. Describe how the thermodynamics of the wind energy conversion system work.
4. Derive the conditions for maximum power generation in a wind conversion system.
5. Derive an equation for the maximum power generated by an ideal horizontal axis wind turbine.
6. Explain the various types of forces acting on the blades of a propeller-type wind turbine.
7. Explain some crucial factors consider for site selection for the wind turbine.
8. What is an advanced wind energy conversion system? How is it classified?
9. How wind energy contributes to sustainability?
10. Enumerate some of the reasons why power generation using wind turbines is considered a significant factor in achieving the goal of sustainability.
11. What are the positive and negative social, environmental, and economic impacts of wind energy generation? – Explain.
12. With the help of recent literature show that wind energy has a significant role in achieving energy sector's sustainability in developing countries.

6.8.3 Numerical problems

1. Suppose that in a windmill, $D = 34.4$ m, $C_p = 0.34$, $V = 8.51$ m/sec, $\rho = 1.135$ kg/m^3, and $P_0 = 90$ KW.
   (i) Calculate P_m.
   (ii) If the turbine operates at C_p at the rated wind speed, then work out the P_{mx}.
   (iii) Calculate η. [Answer: (i) ≈ 324.89 kW, (ii) ≈ 110.46 kW, and (iii) ≈ 81.48%]
2. A wind turbine travels with a speed of 36 kmph and has a blade length of 15 m. Determine the wind power. (Density of air = 1.23 kg/m^3) [Answer: 434.718 kW]
3. Given the following data for a wind turbine operating under certain conditions.
   Blade length = 50 m
   Speed of wind = 54 kmph
   Density of air = 1.225 kg/m^3
   Power Coefficient (C_p) = 0.38

   Calculate the rotational kinetic power in MW produced in a wind turbine at its given rated wind speed. Also, plot the power output P as a function of wind speed v on the interval [0 m/s, 15 m/s]. [Answer: P =6.17 MW]

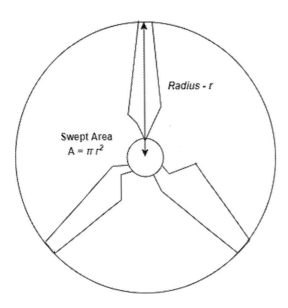

4. If the wind is flowing at a speed of 10 m/s, how much power is the wind striking the wind turbine if the blades are 40 m long and the turbine is located at sea level? How much would more power be needed if the identical wind turbine was operated at a wind speed of 12 m/s?

   [Answer: $P_1 = 3078.76$ kW, $P_2 = 5320.098$ kW, More power required $= 2241.338$ kW].

5. What is the swept area of a wind turbine with 8 blades that are each 50 m long? [Answer: 7853.98 m^2]

6. A wind turbine produces 1000 W at a wind speed of 20 m/s with a blade length of 25 m then calculate:
   (i) Theoretical power produced
   (ii) Power coefficient
   Consider density of air as 1.225 kg/m^3
   [Answer: (i) 9621.128 kW, (ii) 0.104]

7. Find the size of the wind turbine rotor (diameter in m) to generate 20 MW of electrical power at a steady wind speed of 36 kmph. Assume that the air density is 1.225 kg/m^3, power coefficient and efficiency of the wind turbine are 0.549 and 0.45, respectively.

   [Answer: 101.95 m]

Acknowledgment

The authors are grateful to Nigar Sultana for her research assistance.

References

Allaei, D., Andreopoulos, Y., 2014. INVELOX: description of a new concept in wind power and its performance evaluation. Energy 69, 336–344. https://doi.org/10.1016/j.energy.2014.03.021.

References

Andersen, P.H., Drejer, I., Gjerding, A.N., 2017. Industry evolution, submarket dynamics and strategic behaviour among firms in offshore wind energy. Compet. Change 21 (2), 73–93. https://doi.org/10.1177/1024529416689793.

ANEEL, 2014. Micro e minigeração distribuída. Tech. rep., Cadernos Temáticos. Sistema de Compensação de Energia Elétrica. Brasília. Centro de Documentação–Cedoc, Brasil.

Aririguzo, J.C., Ekwe, EB., 2019. Weibull distribution analysis of wind energy prospect for Umudike, Nigeria for power generation. Rob. Comput. Integr. Manuf. 55 (Part B), 160–163.

Arora, D.S., Busche, S., Cowlin, S., Engelmeier, T., Jaritz, H., Milbrandt, A., Wang, S., 2010. Indian Renewable Energy Status Report: Background Report for DIREC 2010. National Renewable Energy Laboratory (NREL), Oak Ridge.

AWEA, 2020. https://www.awea.com/. (Accessed September 2021).

Barla, S., 2020. Global wind turbine supply chain trends 2020. Wood Mackenzie 26 March.

Bell, Tracy, Araujo, Magali, Luo, Zaiming, Tomlinson, James, 2018. Regulation of fluid reabsorption in rat or mouse proximal renal tubules by asymmetric dimethylarginine and dimethylarginine dimethylaminohydrolase 1. American Journal of Physiology 315(1): F74–F78, PMCID: PMC6087787. doi:10.1152/ajprenal.00560.2017.

Bijian, T., Donghai, W., Xiang, Z., Tao, Z., Wenqian, Z., Hong, W., 2017. The observed impacts of wind farms on local vegetation growth in northern China. Remote Sens 9, 332.

BNEF, 2021. "Wind trade and manufacturing: a deep dive", February, www.csis.org/analysis/industrial-policytrade-and-clean-energy-supply-chains.

Cappucci, M., 2018. Explainer: Winds and where they come from. https://www.sciencenewsforstudents.org/article/explainer-winds-and-where-they-come.

Center for Sustainable Systems, 2020. Wind energy factsheet. https://css.umich.edu/factsheets/wind-energy-factsheet

Conkling, Tara, Loss, Scott, Diffendorfer, Jay, 2021. Limitations, lack of standardization, and recommended best practices in studies of renewable energy effects on birds and bats. Conservation Biology 35, 64–76. doi:10.1111/cobi.13457.

Council, G.W.E., 2021. A gust of growth in China makes 2020 a record year for wind energy. https://Gwec.Net/. https://gwec.net/a-gust-of-growth-in-china-makes-2020-a-record-year-for-wind-energy/.

Dai, K., Bergot, A., Liang, C., Xiang, W.-N., Huang, Z., 2015. Environmental issues associated with wind energy–a review. Renew. Energy 75, 911–921.

Daniel, T.C., 2020. The future of wind power is looking bright, IRENA says. https://www.sustainability-times.com/low-carbon-energy/the-future-of-wind-power-is-looking-bright-energy-agency-says/.

Deep, S., Sarkar, A., Ghawat, M., MK, R., 2020. Estimation of the wind energy potential for coastal locations in India using the Weibull model. Renewable Energy 161, 319–339.

Dhar, A., Anne Naeth, M., Jennings, P.D., El-Din, M.D, 2020. Perspectives on environmental impacts and a land reclamation strategy for solar and wind energy systems. Sci. Total Environ. 718 (2020), 134602. https://doi.org/10.1016/j.scitotenv.2019.134602.

Dincer, I., 2018. Energy Production. https://www.sciencedirect.com/topics/engineering/savonius-wind-turbine.

Energy, W., 2020. "China's offshore wind energy industry post-2021", 23 October, www.world-energy.org/article/13292.html.

EPE, 2016. O compromisso do brasil no combate às mudanças climáticas: produção e uso de energia. Tech. rep. Rio de Janeiro. Empresa de Pesquisa Energética, MME, Brasil.

Farina, A., Anctil, A., 2022. Material consumption and environmental impact of wind turbines in the USA and globally. Resourc. Conserv. Recycl. 176 (2022), 105938. https://doi.org/10.1016/j.resconrec.2021.105938.

Gonçalves, S., Rodrigues, T.P., Chagas, A.L.S., 2020. The impact of wind power on the Brazilian labor market. Renewable Sustainable Energy Rev. 128 (2020), 109887. https://doi.org/10.1016/j.rser.2020.109887.

González, M.O.A., Santiso, A.M., de Melo, D.C., de Vasconcelos, R.M., 2020. Regulation for offshore wind power development in Brazil. Energy Policy 145 (2020), 111756. https://doi.org/10.1016/j.enpol.2020.111756.

Gul, M., Tai, N., Huang, W., Nadeem, M.H., Yu, M., 2019. Assessment of wind power potential and economic analysis at Hyderabad in Pakistan: powering to local communities using wind power. Sustainability 11, 1391. https://doi.org/10.3390/su11051391.

GWEC, 2011. Indian Wind Energy Outlook, 2011. Brussels.

GWEC, 2018. Global Wind Report 2018. Global Wind Energy Council, Brussels, Belgium Tech. rep.

GWEC, 2020a. https://gwec.net/gwec-in-2020/. (Accessed September 2021).

GWEC, 2021. Global Wind Report 2021. GWEC, Brussels.

Hartmann, P., Apaolaza-Ibáñez, V., 2012. Consumer attitude and purchase intention toward green energy brands: the roles of psychological benefits and environmental concern. J Bus Res 65, 1254–1263.

Henningsson, M., Jönsson, S., Ryberg, J.B., Bluhm, G., Bolin, K., Bodén, B., et al., 2013. The Effects of Wind Power on Human Interests: A Synthesis. Swedish Environmental Protection Agency, Stockholm.

Hu, W., Liu, Z., Tan, J., 2020. Thermodynamic Analysis of Wind Energy Systems. Wind Solar Hybrid Renewable Energy Syst. doi:10.5772/intechopen.85067. https://doi.org/10.5772/intechopen.85067.

IEA, 2021. https://www.iea.org/reports/world-energy-outlook-2021. (Accessed October 2021).

IRENA, 2021a. Renewable Capacity Statistics 2021. IRENA, Abu Dhabi https://irena.org/publications/2021/March/Renewable-Capacity-Statistics-2021.

IRENA, 2021b. World Energy Transitions Outlook: 1.5°C Pathway. IRENA, Abu Dhabi.

IRENA and ILO, 2021. Renewable Energy and Jobs–Annual Review 2021,. International Renewable Energy Agency, International Labour Organization, Abu Dhabi, Geneva.

Irfan, M., Hao, Y., Panjwani, M.K., Khan, D., Chandio, A.A, Li, H., 2020. Competitive assessment of South Asia's wind power industry: SWOT analysis and value chain combined model. Energy Strategy Rev. 32 (2020), 100540. https://doi.org/10.1016/j.esr.2020.100540.

Jacobson, R., 2016. Where Do Wind Turbines Go To Die? | Inside Energy. http://insideenergy.org/2016/09/09/where-do-wind-turbines-go-to-die/katekar.

Jagadeesh, A., 2000. Wind Energy development in Tamil Nadu and Andhra Pradesh, India Institutional dynamics and barriers—a case study. Energy Pol. 28 (2000), 157–168.

Kaldellis, J.K., 2012. Wind energy - introduction. Compr. Renew. Energy. https://doi.org/10.1016/B978-0-08-087872-0.00201-8.

Katekar, V.P., Asif, M., Deshmukh., S.S., 2021. Energy and Environmental Scenario of South Asia. In: Asif, M. (Ed.), Energy and Environmental Security in Developing Countries. Advanced Sciences and Technologies for Security Applications. Springer, Cham https://doi.org/ https://doi.org/10.1007/978-3-030-63654-8_4.

Katsaprakakis, D.A., 2012. A review of the environmental and human impacts from wind parks. A case study for the prefecture of Lasithi. Crete. Renew. Sustain. Energy Rev. 16 (5), 2850–2863. https://doi.org/10.1016/j.rser.2012.02.041.

Khan, I., 2020. Sustainability challenges for the south Asia growth quadrangle: a regional electricity generation sustainability assessment. J. Cleaner Prod. 243 (118639), 1–13. https://doi.org/10.1016/j.jclepro.2019.118639.

Kondili, E.M., 2021. Environmental Impacts of Wind Power. Compr. Renew. Energy. https://doi.org/10.1016/B978-0-12-819727-1.00158-8.

Koschinski, Sven, Lüdemann, Karin, 2020. Noise mitigation for the construction of increasingly large offshore wind turbines. Semantic Scholar, Corpus ID: 221835230.

Kowalczyk, Szumilas, 2020. Long-term visual impacts of aging infrastructure: Challenges of decommissioning wind power infrastructure and a survey of alternative strategies. Renewable Energy.150 4. doi:10.1016/j.renene.2019.12.143.

Lee, S.Y., Hamilton, S., Barbier, E.B., Primavera, J., Lewis, R.R., 2019. Better restoration policies are needed to conserve mangrove ecosystems. Nat. Ecol. Evol. 3, 870.

Li, D., Miao, S., 2021. Fitting the wind speed probability distribution with Maxwell and power Maxwell distributions: a case study of North Dakota sites. Sustain. Energy Technol. Assess. 47, 101446.

Lintott, P.R., Richardson, S.M., Hosken, D.J., Fensome, S.A., Mathews, F., 2016. Ecological impact assessments fail to reduce risk of bat casualties at wind farms. Curr. Biol. 26, R1135–R1136.

Loss, S.R., Will, T., Marra, PP., 2013. Estimates of bird collision mortality at wind facilities in the contiguous United States. Biol. Conserv. 168, 201–209.

Marques, A.T., Santos, C.D., Hanssen, F., Muñoz, A.-R., Onrubia, A., Wikelski, M., Moreira, F., Palmeirim, J.M., Silva, J.P., 2020. Wind turbines cause functional habitat loss for migratory soaring birds. J. Anim. Ecol. 89, 93–103.

Mathew, S., Philip, G.S., 2012. 2.05 - Wind turbines: evolution, basic principles, and classifications. Compr. Renew. Energy 2, 93–111.

Mathew, S., 2006. Wind Energy Fundamentals, Resource Analysis and Economics. Springer-Verlag, Berlin and Heidelberg.

McNew, L.B., Hunt, L.M., Gregory, A.J., Wisely, S.M., Sandercock, B.K., 2014. Effects of wind energy development on nesting ecology of greater prairie-chickens in fragmented grasslands. Conserv. Biol. 28, 1089–1099.

Mordor Intelligence, 2020. Small Wind Turbine market - Growth, trends, Covid-19 imoact and forecasts, (2021–2026). Available at: https://www.mordorintelligence.com/industry-reports/small-wind-turbine-market. (Accessed: 1 October 2021).

Nadour, M., Essadki, A., Fdaili, M., Nasser, T., 2018. Advanced Backstepping Control of a Wind Energy Conversion System Using a Doubly-Fed Induction Generator. In: Proceedings of 2017 International Renewable and Sustainable Energy Conference, IRSEC 2017.

National Geographic Society, 2021. https://www.nationalgeographic.org/society/. (Accessed November 2021).

Nazir, M.S., Ali, N., Bilal, M., Iqbal, H.M.N., 2020. Potential environmental impacts of wind energy development: a global perspective. Curr. Opin. Environ. Sci. Health 13, 85–90. https://doi.org/10.1016/j.coesh.2020.01.002.

Nazir, M.S., Mahdi, A.J., Bilal, M., Sohail, H.M., Ali, N., Iqbal, H.M., 2019. Environmental impact and pollution-related challenges of renewable wind energy paradigm–a review. Sci. Total Environ. 683, 436–444.

NCTCE, 2019. How is wind energy sustainable? https://nctce.com.au/how-is-wind-energy-sustainable/.

Nordmann, A., 2021. Wind Turbine Design. https://en.wikipedia.org/wiki/Wind_turbine_design.

Nouh, A., Mohamed, F., 2014. Wind energy conversion systems: classifications and trends in application. In: IREC 2014 - 5th International Renewable Energy Congress.

Olson-Hazboun, S.K., Krannich, R.S., Robertson, P.G., 2016. Public views on renewable energy in the Rocky Mountain region of the United States: distinct attitudes, exposure, and other key predictors of wind energy. Energy. Res. Soc. Sci. 21, 1–179. https://doi.org/10.1016/j.erss.2016.07.002.

Paraschiv, L.S., Paraschiv, S., Ion, IV., 2019. Investigation of wind power density distribution using Rayleigh probability density function. Energy Procedia 157, 1546–1552.

Paul, M., 2021. Wind power capacity needs to grow at thrice the current speed to reach net zero: report. https://www.downtoearth.org.in/news/energy/wind-power-capacity-needs-to-grow-at-thrice-the-current-speed-to-reach-net-zero-report-76188.

Peri, E., Tal, A., 2020. A sustainable way forward for wind power: Assessing turbines' environmental impacts using a holistic GIS analysis. Appl. Energy 279 (2020), 115829. https://doi.org/10.1016/j.apenergy.2020.115829.

Renewables, 2021. How windy does it have to be? Windpower Learning Centre. https://www.renewablesfirst.co.uk/windpower/windpower-learning-centre/.

Renewables, N., 2021b. Brazil connects 1.4 GW of wind farms in H1 2021. https://renewablesnow.com/news/brazil-connects-14-gw-of-wind-farms-in-h1-2021-746680/.

Rennkamp, B., Perrot, R., 2016. Drivers and barriers to wind energy technology transitions in India, Brazil and South Africa. Handbook on Sustainability Transition and Sustainable Peace. Springer, Cham, pp. 775–791.

Rodríguez, I., Caldés, N., Garrido, A., De La Rúa, C., Lechón, Y., 2017. Socioeconomic, Environmental, and Social Impacts of a Concentrated Solar Power Energy Project in Northern Chile. In:

Rydell, J., Engström, H., Hedenström, A., Larsen, J.K., Pettersson, J., Green, M. (2012). The effect of wind power on birds and bats. A synthesis. Report, 6511.

Saidur, R., Rahim, N.A., Islam, M.R., Solangi, K.H., 2011. Environmental impact of wind energy. Renew. Sust. Energ. Rev. 15, 2423–2430. https://doi.org/10.1016/j.rser.2011.02.024.

Salameh, Z., 2014. Renewable Energy System Design, 1st Ed Elsevier, Amsterdam, Netherlands. https://doi.org/10.1016/B978-0-12-819727-1.00152-7.

Sanders, Alan, Beecham, Gary, 2017. Genome-Wide Association Study of Male Sexual Orientation. Scientific Report 7, 16950 (2017). https://doi.org/10.1038/s41598-017-15736-4.

Sayed, E.T., Wilberforce, T., Elsaid, K., Rabaia, M.K.H., Abdelkareema, M.A., Chae, K.-J., Olabi, A.G., 2021. A critical review on environmental impacts of renewable energy systems and mitigation strategies: wind, hydro, biomass and geothermal. Sci. Total Environ. 766 (2021), 144505. https://doi.org/10.1016/j.scitotenv.2020.144505.

Sayigh A. (Ed.), Mediterranean Green Buildings & Renewable Energy. Springer, Cham. https://doi.org/10.1007/978-3-319-30746-6_68.

Scheer, H., 1995. Solar energy's economic and social benefits. Sol. Energy Mater. Sol. Cells. 38, 555–568. https://doi.org/10.1016/0927-0248(94)00243-6.

Schmidt, J.H., Klokker, M., 2014. Health effects related to wind turbine noise exposure: a systematic review. PLoS One 9 (12), e114183.

Sebestyén, V., 2021. Renewable and sustainable energy reviews: environmental impact networks of renewable energy power plants. Renewable Sustainable Energy Rev. 151 (2021), 111626. https://doi.org/10.1016/j.rser.2021.111626.

Serban, A., Paraschiv, L.S., Paraschiv, S., 2020. Assessment of wind energy potential based on Weibull and Rayleigh distribution models. Energy Rep. 6, 250–267.

Shahan, Z., 2014. History of Wind Turbines - Renewable Energy World. Renewable Energy World. https://www.renewableenergyworld.com/storage/history-of-wind-turbines/#gref.

Shang, L., 2010. Wind farm's environment impacts on Hexi Corridor. J. Environ. Res. Monit. 1, 3–5.

Sharma, N.K., Tiwari, P.K., Sood, Y.R., 2012. Solar energy in India: strategies, policies perspectives, and future potential. Renew. Sust. Energ. Rev. 16 (1), 933–941. https://doi.org/10.1016/j.rser.2011.09.014.

Simonetti, D.S.L., Amorim, A.E.A., Oliveira, F.D.C., 2018. Doubly fed induction generator in wind energy conversion systems. Adv. Renew. Energies Power Technol 462–489. https://doi.org/10.1016/B978-0-12-812959-3.00015-0.2918281.

Smedley, A.R., Webb, A.R., Wilkins, AJ., 2010. Potential of wind turbines to elicit seizures under various meteorological conditions. Epilepsia 51 (7), 1146–1151.

Soares, I.N., Gava, R., de Oliveira, J.A.P., 2021. Political strategies in energy transitions: exploring power dynamics, repertories of interest groups and wind energy pathways in Brazil. Energy Res. Soc. Sci. 76 (2021), 102076. https://doi.org/10.1016/j.erss.2021.102076.

Sovacool, B.K., 2013. The avian benefits of wind energy: a 2009 update. Renew. Energy. 49, 19–24. https://doi.org/10.1016/j.renene.2012.01.074.

Spellman, F.R., 2014. Environmental Impacts of Renewable Energy. CRC press, Boca Raton, Florida, United States, ISBN 9781482249460.

Statistical Review of World Energy, 2021. The statistical review of world energy analyses data on world energy markets from the prior year. The Review has been providing timely, comprehensive and objective data to the energy community since 1952. https://www.bp.com/en/global/corporate/energy-economics/statistical-review-of-world-energy.html.

Swofford, J., Slattery, M., 2010. Public attitudes of wind energy in Texas: local communities in close proximity to wind farms and their effect on decisionmaking. Energy Policy 38 (5), 2508–2519. https://doi.org/10.1016/j.enpol.2009.12.046.

Tabassum-Abbasi, Premalatha, M., Abbasi, T., Abbasi, S.A., 2014. Wind energy: increasing deployment, rising environmental concerns. Renew. Sust. Energ. Rev. 31, 270–288. https://doi.org/10.1016/j.rser.2013.11.019.

Tang, B., Wu, D., Zhao, X., Zhou, T., Zhao, W., Wei, H., 2017. The observed impacts of wind farms on local vegetation growth in northern China. Remote Sens. 9, 332. https://doi.org/10.3390/rs9040332.

Tasneem, Z., et al., 2020. An analytical review on the evaluation of wind resource and wind turbine for urban application: prospect and challenges. Dev. Built Environ. 4 (November), 1–37. https://doi.org/10.1016/j.dibe.2020.100033.

Taylor, R.H., 1984. Wind power technology. Phys. Bull 35. https://doi.org/10.1088/0031-9112/35/9/023.

Twidell, J., Weir, T., 2021. Wind power technology. Renew. Energy Resour. https://doi.org/10.4324/9781315766416-16.

UN Sustainable Development Goals. 2021. Ensure access to affordable, reliable, sustainable and modern energy for all. https://sdgs.un.org/goals/goal7

US Department of Energy, 2021. https://www.energy.gov/. (Accessed November 2021).

Vasar, C., Prostean, O., Filip, I., Szeidert, I., 2018. Wind energy conversion system-a laboratory setup. In: SACI 2018 - IEEE 12th International Symposium on Applied Computational Intelligence and Informatics, Proceedings, pp. 313–317.

Voicescu, S.A., Michaud, D.S., Feder, K., Marro, L., Than, J., Guay, M., et al., 2016. Estimating annoyance to calculated wind turbine shadow flicker is improved when variables associated with wind turbine noise exposure are considered. J. Acoust. Soc. Am. 139 (3), 1480–1492.

Wang, L., Liu, J., Qian, F., 2021. Wind speed frequency distribution modeling and wind energy resource assessment based on polynomial regression model. Int. J. Electr. Power Energy Syst. 130, 106964.

Wang, Chien, Prinn, Ronald, 2010. Potential Climatic Impacts and Reliability of Very Large-Scale Wind Farms. Atmospheric Chemistry and Physics Discussions. doi:10.5194/acp-10-2053-2010.

Weiss, C.V.V., Ondiviel, B., Guinda, X., Jesus, F.D., Gonzalez, J., Guanche, R., Juanes, J.A., 2018. Co-location opportunities for renewable energies and aquaculture facilities in the Canary Archipelago. Ocean. Coast. Manage. 62–71. https://doi.org/10.1016/j.ocecoaman.2018.05.006.

WNA, 2011. Comparison of Lifecycle Greenhouse Gas Emissions of Various Electricity Generation. World Nuclear Association, London.

WWF, 2015. https://www.worldwildlife.org/. (Accessed September 2021).

Yazdani, S., Deymi-Dashtebayaz, M., Salimipour, E., 2019. Comprehensive comparison on the ecological performance and environmental sustainability of three energy storage systems employed for a wind farm by using an emergy analysis. Energy Convers. Manage. 191 (2019), 1–11. https://doi.org/10.1016/j.enconman.2019.04.021.

Zafirakis, D.P., 2021. Energy yield of contemporary wind turbines. Compr. Renew. Energy. https://doi.org/10.1016/B978-0-12-819727-1.00152-7.

Zhu, Na, Zhang, Dingyu, Wang, Wenling, 2020. A Novel Coronavirus from Patients with Pneumonia in China, 2019. The New England Journal of Medicine, PMCID: PMC7092803 doi:10.1056/NEJMoa2001017.

CHAPTER 7

Substituting coal with renewable biomass for electricity production using co-gasification technique: A short-term sustainable pathway for developing countries

M. Shahabuddin[a,b] and Sankar Bhattacharya[a]

[a]Department of Chemical Engineering, Monash University, Clayton, Australia
[b]Department of Mechanical and Product Design Engineering, Swinburne University of Technology, Hawthorn, Australia

7.1 Introduction

Although gasification of coal is a mature technology, recent research interest has focused on the co-gasification of biomass and coal to substitute coal partially. Because it is a long way to phase out coal-based power generation, especially from developing countries having huge electricity demand (Shahabuddin et al., 2021). The co-gasification of biomass and coal has several advantages over individual gasification of coal or biomass. The advantages include higher carbon conversion, high-quality syngas, less emission, and substitution of coal with renewable biomass (Kumabe et al., 2007). The higher carbon conversion in co-gasification is the result of the synergistic effect due to the catalytic alkali and alkaline earth materials (AAEM) present in biomass ash. However, the catalytic effect depends on the gasifier's operating conditions, such as temperature and pressure (Shahabuddin and Bhattacharya, 2021c). When the gasification temperature is significantly above the melting point of AAEM, the AAEM are evaporated, resulting in an insignificant effect in increasing the gasification rate.

Moreover, if the operating temperature is within the range of the sintering temperature of ash, the AAEM may adversely affect the gasification reactivity by blocking the micropore of the char (Tanner et al., 2016; Ye et al., 1998). Biomass contains high volatile matter but low carbon content compared to coal. Besides, the oxygen content of biomass is significantly higher than that of coal. The biomass properties favor the char reactivity and syngas quality (Taba et al., 2012; Fermoso et al., 2009). Also, the co-gasification decreases tar formation and increases the H_2/CO ratio in the yield gas. The syngas with a higher H_2/CO ratio from co-gasification favors the synthesis of many chemicals (Kumabe et al., 2007; Luque and Speight, 2015).

The concept of co-gasification of biomass with coal has been introduced at the beginning of 21st century, mainly to replace a proportion of coal from gasification and making gasification technology more environmentally friendly. A single study concerning the co-gasification of biomass and coal was found as early as 2000 (de Jong et al., 1999). In that study, biomass and coal were gasified in a 1.5 MW_{th} power plant using a pressurized fluidized bed gasifier. In 2000, the co-gasification behavior of pine chip and low-grade black coal was tested by Pan et al. (2000) using a fluidized bed gasifier with a mixture of air and steam as a reactant. Several co-gasification studies using coal and biomass have been conducted in the laboratory scale in the last one and half decades, predominantly using fixed, and fluidized bed gasifiers (Chmielniak and Sciazko, 2003a; Usón et al., 2004; Vélez et al., 2009; Hernández et al., 2010; Taba et al., 2012; Jeong et al., 2015; Tursun et al., 2016; Zhang et al., 2016; Ali et al., 2017). Despite having some commercial co-gasification plants worldwide using different biomasses, the co-gasification of coal and biomass is not well developed (Fernando, 2009). According to the US world gasification database (Higman, 2017), there is no commercial plant-available globally for the co-gasification of coal and biomass. However, some commercial integrated gasification combined cycle (IGCC) plants are under development to co-gasify coal and biomass to produce power and chemicals (Chmielniak and Sciazko, 2003b).

The low energy density of biomass resulting in the yield of low calorific value (CV) syngas makes them incompatible for the gasification individually. Hence, biomass co-gasification with high-rank coal is a promising approach, making gasification technology environmentally friendly by reducing emissions (Taba et al., 2012). Nevertheless, co-gasification of biomass and coal encounters some challenges, including nonuniform particle size due to the fibrous structure of biomass, a marked difference in gasification temperature between biomass and coal particle and the selection of optimized process and operating conditions (Brar et al., 2012). Co-gasification behavior is further influenced by the type of gasifier, temperature, types and concentration of the gasifying agent, blending ratio of coal and biomass, and physicochemical properties of the feedstocks. However, there are limited review articles in the scientific literature considering the above issues. Hence, these aspects are reviewed and analyzed to propose some recommendations, which will set pathways for the advancement of co-gasification study to partially substitute coal with biomass for power generation.

Coal is the most reliable and cheapest feedstock for electricity generation, meeting 36.4% of the world electricity demand (Energy, 2020). Conversely, renewable energy is still expensive, and its utilizations depend on many factors, including the availability of land and weather. Therefore, it is still a long way to implement renewable energy as a mainstream energy source in densely populated developing countries like Bangladesh to meet the tremendous energy and power demand. Hence, the co-gasification of domestic coal and biomass might be a viable choice, at least in the intermediate term.

7.2 Status of coal and biomass as energy sources

Globally, total energy consumption is about 15.0 Gigatonne of oil equivalent (Gtoe) at present, which is forecasted to be 19.0 Gtoe by 2040 . Coal and biomass collectively meet around 40% of the world energy demand (BP Statistical Review). Out of which, 29% is produced from coal, whereas 9.8% is generated from biomass (Association, 2018). As a dominant source of fossil fuel, coal will continue to play a significant role in meeting the increasing energy demand, at least in the next few decades (Quyn et al., 2002). Likewise, the use of biomass is forecasted to rise by 51.5% by 2040 (Panwar et al., 2011). The global reserve of coal is 841 Billion Tonnes (BT), whereas yearly, about 220 BT of biomass is produced worldwide. Globally, 7.8 BT of coal is consumed per year to produce 228.6 Exajoule (EJ) of energy, in contrast to 19.5 BT of biomass is used to produce 55.0 EJ of energy (World Energy Council Report; Stucley et al., 2004).

7.3 Power and energy from coal and biomass in Bangladesh

Currently, the total installed power generation capacity in Bangladesh is 22,000 MW. However, as of July 2021, the peak generation is less than half (10,300 MW) of the installed capacity (BPDB, 2021). According to the Power system master plan 2016, the Bangladesh government plans to increase coal-based power generation by 25% from its current 2.1%. Hence, by 2040, the total electricity generation from coal will be 25,500 MW from its current production of 1768 MW, which is the share of 8% (BPDB, 2021). Several coal-based power plants are under construction to meet the increasing power demand. Payra 1320 MW Thermal Power Plant (1st Phase) in Patuakhali district has started its commercial production in December 2020. In this plant, about 4.12 million tons of coal will be consumed per year (BPCL, 2021). In the second phase, the same company is building another plant with the same capacity. Several other plants, such as Rampal thermal power plants in Bagerhat districts (IEA clean coal center), Banshkhali power station (1224 MW) in Chattogram and Matarbari coal-fired power plant (1200 MW) in Cox's Bazar district are in the final stage. All these planned power plants will be operated by importing coal from China, Australia and Indonesia. However, due to the country's commitment to the climate vulnerable forum in generating about 40% electricity from different renewable sources by 2041, recently, government of Bangladesh has decided to drop 10 coal-fired power plants from the PSMP (UNB, 2021).

Barapukuria coal mine- so far, the only economically feasible coal mine in Bangladesh has a reserve of 390 million tons. Coal from this mine is predominantly used in the Barapukuria thermal power plant with a production capacity of 400 MW. However, there have been no studies assessing the gasification characteristics of any Bangladeshi coal. Some studies concerning the physicochemical properties of Barapukurian coal and ash have been found in the literature (Howladar and Islam, 2016; Safiullah et al., 2011; Khan et al., 2013; Bhuiyan et al., 2014). In addition, a study conducted by Chowdhury et al. (2014) reported the energy and material balance under partial oxidation conditions for power generation application.

As an agricultural-based country, Bangladesh has considerable biomass potential. Yearly, 213.8 million tons (MT) of biomass is produced in the country, meeting 70% of the total energy (1345 petajoules) demand (Masud et al., 2019a). The primary biomass resources in Bangladesh

TABLE 7.1 Potential biomass resources in Bangladesh (Masud et al., 2019a; Halder et al., 2015).

Biomass sources	Production (MT/year)	Energy content (PJ)	Electricity generation (TWh)
Agricultural residue	94.1	582.3	161.8
Livestock residue	88.9	456.4	126.8
Forest residue	17.4	210.6	58.5
Municipal solid waste	13.4	95.6	26.6
Total	213.8	1344.9	373.7

include agricultural residues (rice husk, straw), wood chips, leaves, and cow dung, mainly used for cooking, particularly in the rural area.

Table 7.1 lists major biomass resources in Bangladesh and their corresponding energy potential. These biomass resources are converted into energy using three main routes: conventional cooking, biochemical and thermochemical conversions (Masud et al., 2019b). The thermochemical conversion includes pyrolysis and gasification. Several pilot-scale pyrolysis and gasification plants are under operation across the country, predominantly using rice husk with a capacity of 100 to 200 kW. The total power capacity from these plants is estimated to be around 50 to 100 MW (Masud et al., 2019a).

7.4 Greenhouse gas emission

According to Paris agreement in the international climate policy conference organized by the United Nations in 2015, the global temperature must hold sufficiently below two degrees above the preindustrial level. Also, effort should be made to limit the increase by 1.5°C above the preindustrial level, potentially reducing the global risk of negative climate change (UNFCCC, 2015). Considering different sectors, power generation contributes 40% global CO_2 emission, this is followed by industry 25%, transport 21.2%, building 8.5% and others 5.6% (Shahabuddin and Bhattacharya, 2021a). As shown in Fig. 7.1, the total CO_2 emission is in the world is 34.8 Bt, accounting 40 % from coal, 32 % from oil, 21% from gas, and 7% from cement, flaring, and others (Friedlingstein et al., 2021). Emission from all fuel types have been rising since 1960 except a drop in 2020 due to the COVID-19 pandemic. The global CO_2 concentration in the atmosphere has increased to 412.5 ppm, which is about 50% higher than preindustrial level (Cozzi et al., 2021). Globally, there are 40 Mt CO_2 capturing capacity from the 22 large-scale carbon capture and sequestration (CCS) plants against the total emission of over 36.2 billion tonnes (Jos et al., 2016; World Energy Council Report).

The cost of CO_2 is associated with its capture, transportation and storage. Furthermore, these costs also depend on the infrastructure, transportation distance, available disposal site, and capture or mitigating technology (Hossain and de Lasa, 2008). Currently, three major technologies are used for different types of plants. These technologies include an amine-based technique for CO_2 capturing from pulverized coal (PC) and Natural Gas Combined Cycle (NGCC) plants, a physical sorbent based system in Integrated Gasification Combined Cycle

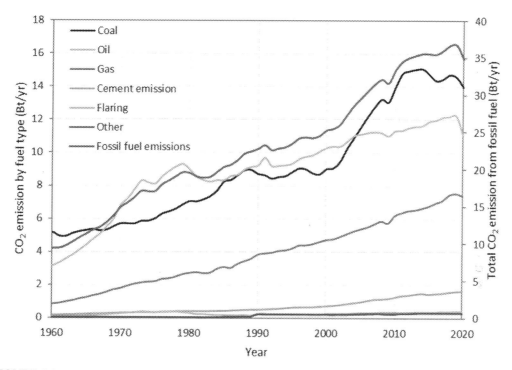

FIGURE 7.1 Global CO$_2$ emissions concerning fuels types (Friedlingstein et al., 2021).

(IGCC) plant, and an oxy-fuel capturing technique in the oxy-combustion system (Rubin et al., 2015). Table 7.2 shows the cost associated with capture, storage and transport using current technologies for the natural gas combined cycle (NGCC), supercritical pulverized coal (SCPC) and integrated gasification combined cycle (IGCC) power plants. The cost in $/tonne CO$_2$ can be calculated by using the following formula (David and Herzog, 2000):

$$\text{The cost of mitigating CO}_2 \text{ (\$/tonne)} = \frac{(COE)_{capture} - (COE)_{reference}}{(E)_{capture} - (E)_{reference}} \quad (7.1)$$

Here COE represents the cost of electricity in US $/kWh, and E represents the production of CO$_2$ per kWh of electricity produced (tonne CO$_2$/kWh).

According to Table 7.2, it can be seen that the levelised cost of electricity (LCOE) for coal-based power plants (SCPC and IGCC) is relatively higher than that of natural gas-based power plants (NGCC). However, the electricity for NGCC and IGCC are quite similar, which is much lower than SCPC.

7.5 Why co-gasification?

The main limitation of coal-based power production is its high greenhouse gas (GHG) emission, resulting in global warming and negative climate change. The CO$_2$ emission from

TABLE 7.2 The cost (in USD) associates with capture, storage, and transport using current technologies for new power plants (Rubin et al., 2015).

Cost and performance parameters	NGCC postcombustion capture	SCPC postcombustion capture	IGCC precombustion capture
Reference plant without CCS: LCOE ($/MWh)	42–83	61–79	82–99
Power plant with CCS:			
Increased fuel requirement/MWh (%)	13–18	21–44	20–35
CO_2 captured (kg/MWh)	360–390	830–1080	840–940
CO_2 avoided (kg/MWh)	310–330	650–720	630–700
% CO_2 captured	88–89	86–88	82–88
Power plant with capture, transport and storage:			
LCOE ($/MWh)	63–122	95–150	112–148
Electricity cost increase for CCS ($/MWh)	19–47	31–71	25–53
% increase	28–72	48–98	26–62

coal power plants ranges from 670 to 850 kg/MWh, depending on the technology used (Moazzem et al., 2012). As of July 2021, Bangladesh's total power generation capacity using coal is 1768 MW (BPDB, 2021). Hence, burning of sub-bituminous/bituminous coal using supercritical technology (BPCL, 2021) emits 1139 tons of CO_2, 115 tons of SO_2, and 7.25 tons of NO_x per hour.

Furthermore, the land and water are being affected by the solid residue of the coal power plants (Hossain et al., 2015). Besides, the limitation of conventional coal-based power plants is their lower thermal efficiency. Globally, 75% of the coal-based power plants are operated under subcritical technology with an efficiency of about 30%.

Thus, it is essential to utilize advanced technology for coal-based power production to increase efficiency and reduce emissions. One such technology is gasification, which can increase the thermal efficiency by up to 45% and simultaneously reduce CO_2 emission by 32% (Bhattacharya et al., 2013). In addition, the most significant advantage of gasification is the generation of synthetic gas, which can be used for the production of a range of chemicals. However, coal gasification still suffers from various issues, including low carbon conversion, mainly using high-rank coal, relatively high greenhouse gas emissions, tar formation, and operational issues due to slagging and high moisture content (Brar et al., 2012).

Several studies have tried to address these issues under catalytic gasification conditions (Wood and Sancier, 1984; Hauserman, 1994; Yeboah et al., 2003; Tang and Wang, 2016). However, this technique has not been commercialized because of the expensive catalyst and

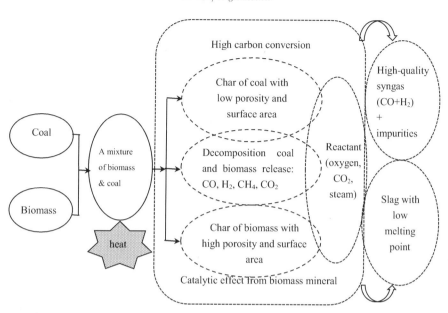

FIGURE 7.2 Principle of co-gasification of biomass and coal.

low lifetime. Hence, recent research interest has focused on co-gasification of coal and biomass, capable of providing synergistic effects, high carbon conversion, improved gas quality, and low carbon footprint (Zhang et al., 2016). Also, biomass is renewable, producing power with up to 97% lower CO_2 emission than coal (Stucley et al., 2004). However, fuel-specific data for the co-gasification of particular coal and biomass are required.

Bangladesh has a significant amount of renewable lignocellulose biomass, which can potentially substitute coal partially. However, fuel-specific data to assess coal and lignocellulose biomass feasibility is essential under practically relevant reactant atmosphere, including O_2, CO_2, and steam (Shahabuddin and Bhattacharya, 2021b). The major steps concerning the co-gasification of biomass and coal are shown in Fig. 7.2, while the major reactions involved in gasification/co-gasification are illustrated in Table 7.3.

Besides the catalytic effect from AAEM in biomass, the higher reactivity in co-gasification is the result of a higher heating rate due to the evolution of combustible pyrolysis products (Mallick et al., 2017). Mallick et al. (2018) showed that the yield of char decreases by 12 to 14%-point due to copyrolysis compared to that of individual pyrolysis of biomasses. The synergistic effect helps to enhance the energy efficiency by increasing the carbon conversion besides altering the syngas composition (Mallick et al., 2017). Also, a significant reduction in tar formation is reported. Shahabuddin et al. (Shahabuddin and Bhattacharya, 2021c) studied the co-gasification characteristics of Bangladeshi Barapukurian Bituminous coal and pine bark biomass in an entrained flow gasifier. Results showed that increasing the biomass concentration in the blend increases the carbon conversion and syngas quality besides reducing emission. The blending of 20% biomass with coal increased the carbon conversion by 21.5, and 4.5%-point at 1000, and 1400°C respectively.

TABLE 7.3 Gasification/co-gasification reactions and corresponding enthalpy (Shahabuddin and Bhattacharya, 2021d).

Reaction	Reaction equation	Enthalpy
Partial combustion	$C_{(s)} + \frac{1}{2}O_2 \rightarrow CO$	$\left(\Delta H = -123 \frac{KJ}{mol}\right)$
Boudouard reaction	$C_{(s)} + CO_2 \rightarrow 2CO$	$\left(\Delta H = +159.7 \frac{KJ}{mol}\right)$
Steam gasification	$C_{(s)} + H_2O \rightleftharpoons CO + H_2$	$\left(\Delta H = +118.9 \frac{KJ}{mol}\right)$
Hydrogen combustion	$C_{(s)} + 2H_2 \rightleftharpoons CH_4$	$\left(\Delta H = -88.4 \frac{KJ}{mol}\right)$
Water-gas shift reaction	$CO + H_2O \rightleftharpoons CO_2 + H_2$	$\left(\Delta H = -41.1 \frac{KJ}{mol}\right)$
Steam methane reforming	$CH_4 + H_2O \rightleftharpoons CO + 3H_2$	$\left(\Delta H = +206.3 \frac{KJ}{mol}\right)$
Dry methane reforming	$CH_4 + CO_2 \rightleftharpoons 2CO + 2H_2$	$\left(\Delta H = -247 \frac{KJ}{mol}\right)$
Sabatier reaction	$CO_2 + 4H_2 \rightleftharpoons CH_4 + 2H_2O$	$\left(\Delta H = +73 \frac{KJ}{mol}\right)$

7.6 Selection of biomass for the co-gasification with coal

A range of biomass has been used in the literature for the co-gasification with coal. The critical properties of the commonly used biomass are shown in Table 7.4. It is observed that there is no significant variation in terms of the properties of biomass. Most biomass possesses high moisture content and volatile matter but relatively low fixed-carbon ash. Also, the calorific value of biomass is significantly lower than coal (Shahabuddin and Bhattacharya, 2021c; Xu, 2018).

Based on the ultimate data, it is clear that biomass is enriched with oxygen, which helps accelerate the gasification rate, resulting in higher carbon conversion. Moreover, high hydrogen content in the biomass feedstock leads to syngas with a higher H_2/CO ratio. In contrast, the low N_2 and S contents in biomass are favorable concerning pollutant emission.

7.7 Gasification technologies

Gasification is the platform of cleaner coal technology, which include integrated gasification combined cycle (IGCC) and zero-emission coal (ZEC) (Yan et al., 2013). The history of gasification is over 200 years. However, the application of gasification technology has increased enormously, particularly from the beginning of the 21st century under different names. For example, gasification technology is used as an Absorption Enhanced Reforming (AER) in Germany (Weimer et al., 2002), Advanced Gasification-Combustion (AGC) (Rizeq et al., 2001; Yan et al., 2013), and the Zero Emission Coal (ZEC) in the USA (Ziock et al., 2001; Boshu et al., 2008; Yan et al., 2012), Hydrogen Production Reaction Integrated Novel Gasification (HyPr-RING) process in Japan (Lin et al., 2005). Based on the configurations and operating conditions, these gasification technologies use generic gasifier technology such as fixed/moving bed,

7.7 Gasification technologies

TABLE 7.4 Proximate and ultimate data for commonly used biomasses in the literature (Kamble et al., 2019; Dass and Jha, 2015; Raveendran et al., 1995).

Feedstock	Proximate analysis (wt.%)				Ultimate analysis (*daf., wt.%)					CV (daf) MJ/kg
	Moisture	Volatile matter (daf)	Fixed carbon (daf)	Ash	C	H	N	S	O	
Baggage	28.1	48.1	35.1	11.8	46.8	5.9	0.1	0.2	35.2	3.90
Bamboo	12.6	72.1	24.9	3.0	47.4	5.8	0.9	0.4	42.5	4.33
Bamboo dust	13.2	71.6	16.6	7.9	43.9	8.2	0.4	0.5	39.1	4.13
Bark	18.0	76.0	24.0	0.7	48.6	5.8	0.4	0.1	38.1	3.87
Coconut coir	9.4	69.6	13.2	17.2	42.1	5.5	0.2	0.4	34.6	4.32
Coconut fiber	12.5	82.1	13.2	4.7	52.4	6.7	0.3	0.5	35.4	4.33
Coconut shell	6.2	79.9	13.2	1.9	46.6	6.2	0.3	0.4	44.6	4.65
Corn cob	10.2	78.5	18.6	2.9	46.5	6	0.4	0.1	44.1	4.06
Corn stalks	8.2	73.2	19.3	8.1	41.9	5.6	0.6	0.1	43.7	4.01
Cotton gin waste	7.4	75.8	20.2	3.7	46.1	5.7	0.5	0.2	43.8	4.32
Cotton shell	5.6	72.2	23.3	4.6	38.5	6.1	0.3	0.2	50.3	4.36
Mustard shell	8.2	75.4	16.2	6.3	48.7	6.3	0.4	0.7	37.6	4.23
Mustard stall	10.6	70.9	15.5	10.8	42.3	7.9	0.8	0.5	37.7	4.22
Needles pine	3.6	72.4	26.1	1.5	50.6	0.2	0.3	0.3	43.1	4.75
Rice bran	5.8	75.7	11.2	13.1	36.2	6.1	0.4	0.2	44.0	3.95
Rice husk	5.4	65.9	26.7	18.9	40	5.3	0.3	0.4	35.1	3.43
Rice straw	4.8	68.3	16.2	19.8	36.7	5.5	0.6	0.3	37.1	3.73
Sugarcane leaves	5.3	73.5	16.9	7.9	35.8	6.1	0.4	0.2	49.6	4.22
Wheat stalk	7.2	78.7	15.6	5.7	39.8	5.3	0.7	0.4	48.1	3.91
Wheat straw	6.8	74.1	20.8	7.4	43.9	5.6	1.4	0.6	41.1	4.32
Wood	11.3	83	17	1.8	5.5	6.1	0.3	0.1	41.2	4.47

* daf: dry-ash-free.

fluidized bed and entrained flow (Ratafia-Brown et al., 2002a). Some modern gasifiers are operated under different names. For example, modern moving beds and transport gasifiers are modified versions of fixed and fluidized bed gasifiers.

In a fixed or moving bed gasifier, solid fuels (i.e., coal, biomass) fall slowly due to the gravity through the bed and gasification reactants flow upward (Collot, 2006). Due to the counter-flow arrangement, solid particles absorb heat from the flowing syngas before reaching the gasification zone (Phillips, 2006). The significant drawbacks of a fixed or moving bed gasifier are incomplete carbon conversion, nonuniform temperature, higher residence time, lower syngas temperature, and unwanted tar and oil yield.

A fluidized bed is a back-mixing type gasifier, where fresh solid particles continuously mix with the existed fully or partially gasified particles. Coal or biomass particles become lighter and smaller during fluidized bed operation and thus entrain out from the bed. Therefore, a cyclone separator is commonly incorporated to recycle the entrained particles in the bed, which leads to an increase in the cost. In order to overcome this limitation, several attempts have been made in several investigations (Johansson et al., 2006; Huseyin et al., 2014). Xu et al. (2009) and Virginie et al. (2012) employed a two-stage dual fluidized bed gasifier for biomass gasification, whereas Xie et al. (2010) used a circulating spout-fluid bed reactor for the same application. Another limitation of fluidized bed gasifiers is their inability at high-temperature applications due to agglomeration. However, the residence time is shorter than that of the moving bed gasifier (Phillips, 2006).

Entrained flow gasifier is believed to be the most advanced gasifier technology. In entrained flow gasifier, pulverized fuel particles and reactants are simultaneously supplied to the gasifier, resulting in the flow being a dense cloud (Ratafia-Brown et al., 2002a). Fuel particles are fed in the reactor either in dry or in the form of a slurry. Entrained flow gasifiers are operated under highly turbulent conditions, resulting in the residence time being a few seconds. High temperatures (1200–1600°C) and high pressure (2–8 MPa) application of entrained flow gasifier result in high carbon conversion (Phillips, 2006; Collot, 2006). Moreover, due to the high-temperature application, the unwanted devolatilized products (tar, phenol, etc.) transform into clean syngas and increase the quantity of syngas thereby. The high temperature in the gasifier also helps to melt the ash of coal and biomass, resulting in the formation of slag, which is crucial for the uninterrupted operation of the gasifier.

The raw syngas obtained from the gasifier is quenched and cleaned up using water or other coolers (Collot, 2006). About 25% of the heat from high-temperature syngas can be recovered and reused to preheat the fuel or generate steam (Ratafia-Brown et al., 2002a). The advantages of entrained flow gasifiers over others include fuel flexibility, higher carbon conversion, uniform temperature profile, high-temperature application, shorter residence time, and high-quality syngas (Simbeck et al., 1993; Ratafia-Brown et al., 2002a; Phillips, 2006; Collot, 2006). In contrast, the challenges of entrained flow gasifiers include a high volume of reactant, heat loss from the syngas and slag handling. Table 7.5 lists the critical comparison of available gasifier technologies.

7.7.1 Commercial gasifiers

Based on the configuration, commercially available entrained flow gasifiers include Texaco, E-Gas, Shell and PRENFLO (Ratafia-Brown et al., 2002a; Boot-Handford, 2015). Table 7.6

TABLE 7.5 Key characteristics of generic gasifier technologies (Tanner, 2015, Rataña-Brown et al., 2002b; Higman, 2014; Fernando, 2008; Shahabuddin et al., 2019).

Operating conditions	Fixed bed		Fluidized bed		Entrained flow	
Ash conditions	Dry ash	Slagging	Dry ash	Agglomerating	Slagging	Slagging
Fuel feeding conditions	Dry feeding	Dry feeding	Dry feeding	Dry feeding	Dry feeding	Slurry feeding
Reactant type	Air/O_2	Air/O_2	Air/O_2	Air/O_2	O_2	O_2
Reactant requirement	Low	Low	Medium	Medium	High	High
Syngas flow direction	Up	Up	Up	Up	Up or down	Up or down
Typical reactor temp (°C)	1000	1500–1800	900–1050	900–1050	1200–1600	1200–1600
Syngas temperature (°C)	425–650	425–650	925–1040	925–1040	1400–1600	1200–1400
Syngas cooling	Water	Water	Coolant	Coolant	Coolant	Water/syngas coolant
Pressure (MPa)	3.0	2.5	Up to 3.0	1.0–3.0	2.5–3.0	2.5–3.0
Feedstock preference	Low to high-rank coals and waste	Medium to high-rank coals, petcoke and waste	Low to medium rank coals and waste	Low to medium rank coals, biomass and waste	Low to high-rank coals, biomass, petcoke and waste	Low to high-rank coals, biomass, petcoke and waste

(continued on next page)

TABLE 7.5 Key characteristics of generic gasifier technologies (Tanner, 2015, Ratafia-Brown et al., 2002b; Higman, 2014; Fernando, 2008; Shahabuddin et al., 2019)—cont'd

Operating conditions	Fixed bed	Fluidized bed		Entrained flow	
Typical particle size (mm)	5–80	5–80	< 6	< 6	<0.1	<0.1
Residence time (s)	900–3600	900–3600	10–100	10–100	1.5–4	1.5
Moisture (%)	No limit	<28	No limit	No limit	Possible to use coal with high moisture	limited
Ash content limit (%)	<15	<25	<40	<40	2–25	<25
Ash Fusion Temp limit (°C)	Any	Any	>1100	>1100	Generally <1300	Generally <1300
Commercial Gasifier	Lurgi	BGL	IDGCC, HTW and KBR	KRW and U-Gas	Shell, PRENFLO, EAGLE, Siemens, MHI	GE, E-Gas
Conversion	>99	>99	96	95	98–99	100
Typical Cold gas efficiency (%)	~88	~88	~85	70–80	~80	74–77
Unit capacity (MWth)	10–350	10–350	100–700	20–50	Up to 700	Up to 700
Key technical issues	Agglomeration and use of hydrocarbon liquid	Agglomeration and use of hydrocarbon liquid	Lower carbon conversion and, agglomeration	Lower carbon conversion and, agglomeration	Syngas cooling and slagging	Syngas cooling and slagging

TABLE 7.6 Key features of commercially available entrained flow gasifiers (Ratafia-Brown et al., 2002a; Shahabuddin et al., 2020b,2020a).

Parameter	ChevronTexaco	E-Gas	Shell	PRENFLO
Fuel type	Bituminous coal	Bituminous coal	Bituminous coal	Petroleum coke and bituminous coal
Gasification process	Single stage entrained flow	Two-stage entrained-flow	Single-stage updraft entrained flow	Single-stage updraft entrained flow
Fuel feeding	Slurry feeding	Slurry feeding	Dry feeding	Dry feeding
Reactant	95% pure oxygen	95% pure oxygen	95% pure oxygen	95% pure oxygen
Syngas cooler type	Downflow radiant, water-tube and fire-tube	Downflow fire-tube	Downflow water-tube	Downflow or up-flow radiant water-tube and convective water-tube
Controlling particles	Water-scrubber	Metallic candle-filter and water-scrubber	Candle-filter	Candle-filter
Chloride, Fluoride, and Ammonia Control	Water-scrubber	Water-scrubber	Water-scrubber	Water-scrubber
Sulfur recovery (%)	98%	99%	99%	99%
Air separation	Cryogenic distillation	Cryogenic distillation	Cryogenic distillation	Cryogenic distillation
Combustors	Multiple cans	Multiple cans	Twin vertical silos	Twin horizontal silos
Firing temperature, °C	1287	1287	1100	1260
Heat recovery steam generator	Triple-pressure reheat and natural circulation	Triple-pressure reheat and natural circulation	Triple-pressure reheat and natural circulation	Triple-pressure reheat and natural circulation
Slag removal	Lock-hopper	Continuous	Lock-hoppers	Lock-hoppers

outlines an overview of the commercially available entrained flow gasifier used in IGCC plants, whereas Fig. 7.3 shows the configuration of the major commercial entrained flow gasifiers.

7.8 Present status of co-gasification of biomass and coal

Most of the co-gasification studies in the literature are based on the thermogravimetric study under various operating conditions such as temperature, reactant type and concentrations, particle size, and blending ratio. Kumabe et al. (2007) studied the co-gasification characteristics of Japanese cedar biomass with Mulia coal under a fixed bed downdraft gasifier at a temperature of 900°C using air and steam as the reactants. It was observed that the yield of syngas increased with the increasing biomass ratio in the blend. The cold gas efficiency (CGE) was 65-85% under different operating conditions. Table 7.7 summarizes the key findings for

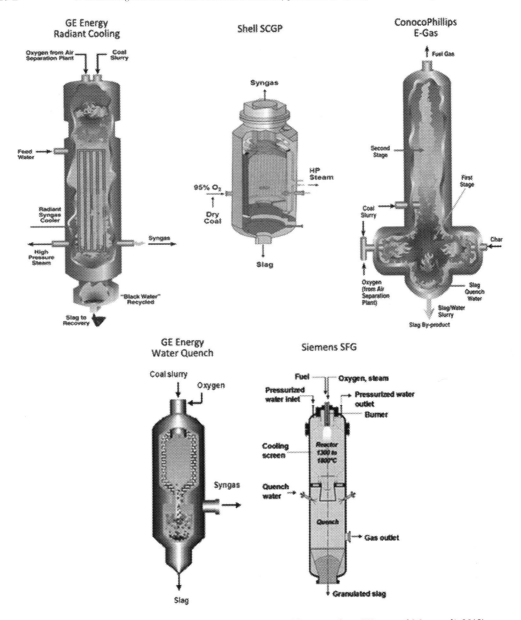

FIGURE 7.3 Major configuration of the commercial entrained flow gasifiers (Wang and Massoudi, 2013).

co-pyrolysis and co-gasification of biomass and coal using a thermogravimetric analyzer (TGA) or fixed bed reactor. Almost all studies found a synergistic effect in the yield of devolatilized syngas and reactivity due to co-pyrolysis and co-gasification. In contrast, some studies did not find any synergistic effect, which might be due to the operating conditions such as particle size and heating rates.

TABLE 7.7 Summary of the co-pyrolysis and co-gasification of coal and biomass in TGA/fixed bed reactor.

References	Operating conditions	Major findings
Jeong et al. (2015)	• Temperature: 750–850°C • Pressure: 1 atm • Reactants: H_2O • Coal/biomass ratio: 4:1,1:1, 1:4 • d_p coal: 60-700 µm • d_p biomass: 1000 µm	• Carbon conversion of pine biomass was much higher than that of bituminous coal used in the study. • The synergistic effect from co-gasification of biomass and coal was observed, which increased with increasing biomass ratio. • It is impossible to predict the reactivity of co-gasified char from the individual reactivity of biomass and coal.
Ellis et al. (2015), Masnadi et al. (2015)	• Temperature: 850°C • Pressure: 1 atm • Reactants: CO_2 • Coal/biomass ratio: 3:1, 1:1,1:3. • d_p: 0.6-1.0 mm	• Reactivity of char produced from individual pyrolysis was higher than that of co-pyrolyzed char. • The mineral matter transformation/interaction during co-pyrolysis is responsible for decreasing the reactivity of char in gasification. • The key mineral matter formed during co-pyrolysis was gehlenite crystals, as confirmed by the XRD test.
Masnadi et al. (2014)	• Temperature: 750–900°C • Pressure: 1 atm • Reducing gas: N_2 • Heating rate: 25°C/min • Coal/biomass ratio: 3:1, 1:1, 1:3 • d_p: 300-355 µm	• The pyrolysis temperature dramatically influences the surface area of the char. • The co-pyrolysis of biomass and coal did not show any influence on the char and devolatilization behavior. • AAEM began to evaporate with increasing temperature, and hence no synergy was observed.
Ulloa et al. (2009)	• Maximum temperature: 1200°C • Pressure: 1 atm • Reducing gas: N_2 • Heating rate: 10, 30, 50°C/min • Coal/biomass ratio: 1:1 • d_p: 1000 µm	• Synergistic effect of co-pyrolysis of biomass and coal on char and devolatilized gas. • A higher devolatilization was detected during co-pyrolysis compared to the independent pyrolysis of pure coal and biomass.

Li et al. (2010) gasified bituminous coal with pine and rice straw biomass in a fluidized bed gasifier at a temperature between 921 and 1033°C using oxygen-rich air and steam. It was found that increasing the Equivalent ratio (ER) of air and steam decreases the yield of syngas; steam carbon ratio above 0.5 leads to decreasing the bed temperature and the yield of syngas, and increasing biomass ratio increases the yield of H_2 but decreases CO.

Vélez et al. (2009) investigated the co-gasification behavior of three different biomass, namely coffee husk, sawdust and rice husk with Columbian coal under fluidized bed conditions at a maximum temperature of 900°C using air-steam reactants. Results revealed that increasing biomass ratio and steam concentration increase the yield of H_2 but decreases energy efficiency. The optimum ratios of biomass and steam for suitable gas quality varied with biomass types. Also, bed temperature played an essential role in the gas yield, which varied with the biomass ratio in the feed. As observed, the bed temperature dropped by around 30°C with an increasing biomass ratio from 6 to 15%. In terms of gas production, the yield of H_2 decreases, but CO increases with increasing temperature due to the dominance of reverse water gas shift reaction and Boudouard reaction, respectively.

Pan et al. (2000) conducted a co-gasification study for pine chips with a mixture of low-grade black coal, and Sabero refused coal in a superficial fluidized bed gasifier at a temperature between 840 and 910°C using air and steam as the reactants. It was found that the carbon conversion increased by 63% to 83.4%, while the thermal efficiency increased by 40% to 68% with increasing biomass ratios. Pinto et al. (2003) gasified coal with plastic and pine bark biomass using a fluidized bed gasifier at a temperature between 750 and 890°C under the air-stream mixture. It was revealed that increasing temperature decreases CH_4 and other hydrocarbons by 30 to 63% but increased the yield of H_2 by about 70%. The increase of steam concentration was favorable for the production of H_2 while unfavorable for CO. Table 7.8 presents a summary of the co-gasification behavior of biomass and coal in fluidized bed gasifier (FBG).

To date, most of the co-gasification studies have been conducted using TGA and laboratory-scale fixed/fluidized bed gasifiers. Limited studies are available in the literature using high-temperature entrained flow co-gasification of coal and biomass. A study conducted by Sripada et al. (2017) compared the individual gasification behavior of pine bark biomass with Loy Yang Victorian brown coal at a temperature between 1000 and 1400°C using the CO_2 reactant. They observed that pine bark's carbon conversion was 4 to 10% higher than that of Loy Yang coal under different temperatures and CO_2 concentrations. However, the co-gasification behavior of biomass and coal was not analyzed in that study. Tilghman and Mitchell (2016) studied the co-pyrolysis behavior of Wyodak subbituminous coal and corn stover biomass at a temperature of 1277°C using an entrained flow gasifier under atmospheric pressure. After that, the pyrolysis char produced was studied using a TGA to analyze the gasification reactivity. The experimental study conducted by Hernández et al. (2010) investigated the co-gasification behavior of dealcoholized grape biomass with a low-rank coal-coke blend at a temperature between 750 and 1150°C using an air-blown entrained flow gasifier. The results showed that increasing biomass content in the blend improved the syngas quality and CGE. The synergistic effect was reported under the co-gasification condition due to the dominant percentages of Ca and K in biomass ash.

7.9 Challenges of co-gasification

Despite having numerous advantages of co-gasification of biomass and coal, co-gasification suffers from the following challenges (Luque and Speight, 2015; Brar et al., 2012; Shahabuddin and Bhattacharya, 2021c):

- It is hard to make a uniform particle size for biomass due to its fibrous structure. Therefore, size reduction and uniform mixing of coal and biomass particles are challenging.
- There is a significant difference in gasification temperature between biomass and coal particles. Therefore, reactor compatibility is an issue.

TABLE 7.8 Summary of the co-gasification behavior in fluidized bed gasifier using coal and wood biomass.

References	Reactor configuration/Operating conditions	Major findings
Pan et al. (2000)	• H x ID: 2000 × 43 mm • Temperature: 840–910°C • Pressure: 1 atm • Superficial gas velocity: 0.7–1.4 m/s • Reactants: air and steam • Coal: pine: 3:1 and 2:3 • d_p: 750–1200 µm	• The carbon conversion of pine chips was 20% higher than that of black coal. • Gas composition was (mol%): CO:17.55, H_2: 11.58, CO_2: 8.44, CH_4:1.88 and C_2H_2: 0.37. • The lower heating value (LHV) of the product gases from biomass increases with increasing biomass ratio. • Overall thermal efficiency decreased by 13% with increasing biomass ratio by 15%.
Pinto et al. (2003)	• H x ID: 3200 × 504 mm • Temperature: 750–890°C • Reactants: Air, steam and mixture of both • O_2/fuel ratio: 0.03–0.33 • O_2/steam ratio: 0.02–0.28 • Steam flow rate: 2–5 kg/h • Coal: pine: plastic: 3:1:1 • d_p: 1250–2000 µm	• Temperature is found to be the most influential operating parameter affecting co-gasification performance. • Increasing gasification temperatures leads to a decrease in tar formation and increases H_2 by a maximum of 70%. However, the concentration of CH_4 decreased by 30–63%. • Increasing air to fuel flow rate helped to decrease the tar and other hydrocarbons. However, access air led to an increase in the CO_2 emission and simultaneously decreased the product's heating value as CH_4 decreases by 40–54%.
Li et al. (2010)	• H x ID: 1578 × 120 mm • Temperature: 948–1026°C • Reactants: Oxygen and steam • ER: 0.31–0.47 • Steam-carbon ratio (Fs/Fc): 0.26–0.88 • Coal: pine: 4:1 and 2:1 • d_p pine: 420 µm • d_p coal: 250, 750 µm	• The lower ER favors the yield of CO and H_2 • The yield of H_2 increases, but CO decreases with increasing biomass ratio in the blend. • Increasing biomass ratio from 0 to 33% increased the bed temperature from 912 to 976°C. • Gas composition (vol.%): CO: 33.5, H_2: 21.57, CO^2: 19.74, CH_4:2.44 using ER of 0.3, Fs/Fc of 0.41 and operation time of 300 min. • The syngas yield and carbon conversion were determined to be 1.0 Nm3/kg fuel and 88.9%. A maximum gasification efficiency was calculated to be 60.9%.
Tursun et al. (2016)	• H x ID: 220 × 24 mm • Temperature: 700–850°C • Pressure: 1 atm • Reactants: steam • Fs/Fc: 0.13–1.9 • Fuel feeding rate (kg/h): 0.20 • Coal: pine: 3:1, 1:1, and 1:3 • d_p of coal: 380–830 µm	• Increasing biomass ratio in the blend increases the yield of H_2, CO, and CH_4 within the range of 3–7%. Whereas the LHV increases by up to 65%. • Increasing biomass ratio in the blend increased the chemical efficiency and carbon conversion over 200%. • Increasing gasifier temperature from 700 to 850°C augments the chemical efficiency and carbon conversion about 5%.

(continued on next page)

TABLE 7.8 Summary of the co-gasification behavior in fluidized bed gasifier using coal and wood biomass—cont'd

References	Reactor configuration/Operating conditions	Major findings
Seo et al. (2010)	• W × B× H: 40 × 285 × 2130 mm • Temperature: 750–900°C • Pressure: 1 atm • Reactants: steam • Fs/Fc: 0.5–0.8 • Fuel feeding rate (kg/h): 6.17–10.3 • Coal: pine: 3:1, 1:1, and 1:3 • d_p coal: 348 μm • d_p biomass: 1438 μm	• Syngas yield, carbon conversion and CGE of biomass were found to be higher than that of coal. • Maximum syngas yield was determined at a 50/50 ratio of biomass and coal. • The heating value of product gases was calculated to be pure coal: 9.89–11.15, pure biomass: 12.10–13.19 and 50/50 blend: 13.77–14.39 MJ/m^3.

- High moisture content in biomass requires pretreatment or torrefaction, which associates with the energy penalty. Furthermore, the high moisture content in biomass decreases the gasifier temperature.
- There is a difference in impurities from biomass and coal during co-gasification.
- The mineral matter present in wood biomass possesses a low melting point, which causes agglomeration and thus creates problems for fluidized bed gasification.

7.9 Conclusion

Co-gasification of coal and renewable biomass is an attractive technology for power and chemical production. Predominantly, the potentiality, present status, available technologies, and challenges of co-gasification of coal and biomass are discussed. Although this method of electricity generation has not been commercialized yet, it has huge potential in the developing economies due to the availability of low cost coal. However, this would not be a complete sustainable electricity generation solution rather a short-term pathway to reduce GHG emissions and to slow down the negative climate change. In addition, the sustainability assessment of coal-biomass co-gasification technique to produce electricity should be carried out to know its sustainability status.

Co-gasification studies in the literature are based on three major configurations, such as (i) the thermogravimetric study, which was predominantly used for kinetic and reactivity analysis, (ii) laboratory-scale fixed-bed study to test the effect of operating conditions such as blending ratio, temperature and reactant to fuel ratio on gasification performance, (iii) laboratory-scale fluidized-bed study to test the gasification performance under various operating conditions. Very few studies on co-gasification are found using entrained flow gasifiers. Based on the literature review, the effects of co-gasification on gasification performance include:

- Increasing temperature increases the carbon conversion, which is similar to individual gasification of coal or biomass. The reactivity of the co-gasification cannot be predicted from the individual gasification of coal and biomass.

- The synergistic effect of co-gasification is observed mostly at lower temperatures. The AAEM of biomass ash governs the catalytic effect during co-gasification.
- Syngas yield, LHV and CGE increase with increasing biomass ratio up to a certain blending ratio. The blending ratio can be controlled by modifying other operating parameters.
- Co-gasification significantly reduces pollutant emission from the gasification

The following recommendations are made to overcome the challenges of co-gasification:

- Torrefaction of biomass using waste energy.
- Maintaining an optimum blending ratio is crucial. The optimum biomass ratio predominantly ranges over 10 to 50%, depending on the coal and biomass properties.
- Practically relevant operating conditions, including the use of gasification reagent: air/oxygen, CO_2, steam and a mixture of them, should be tested considering a wide variety of biomass.
- It is crucial to understand the effect of steam in the gasifier and downstream process to improve the syngas quality applicable for the turbine and chemical synthesis.
- It is imperative to understand the slagging behavior of co-gasification.
- More pilot-scale studies to generate electricity through this process, including process modeling and techno-economic analysis, sustainability assessment are critical to scale up.

Self-evaluation questions

1. What do you mean by co-gasification?
2. What are the advantages and disadvantages of co-gasification?
3. What are the gasification technologies available commercially that can be used for co-gasification of coal-biomass?
4. What are the factors that need to take into account for coal-biomass co-gasification?
5. What are the challenges of co-gasification technologies?
6. How co-gasification of coal and biomass might be a short-term sustainable electricity generation solution for developing countries? – Explain.
7. Conduct a literature survey to identify the latest studies dealt with coal-biomass co-gasification and evaluate their progress in terms of commercial maturity and practical application.

References

Ali, D.A., Gadalla, M.A., Abdelaziz, O.Y., Hulteberg, C.P., Ashour, F.H, 2017. Co-gasification of coal and biomass wastes in an entrained flow gasifier: modelling, simulation and integration opportunities. J. Nat. Gas Sci. Eng. 37, 126–137.

Association, W. B., 2018. WBA Global Bioenergy Statistics 2018. World Bioenergy Association, Stockholm.

Bhattacharya, S., Kabir, K.B., Hein, K., 2013. Dimethyl ether synthesis from Victorian brown coal through gasification–current status, and research and development needs. Prog. Energy Combust. Sci. 39, 577–605.

Bhuiyan, M.M.H., Islam, M.A., Hossain, M.I., Bagum, M., Akter, Y., Chittagong, B., 2014. Analysis and comparison of different coal fields and imported coal in Bangladesh.

Boot-Handford, M., 2015. The utilisation of biomass as a fuel for chemical looping combustion.

Boshu, H., Mingyang, L., Xin, W., Ling, Z., Lili, W., Jiwei, X., Zhenxing, C., 2008. Chemical kinetics-based analysis for utilities of ZEC power generation system. Int. J. Hydrogen Energy 33, 4673–4680.

BP Statistical Review, 2017. BP Statistical Review of World Energy 2017. Available at: https://www.bp.com/content/dam/bp/business-sites/en/global/corporate/pdfs/energy-economics/statistical-review/bp-stats-review-2019-full-report.pdf. Accessed on 19.09.2020.

BPCL, 2021. Bangladesh China Power Company (BCPCL), Payra 1320 MW Thermal Power Plant (1st Phase). Availabe at: https://www.bcpcl.org.bd/power-plant. Accessed on 20-04-2021.

BPDB, 2021. Power Generation Units (Fuel Type Wise), Bangladesh power development board. Available at: https://www.bpdb.gov.bd/bpdb_new/index.php/site/power_generation_unit. Accessed on 04-07-2021.

Brar, J., singh, K., Wang, J., Kumar, S., 2012. Cogasification of coal and biomass: a review. Int. J. Forest. Res. 2012, 1–10.

Chmielniak, T., Sciazko, M., 2003a. Co-gasification of biomass and coal for methanol synthesis. Appl. Energy 74, 393–403.

Chmielniak, T., Sciazko, M., 2003b. Co-gasification of biomass and coal for methanol synthesis. Appl. Energy 74, 393–403.

Chowdhury, S., Inayat, A., Abdullah, B., Omar, A.A., Ganguly, S., 2014. Hydrogen and syngas generation from gasification of coal in an integrated fuel processor. Applied Mech. Mater. 625, 644–647 Trans Tech Publ.

Collot, A.-G., 2006. Matching gasification technologies to coal properties. Int. J. Coal Geol. 65, 191–212.

Cozzi, L., Erdogan, M., Goodson, T., Arsalane, Y., Bahar, H., Barret, C., Zavala, P.B., Couse, J., Criswell, T., Dasgupta, A., Daugy, M., Baptiste, J., Carlos, D., Alvarez, F., Rosa, L.F., Guyon, J., Lorenczik, S., Molnar, G., Losz, A., Varro, L., Wanner, B., Tonolo, G., 2021. Global Energy Review 2021- Assessing the effects of economic recoveries on global energy demand and CO2 emissions in 2021. International Energy Agency (IEA).

Dass, B., Jha, P., 2015. Biomass characterization for various thermo-chemical applications. Int. J. Curr. Eng. Scientific Res. 2, 59–63.

David, J., Herzog, H., 2000. The cost of carbon capture. In: Fifth international conference on greenhouse gas control technologies, Cairns, Australia, pp. 13–16.

De Jong, W., Andries, J., Hein, K.R, 1999. Coal/biomass co-gasification in a pressurised fluidised bed reactor. Renewable Energy 16, 1110–1113.

Ellis, N., Masnadi, M.S., Roberts, D.G., Kochanek, M.A., Ilyushechkin, A.Y, 2015. Mineral matter interactions during co-pyrolysis of coal and biomass and their impact on intrinsic char co-gasification reactivity. Chem. Eng. J. 279, 402–408.

Energy, B.S.R.O.W., 2020. Share of global electricity generation by fuel (percentage). Available at: https://www.bp.com/en/global/corporate/energy-economics/statistical-review-of-world-energy/electricity.html. Accessed on 20-04-2021.

Fermoso, J., Arias, B., Plaza, M., Pevida, C., Rubiera, F., Pis, J., García-Peña, F., Casero, P., 2009. High-pressure co-gasification of coal with biomass and petroleum coke. Fuel Process. Technol. 90, 926–932.

Fernando, R., 2008. Coal gasification. IEA Clean Coal Centre, London, UK.

Fernando, R., 2009. Co-gasification and indirect cofiring of coal and biomass. IEA Clean Coal Centre.

Friedlingstein, P., Jones, M.W., O'Sullivan, M., Andrew, R.M., Bakker, D.C.E., Hauck, J., Le Quéré, C., Peters, G.P., Peters, W., Pongratz, J., Sitch, S., Canadell, J.G., Ciais, P., Jackson, R.B., Alin, S.R., Anthoni, P., Bates, N.R., Becker, M., Bellouin, N., Bopp, L., Chau, T.T.T., Chevallier, F., Chini, L.P., Cronin, M., Currie, K.I., Decharme, B., Djeutchouang, L., Dou, X., Evans, W., Feely, R.A., Feng, L., Gasser, T., Gilfillan, D., Gkritzalis, T., Grassi, G., Gregor, L., Gruber, N., Gürses, Ö., Harris, I., Houghton, R.A., Hurtt, G.C., Ii Da, Y., Ilyina, T., Luijkx, I.T., Jain, A.K., Jones, S.D., Kato, E., Kennedy, D., Klein Goldewijk, K., Knauer, J., Korsbakken, J.I., Körtzinger, A., Landschützer, P., Lauvset, S.K., Lefèvre, N., Lienert, S., Liu, J., Marland, G., Mcguire, P.C., Melton, J.R., Munro, D.R., Nabel, J.E.M.S., Nakaoka, S.-I., Niwa, Y., Ono, T., Pierrot, D., Poulter, B., Rehder, G., Resplandy, L., Robertson, E., Rödenbeck, C., Rosan, T.M., Schwinger, J., Schwingshackl, C., Séférian, R., Sutton, A.J., Sweeney, C., Tanhua, T., Tans, P.P., Tian, H., Tilbrook, B., Tubiello, F., Van der Werf, G., Vuichard, N., Wada, C., Wanninkhof, R., Watson, A., Willis, D., Wiltshire, A.J., Yuan, W., Yue, C., Yue, X., Zaehle, S., AndZeng, J., 2021. Global carbon budget 2021. Earth Syst. Sci. Data 12, 3269–3340.

Halder, P., Paul, N., Joardder, M.U., Sarker, M., 2015. Energy scarcity and potential of renewable energy in Bangladesh. Renewable Sustainable Energy Rev. 51, 1636–1649.

Hauserman, W.B., 1994. High-yield hydrogen production by catalytic gasification of coal or biomass. Int. J. Hydrogen Energy 19, 413–419.

Hernández, J.J., Aranda-Almansa, G., Serrano, C., 2010. Co-gasification of biomass wastes and coal— coke blends in an entrained flow gasifier: an experimental study. Energy Fuels 24, 2479–2488.

References

Higman, C., 2014. State of the gasification industry: worldwide gasification database 2014 update. In: Gasification Technologies Conference. Washington, DC.

Higman, C., 2017. GSTC Syngas Database: 2017 Update. In: Gasification & Syngas Technologies Conference. Colorado Springs.

Hossain, M.M., De Lasa, H.I., 2008. Chemical-looping combustion (CLC) for inherent CO_2 separations—a review. Chem. Eng. Sci. 63, 4433–4451.

Hossain, M.N., Paul, S.K., Hasan, M.M, 2015. Environmental impacts of coal mine and thermal power plant to the surroundings of Barapukuria, Dinajpur, Bangladesh. Environ. Monit. Assess. 187, 202.

Howladar, M.F., Islam, M.R, 2016. A study on physico-chemical properties and uses of coal ash of Barapukuria Coal Fired Thermal Power Plant, Dinajpur, for environmental sustainability. Energy Ecol. Environ. 1, 233–247.

Huseyin, S., Wei, G.-Q., Li, H.-B., He, F., Huang, Z., 2014. Chemical-looping gasification of biomass in a 10 kWth interconnected fluidized bed reactor using Fe_2O_3/Al_2O_3 oxygen carrier. J. Fuel Chem. Technol. 42, 922–931.

IEA Clean Coal Centre. Bangladesh: 13 coal-fired power plants to start generation by 2023. Available at: https://www.iea-coal.org/bangladesh-13-coal-fired-power-plants-to-start-generation-by-2023/. Accessed on 20/12/2019 [Online]. [Accessed].

Jeong, H.J., Hwang, I.S., Hwang, J., 2015. Co-gasification of bituminous coal–pine sawdust blended char with H_2O at temperatures of 750–850°C. Fuel 156, 26–29.

Johansson, M., Mattisson, T., Lyngfelt, A., 2006. Creating a synergy effect by using mixed oxides of iron-and nickel oxides in the combustion of methane in a chemical-looping combustion reactor. Energy Fuels 20, 2399–2407.

Jos, G.J.O, Greet, J.-.M., Marilena, M., Peters, J.A.H.W, 2016. Trends in global CO_2 emissions: 2016 Report. PBL Netherlands Environmental Assessment Agency.

Kamble, A.D., Saxena, V.K., Chavan, P.D., Mendhe, V.A, 2019. Co-gasification of coal and biomass an emerging clean energy technology: status and prospects of development in Indian context. Int. J. Min. Sci. Technol. 29, 171–186.

Khan, M.A.A., Saha, M.S., Sultana, S., Ahmed, A.N., Das, R.C., 2013. Coal Fly Ash of Barapukuria Thermal Power Plant, Bangladesh: Physico Chemical Properties Assessment and Utilization.

Kumabe, K., Hanaoka, T., Fujimoto, S., Minowa, T., Sakanishi, K., 2007. Co-gasification of woody biomass and coal with air and steam. Fuel 86, 684–689.

Li, K., Zhang, R., Bi, J., 2010. Experimental study on syngas production by co-gasification of coal and biomass in a fluidized bed. Int. J. Hydrogen Energy 35, 2722–2726.

Lin, S., Harada, M., Suzuki, Y., Hatano, H., 2005. Process analysis for hydrogen production by reaction integrated novel gasification (HyPr-RING). Energy Convers. Manage. 46, 869–880.

Luque, R., Speight, J., 2015. Gasification and synthetic liquid fuel production: an overview. Gasification for Synthetic Fuel Production, Cambridge, UK. Elsevier, pp. 3–27.

Mallick, D., Mahanta, P., Moholkar, V.S, 2017. Co-gasification of coal and biomass blends: chemistry and engineering. Fuel 204, 106–128.

Mallick, D., Poddar, M.K., Mahanta, P., Moholkar, V.S, 2018. Discernment of synergism in pyrolysis of biomass blends using thermogravimetric analysis. Bioresour. Technol. 261, 294–305.

Masnadi, M.S., Grace, J.R., Bi, X.T., Lim, C.J., Ellis, N., 2015. From fossil fuels towards renewables: inhibitory and catalytic effects on carbon thermochemical conversion during co-gasification of biomass with fossil fuels. Appl. Energy 140, 196–209.

Masnadi, M.S., Habibi, R., Kopyscinski, J., Hill, J.M., Bi, X., Lim, C.J., Ellis, N., Grace, J.R, 2014. Fuel characterization and co-pyrolysis kinetics of biomass and fossil fuels. Fuel 117, 1204–1214.

Masud, M., Ananno, A.A., Arefin, A.M., Ahamed, R., Das, P., Joardder, M.U, 2019a. Perspective of biomass energy conversion in Bangladesh. Clean Technol. Environ. Policy 21, 719–731.

Masud, M.H., Nuruzzaman, M., Ahamed, R., Ananno, A.A., Tomal, A.A, 2019b. Renewable energy in Bangladesh: current situation and future prospect. Int. J. Sustainable Energy 39, 1–44.

Moazzem, S., Rasul, M., Khan, M., 2012. A review on technologies for reducing CO_2 emission from coal fired power plants, chapter.

Pan, Y., Velo, E., Roca, X., Manya, J., Puigjaner, L., 2000. Fluidized-bed co-gasification of residual biomass/poor coal blends for fuel gas production. Fuel 79, 1317–1326.

Panwar, N., Kaushik, S., Kothari, S., 2011. Role of renewable energy sources in environmental protection: a review. Renewable Sustainable Energy Rev. 15, 1513–1524.

Phillips, J., 2006. Different types of gasifiers and their integration with gas turbines. The Gas Turbine Handbook, National Energy Technology Laboratory (NETL), The USA 67–75.

Pinto, F., Franco, C., Andre, R.N., Tavares, C., Dias, M., Gulyurtlu, I., Cabrita, I., 2003. Effect of experimental conditions on co-gasification of coal, biomass and plastics wastes with air/steam mixtures in a fluidized bed system. Fuel 82, 1967–1976.

Power Generation Units, Bangladesh Power development board. Available at: http://www.bpdb.gov.bd/bpdb/index.php?option=com_content&view=article&id=150&Itemid=16. Accessed on 10/02/17.

Power system master plan-2016, Power division, Ministry of Power, Energy and Mineral Resources, The people's republic of Bangladesh. Available at: http://www.bpdb.gov.bd/bpdb/index.php?option=com_content&view=article&id=12&Itemid=126. Accessed on 15/04/2019.

Quyn, D.M., Wu, H., Li, C.-Z, 2002. Volatilisation and catalytic effects of alkali and alkaline earth metallic species during the pyrolysis and gasification of Victorian brown coal. Part I. Volatilisation of Na and Cl from a set of NaCl-loaded samples. Fuel 81, 143–149.

Ratafia-Brown, J., Manfredo, L., Hoffmann, J., Ramezan, M., 2002a. Major environmental aspects of gasification-based power generation technologies. Final Report prepared by Science Application International Corporation for US Department of Energy. Office of Fossil Energy, National Energy Technology Laboratory http://www.netl.doe.gov/.

Ratafia-Brown, J.A., Manfredo, L.M., Hoffmann, J.W., Ramezan, M., Stiegel, G.J, 2002b. An environmental assessment of IGCC power systems. In: Nineteenth annual Pittsburgh coal conference, pp. 23–27.

Raveendran, K., Ganesh, A., Khilar, K.C, 1995. Influence of mineral matter on biomass pyrolysis characteristics. Fuel 74, 1812–1822.

Rizeq, G., Kumar, R., West, J., Lissianski, V., Widmer, N., Zamansky, V., 2001. Fuel-Flexible Gasification-Combustion Technology for Production of H_2 and Sequestration-Ready CO_2. National Energy Technology Lab., Pittsburgh, PA (US); National Energy Technology Lab., Morgantown, WV (US).

Rubin, E.S., Davison, J.E., Herzog, H.J, 2015. The cost of CO_2 capture and storage. Int. J. Greenhouse Gas Control 40, 378–400.

Safiullah, S., Khan, M., Sabur, M., 2011. Comparative study of Bangladesh Barapukuria coal with those of various other countries. J. Bangladesh Chem. Soc. 24, 221–225.

Seo, M.W., Goo, J.H., Kim, S.D., Lee, S.H., Choi, Y.C, 2010. Gasification characteristics of coal/biomass blend in a dual circulating fluidized bed reactor. Energy Fuels 24, 3108–3118.

Shahabuddin, M., Alam, M.T., Krishna, B.B., Bhaskar, T., Perkins, G., 2020a. A review of producing renewable aviation fuels from the gasification of biomass and residual wastes. Bioresour. Technol., 123596.

Shahabuddin, M., Bhattacharya, S., 2021a. Co-gasification characteristics of coal and biomass using CO_2 reactant under thermodynamic equilibrium modelling. Energies 14, 7384.

Shahabuddin, M., Bhattacharya, S., 2021b. Effect of reactant types (steam, CO_2 and steam+ CO_2) on the gasification performance of coal using entrained flow gasifier. Int. J. Energy Res. 45, 9492–9501.

Shahabuddin, M., Bhattacharya, S., 2021c. Enhancement of performance and emission characteristics by co-gasification of biomass and coal using an entrained flow gasifier. J. Energy Inst. 95, 166–178.

Shahabuddin, M. Bhattacharya, S. 2021d. Process modelling for the production of hydrogen-rich gas from gasification of coal using oxygen, CO_2 and steam reactants. Int. J. Hydrogen Energy 46 (47), 24051–24059.

Shahabuddin, M., Kibria, M.A., Bhattacharya, S., 2021. Evaluation of high-temperature pyrolysis and CO_2 gasification performance of bituminous coal in an entrained flow gasifier. J. Energy Inst. 94, 294–309.

Shahabuddin, M., Krishna, B.B., Bhaskar, T., Perkins, G., 2019. Advances in the thermo-chemical production of hydrogen from biomass and residual wastes: Summary of recent techno-economic analyses. Bioresour. Technol. 299, 122557.

Shahabuddin, M., Krishna, B.B., Bhaskar, T., Perkins, G., 2020b. Advances in the thermo-chemical production of hydrogen from biomass and residual wastes: summary of recent techno-economic analyses. Bioresour. Technol. 299, 122557.

Simbeck, D., Korens, N., Biasca, F., Vejtasa, S., Dickenson, R., 1993. Coal gasification guidebook: status, applications, and technologies. Electric Power Research Institute Final Report No. TR-102034, Palo Alto, CA.

Sripada, P.P., Xu, T., Kibria, M., Bhattacharya, S., 2017. Comparison of entrained flow gasification behaviour of Victorian brown coal and biomass. Fuel 203, 942–953.

Stucley, C., Schuck, S., Sims, R., Larsen, P., Turvey, N., Marino, B., 2004. Biomass Energy Production in Australia. Rural Industries Research and Development Corporation, Canberra *Revised Edition*.

Taba, L.E., Irfan, M.F., Daud, W.A.M.W., Chakrabarti, M.H, 2012. The effect of temperature on various parameters in coal, biomass and CO-gasification: a review. Renewable Sustainable Energy Rev. 16, 5584–5596.

Tang, J., Wang, J., 2016. Catalytic steam gasification of coal char with alkali carbonates: a study on their synergic effects with calcium hydroxide. Fuel Process. Technol. 142, 34–41.

Tanner, J., 2015. High Temperature, Entrained Flow Gasification of Victorian Brown Coals and Rhenish Lignites Ph.D Thesis. Monash University, Australia.

Tanner, J., Bläsing, M., Müller, M., Bhattacharya, S., 2016. High temperature pyrolysis and CO_2 gasification of Victorian brown coal and Rhenish lignite in an entrained flow reactor. AlChE J. 62, 2101–2211.

Tilghman, M.B., Mitchell, R.E, 2016. Impact of co-firing coal and biomass on mixed char reactivity under gasification conditions. Energy Fuels 30, 1708–1719.

Tursun, Y., Xu, S., Wang, C., Xiao, Y., Wang, G., 2016. Steam co-gasification of biomass and coal in decoupled reactors. Fuel Process. Technol. 141, 61–67.

Ulloa, C.A., Gordon, A.L., García, X.A, 2009. Thermogravimetric study of interactions in the pyrolysis of blends of coal with radiata pine sawdust. Fuel Process. Technol. 90, 583–590.

UNB, 2021. Govt officially cancels 10 coal power plant projects. Available at: https://www.thedailystar.net/news/bangladesh/governance/news/govt-officially-10-coal-power-plant-projects-2119405. Accessed on 11-11-2011. *The Daily Star*.

UNFCCC, 2015. Adoption of the Paris Agreement. United Nations New York, NY.

Usón, S., Valero, A., Correas, L., Martínez, Á., 2004. Co-gasification of coal and biomass in an IGCC power plant: gasifier modeling. Int. J. Thermodyn. 7, 165–172.

Vélez, J.F., Chejne, F., Valdés, C.F., Emery, E.J., Londoño, C.A, 2009. Co-gasification of Colombian coal and biomass in fluidized bed: an experimental study. Fuel 88, 424–430.

Virginie, M., Adánez, J., Courson, C., De Diego, L., García-Labiano, F., Niznansky, D., Kiennemann, A., Gayán, P., Abad, A., 2012. Effect of Fe–olivine on the tar content during biomass gasification in a dual fluidized bed. Appl. Catal. B 121, 214–222.

Wang, P., Massoudi, M., 2013. Slag behavior in gasifiers. Part I: Influence of coal properties and gasification conditions. Energies 6, 784–806.

Weimer, T., Specht, M., Baumgart, F., Marquard-Möllenstedt, T., Sichler, P., 2002. Hydrogen/syngas generation by simultaneous steam reforming and carbon dioxide absorption. Gasification, the Clean Choice for Carbon Management. *Institution of Chemical Engineers*, Rugby, UK.

Wood, B.J., Sancier, K.M, 1984. The mechanism of the catalytic gasification of coal char: a critical review. Catal. Rev. Sci. Eng. 26, 233–279.

World Energy Council Report World Energy Resources 2016.

Xie, Y., Xiao, J., Shen, L., Wang, J., Zhu, J., Hao, J., 2010. Effects of Ca-based catalysts on biomass gasification with steam in a circulating spout-fluid bed reactor. Energy Fuels 24, 3256–3261.

Xu, G., Murakami, T., Suda, T., Matsuzaw, Y., Tani, H., 2009. Two-stage dual fluidized bed gasification: its conception and application to biomass. Fuel Process. Technol. 90, 137–144.

Xu, T., 2018. Entrained flow gasification of Victorian brown coals using CO_2 as reactant. Monash University.

Yan, L., He, B., Ma, L., Pei, X., Wang, C., Li, X., 2012. Integrated characteristics and performance of zero emission coal system. Int. J. Hydrogen Energy 37, 9669–9676.

Yan, L., He, B., Pei, X., Wang, C., Liang, H., Duan, Z., 2013. Computational-fluid-dynamics-based evaluation and optimization of an entrained-flow gasifier potential for coal hydrogasification. Energy Fuels 27, 6397–6407.

Ye, D.P., Agnew, J.B., Zhang, D.K, 1998. Gasification of a South Australian low-rank coal with carbon dioxide and steam: kinetics and reactivity studies. Fuel 77, 1209–1219.

Yeboah, Y.D., Xu, Y., Sheth, A., Godavarty, A., Agrawal, P.K, 2003. Catalytic gasification of coal using eutectic salts: identification of eutectics. Carbon 41, 203–214.

Zhang, Y., Zheng, Y., Yang, M., Song, Y., 2016. Effect of fuel origin on synergy during co-gasification of biomass and coal in CO_2. Bioresour. Technol. 200, 789–794.

Ziock, H., Lackner, K., Harrison, D., 2001. Zero emission coal power, a new concept. In: First National Conference on Carbon Sequestration.

CHAPTER 8

Waste-to-energy (WtE): A potential renewable source for future electricity generation in the developing world

Zobaidul Kabir[a], Mahfuz Kabir[b] and Nigar Sultana[c]

[a] School of Environmental and Life Sciences, University of Newcastle, Ourimbah Campus, Australia [b] Bangladesh Institute of International and Strategic Studies (BIISS), Dhaka, Bangladesh [c] Knowledge for Development Management (K4DM) Project, UNDP, Bangladesh

8.1 Introduction

The quantity of municipal solid waste (MSW) is increasing due to rapid urbanization with high population growth and improving lifestyle of city dwellers. Globally, the population in cities is growing rapidly. In 1980, urban population was 1.73 billion (39% of total population) and the number of urban populations had climbed to 3.96 billion (54% of total population) in 2015 (United Nations, 2019). It has been predicted that by 2050, the urban share of total population may increase up to 6.4 billion (66% of total population globally) where most (around 90%) of the urban and total population will increase in developing regions such as in Asia and Africa. Currently, the estimated generation of MSW globally is around 2 billion tons per year (United Nations, 2014; United Nations, 2020). Given the increasing urban population and total population globally, it is reasonable to think that the quantity of MSW generation will increase significantly. The increasing consumption of goods in urban areas, changing lifestyle and social culture will drive the increase of the quantity of MSW generation. It has been estimated that the rate of waste generation will be more than double in the next two decades in developing countries (Hoornweg and Bhada-Tata, 2012). The management of considerable amount of waste therefore will be a great challenge to many local urban authorities, especially in the developing economies.

While the generation of MSW has been increasing in developing countries, literature shows that the management of MSW in developing economies provides a despondent picture. To date, most of the MSW generated in developing countries either goes to uncontrolled landfills or open dump sites. The burning of waste in open dump sites and the releases of gases from uncontrolled landfills pollute the air, water, and soil and thereby the ecosystem including human health can be affected in many ways (Mavropoulos et al., 2012). Also, open dumping and landfillable require a large area of land that may not be possible to provide for some countries (e.g., Bangladesh) where open land is limited (Kabir and Khan, 2020). Given the negative environmental impacts, methods for waste management such as uncontrolled landfilling and open dumping are not preferable options. Although sanitary or controlled landfills are modern version of uncontrolled landfills and possible to collect gas (e.g., methane) from sanitary landfills for generating electricity, the adoption of this option is also decreasing especially in developed countries. The reason behind this is the advancement of technologies for MSW management where the technologies may substantially reduce the environmental and social impact as well as the environmental pressure on cities. Furthermore, these technologies can be used to recover materials and generate electricity.

For safe and livable cities for residents, it is imperative to promote cities resilient to any disasters or environmental changes. In developing countries, the increasing pressure of population growth, expansion of urban areas and unsustainable management of MSW are making cities unlivable. Therefore, one of the sustainable development goals (SDGs) is to make the cities sustainable (Goal-11) socially, environmentally, culturally, and economically.

The improvement of MSW management can be one of the key potential solutions for making cities livable as well as to achieve the target (target-11.6) of Goal-11 by 2030. The trend of urbanization in developing countries indicates that there is a long way to go to make the urban life livable. Therefore, financial, and technical support from developed countries is desirable for the development of sustainable urban infrastructure as well as sustainable waste management. It is to be noted that about 95% of developing economies will experience urban expansion in the next decades (United Nations, 2019). Although only 3 percent of total land of the Earth is covered by city dwellers, the environmental footprint of cities is very high where cities are responsible for 60 to 80% of total consumption and 75% of human-induced carbon emission (UN-HABITAT, 2020). Undoubtedly, the generation of huge quantity of waste is the result of high consumption of food and other materials. As mentioned earlier, the sustainable management of MSW is one of the key aspects of a livable city. One of the key issues of sustainable MSW management is to generate energy from MSW using advanced and appropriate technologies. It is to be noted that one of the important targets of sustainable development Goal-7 (target-7.2) is to achieve the use of renewable energy globally by 2030. Thereby, the management of MSW through the generation of energy is relevant.

The advanced technologies for MSW management are going to be popular instead of traditional methods such as landfilling and open dumping. Although developing countries have been putting efforts to recover materials from MSW using recycling, this (recyclables) is only a small percentage of total MSW. It is necessary to go beyond recycling where materials will be recovered, and energy will be generated from MSW. The waste-to-energy (WtE) technologies not only recover materials for recycling but also offer the opportunity to generate electricity (GIZ, 2017). This means the WtE technologies may have the potential to add more value to the economy. Therefore, given the rapidly growing urban population in developing countries and

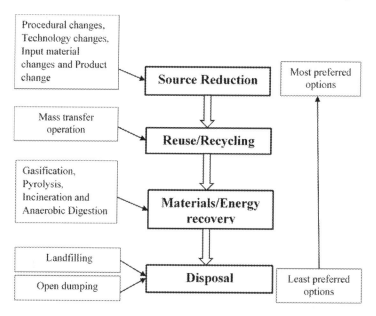

FIGURE 8.1 Showing tools and techniques for waste management.

generation of significant quantity of MSW, the use of WtE could be the best available solution to harness the benefits of waste management including electricity generation. The process of generating energy in different forms (such as heat, electricity, fuel) through primary treatment of the waste using different technologies is known as waste-to-energy generation.

The aim of this chapter is to provide an overview on renewable energy technologies (WtE technologies) as a source of sustainability in developing economies. This chapter is divided into seven sections. The second section offers a brief on foundational content of waste management tools and techniques following an introduction. The third section provides a description of WtE technologies with a particular focus on electricity generation. The fourth section provides the status of MSW management in developing countries. The fifth section is on the contribution of WtE to sustainability in developing countries. The sixth section is about the prospects of the developing economies relating to the use of WtE technologies for waste management. The seventh section is a case study where Bangladesh has represented the developing countries to understand the situation of waste management and prospect of WtE energy application. This is followed by a conclusion.

8.2 Waste management tools and techniques

Several tools and techniques are available for management interventions of solid waste. The preference in the use of operational tools and techniques for solid waste management can be arranged as a minimization hierarchy as shown in Fig. 8.1, where source reduction is the most preferred option and disposal is the least. The fundamental argument for waste minimization is that it is far better to reduce waste in the first place than to cope inadequately with their aftermaths (Holmes, 1997). The philosophy behind the waste minimization hierarchy is that it is crucial in developing policies for using appropriate tools and techniques and that the solutions are easier to deal with at the front end of the situation rather than at the back end.

Fig. 8.1 showing the reduction and reuse including recycling are preferred options to WtE technologies and disposal of waste. However, the reduction or reuse (including recycling) of products depend on the behavior of consumers and it is difficult to make effective these options particularly in developing countries where people are not much aware of the negative impacts of waste. Furthermore, it is not possible to make the generation of waste at 'zero waste' level even if these techniques are used. To address this limitation or to harness the benefits for example, electricity generation from increasing MSW, some advanced technologies have been developed and found to be useful to recover materials as well as to generate electricity. These may include, for example, controlled landfilling, gasification, pyrolysis, anaerobic digestion, and incineration. The application of these technologies may depend on the geographic size and location of a country, the quantity of waste generated in a municipality area, the physical and chemical characteristics of MSW generated in a city.

8.3 Available WtE generation technologies

Literature shows that there are many methods for WtE development and thereby electricity generation. In general, the advanced technologies used for recovery of resources and generation of energy or electricity from waste can be broadly categorized in two types. This includes (i) thermochemical process that involves gasification, incineration, and pyrolysis, and (ii) biological process that involves anaerobic digestion, composting, and landfill (Kabir and Khan, 2020). Fig. 8.2 shows a schematic view of some advanced methods for MSW management.

8.3.1 Sanitary landfilling

Sanitary landfilling is an improved waste disposal technique, which refers to waste compaction, exclusion of moisture and leachate and extraction of gas from waste, and daily application of soil cover on the dumped waste to reduce air pollution and spread of diseases (Blight and Mande, 1996). The requirement depends on the treatment of solid waste prior to the disposal depends on its composition. For example, sanitary landfills with impermeable liners, constructed of clay or polyethylene can be a good option to avoid groundwater contamination by leaching waste minerals (Miller, 2000). Furthermore, it is apparent that high and upper-middle-income countries are operating controlled landfills and going beyond recycling adopting advanced technologies (Kabir and Khan, 2020). In general, countries with lower income rely on open dumping and this may account for 93% of total waste generated (Goldberg, 2018). Fig. 8.3 shows a systematic flowchart of generating electricity from landfill gases.

8.3.2 Gasification

Gasification is another WtE generation technology and this process is undertaken in the absence of oxygen (Seo et al., 2018). The process depends on temperature which is above 650°C, vapor pressure, heating rate, and concentration of O_2 in the reactor. Heat production

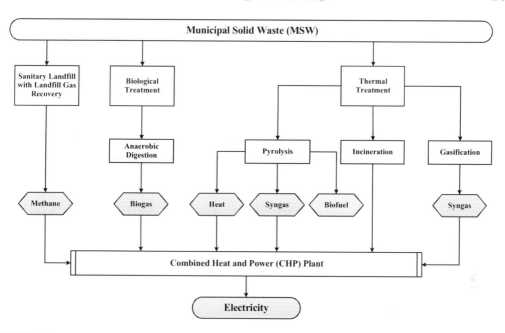

FIGURE 8.2 A schematic basic view of waste-to-energy generation technologies for electricity production.

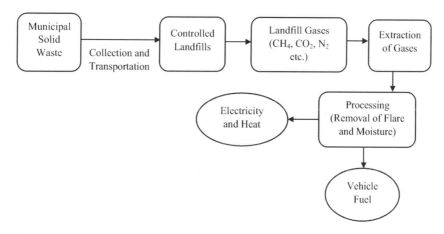

FIGURE 8.3 A systematic flowchart of generating electricity and fuel from controlled landfills gases.

is the character of gasification through reaction. Overall, there are three stages of gasification. First, syngas is produced at the first stage through a gasifier. Second, the syngas is cleaned by removing the pollutants and harmful compounds such as tar. At the third stage, the syngas is used to recover energy using a gas engine (Seo et al., 2018). Gasification technology may produce about 1000 kWh electricity from one ton of MSW. Fig. 8.4 shows a schematic process of gasification. It is to be noted that there are different types of gasifiers used to produce syngas

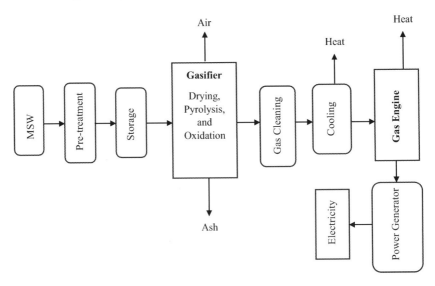

FIGURE 8.4 A schematic showing the process of gasification.

given the nature of waste. For example, some gasifiers are designed to gasify construction and demolition debris. Other gasifiers are used for MSW. In general, gasification requires preprocessing to remove inorganic materials such as glass and metals because these materials are not possible to gasify.

Syngas is the key gas produced from gasification. Syngas is the combination of hydrogen, carbon monoxide, carbon dioxide, and methane. Also, other materials such as oils, inert gases, tars, slag, water, and char gas pollutants are produced from gasification (Seo et al., 2018). The structure and amount of syngas produced from gasification usually depend on the operational parameters of the gasifier. These parameters may include the features of feedstock, level of air, steam, CO_2, plasma, temperature, pressure, and catalyst type. Ammonia, methanol, and hydrogen are produced from syngas as major chemicals (Indrawan et al., 2020).

8.3.3 Incineration

This is a common WtE generation technology around the world to recover energy from MSW. An incineration facility for the treatment of MSW may go through several stages. These may include (i) collection and storage of MSW used as feedstock, (ii) the feedstock is combusted in a furnace to generate hot gases in addition to residue such as bottom ash, (iii) heat is recovered through the generation of steam at continuously reduced gas temperature, (iv) air pollutants are removed through the treatment of cooled gas, (v) disposal of residuals produced from the process, and (vi) releasing of treated gas to the atmosphere through a stack (Liu et al., 2020). The gases produced and released after treatment are mostly GHGs. Fig. 8.5 shows a schematic view of the incineration process.

The incineration process is the combination of pyrolysis, gasification, and combustion phases. Waste is used as feedstock without any pretreatment for the incineration process.

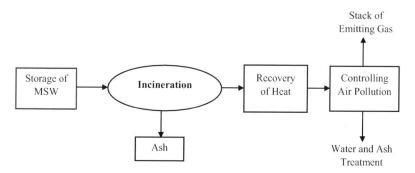

FIGURE 8.5 Schematic view of incineration process.

To oxidize the feedstock, enough air is needed in this process. Also, the conversion of waste materials into products requires a high temperature of 850°C. Carbon dioxide (CO_2), water and solid residue called bottom ash are the products produced from the conversion of waste materials (Zaman, 2010). Bottom ash is useful to the construction of infrastructure such as roads and bridges (Sahin et al., 2018). Although variation exists to the incineration process, most facilities have common units of operation. Table 8.1 shows the products and bi-products of MSW treatment using incineration.

8.3.4 Pyrolysis

Pyrolysis is another thermal process where various materials are recovered from the conversion of MSW, and the materials are used to generate heat and electricity. About 80% of energy can be recovered from the treatment of MSW using pyrolysis process (Chen et al., 2015; Ouda et al., 2016). There are three different types of chemical reactions in pyrolysis depending on the transfer rate of heat, residence time, particle size and temperature. These chemical reaction processes comprise slow pyrolysis that generates charcoal, fast pyrolysis that generates bio-oil, and flash pyrolysis that generates gas such as syngas (Qazi et al., 2018). There are different types of reactors in the pyrolysis process. In general, the fluidized bed reactor-based pyrolysis reaction is more popular than others among the different reactors. Temperature required for the degradation of MSW in the pyrolysis process may range from 300 to 800°C (Qazi et al. 2018b).

Separation of recyclables such as glasses, metals (e.g., copper and zinc), inert materials such as sand, soil, concrete, rock from MSW is required before sending the MSW at the inlet point of the pyrolizer. The process initially begins with the break-down of pretreated MSW at the chamber heated with 300°C (Qazi et al., 2018). The temperature of the chamber is then elevated up to 800°C. The main products produced from pyrolysis are biochar and bio-oil. The production of these products depends on composition, moisture, and particle size of biomass or MSW, reaction time, feed rate, gas flow rate, temperature, and heating rate during chemical reaction (Guedes et al., 2018; Bach and Chen, 2017). Also, some gaseous products including carbon monoxide (CO), methane (CH_4), hydrogen (H_2), and carbon dioxide (CO_2) are generated. Fig. 8.6 displays the schematic representation of this technology.

TABLE 8.1 The products and bi-products generated from MSW treatment through incineration.

Product type	Quantity
Electricity (kWh/t)	519
Heat (MJ/t)	1785
Recovered materials	
Ferrous metals (kg/t)	22.1
Bottom ash (kg/t)	219.4
Hazardous pollutants	
Air pollution residue (kg/t)	31.3
Carbon dioxide (CO_2) (kg/t)	452
Carbon monoxide (CO) (g/t)	48.7
Sulphur dioxide (SO_2) (g/t)	94.9
Nitrous oxide (N_2O) (kg/t)	1.7
Hydrogen chloride (HCl) (g/t)	32.3
Ammonia (NH_3) (g/t)	17.9
Particulates PM_{10} (g/t)	17.9
Dioxin (Furans) (ng/t)	92.3

Source: Adapted from (Kabir and Khan, 2020).

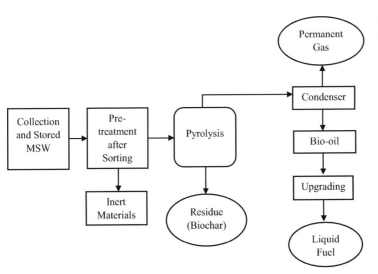

FIGURE 8.6 Flowchart of pyrolysis process for MSW (partially adapted from Qazi et al., 2018).

Bio-oil is dark-brown liquid containing gases (hydrogen, syngas), chemicals (resin, fertilizer, flavors, adhesive, acetic acid), heat from cofiring of boiler and furnace and power from diesel engine and turbine. Biodiesel can be produced from bio-oil. Thereby, bio-oil can be used as liquid fuel for diesel engines and to generate electricity through gas turbines. Biogas (e.g., syngas) produced from pyrolysis is mainly the mixture of methane, carbon monoxide, hydrogen in addition to a wide variety of Volatile Organic Compounds (VOCs) and can be used to produce electricity. The syngas is cleaned for production of electricity by removing particulates, hydrocarbons or tar and soluble matter. Finally, the residue generated from pyrolysis is called "char" consists of noncombustible materials and carbon (Qazi et al. 2018; Guedes et al., 2018) and used as solid fuel, adsorbent and to increase soil quality.

8.3.5 Anaerobic digestion

In anaerobic digestion (AD) process, the organic fraction of MSW is separated from other materials to be used as feedstock. This process decomposes the organic waste with the help of micro-organisms and the process occurs in closed spaces in the absence of oxygen. The AD process consists of different stages namely acidogenesis, acetogenesis, and methanogenesis and a suitable atmosphere to produce desired products is created involving microorganisms (Khan, 2020; Farooq et al., 2021). Appropriate conditions including well-maintained temperature and pH level in the reactor is necessary for the AD process (Qazi et al., 2018). According to the use of microorganisms in the corresponding stage of the process and different ranges for temperature at the same time, the pH level may range from 6.7 to 7. Considering the economic feasibility, mesophilic or thermophilic temperature are followed in most cases (Mutz et al., 2017). The organic feedstock is placed in the digester for 5 to 10 days. Reactor's type varies given the nature of feedstock to generate and recover energy from MSW. For example, a continuously stirred tank reactor is used for food waste, while plug-flow and batch reactors are used for other types of organic waste (Mutz et al., 2017). Fig. 8.7 shows a schematic view of WtE generation through AD process.

Biogas, fiber and liquid digestate are key products produced from the AD process. Among the products, biogas is the dominating product that contains 50 to 80% methane (CH_4), 20 to 50% carbon dioxide (CO_2), and little amount of sulfide and ammonia (Qazi et al., 2018). Biogas can be used as an alternative to natural gas to generate combined heat and power (CHP). However, heat and power production efficiency of biogas (around 5.5–7.5 kWh/m^3) is less than that of natural gas. This is due to the lower calorific value of biogas (Qazi et al., 2018; Mutz et al., 2017; Khan, 2020). The other two products (liquid digestate and fiber) are rich in nutrients and are used as fertilizer for crops.

8.4 Present status of MSW in developing countries

8.4.1 Collection of MSW

In low-income countries around 33 to 40% of total MSW are collected and of which 93% are burnt in open space (Goldberg, 2018). Municipalities collect MSW through public-private partnership in many developing countries. Also, MSW is collected by informal sectors such

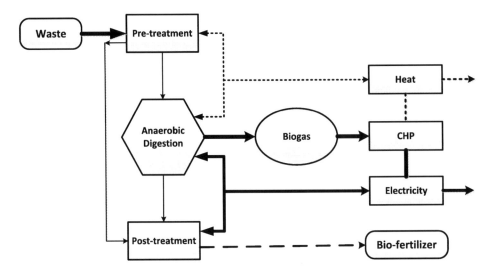

FIGURE 8.7 WtE generation process through anaerobic digestion. CHP, combined Heat and Power.

as scavengers who usually collect the recyclables from dumpsites and supply to factories for recycling. In most of the major cities in Bangladesh, for example, there is no door-to-door MSW collection system. On the other hand, there are improved waste collection systems (for example, door to door collection) in developed countries and almost hundred percent of total waste is collected from sources and only 2% of waste goes to dumpsites. The coverage rate of MSW collection in the middle-income countries is between 50% and 80% (World Bank, 2018).

8.4.2 Composition of waste

The composition of waste varies substantially from one jurisdiction to another though the quantity of MSW is increasing rapidly. This variation in composition depends on the income level, quality of life and living standard, lifestyle, social culture, geographic location, and weather conditions of a country (World Bank, 2012; Couth and Trois, 2011). There is a clear difference in MSW composition between developed and developing countries. The composition of MSW varies among the countries with high, middle, and low-income status as shown in Table 8.2. The organic part of MSW in low- and middle-income countries is higher than that of in countries with high income. In general, the composition of MSW in developing countries is characterized by high portion of organic materials (56–64%), high moisture (40–80%), high density (250–500 kg/m^3), and low calorific value (CV) (800–1,100 kcal/kg) (Khan, 2020). In contrast, countries with high income generate MSW with relatively lower portion of organic materials (20–30%), lower moisture and high calorific value (Habib et al., 2021; Tun and Juchelková, 2019; Ozcan et al., 2016) than those of in developing countries. Importantly, the application of WtE technology is largely dependent on these characteristics of MSW irrespective of socioeconomic and geographic context of a country (Khan, 2020).

TABLE 8.2 Average waste composition in various income group of countries.

Countries by Income	Composition of materials				
	Organic (%)	Paper (%)	Plastic (%)	Metals and glass (%)	Others (%)
Low-income countries	64	6	9	6	15
Middle-income countries	56	12	13	7	12
High-income countries	28	30	11	13	18

Source: (Kumar and Samadder, 2017).

TABLE 8.3 MSW disposal methods by countries.

Countries	Open dumping	Landfill	Composting	Recycling	Incineration	Other advanced method
HI	2%	39%	6%	29%	22%	2%
UMI	30%	54%	2%	4%	10%	-
LMI	66%	18%	10%	6%	1%	-
LI	93%	3%	0.3%	3.7%	-	-

Source: (World Bank, 2018).
Note: HI, high income; UMI, upper middle income; LMI, lower middle income; LI, low income.

8.4.3 Disposal of MSW

Table 8.3 indicates a representative situation of the disposal of MSW in countries by the level of income. It is clear that countries with more income have a lower percentage of open dumping of MSW. This is completely opposite to the countries with low and lower-middle income where the practice of open dumping of MSW is very common. The Table 8.3 indicates that the percentage of open dumping of MSW usually lies between 66% and 93%. On the other hand, countries with high and upper middle income show much lower percentage of open dumping of MSW; only 2% and 30% respectively than that of countries with low and lower middle income (World Bank, 2018).

Almost all developing countries depend on uncontrolled landfills and open dumping combining 96% and only around 4% on recycling (World Bank, 2018). On the other hand, high income countries manage MSW using controlled or sanitary landfills and recycling in addition to advanced technologies such as incineration. In South Asia, for example, four out of the eight countries have recycling systems of MSW, and the percentage of recycling of MSW ranges from 1 to 13%. To manage organic waste, seven out of the eight countries in South Asia have begun composting programs (World Bank, 2017). One of the key reasons for the low level of waste recycling in South Asia is that 44% of total waste is burnt that may carry energy (Haas et al., 2015). Therefore, recycling is undertaken at a limited scale and cannot add value to the economy adequately. This scenario of waste disposal is available in many other developing countries.

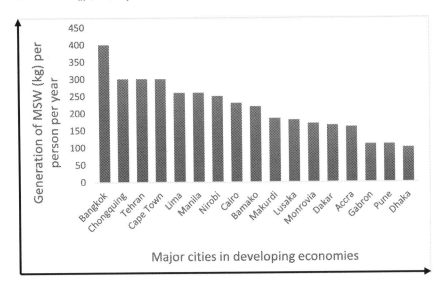

FIGURE 8.8 Generation of MSW (kg/per-capita/year) in some selected major cities of developing countries (adapted from EAWAG, 2008).

In developing countries, an urban citizen generates on average between 100 and 400 kg of MSW per year (EAWAG, 2008). This wide range of variation in generation of per capita MSW per year is due to variation in per capita income, GDP growth rate and consumption pattern. Fig. 8.8 shows the MSW generation per person per year in some of the cities in developing countries.

8.4.4 WtE technology status

Study shows that as of 2018, there are 1700 WtE plants worldwide (GIZ, 2017). Although WtE technology as the source of electricity is popularly increasing globally, only 20% of total WtE technologies are available in developing countries and 80% are in the developed world (GIZ, 2017). For MSW management, landfill, and open dumping are still prevalent in developing countries. Only a few countries have taken initiative to install advanced technologies such as incineration or anaerobic digestion. This shifting from landfilling, open dumping and burning practices to WtE plants may support the management of MSW smartly (UNEP, 2017).

Most developing countries are still depending on landfills and open dumping as mentioned earlier. Study shows that there are hardly any WtE technologies such as anaerobic digestion for large scale operation in developing countries (World Bank, 2018). There are a few incinerators under continuous operation in developing countries in Asia and Africa. Some countries have taken initiatives to install and operate pyrolysis and gasification technologies for MSW management but could not move beyond pilot scale. One of the key reasons could be the framework conditions of developing countries are different from the developed ones where the application of WtE technologies is on the rise. Given the difference in socioeconomic and cultural context, the transfer of technology directly from developed to developing countries may not be possible. This requires meeting the satiation of developing countries such as financial requirements, composition of input materials, and local capacities. Nonetheless, WtE technologies are essential in fast-growing cities of developing countries and there is a potential to advance the waste management system for the generation of electricity. Overall, the objective of the use of WtE is not only to reduce the volume of MSW and environmental pollution but also to contribute to the economy through recovery of materials and generation of electricity (GIZ, 2017).

8.5 Future prospect in the developing world

While many of the developing countries are at their infancy in the application of WtE technologies, the future prospect in the developing world is bright. For example, countries like China, Brazil and India have installed renewable energy capacity including 895, 150, and 134 GW, respectively (Statista, 2020). In developing countries, the amount of waste will increase due to population growth, urbanization, and economic growth. Therefore, WtE can be one of the best solutions for the management of MSW as well as to generate electricity. At present, developing countries send 93% of MSW to landfill and open fields where hardly any gases from landfill are extracted for generation of electricity (Goldberg, 2018). The use of WtE is an opportunity to modernize the landfills to extract gases for electricity generation.

For continuous economic growth and industrial development, developing countries are concerned about the increasing need for energy and therefore, looking for alternative sources of energy. For example, South and Southeast Asian countries need increasing energy for economic growth and now heavily dependent on fossil fuel such as Bangladesh (Khan, 2020a; Khan, 2019). The countries of this region are now looking for renewable energy not only to reduce the dependency on fossil fuel but also to utilize the huge amount of MSW and other sources of renewable energy. Many member countries in this region proposed the utilization of WtE technologies to recover energy from MSW (Rohatgi 2017) as much as possible. The need for sustainable energy development has attracted these countries to apply WtE technologies in addition to the management of MSW smartly. It is to be noted that many countries of this region are now recovering energy by converting existing landfill stations into sanitary landfills in addition to the establishment of other WtE technologies such as incineration. In Malaysia, the adoption of WtE technologies resulted in the installation of an incineration plant at Langkawi and establishment of biogas power plants to convert MSW to electricity (Rohatgi, 2017). There are four mini-incinerators under implementation in four locations of the country and recently another WtE plant is under construction. In general, energy is recovered from MSW in Malaysia from thermal, biological, and landfilling via LFG (Wong et al. 2019).

Another example is Thailand, where initiative has been taken to recover energy from MSW. Thailand's target to collect recoverable MSW fraction has been estimated to be 83.4% by 2025 where 58.2% of MSW is compostable and 25.2% is recyclable (Shapkota, 2006). The government is endorsing recycling as one of the key activities of sustainable waste management. The government has been promoting recycling as one of the major techniques to manage the increasing MSW in a sustainable way and advanced meaningfully. For recycling, the government has developed good practice of waste management including sorting, collection from sources, separation, sorting, and pretreatment where and when applicable to convert the MSW into useful products. In addition to recycling, the government has planned to use advanced technologies for MSW management. It has been estimated that around 5.1 Mt (19.5% of total MSW generated) of MSW would be managed through recycling, 3.9 Mt through composting, 1.1 Mt using anaerobic digestion technology, and 0.1 Mt using incineration in 2013 (PCD, 2014). While municipality authorities are responsible to collect the MSW, there are also public-private partnership mechanisms to collect MSW.

One of the key prospects of WtE as a source of renewable electricity in developing countries is the commitment of the Abu Dhabi Fund for Development (ADFD) to deliver USD 350 million to developing countries for the execution of renewable energy projects supported by governments. Since January 2014, the IRENA/ADFD Project Facility have selected a total of 21 renewable energy projects and received USD 214 million loans with concession. Among the 21 nominated projects, 16 are now progressive over some steps of execution (IRENA, 2019).

Donor agencies, be it multilateral or bilateral, for example, the Asian Development Bank (ADB) or GIZ have accentuated the prospect of WtE in developing countries. The ADB has been supporting the implementation of WtE technologies in Asia and the Pacific for more than a decade. It is interesting to know that the agency's first initiative was the approval of a project in China in 2007. The aim of the project was to generate power from biomass in rural China. The agency has proposed a total of 27 WtE projects in 2018. Among the 27 projects, six projects are active, and three projects are already approved. "With nearly half of the projects in the proposed stage, there is great potential of WtE in the ADB portfolio in the coming years" (ADB, 2020). Countries such as China, Malaysia, Brazil, and India have already developed plans for WtE technology and are acting accordingly. For example, there are four incineration plants in Malaysia and another plant was built and completed in 2018 under 10th Malaysia Plan (Wong et al., 2019).

While there is huge potential of generation of electricity from MSW in the developing world, there are some challenges that need to be addressed to harness the benefits of the application of WtE technologies. The characteristics of MSW in developing countries include lower calorific value and high moisture content compared to those of in developed countries. This may remain a challenge especially for thermal WtE technologies where energy content or low calorific value (LCV) of waste should be 7 MJ/kg and never fall below than 6 MJ/kg on an average over a year (Blight and Mande, 1996). It is to be noted that the LCV of one kg fuel is about 40 MJ/kg. To maintain the LCV not below the 6 MJ/kg and make it suitable for WtE technology, it is necessary to sort the MSW properly. Other challenges include whether the waste minimizations hierarchy (e.g., avoid, reduce, reuse/repair and recycling) is followed properly. Also, the prerequisites for a favorable environment of WtE technologies include information on the composition and quantity of waste, sources of financing, skilled personnel, globally accepted emission standards, and the selection of more suitable WtE technology given the context of the country and characteristics of MSW (GIZ, 2017).

8.6 Contribution to sustainability in developing countries

There is a potential of WtE plants to contribute towards sustainability in developing countries considering social, economic, and environmental aspects compared to fossil-fuel based electricity generation. A well-managed WtE technology may reduce storage of physical

TABLE 8.4 Sustainable development goals' targets relating to renewable energy.

SDGs	Targets
SDG-7	(i) To make sure the inexpensive and consistent supply of services relating to energy. (ii) To enhance the portion of renewable energy to universal energy mix by 2030 substantially. (iii) To make the rate of improvement of energy proficiency twice worldwide by 2030. (iv) To make available technologies for renewable energy generation and encourage investment for infrastructural facilities by 2030; and (v) To intensify the availability of facilities relating to sustainable renewable energy globally by 2030.
SDG-11	(i) To reduce environmental impact of cities by enhancing the collection of MSW and managing the MSW generated by cities in a sustainable way by using advanced technologies by 2030. (ii) To reduce or maintain the level of suspended particulate matters (SPM) (e.g., $PM_{2.5}$ and PM_{10}) in cities to avoid the impact on human health by 2030. (iii) To make the cities livable by enhancing safety, resilient to disasters and inclusiveness by 2030.

Source: United Nations, Sustainable Development Goals (2015) (https://sdgs.un.org/goals).

waste, emit relatively lower amounts of carbon, contaminate minimum land, and recover more energy (Zhang, 2015; Nwokolo et al., 2017). In developing countries, currently landfills are used for MSW management and these require huge areas of land. The WtE technologies are good for developing countries with a scarcity of land for landfills, for example, Bangladesh (International Business Publications, 2015) where more than 1000 people live per square kilometer. The WtE as a source of renewable energy may play a vital role in the achievement of a sustainable energy ecosystem in addition to an efficient substitute for fossil fuel for generating electricity. Also, biofuel produced using WtE technologies, can play an increasing role in reducing carbon emissions from marine vessels and airplanes since electric vehicles and fuels cells are not suitable for these transport modes.

For electricity generation, developing countries are heavily dependent on fossil fuel such as natural gas and diesel fuel and only 2 to 3% of electricity is generated from renewable energy sources (UNEP, 2017). WtE technologies used for MSW treatment may offer several options for producing alternatives for gasoline and diesel. For example, production of ethanol from MSW may replace gasoline and biodiesel may replace traditional fossil-fuels used for transport (Swaraz et al., 2019). Similarly, the reduction of air pollution and impact on public health is possible to a great extent using WtE technologies (ADB, 2020). It is to be noted that 93% of the MSW in developing countries either goes to landfill or open dumping stations. Landfills without gas capture can erupt fire, for example in Myanmar in 2018 (Goldberg, 2018) and release toxic pollutants in the air. Also, it is common to burn solid waste in landfills in developing countries and these fires generate health risks. Similarly, the burning of dumped MSW in the open air directly releases toxic pollutants such as suspended particulate matters (SPMs). The adoption of a better waste management system through WtE technologies may reduce the air pollution.

The WtE technologies may contribute to the sustainability in developing countries through the achievement of SDG targets specially SDG-7 and SDG-11 (Table 8.4). While the achievement of all the targets depicted in Table 8.4 are important, the enhancement of the share of renewable energy using MSW is relatively more important than others. This will play a key role to make the cities livable by reducing environmental pollution. It is well recognized that the production of renewable energy may significantly reduce the release of GHGs. This means the conversion of MSW into clean electricity using WtE technology will not only reduce the amount of MSW in a sustainable way but also make the cities livable (Tyagi, 2013; Kothari, 2010).

One example of MSW management using advanced technology is Delhi in India, a megacity. The East Delhi municipal corporation is operating a landfill including a WtE plant through partnership with infrastructure leasing and financial services environment (IL&FS Environment). The WtE technology generates 12 MW of power annually using the gas from the landfill (UNEP, 2017). In addition to energy, the plant also produces 127 tons of fuel annually. It has been predicted that the WtE plant will save approximately 8.2 million tons of carbon emissions or will avoid the emission of 8.2 million tons of methane emissions over its 25-year life span. The benefits of the project will go beyond the avoidance of methane emissions. The plant saved 200 acres of land that was used for dumping of MSW/from being consumed by the landfill. The value of the land is US$308 million.

Among the outputs generated using WtE technologies, biofuel is an important option in the transport sector for the reduction of GHGs. Biofuel can be produced from syngas where syngas is generated from gasification and pyrolysis. Biofuel also may enhance the fuel efficiency and electrification transport. Also, biofuel can play a role in reduction of air pollution and mitigation of climate change impacts. It has been found that the first-generation biofuels such as from corn-based ethanol may address these challenges. Furthermore, the second-generation biofuels from agricultural waste and MSW have resulted in a variety of substitutes to gasoline or diesel (ADB, 2020). Agricultural biomass for example, residue from paddy field or sugarcane field can be used for pyrolysis. Biochar is generated from pyrolysis and may be used for the improvement of soil fertility. This may reduce the acidity of the soil and increase water nutrient retention due to its high porosity and thereby help boosting agricultural productivity. Biochar is efficient to bind CO_2 from the atmosphere to the soil. Although, biochar can be used as fuel it provides a good support for microorganisms to balance the soil nutrients. It can be used as fuel but is best as a high-surface area scaffold for microbes to fix nutrients into soils.

The transferring of WtE technologies from developed to developed countries is not possible directly given the variation of characteristics of MSW and different socioeconomic and cultural context of the developing countries. Also, the acceptance of technology depends on the local community who will be affected by the technology as well as benefited. Therefore, participation of community people at the beginning of adoption of WtE is important in addition to research (Intharathirat et al., 2015).

To promote sustainability using WtE technology, the financial aspects of the establishment and operation of WtE technologies is a key issue. Sufficient resources in many developing countries may not be available to install the plants and the operation of plants. To secure continuous operation and maintenance in addition to establishment of a plant, the government of developing countries may go for public private partnership. Furthermore, donor agencies may provide financial support in this regard. The initial investment for WtE technologies may be available in developing countries but financing for the long-term operation of plants may not be adequate (Intharathirat and Salam, 2015). Comprehensive financial feasibility for each of the potential WtE technologies is necessary to compare and select the most suitable one. Study found that for example, for an incineration technology with the treatment capacity of

TABLE 8.5 Estimated costs (in Euro) for incineration in industrialized (developed) and emerging (developing) countries.

Capacity: 150,000 t/a	Installation costs	Capital costs per ton of waste input	Operation and management costs/ton	Total cost/ton	Revenues from energy sales/ton	Costs to be covered per ton waste input
Developed countries (EU countries)	135–185 million Euro	80–115 Euro/t	180 Euro/t	260–295 Euro/t	60 Euro/t (Heat and electricity) 27 Euro/t (Electricity)	200–235 Euro/t
Developing countries	30–75 million Euro	22–55 Euro/t	20–35 Euro/t	42–90 Euro/t	2–10 Euro/t (Electricity)	40–80 Euro/t

Source: (GIZ, 2017).

150,000 metric tons of waste annually the market revenues from selling recovered energy and materials only may not cover the full annual cost of the plant. The expected net costs per metric ton of waste must be covered by other financing means such as additional revenues from gate fees, subsidies from government so that operations can be financed sustainably for the life of the plant. It is to be noted that in developed countries the cost for installation and operation of a WtE is well established. On the other hand, it is sometimes difficult to provide demonstrative costing information in developing countries. Table 8.5 provides a comparative picture of estimated cost of incineration plants for MSW treatment. It can be seen that in both developed and developing countries the cost to be recovered per ton of waste input is proportionately similar comparing with total cost. In developing countries, the recovery of costs from per ton of waste is 40 to 80 Euro against the total cost 42-90 Euro. In developed countries the cost to be recovered from one ton of waste is 200 to 235 Euro against the total investment. However, the revenue generation from WtE plants in developing countries is relatively lower than that of in developed countries.

8.7 Case study: Bangladesh

8.7.1 Generation and composition of MSW

The generation of waste in Bangladesh is increasing due to population growth (rate: 1.31%), rural-urban migration, rapid urbanization, and improved lifestyle due to economic growth. It is to be noted that Bangladesh is graduating from least developed country (LDC) to developing country due to continuous economic growth and the growth rate is around 7% over the last decade. The country has targeted to be a developed nation by 2041. Therefore, more energy (electricity) generation is required to meet the basic need of growing population and the demand of electricity for both rural and urban economies including industries. In future, the country will face challenges with the emission of greenhouse gases (GHGs) in addition to more waste. At present average solid waste generation per capita in Bangladesh is 0.7 kg. However, per capita waste will increase in the future in Bangladesh because per capita income is gradually increasing. Due to mass consumption and higher purchasing capacity, already major cities are facing challenges with generation of more waste. For example, Dhaka one of the megacities in the world is generating around 4334.52 kg per person per year the highest amount of waste among the cities in Bangladesh and the lowest amount of waste is generated in Barisal city with 134.38 kg per person per year (Khan et al., 2020; UNDP, 2017). It is to be noted that in 336 municipalities including 8 cities the generation of waste varies from 0.25 to 0.56 kg per person per day. In Dhaka city, the current generation of total waste is likely to be 120,000 ton per year in 2021–2022 (UNDP, 2019).

In Bangladesh, the composition of MSW includes organic waste such as kitchen or household waste, glasses, metals, papers, wood, plastics, textile, glass, rubber, and debris. The density of these materials varies. It has been estimated that per cubic meter of MSW include compostable 240 kg, paper 85 kg, plastic 65 kg, metal 320 kg, wood 240 kg, textile 65 kg, glasses 195 kg and, wreckage 480 kg (UNDP, 2017). A recent study on the composition of waste in four major cities in Bangladesh indicates that most of the MSW is organic in nature where 70% of total MSW is compostable (food and vegetable waste). The other materials include paper 4.5%, plastics 5.3%, metals 0.7%, wood grasses and leaves 4.5%, rags, textile, and jute 3.7%, glasses 0.3%, organic noncompostable 7%, and others 3% (Habib et al., 2021). In general, the organic or compostable part of MSW in developing countries is higher than that of developed countries (World Bank, 2017). Furthermore, the waste generated in Bangladesh contains a relatively high amount of moisture than that of developed countries.

8.7.2 WtE and institutional policy

The generation of electricity in Bangladesh depends on gas, imported coal and diesel. This fossil fuel-based electricity generation is contributing to the generation of GHGs. Given the realization of limited stock of fossil fuels globally and the impacts of the use of fossil fuel for electricity generation on the environment and society, the Government of Bangladesh (GoB) has adopted a renewable energy policy in 2008. The policy targeted to generate 10% of total electricity from renewable energy sources by 2020 although the contribution of electricity generated from renewable energy sources was around 3% only (British Petroleum, 2019). The policy identified MSW as one of the potential sources of renewable energy in addition to biomass, solar, hydro, and windmills. Importantly, the National Energy Policy (2004) has mentioned the requirement of waste-to-energy technologies to manage MSW in a sustainable way. Accordingly, the GoB has established a separate organization called Sustainable and Renewable Energy Development Authority (SREDA) to promote renewable energy (GOB, 2008).

8.7.3 Potential of WtE technologies in Bangladesh

Although the energy policies in Bangladesh underscores the generation of electricity from renewable energy sources such as MSW, there is an inadequate initiative to generate electricity from MSW though there is a huge potential to generate electricity from MSW (Kabir and Khan, 2020; Khan et al., 2020). Currently, there are few WtE plants under operation in Bangladesh to generate electricity.

For example, one WtE plant based on anaerobic digestion (AD) has been established in Jashore located in the south-western region of Bangladesh. This is the first kind of WtE plant

TABLE 8.6 Energy recovery potential using anaerobic digestion in Rajshahi city, Bangladesh.

Total MSW generation (tons/day)	MSW from food and vegetables (tons/day)	Total biogas generation* (m^3)	Compost fertilizer** (tons/day)
358	259 (72.29%)	38,850	65

(Source: Rana, 2016).
* From 1 ton of MSW, 150 m^3 biogas can be produced by anaerobic digestion.
** 250 tons of compost fertilizer can be produced from 1000 tons of MSW by anaerobic digestion.

in Bangladesh established in 2017 with the support of Asian Development Bank. Previously, the location of the plant was an open dumping site where MSW from Jashore was dumped. Now, this place has been converted into the site for the WtE plant. It is expected that the production of compost will be used for agricultural production where 2 to 2.5 tons of compost are produced per day. Currently the total generation of MSW in this area is around 50 tons every day. However, only 20 tons of total waste are collected, and this amount of waste is not adequate given the capacity of the plant. The municipal authority is concerned about the demand of the waste for the plant but cannot supply adequate MSW due to poor collection system. It is expected that the collection system of MSW will be improved to meet the demand of the plant at its full capacity. This will enable the plant to produce 450 cubic meter per day and generate 200 kWh electricity.

The installation of another WtE plant is under consideration nearby Dhaka city. Certainly, there is a great potential of generating electricity from waste generated in Bangladesh if the collection system and sorting of MSW can be improved. A study on available WtE technologies for MSW treatment in Rajshahi city shows the potential of recovering materials and energy (Habib et al., 2020). It was found that the AD process is more suitable technology for MSW treatment in the city. It was calculated that approximately 159.40MWh energy can be generated from MSW of Rajshahi municipality area. Furthermore, recyclables can be separated from collected MSW and can be used to produce various recycled and saleable products. Thereby, the WtE technologies show the prospect of sustainability in the cities in Bangladesh. For example, Table 8.6 shows the total generation of MSW in Rajshahi city and potential production of biogas and fertilizer.

In addition to MSW, there is supply of feedstock from the agricultural sector. It is to be noted that Bangladesh is an agricultural country and there is a huge amount of agricultural residue generated that can be used as feedstock for WtE plants to generate electricity. Overall, Bangladesh is still at the infancy stage of using WtE technologies for generating energy from waste. The barriers or challenges behind this may include lack of awareness about waste management with proper procedure, poor capacity of collection of waste, absence of mechanical mechanism for collection and sorting of waste, lack of financing for WtE technologies, lack of action plan for waste management, and inadequate people with technical knowhow about WtE technologies and their operations (Habib et al., 2021; Zaman, 2010).

One of the key issues relating to the characteristics of MSW in Bangladesh is that the MSW in general consists of ash (10–20%), combustible (20–30%), and moisture or water 57 to 68% (UNDP, 2017). Because of relatively more water content (average around 70%) of MSW in Bangladesh than the standard where water or moisture content should be less than 50% for

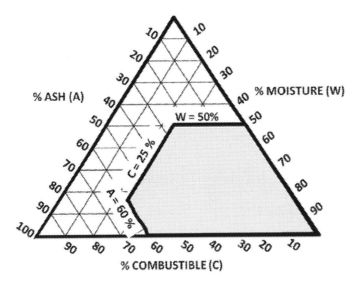

FIGURE 8.9 Tanner triangle showing the combustibility of fuel (MSW). Grey colored area indicates the capability of any fuel (MSW) to be combusted without any auxiliary fuels. *(Reproduced from Kabir and Khan, 2020).*

WtE conversion, an additional auxiliary fuel is needed for WtE technologies such as pyrolysis, incineration, and gasification. This can be explained by the "tanner triangle" (Fig. 8.9) which shows that the MSW in Bangladesh as a source of energy fall outside the grey area of the triangle (Kabir and Khan, 2020). That means, due to higher moisture than the standard level, AD process is more suitable in Bangladesh than any other WtE technologies. If the fuel would fall in the grey area of the triangle, no additional fuel is required to convert MSW into energy and therefore no incurred costs (Kabir and Khan, 2020).

8.8 Conclusion

The aim of this chapter was to provide an overview on the potentials of WtE technologies as a source of electricity generation from municipal solid waste in the context of developing economies. Due to increasing population, rural urban migration and improved lifestyle, the quantity of MSW is rapidly increasing in developing countries and this is posing a challenge to make cities leviable. To manage the MSW there are tools and techniques used in all developed countries globally. The developed nations are well ahead of the management of MSW in a sustainable fashion using advanced technology. The use of advanced technology in developing countries is very limited although examples show that many of the developing countries have taken initiative to adopt advanced technologies. Given the increasing amount of MSW in developing countries, there are huge potentials to generate electricity using WtE technologies. To advance the adoption of these technologies, some challenges in developing countries were identified. These include, financial support, awareness among people, improved waste collection system, enforcement of policy and legislations. It is expected that developing countries will be able to use WtE to recover materials and generate electricity through addressing these challenges.

Self-evaluation questions

1. What do you mean by waste-to-energy generation? What are the basic tools and techniques of waste management? - Explain with proper diagram.
2. What are the available WtE generation technologies? Describe them briefly.
3. In your opinion, what is the prospect of generating electricity from MSW in developing countries?
4. What are the barriers and opportunities for the generation of electricity from MSW in developing countries?
5. Do you think that WtE generation in the developing world could be a potential energy source? – Justify your answer.
6. Explain the use of the "tanner triangle" in determining the combustibility of a fuel (MSW).
7. Considering the waste composition characteristics of developed and developing countries propose a WtE generation technology for each country and justify your choice.
8. Conduct a literature survey on WtE generation technologies and present a comparative study considering their technical maturity, social, environmental, and economic aspects with respect to developed and developing countries.

References

Asian Development Bank (ADB), 2020. Waste to energy in the age of the circular economy: a good practice handbook. ADB, Manilla, Philippines.

Bach, Q.V., Chen, W.H., 2017. Pyrolysis characteristics and kinetics of microalgae via thermogravimetric analysis (TGA): a state-of-the-art review. Bioresour. Technol. 246, 88–100. http://doi.org/10.1016/j.biortech.2017.06.087.

Blight, G.E., Mbande, C.M., 1996. Some problems of waste management in developing countries. J. Solid Waste Technol. Manag. 23 (1), 19–27.

British Petroleum, 2019. BP Statistical Review of World Energy, 68th edition, Available at https://www.bp.com/content/dam/bp/business-sites/en/global/corporate/pdfs/energy-economics/statistical-review/bp-stats-review-2019-full-report.pdf.

Chen, D., Yin, L., Wang, H., He, P., 2015. Reprint of: pyrolysis technologies for municipal solid waste: a review. Waste Manage. (Oxford) 37, 116–136. http://doi.org/10.1016/j.wasman.2015.01.022.

Couth, R., Trois, C., 2011. Waste management activities and carbon emissions in Africa. Waste Manage. (Oxford) 31 (1), 131–137. http://doi.org/10.1016/j.wasman.2010.08.009.

EAWAG, 2008. Global Waste Challenge, Situation in Developing Countries. Swiss Federal Institute of Aquatic Science and Technology and Sandec, Dübendorf, Switzerland.

Farooq, A., Haputta, P., Silalertruksa, T., Shabbir, H., Gheewala, SH., 2021. A framework for the selection of suitable waste to energy technologies for a sustainable municipal solid waste management system. Front. Sustain. 2, 681690. http://doi.org/10.3389/frsus.2021.681690.

GIZ, 2017. Waste-to-Energy Options in Municipal Solid Waste Management: A Guide for Decision Makers in Developing and Emerging Countries. Eschborn, Germany Available at: https://www.giz.de/en/downloads/GIZ_WasteToEnergy_Guidelines_2017.pdf.

Goldberg, J. 2018. Yangon's two-week landfill fire raises burning questions for authorities. The Guardian. Available at: https://www.theguardian.com/cities/2018/may/17/yangon-two-week-landfill-fire-raises-burning-questions-for-authorities-myanmar.

GOB, 2008. Renewable Energy Policy of Bangladesh. Power Division, Ministry of power, energy and mineral resources. Government of the People's Republic of Bangladesh, Dhaka Available at: https://policy.thinkbluedata.com/sites/default/files/REP_English.pdf .

Guedes, R.E., Luna, A.S., Torres, A.R., 2018. Operating parameters for bio-oil production in biomass pyrolysis: a review. J. Anal. Appl. Pyrolysis 129, 134–149. https://doi.org/10.1016/j.jaap.2017.11.019.

Haas, W.F., Krausmann, D.W., Heinz, M., 2015. How circular is the global economy? An assessment of material flows, waste production, and recycling in the European Union and the world in 2005. J. Ind. Ecol. 19 (5), 765–777. https://doi.org/10.1111/jiec.12244.

Habib, M.A., Ahmed, M.M., Aziz, M., Beg, M.R.A., Hoque, M.E., 2021. Municipal solid waste management and waste-to-energy potential from Rajshahi City Corporation in Bangladesh. Appl. Sci. 11 (9), 3744. https://doi.org/10.3390/app11093744.

Hoornweg, D., Bhada-Tata, P., 2012. What a Waste: A Global Review of Solid Waste Management. Urban Development Series; Knowledge Papers no. 15. World Bank, Washington, D.C. Available at: https://openknowledge.worldbank.org/handle/10986/17388.

Holmes, D., 1997. Waste minimization in the small island developing states: implications for strategy and policy. In: Paper presented in Regional Workshop on Waste Management in Small Island Developing States In The South Pacific. Canberra.

Indrawan, N., Kumar, A., Moliere, M., Sallam, K.A., Raymond, L., Huhnke, R.L., 2020. Distributed power generation via gasification of biomass and municipal solid waste: a review. J. Energy Inst. 93, 2293–2313. http://doi.org/10.1016/j.joei.2020.07.001.

International Business Publications, Inc., 2015. Malaysia energy policy, laws and regulations handbook. Strategic Information and Basic Laws, Volume 1. World Business and Investment Library, Washington, DC.

Intharathirat, R., Salam, P.A., 2015. Valorization of MSW-to-Energy in Thailand: status, challenges and prospects. Waste Biomass Valor 7, 1–27. http://doi.org/10.1007/s12649-015-9422-z.

IRENA, 2019. Advancing Renewables in Developing Countries: Progress of projects supported through the IRENA/ADFD Project Facility, Abu Dhabi, UAE. Available at: https://www.irena.org/-/media/Files/IRENA/Agency/Publication/2019/Jan/IRENA_ADFD_Advancing_renewables_2019.pdf.

Kabir, Z., Khan, I., 2020, Environmental impact assessment of waste to energy projects in developing countries: A guideline in the context of bangladesh, Sustain. Energy Technol. Assess. 37, 1–13.

Khan, I., 2020. Waste to biogas through anaerobic digestion: Hydrogen production potential in the developing world - a case of Bangladesh. Int. J. Hydrogen Energy Vol. 45, 15951–15962. http://doi.org/10.1016/j.ijhydene.2020.04.038.

Khan, I., 2020a. Sustainability challenges for the south Asia growth quadrangle: a regional electricity generation sustainability assessment. J. Cleaner Prod. 243 (118639), 1–13. http://doi.org/10.1016/j.jclepro.2019.118639.

Khan, I., 2019. Power generation expansion plan and sustainability in a developing country: a multi-criteria decision analysis. J. Cleaner Prod. 220, 707–720. http://doi.org/10.1016/j.jclepro.2019.02.161.

Kothari, R., Tyagi, V.V., Pathak, A., 2010. Waste-to-energy: a way from renewable energy sources to sustainable development. Renew. Sustain. Energy Rev. 14 (9), 3164–3170. http://doi.org/10.1016/j.rser.2010.05.005.

Kumar, A., Samaddar, S.R., 2017. A review on technological options of waste to energy for effective management of municipal solid waste. Waste Manage. (Oxford) 69, 407–422. http://doi.org/10.1016/j.wasman.2017.08.046.

Liu, C., Nishiyama, T., Kawamoto, K., Sasaki, S., 2020. CCET Guideline Series on Intermediate Municipal Solid Waste Treatment Technologies: Waste-to-Energy Incineration. IGES, Japan Report for UNEP (45 pages).

Mavropoulos, A., Wilson, D., Velis, C., Cooper, J., Appelqvist, B., 2012. Globalization and Waste Management. Phase 1: Concepts and Facts. International Solid Waste Association, Wien.

Miller, G.T., 2000. Living in the Environment: Principles, Connections, and Solutions, 11th ed Brooks/Cole Publishing Company, Pacific Grove, California, USA Available at: http://www.mtcarmelacademy.net/uploads/1/1/7/5/11752808/living_in_the_environment_16th_edition_-_miller.pdf.

Mutz, D., Hengevoss, D., Christoph, H., Gross, T., 2017. Waste-to-Energy Options in Municipal Solid Waste Management-A Guide for Decision Makers in Developing and Emerging Countries. GIZ, Germany, p. 57 Technical Reportpages).

National Energy Policy, 2004. Government of the people's republic of Bangladesh. Ministry of power, energy and mineral resources, Dhaka, Bangladesh.

Nwokolo, N., Mamphweli, S., Makaka, S., 2017. Analytical and thermal evaluation of carbon particles recovered at the cyclone of a downdraft biomass gasification system. Sustainability 9 (4), 645. https://doi.org/10.3390/su9040645.

Ouda, O.K.M., Raza, S.A., Nizami, A.S., Rehan, M., Al-Waked, R., Korres, N.E., 2016. Waste to energy potential: a case study of Saudi Arabia. Renew. Sustain. Energy Rev. 61, 328–340. http://doi.org/10.1016/j.rser.2016.04.005.

Ozcan, H.K., Guvenc, S.Y., Guvenc, L., Demir, G., 2016. Municipal solid waste characterization according to different income levels: a case study. Sustainability 8 (10), 1044. https://doi.org/10.3390/su8101044.

PCD, 2014. Annual Thailand State of Pollution Report 2013. Pollution Control Department, Thailand.

Qazi, W.A., Abushammala, M.F.M., Azam, M.H., Younes, M.K., 2018. Waste-to-energy technologies: a literature review. J. Solid Waste Technol. Manag. 44, 387–409. http://doi.org/10.5276/JSWTM.2018.387.

Rana, M.S., 2016. Feasibility of Study Waste to Energy and Power Generation of Dhaka City Master's Thesis. University of Dhaka, Dhaka, Bangladesh.

Rohatgi, A., 2017 Southeast Asia Renewables Market Outlook; Bloomberg New Energy Finance: UK.

Sahin, O., Kirim, Y., 2018. Material recycling. J. Compr. Energy Syst. 2, 1018–1042. http://doi.org/10.1016/B978-0-12-809597-3.00260-1.

Seo, Y., Alam, M.T., Yang, W., 2018. Chapter-7: Gasification of Municipal Solid Waste. In: Yun, Y. (Ed.), Gasification for Low-Grade Feedstock. IntechOpen, London, UK, pp. 115–141. http://doi.org/10.5772/intechopen.73685.

Shapkota, P., Coowanitwong, N., Visvanathan, C., and Traenkler, J., 2006. Potentials of Recycling MSW in Asia vis-a-vis Recycling in Thailand, Available at: https://www.researchgate.net/publication/237476530_Potentials_of_Recycling_Municipal_Solid_Waste_in_Asia_vis-a-vis_Recycling_in_Thailand.

Statista, 2020, Leading countries in installed renewable energy capacity worldwide in 2020, available at https://www.statista.com, accessed on 9/7/20/2021.

Swaraz, A.M., Satter, M.A., Rahman, M.M., Asad, M.A., Khan, I., Amin, M.Z., 2019. Bioethanol production potential in Bangladesh from wild date palm (*Phoenix sylvestris*Roxb.): an experimental proof. Ind. Corps Products 139 (111507), 1–9. http://doi.org/10.1016/j.indcrop.2019.111507.

Tun, M.M., Juchelková, D., 2019. Drying methods for municipal solid waste quality improvement in the developed and developing countries: a review. Environ. Eng. Res. 24 (4), 529–542. https://doi.org/10.4491/eer.2018.327.

Tyagi, V.K., Lo, S.L., 2013. Sludge: a waste or renewable source for energy and resources recovery? Renew. Sustain. Energy Rev. 25, 708–728. https://doi.org/10.1016/j.rser.2013.05.029.

United Nations, 2020. Sustainable Development Goals, Available at: https://www.un.org/sustainabledevelopment/cities/.

Nations, U., 2014. Department of Economic and Social Affairs, Population Division. World Urbanization Prospects: The 2014 Revision, Highlights Available at: https://population.un.org/wup/Publications/Files/WUP2014-Report.pdf.

United Nations, 2019. Department of economic and social affairs, population division. World Urbanization Prospects: The 2018 Revision (ST/ESA/SER.A/420) Available at: https://population.un.org/wup/Publications/Files/WUP2018-Report.pdf.

United Nations Environment Programme (UNEP), 2015. Global Waste Management Outlook. Available at: https://www.uncclearn.org/sites/default/files/inventory/unep23092015.pdf.

UNEP, 2017. Renewable Energy and Energy Efficiency in Developing Countries: Contributions to Reducing Global Emissions. Paris, France. Available at: https://www.unep.org/resources/report/renewable-energy-and-energy-efficiency-developing-countries-contributions-0.

UN-Habitat, 2020, Strategic Plan 2020-2023. Nairobi, Kenya. https://unhabitat.org/sites/default/files/documents/2019-09/strategic_plan_2020-2023.pdf.

Wong, Z.J., Bashir, M.J.K., Ng, C.A., Sethupathi, S., Lim, J.W., Show, P.L., 2019. Sustainable waste-to-energy development in Malaysia: appraisal of environmental, financial, and public issues related with energy recovery from municipal solid waste. Processes 7 (10), 676. https://doi.org/10.3390/pr7100676.

World Bank, 2012. What a waste: a global review of solid waste management. The Urban Development Series Knowledge papers Available at: https://openknowledge.worldbank.org/handle/10986/17388.

World Bank, 2017. Rapid Assessment of Kabul Municipality's Solid Waste Management System. World Bank, Washington DC Report No. ACS19236.

World Bank, 2018. What a waste 2.0: a global snapshot of solid waste management to 2050.

The Urban Development Series Knowledge papers. Washington, DC, USA. Available at: http://hdl.handle.net/10986/30317.

Zaman, A.U., 2010. Comparative study of municipal solid waste treatment technologies using life cycle assessment method. Int. J. Environ. Sci. Technol. 7 (2), 225–234. https://doi.org/10.1007/BF03326132.

Zhang., S., 2015. All about biofuels. Wixsite.com http://allaboutbiofuels.wixsite.com/biofuels/thermalconversion.

CHAPTER 9

Geothermal energy in developing countries–The dilemma between renewable and nonrenewable

Nurdan Yildirim[a] and Emin Selahattin Umdu[b]

[a]Mechanical Engineering Department, Yasar University, Bornova, İzmir-Turkey
[b]Energy Systems Engineering Department, Yasar University, Bornova, İzmir-Turkey

9.1 Introduction

The word geothermal is formed by the combination of the Greek words geo (earth) and therme (heat). Basically, geothermal energy, resulting from the molten interior of the earth and the decay of radioactive materials in underground rocks, is generally defined as the energy coming from the depths of the earth. The temperature at the center of the Earth, around 6500 km-depth, is about 6000°C and is almost the same as the surface temperature of the sun. Magma, rising in places on the buoyancy forces, pushes the plates towards the Earth's crust and in this way a large amount of heat has been emitted from the Earth's core for about 4.5 billion years. The temperature gradient of the Earth and the total heat flux from the Earth's interior are normally 2 to 3°C/100 m and around 80 mWth/m^2, respectively (Dumas, 2017). Since the radioactive processes in the core of the Earth are continuous, geothermal energy can be classified as a renewable energy source (G. Systems, 2016; Yildirim, 2010; Petroski, 2013; Miller, 2006).

Rainwater and melted snow seepage into the ground with the help of faults and cracks in the earth, and then they are heated up to above boiling temperature such as 260°C or more by hot rocks (Nemzer, 2005). If hot rocks are permeable, the percolating surface waters naturally circulate back to the surface, returned as steam forming hot springs, geysers, and fumaroles. When the rocks are not permeable, the water is trapped by impermeable rocks, filling the rock pores and cracks, geothermal reservoir is formed (Burgess, 1989).

Geothermal systems and reservoirs are classified according to different aspects such as their nature and geological setting, reservoir temperature, enthalpy, and physical state (Fig. 9.1). There are mainly six classifications depending on the nature and geological

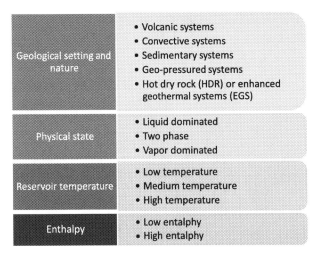

FIGURE 9.1 Classification of geothermal resources.

settings of geothermal resource as volcanic systems, convective systems, sedimentary systems, geo-pressured systems, engineered (enhanced) geothermal systems (EGS) or hot dry rock (HDR), and shallow resources. In volcanic systems the flow of water is controlled by permeable fractures and fault zones. Geothermal water has been circulated mostly through vertical cracks to deeper than 1 km, to extract heat from rocks in convective systems. Sedimentary systems are conductive in nature rather than convective and owe these systems are owed by the formation of permeable sedimentary layers at gradients above 30°C/km and at depths of more than 1 km. Geo-pressured systems are generally quite deep and similar to geo-pressured oil and gas reservoirs. EGS or HDR consist of volumes of rock that has low permeability or are virtually impermeable have been heated to by abnormally high heat flow or volcanism. In these systems to extract heat, mostly a closed loop is used, not conventional methods. Shallow resources, where ground source heat pumps are commonly used, refer to normal heat flow through formations, thermal energy stored in rocks and in hot groundwater systems near the surface the Earth's (Saemundsson et al, 2009). Geothermal resources are classified in three groups depending on reservoir temperature as low temperature (LT), medium or moderate temperature (MT), and high temperature (HT). If the reservoir temperature at 1 km depth is less than 150°C, the geothermal resources are called as LT, MT if it is between 150 and 200°C, and HT if more than 200°C. These temperature ranges are widely used temperatures for classifications, but they may differ based on the characteristics of geothermal reservoirs. In low enthalpy geothermal systems, the reservoir fluid enthalpy is less than 800 kJ/kg, which corresponds to temperatures below 190°C. Otherwise, it is defined as a system with high enthalpy. Depending on the physical states, geothermal resources can be liquid dominated, vapor-dominated or two-phase (Saemundsson et al., 2009, Bödvarsson, 1961; Axelsson and Gunnlaugsson, 2000).

Geothermal energy uses can be grouped into two main categories as direct use and power generation as can be seen from the Lindal diagram (Fig. 9.2), which summarizes the temperature requirements of geothermal applications, proposed by Icelandic engineer Baldur Lindal for the first time (CanGEA, 2014). Direct use of geothermal energy is a utilization

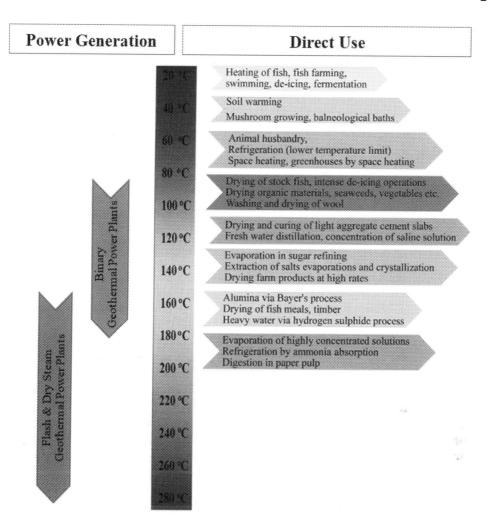

FIGURE 9.2 Lindal diagram (Data from Gupta and Roy, 2007).

in which thermal energy of geothermal resources is transferred without changing the form of energy. District and space heating/cooling, aquacultural heating, greenhouse heating, industrial process heating, food drying, snow melting, heat pumps are some applications of direct use of geothermal energy.

Geothermal resources have been used in many ways throughout history, including cooking, healing and physical therapy, bathing, and underfloor heating. More than 10,000 years ago, Paleo-Indians settling at hot springs for bathing and cooking is considered one of the first known human use of geothermal resources. While the direct use of geothermal energy dates back thousands of years, electricity generation took place in the 1900s. In 1904, the first dry steam Geothermal Power Plant (GPP) was invented by Prince Piero Ginori Conti in Italy at the Larderello field (Dickson and Fanelli, 2013). It was followed by a small GPP that started operating in 1958 in New Zealand. The first geothermal power plant of the United States was commissioned at the Geysers in California in 1960 with a capacity of 11 MW (WEC, 2010; Salmon et al., 2011).

9.2 Available technologies

Geothermal resources are not alike, they have their own characteristics. One of the most important features that will determine the utilization area of a geothermal resource is its temperature. Generally, geothermal resources at high temperatures (>120°C) are used primarily for electricity generation. Other properties of the geothermal resource such as physical state, chemical compositions, pressure, thermal gradient, etc. are also effective in determining the components and technologies to be used in the geothermal system. In this section, basic/available systems and technologies in direct use and power generation of geothermal energy will be introduced.

9.2.1 Geothermal power generation

A geothermal resource can have geothermal fluid in one of these three physical states: (1) vapor dominant (dry steam), (2) two-phase (saturated liquid-vapor mixture), and (3) liquid dominant (brine). Different technologies are used to generate power depending on the physical state of the extracted geothermal fluid. GPPs can be divided into four main categories: dry steam, flash steam, binary, and kalina power plants (Kanoglu, 1999).

The geothermal fluid required for electricity generation is located in reservoirs hundreds or thousands of meters deep. To reach them, wells are drilled, and geothermal fluid is carried to the surface through these wells. Geothermal fluid can be produced entirely in vapor phase without containing any liquid phase from a few geothermal reservoirs in the world. The power plants that operate on this geothermal steam are called as dry steam GPPs, which was the first type of GPPs built in the World. Dry steam GPPs are the simplest and cheapest GPPs and they don't require a separator (Fig. 9.3A). Steam from a geothermal production well is passed through a steam turbine-generator couple that converts thermal energy into electricity. Condenser and cooling towers are the other main components of dry steam GPPs. Dry steam GPP was first used in 1904 at Lardarello in Italy. The Geyser GPP, which is the largest geothermal power generation facility in the World fed by a single geothermal source, also includes dry steam GPP technology currently (EIA, 2021).

Hot water reservoirs with fluid temperature generally above 180°C are used in flash steam GPP. The pressure of the geothermal fluid brought to the surface through the production wells is reduced and then the liquid and vapor phases are separated using a separator (Fig. 9.3B). The steam is directed to the turbine to produce electricity. After it is cooled and condensed, it can inject back into the reservoir or used in the other low thermal energy required processes of the plant. Since the majority of geothermal reservoirs in the World consist of liquid-dominated systems, the most widely used GPP type in the World is flashed-steam GPP.

The geothermal steam circulating in the power cycle of dry or flash steam GPPs is not pure and contains various noncondensable gases (NCGs) such as CO_2, NH_3, H_2S, CH_4, and N_2. The

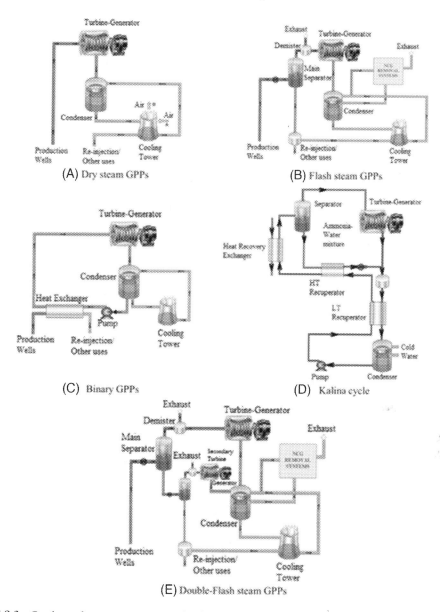

FIGURE 9.3 Geothermal power generation technologies.

fraction of NCGs in the World varies from almost zero to 25% by weight, depending on the source (Hall, 1996; Coury et al., 1996). NCGs in the steam increase the condenser pressure and total auxiliary power consumption, reducing the net power output of the facility. Hence, NCGs should be withdrawn by the help of NCG removal system for better the performance of GPPs. Steam jet ejector (SJE)s, centrifugal compressors (CS)s, and liquid ring vacuum pumps (LRVP)s are commonly used conventional NCG removal systems. Also, upstream reboiler systems

TABLE 9.1 NCG removal systems used in some flashed-steam GPPs around the world.

GPP name	Location	Type of flashed – steam GPP	NCG removal system type
Zorlu-Kizildere	Turkey	Single	3 Compressors
Gurmat	Turkey	Double	SJE and LRVP
Miravalles Unit-3	Costa-Rica	Single	2 SJEs and LRVP
Olkaria-1	Kenya	Single	2 SJEs
Kawerau	New Zealand	Double	2 SJEs and LRVP

Source: Wallace et al., 2009; Horie et al., 2010; Kwambai, 2010; Moya and DiPippo, 2010.

(RSs) are another innovative approach in which NCGs are extracted from geothermal steam before entering the turbine. Hybrid NCG removal systems (HS), usually SJE-LRVP (Steam Jet Ejector-Liquid Ring Vacuum Pump) combination, are the most preferred systems in GPPs recently. NCG removal systems used in some flashed-steam GPPs around the world are listed in Table 9.1.

Generally, geothermal reservoirs with low-medium temperature geothermal fluid between 85°C and 150°C are not suitable for use in flash steam GPP. But these reservoirs can be used to generate electricity in a binary power plant called also Organic Rankine Cycle (ORC). In a binary GPP, the geothermal fluid is not used directly in the power cycle, but through a heat exchanger the geothermal fluid transfers its energy to a binary (secondary) fluid used in the power cycle with low boiling points such as propane, iso-pentane, ammonia, iso-butane, etc. (Fig. 9.3C). By using geothermal energy in this way, the secondary fluid flashes into vapor and drives the turbines. Then the vapor is condensed, and the power cycle is repeated (Gupta and Roy, 2007). The geothermal fluid is pumped back into the reservoir after transferring its energy.

Another cycle used in GPPs is the Kalina cycle, which can be considered a modified Rankine cycle (Fig. 9.3D) (Bombarda et al., 2010). Unlike the Rankine cycle, which uses pure water, a mixture of ammonia and water is used as working fluid in a Kalina cycle. Kalina GPPs have key advantages such as producing less emissions, requiring less energy and higher thermodynamic efficiency compared to Binary GPPs. Kalina cycle technology is currently being actively used at the Husavik GPP in Iceland with an installed capacity of 2 MW and a working fluid of 82 wt% ammonia (DiPippo, 2004; A.N.M Nihaj Uddin Shan, 2020).

Some geothermal resources require more complex energy conversion systems than the basic power cycles previously described. Furthermore, in order to benefit more energy from geothermal fluid, combined cycles such as double flash, combined flash, and dual, combined flash and kalina have been developed by combining different types of power plants and using them in an integrated manner. A simplified schematic diagram of a double flash facility as an improvement of the single flash plant design is shown in Fig. 9.3E as an example. Low- and high-pressure separator-, low- and high-pressure turbine, condenser, cooling tower, and circulation water pumps and NCG discharge system are the main equipment of double flash GPPs. In the double flash facility, a two-phase mixture is obtained by reducing the pressure of the liquid coming from the high-pressure separator. In the low-pressure separator, the mixture is separated into liquid and vapor phases. The additional low-pressure steam is directed to

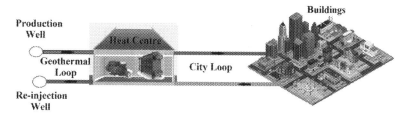

FIGURE 9.4 Schematic of a geothermal district heating system.

the low-pressure turbine to generate electrical energy and then the fluid goes to the condenser (DiPippo, 2005; Swandaru, 2009).

9.2.2 Direct use of geothermal energy

9.2.2.1 District heating

The systems that provide thermal energy and/or domestic hot water need of a group of processes, building by using one or more geothermal production areas consisting of steam or hot water as the heat source are defined as geothermal district heating system (GDHS). A GDHS consists of three main parts (Fig. 9.4). One of the parts is heat production. Geothermal production and recharge fields, conventional peaking station and wellhead equipment are involved in heat production part. The other part is piping system, in which geothermally heated water and/or geothermal fluid is delivered to consumers through insulated pipes consisting of supply and return networks. The last part is heat center and in-building equipment consists of circulation pumps, heat exchangers and may also thermal storage tanks to meet the dynamic heating demand. Depending on the temperature of the geothermal fluid, heat can be extracted by two basic methods: direct heat exchange with heat exchangers and heat pumps.

9.2.2.2 Geothermal greenhouse heating

Geothermal energy has been commonly used in agriculture especially for greenhouse heating. In many countries, geothermal energy is preferred for the off-season production of commercial vegetables, flowers, and fruits. Heating of a greenhouse environment by geothermal energy reduces operating costs, which accounts for 35% of the product cost, and allows the operation of a commercial greenhouse in colder climates that would normally not be economical if fossil fuel is used for heating. Greenhouse construction type and material, technological level of production, climate, location, the temperature of the heating medium and the geothermal water, the particular requirements of the plants in the greenhouse are the main factors dictate the design of geothermal heating system for a greenhouse. There are basically six different heating systems for greenhouses heated by geothermal energy which are finned pipe, fan coil units, low temperature unit heaters, standard unit heaters, bare tube, and soil heating. Heat exchanger is preferred to provide heat in greenhouses because of the corrosion and scaling associated with most geothermal fluids (Boyd and Rafferty, 2008).

9.2.2.3 Geothermal aquacultural heating

Aquaculture involves the raising of fish and various aquatic animals in a controlled environment to ensure faster growth of the creatures and increase their survival rates. Geothermal hot water is used to heat freshwater in heat exchangers or is mixed with fresh water to heat pond and raceway water up to suitable temperature ranging from 13 to 30°C based on species involved. Using geothermal resources in this way not only increases the growth rate of aquatic species by 50% to 100%, but also allows aquaculture activities to be sited in colder climates or closer to markets where conventional heating may not be economical. In the design of geothermally heated pools and channels, for the required thermal capacity, it is necessary to determine the optimum water temperature for living things as well as heat losses through evaporation, convection, radiation, and conduction.

Catfish, carp, tilapia, salmon, mullet, frogs, eels, shrimps, sturgeon, crayfish, lobsters, oysters, crabs, clams, alligators, scallops, abalone, and mussels are main species raised in aquacultural environment. The use of geothermal energy in fish farming worldwide is increasing rapidly. Iceland, Greece, France, New Zealand, Hungary, and the United States are leading countries in this area (Boyd and Lund 2006; Ragnarsson, 2014; Öz and Dikel, 2016).

9.2.2.4 Food drying

The drying process, which is a heat and mass transfer event that involves the transport of water in the product from the interior of the product to the surface and evaporating from the surface, is one of the processes commonly used in the food sector and industry. With drying, the shelf life of the products is extended, the mass and volume of the products are reduced, and thus the transportation and storage of the products becomes easier, more efficient, and economic. Geothermal energy as a sustainable and continuous energy source without depending on meteorological conditions provides many benefits such as reducing the cost and time in drying processes. Based on the properties of the product to be dried (type, current water content, size, desired moisture content after drying, drying time, etc.), mainly chamber, tunnel (corridor), conveyor, drum, pneumatic type dryers are used. Mostly, drying processes take place by direct contact of products (cereal, vegetable, fruit, etc.) with warm air. Drying energy requirements are linked to heating the product to evaporation temperature and evaporating a certain percentage of moisture. All or part of required drying energy for conditioning the air to be used in a drying process can be provided by geothermal energy with the help of a heat exchanger. Generally, geothermal fluid with low temperatures such as 65 to 95°C can be used for drying agricultural products (fruits, vegetables etc.), but higher temperatures are required in industrial drying applications. For example, 100 to 120°C for drying leather and fabric, 125 to 160°C for drying cement, fish meat (Lund et al., 2010; Helvacı et al., 2013; Popovska, 2003),

9.2.2.5 Snow melting

The covering of roads and pavements with snow and ice creates serious disturbance for pedestrians and vehicles and causing various major accidents and traffic problems. In regions with very cold climates and heavy snowfall for a long time, outdoor life is very difficult for people. For this reason, systems called snow melting systems are needed for the rapid removal of snow and ice from the pavements, roads, entrance of important buildings (hospitals, schools, etc.). There are mainly two main methods in snow melting systems as conventional and novel snow melting systems. Mechanical and chemical methods are commonly used in conventional methods. Mechanical methods, cannot remove the snow at a time, are time consuming, sometimes inefficient. Chemical methods, which used chemical solvents such as potassium acetate and chlorine salt, corrode the road, pollute the soil, and can be harmful to the freshwater system and the environment. For example, the use of salts and chemicals on airport pavements for snow melting is prohibited, as they are harmful to both pavement materials and metallic materials of aircraft. Electrical snow melting method, heat pipe snow melting method, phase change snow melting method and hydronic snow melting method are the main environmentally friendly and sustainable novel removal methods. The electrical snow melting method, which removes the snow by electric mats, infrared heaters, or microwave inductors, is generally less energy efficient with high electricity consumption. Heat pipe snow melting method, which melts snow with geothermal energy, may be insufficient at air temperatures below −10°C. Studies continue to investigate the durability, reliability and stability of the phase change snow melting method, which cleans the snow with thermal energy released from the phase change material (PCM). Hydronic snow melting method is the most practical method of the novel snow removal methods and has better energy efficiency. In the hydronic snow melting method, snow and ice are melted with energy transferred from a circulation fluid such as heated water, brine, ethylene glycol or oil circulated in a series of pipe circuits placed under the surface. In the hydronic snow melting method, the heat source is more reliable as it can be controlled through the circulation fluid, it can be diversified into electrical energy and/or renewable energy. Systems, where the thermal energy of geothermal fluid or ground is used as a heat source in hydronic snow melting method, are called geothermal assisted snow melting systems. A ground heat exchanger, a geothermal heat pump, and a piping circuits for different utilizations are main components of a shallow geothermal system. The depth of buried pipes is generally 50 to 100 mm, the distance between the pipes can be 150 to 350 mm and the diameter of the pipes is between 18 and 35 mm. In geothermal assisted snow melting systems, low temperature geothermal resources can be used. On the other hand, snow melting systems can be used in integration with processes such as electricity generation and industrial drying by geothermal energy. The high temperature geothermal fluid, which uses its energy primarily in processes such as electricity generation and industrial drying, then can be used for snow melting (Zhao, et al, 2016; Akrouch et al, 2016; Liu et al., 2018; Zhao, et al, 2020).

9.2.2.6 *Heat pumps*

Heat pumps are devices that transfer thermal energy from low temperature environment to high temperature environment. Basic heat pumps have 4 main components: compressor, condenser, throttling/expansion valve, and evaporator, and operate on the basis of the vapor-compression refrigeration cycle. Reversing valve, fans, pipes, and control equipment can be added to the heat pump system as auxiliary equipment if needed. A heat pump has a closed cycle operating in 2 pressure ranges, and generally refrigerants are used as working fluid in this cycle. The working principle of a heat pump can be briefly summarized as follows: Working liquid is evaporated in the evaporator with the heat transferred to the evaporator from the low temperature environment. The evaporator is located on the low-pressure side of the heat pump. At the evaporator outlet, the working fluid can be in the saturated vapor or mostly superheated vapor phase. Then, the pressure of the working fluid is increased to the condenser pressure with the compressor operating with electrical energy. At the same time, the temperature of the working fluid increases due to compression. Working fluid condenses in the condenser. During condensation, heat is rejected to the outside (high temperature environment). At the condenser outlet, the working fluid may be in a saturated liquid or in some cases in a subcooled (compressed liquid) phase. In order to complete the cycle, the fluid at high pressure at the condenser outlet is passed through the throttle valve, and the pressure is reduced to the evaporator pressure while the temperature decreases.

The low-temperature environment where the heat pump receives heat can be called as "heat source," and the high-temperature environment where the heat rejection occurs can be defined as "heat sink." Generally, air, water, and ground/soil can be used as heat sources. For this reason, heat pumps are called air, water, or ground source heat pumps, depending on the type of heat source they use. Water and ground source heat pumps are also classified as geothermal heat pumps.

Geothermal heat pump (GHP) systems, which were first developed in the late 1940s, are widely used today to meet the heating, domestic hot water and even cooling needs of buildings. If the heat pump will also be used for cooling, a reversing valve is added to the system and the system is operated in cooling mode by changing the flow direction of the working fluid in the cycle.

Comparing air source and geothermal heat pumps, geothermal heat pumps are generally more economical and environmentally friendly. Main advantages of geothermal heat pumps:

- It uses an energy source that is more stable than air. Therefore, it can be operated without the need for additional energy even in periods of very low outside air temperatures.
- Since they have a higher performance coefficient, they consume less energy and use less working fluid.
- They are simpler in design and require less maintenance.

In geothermal source heat pumps, vertical and/or horizontal piping is required to obtain the energy of ground or water source. Due to these piping and drillings, the capital investment costs of geothermal heat pumps are 30 to 50% higher than air source heat pumps. Despite this, they are still more economical throughout their working life due to their low operating costs.

Electrical energy is used to drive the compressor in the heat pump. There are also thermally driven absorption heat pumps, mostly of high capacity, that are rarely used. The performance of a heat pump is measured by its COP value. COP is calculated by dividing the thermal energy obtained by the electrical energy used. Therefore, the higher the COP value, the better the performance of the system. In general, COP = 1.5 to 3.9 in air-to-air heat pumps; COP = 2.0 to 4.0 in water-to-air heat pumps; COP = 3.0 to 5.0 in water-to-water heat pumps. The COP value increases with using feeding fluid at lower temperature (Hepbasli and Balta, 2007; Milenić et al., 2010; Self et al., 2013; Ghoreishi and Kuyuk, 2017; Maddah, et al., 2020).

9.2.2.7 Other direct-use applications

Geothermal energy is also used for various industrial applications such as preheating, washing, peeling and planching, evaporation and distillation, sterilization, pasteurization, drying, heap leaching, food production (beer, milk powder, etc.), and cooling. Heat exchangers and heat pumps are commonly used equipment for these thermal applications. Heat pump system is generally preferred for cooling purposes, but vapor absorption refrigeration systems are also used rarely. These systems are economical where there is an inexpensive thermal energy with a temperature range of 100 to 200°C. Geothermal sources has gaining interest for adsorption refrigeration by being renewable and low emission sources with stable and low costs. Ammonia – water systems are the most common adsorption refrigeration systems, yet water lithium bromide and chloride systems are also available for low temperature applications above freezing point of water, such as air conditioning (Cengel and Boles, 2015).

9.2.3 Engineered (enhanced) geothermal systems or hot dry rock systems

In some high-thermal gradients regions, there is no reservoir containing fluid or the flow rate of the fluid produced from these reservoirs is not sufficient to use in thermal processes. In such cases, deep wells are drilled, and various fluids are injected into these wells. These drilled wells can be a separate production and re-injection well, or they can be a single well with different geometry such as U-type, multi-tube, coaxial. The fluid whose temperature increases with the heat of the ground is circulated to the surface and is mostly used in various direct use applications. After harvesting its useful energy, the fluid is injected back into the wells and the cycle is completed. Although this is a new technology, in which electricity generation is possible, has a pilot scale application in the USA, and studies are ongoing for its development (Dincer, 2018).

9.2.4 Combined (hybrid) geothermal systems

Geothermal energy, comes from the depths of the earth, is a renewable energy source that is very difficult, time consuming and costly to discover and transfer it to the surface. A resource obtained with such difficulties must be very-well analyzed in detail before it can be used. During the analysis, the temperature, pressure, flow rate, chemical composition of the source, energy capacities of the source, energy requirements of the possible utilization, the distance to the utilization area, the initial investment, operation, maintenance, and lifetime costs should be taken into account. Consequently, if it is understood that the considered application of this resource is not feasible, the integration of the resource with other possible applications should be investigated. The biggest advantage of geothermal energy is that it can be used in a combined system for multiple processes depending on thermodynamics properties of the resource.

For example, as shown in Fig. 9.5, after using some parts of its energy for electricity generation, the geothermal fluid can be used for district / space heating and cooling. Additionally, the heating of a greenhouse, fish farm or various industrial thermal processes can be integrated into the system.

9.3 Current status

The total primary energy consumption of OECD countries and the world in 2019 was 233.43 EJ and 583.9 EJ, respectively. Five percent of global consumption comes from renewable energy sources excluding hydro. Global primary energy consumption per capita was 75.7 GJ, while it was 178.5 GJ for OECD countries. Qatar was the highest global primary energy consuming

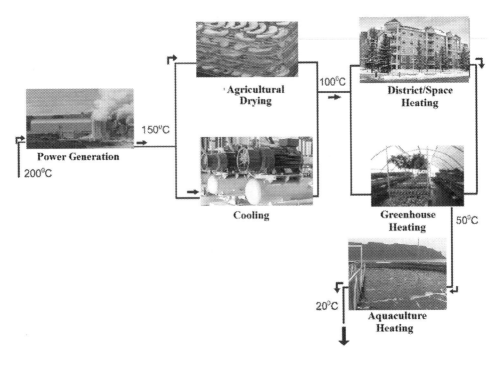

FIGURE 9.5 An example of a combined geothermal system (adapted from Dincer and Ezan, 2020).

country per capita at 714.3 GJ. Singapore, Trinidad & Tobago, and the United Arab Emirates followed Qatar with 611.6 GJ, 511.6 GJ, and 494.4 GJ respectively (BP, 2020).

Global geothermal power generation development is shown in Fig. 9.6 from 1950 to 2019. As can be clearly seen from Fig. 9.6, over the years, the installed power of the geothermal power plant and in parallel, the annual electricity production has increased significantly. The installed power capacity has doubled in the last 19 years and reached the level of 15,950 MWe in 2019.

In 2019, OECD, non-OECD countries and the World's electricity generation was 11,136.0 TWh, 15,868.7 TWh and 27004.7 TWh, respectively. 2805.5 TWh (10.38% of global electricity generation) was generated from renewable energy sources excluding hydro (BP, 2020). Geothermal electricity generation with approximately 95 TWh had the share of 3.4% among the renewable sources in 2019 (Huttrer, 2020).

The tectonic plate boundaries and global average heat flow in mW/m^2 is shown in Fig. 9.7. The installed geothermal power capacities of the countries in 2019 are also written on the map. According to the installed power capacity, the top five countries are US, Indonesia, Philippines, Turkey, and Kenya. With an installed capacity of 3,700 MWe, 24% of the geothermal electricity installed capacity in the world belongs to the USA. Indonesia follows with the share of 14% and the Philippines with 12%, respectively (Fig. 9.8). According to the data of 2019, 7 countries with 1000 MWe and above geothermal electricity installed capacity correspond to 74% of global installed geothermal power.

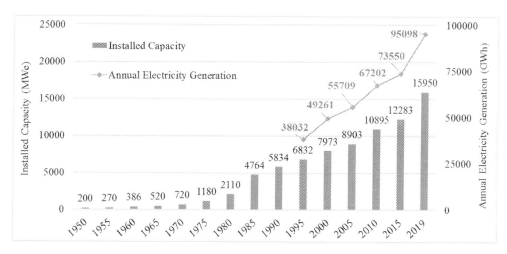

FIGURE 9.6 Global geothermal power development (Data from Bertani, 2016; Huttrer, 2020).

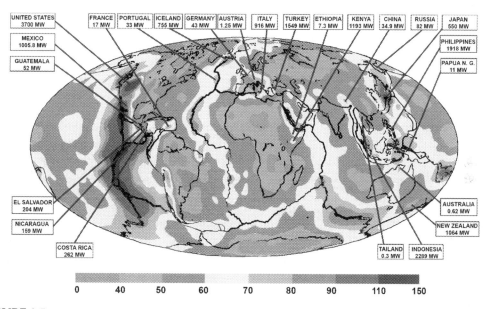

FIGURE 9.7 Total installed capacity of the GPPs for some countries in 2020 (Figure from Hamza et al., 2008; Ipcc 2021; data from Huttrer, 2020).

Among the developing countries, when the changes in the installed power capacities of the geothermal power plants in 2015 and 2019 are examined, it is seen that the biggest development is in Turkey. Turkey's installed capacity of 397 MWe in 2015 increased by approximately 290% in 2019 and reached the levels of 1549 MWe with a capacity increase of 1152 MWe. Turkey is followed by Indonesia with a capacity increase of 949 MWe (70.8%), followed by Kenya with a capacity increase of 599 MWe (100.8%). According to the changes in capacity increases in the last 5 years, the capacity expectations in 2025 have been determined

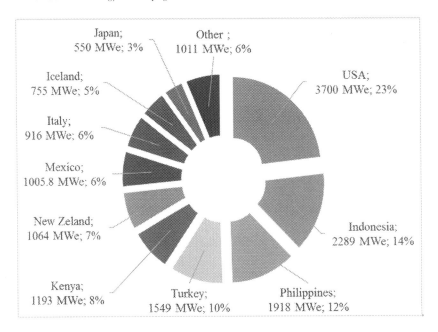

FIGURE 9.8 Leading countries by total installed capacity of GPPs in 2019.

and it is predicted that Indonesia's installed power will increase to 4362 MW, exceeding the USA, which has the highest installed power in the world. It is estimated that Turkey will rank 3rd in the world with its 2600 MW installed power projection (Huttrer, 2020).

Bertani (2016) stated that the total number of units of GPPs in the world in 2015 is 613. Among the technologies of GPPs, binary GPPs have the biggest share with 46% and single flash GPPs with 28% according to their total installed power capacities (Fig. 9.9).

The development of direct use of geothermal energy between 1995 and 2019 according to the installed capacities is shown in Fig. 9.10. The total installed capacity, which was 8664 MWt in 1995, increased 11.4 times in approximately 25 years. According to the data of 2019, the total installed capacity of geothermal direct use applications in 88 countries in the World was 107,727 MWt, with an annual energy amount of 283,580 GWh (1,020,887 TJ) (Lund and Aniko, 2020) could only meet 1.75/1000 of the global primary energy consumption.

China (40,610 MWt), United States (20,713 MWt), Sweden (6680 MWt), Germany (4,806 MWt), and Turkey (3,488 MWt) were the leading countries based on total installed capacity of geothermal direct use applications in 2019. Although the first three countries are the same in terms of annual energy use of geothermal direct use applications, Turkey and Japan were ranked 4th and 5th, respectively.

The distribution of installed capacities of direct use applications in 2019 is presented in Fig. 9.11. It is seen that geothermal heat pump applications are the most widely used direct use application with the largest share in total capacity with 72%. Space heating includes district and individual heating, accounting for 12% of total capacity. This is followed by bathing and

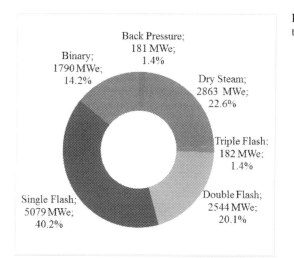

FIGURE 9.9 Total install capacities in 2015 based on type of GPPs (Data from Bertani, 2016).

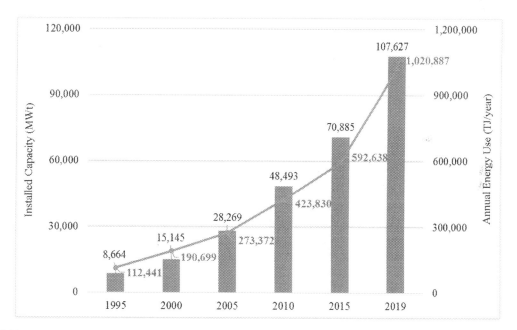

FIGURE 9.10 Global geothermal direct use development (Data from Lund and Aniko, 2020).

swimming utilizations with a share of 11%. As a result, the total installed capacity of these three utilization types corresponds to 95% of the total.

Fifty-four countries had geothermal heat pump applications in 2019 with 77,547 MWt total capacity. This capacity can be considered as equivalent to 6.46 million GHP systems with 12 kW capacity. The leading countries are China, the United States, Sweden, Germany, and Finland have 83.6% of the total annual energy for direct use applications (Table 9.2).

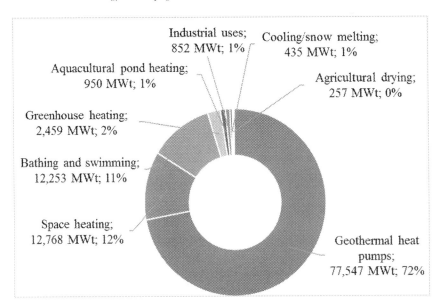

FIGURE 9.11 Distribution of direct use applications of geothermal energy based on installed capacity for the year of 2019 (Data from Lund and Aniko, 2020).

TABLE 9.2 The capacities and leading countries of direct use applications of geothermal energy in 2019 (Data from Lund and Aniko, 2020).

	Capacity		Leading countries based on annual energy usage				
Utilization	MWt	TJ/year	1st	2nd	3rd	4th	5th
Geothermal heat pumps	77,547	599,981	China	US	Sweden	Germany	Finland
Space heating	12,768	162,979	Turkey	Japan	Russia	US	Switzerland
Bathing and swimming	12,253	184,070	China	Japan	Turkey	Brazil	Mexico
Greenhouse heating	2549	35,826	Turkey	China	Netherlands	Russia	Hungary
Aquacultural pond heating	950	13,573	China	US	Iceland	Italy	Israel
Industrial uses	852	16,390	China	New Zeland	Iceland	Russia	Hungary
Snow melting	415	2389	Iceland	Japan	Argentina	US	Slovenia
Cooling	19.9	200.1	Bulgaria	Brazil	Australia	Slovenia	Algeria
Agricultural drying	257	3529	China	France	Hungary	US	Japan
Other	106	1950	New Zeland	Japan	Kenya		

About 162,979 TJ of energy was generated with an installed capacity of 12,768 MWt in 2019 for geothermal space heating, including space heating and district heating. Ninety-one percent of the total space heating capacity belongs to district heating. Among the 29 countries with geothermal space heating applications around the world, Turkey, Japan, Russia, the United States and Switzerland, are the leading countries in terms of annual energy use.

The leaders among 32 countries for geothermal greenhouse heating are Turkey, China, Netherlands, Russia, and Hungary, using about 83% of 35,826 TJ/year total energy use of the World in 2019. It can be said that 35,826 TJ/year of energy means heating a total of 1791 hectares of greenhouse, with an average of 20 TJ/year/ha of energy.

The total installed capacity of geothermal systems, used to obtain energy needed in processes such as milk pasteurization, chemical extraction, bottling of water and carbonated beverages, concrete curing, leather industry, borate and boric acid production, CO_2 extraction, iodine and salt extraction, pulp, and paper processing, is 852 MWt.

It is estimated that a total of 2.5 million square meters of pavement in the world is heated by geothermal snow melting systems, 74% of which are in Iceland. Japan, Argentina, the United States of America, and Slovenia are other countries with significant capacity geothermal snow melting systems. In 2019, the total installed power of the existing geothermal snow melting systems is assumed to be 415 MWt.

According to the direct use capacities of developing countries in 2019 (Table 9.3), China is the country with the highest use in all direct use types except greenhouse heating. Turkey is the leading country in greenhouse heating. The biggest direct use application in terms of capacity in China is heat pumps with a capacity of 26,450 MWt. Turkey leads the world in geothermal greenhouse heating with a capacity of 820 MWt. When the total installed capacity of direct use utilization of developing countries, including the heat pump application, is examined, China is in the 1st place with 40,610 MWt, Turkey is in the 2nd place with 3488.4 MWt, Ukraine is in the 3rd place with 1607 MWt and Hungary is in the 4th place with 1023.7 MWt and Poland ranks 5th with 576 MWt. As can be seen, except China, which is the first in the world in direct usage capacity, all other countries are located in the European continent.

9.4 The dilemma: Renewable or nonrenewable

Geothermal energy is a renewable source and related technologies proves themselves sustainable for decades if not a century of utilization for different applications. But use of geothermal energy becomes a controversial issue for about two decades. The main reasons for it based on its effects on the local communities. The primary energy challenge for developing countries is to meet growing demand for electricity while diversifying the energy generation mix. This requires increase in significant generation capacity which can be supplied by renewables significant drop in the prices. Geothermal comprised only a small fraction in increasing renewable share since both solar and wind energy have relatively lower risks, rapid deployment times, and lower investment requirements compared to geothermal. Yet the primary benefit of geothermal in power generation without fluctuations can play an important part in diversifying the power generation mix for developing countries especially for base-load power generation. And importance of geothermal power generation is increasing as other renewable sources such as hydropower is increasingly risky due to possible droughts and

TABLE 9.3 The utilization capacities of direct use applications of geothermal energy in 2019 for developing countries (Data from Lund and Aniko, 2020).

Country	Space / District heating MWt	TJ/y	Greenhouse heating MWt	TJ/y	Aquacultural heating MWt	TJ/y	Swimming and bathing MWt	TJ/y	Air conditioning MWt	TJ/y	Heat pumps MWt	TJ/y	Food drying MWt	TJ/y	Other uses MWt	TJ/y	Total MWt	TJ/y
Algeria	2.0	26.0	1.2	5.0	15.0	300.0	58.3	1955.4	0.5	3.2	0.7	85.0					77.7	2376.1
Argentina	22.4	50.0	21.5	40.1	7.0	13.1	137.2	1029.6							15.3	44.6	204.8	1209.1
Belarus																	10.0	137.0
Bosna & Her.	16.5	170.1					11.8	61.8			7.2	71.2					36.0	306.7
Brazil					1.0	20.0	355.9	6545.4	2.3	40.0	0.1	0.3	0.5	3.6			363.5	6682.7
Bulgaria	19.2	150.4	1.7	25.5			65.7	994.0	3.3	50.0	10.0	47.3			4.2	77.0	109.4	1327.0
Chile							14.7	228.9			7.9	50.0					22.6	278.9
China	7011.0	90,650.0	346.0	4255.0	482.0	5016.0	5747.0	86,993.0			26,450.0	246,212.0	179.0	2145.0	395.0	8221.0	40,610.0	443,492.0
Columbia							18.0	300.0			1.0	20.0					20.0	340.0
Crotia	29.1	177.6	2.3	87.4			44.1	95.1			3.8	30.0					79.3	390.6
Czech Republic							4.5	90.0			320.0	1700.0					324.5	1790.0
Ecuador							5.1	102.4			0.0	1.1					5.2	103.5
Georgia	13.6	433.2	18.1	571.0	0.0	1.1	37.5	1180.7			0.0	0.2					69.2	2186.2
Hungary	300.6	2587.0	358.1	2891.0			249.0	3684.0			72.0	1022.0	25.0	297.0	19.0	220.6	1023.7	10701.6
India							357.5	4004.0			0.1	3.8					357.6	4007.8
Iran							81.7	2576.9			0.5	6.4					82.2	2583.2
Kenya			5.3	185.0	0.2	6.5	8.7	275.5					0.3	9.9	4.0	125.5	18.5	602.4
Lithuania																	125.5	1044.0
Macedonia	43.4	525.0	2.8	61.1							1.3	37.5	0.5	13.2			47.4	623.6
Mexico	0.1	3.6					155.3	4166.5			0.1	0.5					156.1	4183.9
Mongolia	3.5	54.4					16.0	261.9			3.3	82.4					22.7	398.7
Poland	85.1	912.9			2.1	17.8	17.0	137.2			650.0	3100.0			1.0	5.5	756.0	4176.0
Romania	107.9	823.5	15.7	80.5	4.8	9.5	66.7	492.3			40.0	480.0	6.3	12.7	3.8	6.8	245.1	1905.3
Saudi Arabia																	45.0	172.9
Slovakia	56.5	432.7	45.0	229.3	0.2	0.4	127.0	1325.0	3.2	32.9	1.6	13.5					230.3	2000.9
Slovenia	19.7	179.5	10.8	111.6			23.4	197.3			203.1	1031.8					265.6	1610.5
Thailand							127.5	1168.9			1.0	12.0					128.5	1181.2
Turkey	1453.0	16,037.0	820.0	155,160			1205.0	22,800.0	0.4	10.0	8.5	171.0	0.0	0.3			3488.4	54,584.0
Ukraine							7.0	96.0			1600.0	4990.0	1.5	50.0			1607.0	5086.0
Vietnam					0.0	0.1					0.0	1.5					18.2	188.5

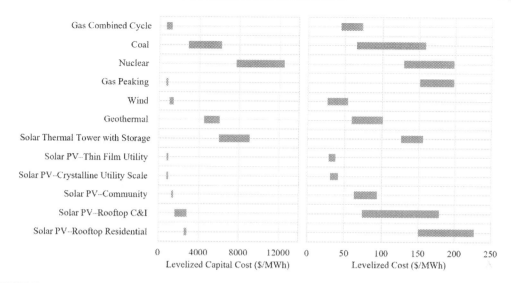

FIGURE 9.12 Levelized costs of energy generation (data from Lazard, 2020).

global warming in developing countries. Costa Rica aims to increase its geothermal capacity to decrease variability risks of its renewable energy mix generating over 90% of its electricity (World Bank, 2018).

Geothermal energy has significant advantages compared to other energy harvesting and utilization technologies. First and most importantly it is continuously available and not dependent on weather conditions, seasons, or time of the day. Its initial investments are high compared to other renewable energy technologies as shown in Fig. 9.12. Yet lower operational and maintenance costs result in lower levelized costs than many energy technologies (Lazard, 2020). Geothermal power generation mainly restricted to regions with medium to high temperatures and its both more cost effective and environmentally friendly than many nonrenewable technologies. Further it can help to reduce energy heating bills down to one quarter of their current level with direct heating and cooling networks (Garabetian, 2019).

The reason of debate about geothermal energy mostly focused on emissions and its effects on local community. Global average GHG emissions from geothermal power production were 122 gCO_2/kWh, but there are extreme cases reported. As in the case of Menderes and Gediz grabens in South West Turkey which is 900 to 1300 g CO_2/kWh. Such high GHG emissions seem to be restricted to high temperature geothermal reservoirs located in carbonate rich rocks, which are a rare occurrence, and not using reinjection systems for NCGs. Thus, this removes all benefits of geothermal power generation compared to coal. There are several technologies that have been developed to capture and treat NCGs from geothermal power plants. And captured CO_2 can be commercialized for dry ice production and for production of carbonated beverages as in some of the geothermal power plants in Turkey, including Kizildere, Dora I and II, and Gumuskoy. These power plants decrease their emissions and use disadvantage of abundance of carbonate sedimentary and metamorphic rocks rich in the region to an advantage. Reaching near zero real emissions through reinjection costs 10.3 USD/$tonCO_2$ (Fridiksson et al., 2016).

However, capture and treatment of geothermal CO_2 is currently uncommon worldwide. At present, NCG reinjection has been applied in few cases; Hellisheiði (Iceland) geothermal power plant, Umurlu geothermal reservoir (Turkey) (Bonafin et al., 2019).

9.5 Contribution to sustainability

Geothermal is a renewable resource as the extracted heat from an active reservoir is continuously restored by natural heat production, conduction, and convection from surrounding hotter regions, and the extracted geothermal fluids are replenished by natural recharge and by injection of the depleted (cooled) fluids. If managed properly, geothermal systems can be sustainable for the long term (Edenhofer et al., 2010). Further, geothermal heat sources will not be impacted by climate change. Although there are clear challenges to realizing the massive potential of geothermal energy, current research and development is mainly aimed for expending its applications for different areas beyond power generation. Geothermal energy is uniquely positioned to play a key role in climate change mitigation strategies by stable cost and supply unlike other renewable energy options such as wind and solar (Edenhofer et al., 2010).

Operational sustainability of geothermal operations are already proven for both direct use and power generation from geothermal energy. There are clear indicators that direct use has been practised at least since the Middle Palaeolithic when hot springs were used for ritual or routine bathing, and industrial utilization began in Italy by exploiting boric acid from the geothermal zone of Larderello. The area is also important since in 1904 the first geothermal power generation for 15 kilowatts of electric energy were generated and in 1913 the first 250 kW commercial geothermal power unit was installed. For more than 100 years, geothermal energy has provided safe, reliable, environmentally benign energy used in a sustainable manner to generate electric power and provide direct heating services on both large and small scales. Approximately 40% of the present-day installed electricity capacity has been in operation for more than 25 years, which shows sustainability of operations with reliability (Edenhofer et al., 2010). Yet general term sustainability covers a wider meaning which includes many aspects where environmental and social is main issues. Geothermal energy utilization can be made nearly climate neutral, and its many other positive environmental attributes enable it to operate in an environmentally sustainable manner. With its natural thermal storage capacity, geothermal is especially suitable for supplying dispatching base-load power. Thus, geothermal could function in a portfolio approach to increase the effectiveness of intermittent renewable energy sources such as hydro, wind and solar, resulting in a much larger net impact for mitigating climate change (Edenhofer et al., 2010). In recent years the use of geothermal resources increased with the technological developments. This also brings unwanted effects on the environment because of incorrect applications. This cases also negatively affected businesses that have good practices. The environmental impacts of geothermal power plants are generally in the form of discharge of geothermal fluid to the surrounding receiving environments without re−injection, leaks and leaks caused by deformations in plant components due to corrosion and scaling, gas emission, visual pollution, micro-seismicity, collapses, thermal and chemical pollution is taking place. Such environmental problems can be minimized by innovative methods.

Potential environmental impact of geothermal use starts from the drilling operations and follows plant life cycle till land reclamation. Most of the time construction activities underestimated compared to operational effects of a facility, but most of the significant effect on local environment and also social disturbances are risen during drilling operations for geothermal applications. These operations increase local traffic and cause excessive noise up to 120 dB (Bošnjaković et al., 2019). Further geothermal fluid can contain high concentrations of heavy metals such as arsenic or boron. Explosions or jets that may occur in the well during or after geothermal well drilling affect both soil and water resources as a result of the discharge of this fluid or its improper application (Dinçer and Ezan, 2020). This also brings issues on possible toxic effect to environment. These problems are not faced during plant operations where geothermal fluids used for electricity are injected back into geothermal reservoirs using wells. Measures to prevent cross-contamination of brines with groundwater systems are taken for all plants. But due to the acidic or basic nature of geothermal fluids, they can react with the metallic material used in the construction of equipment resulting in corrosion (IRENA, 2019).

Developing countries benefit geothermal energy and especially Asian and African countries have a better performance in exploiting geothermal resources. There is significant development in Latin America and Caribbean Region yet their geothermal resource lower than 15% of estimated geothermal resource potential with utilized capacity of 1.67 GW (World Bank, 2018).

The visible plumes seen rising from some geothermal power plants are actually water vapor since these plants do not use fossil fuels. Yet air pollution is more related with long term operations of geothermal power plants where closed cycle or re-injection systems are not utilized. Carbon dioxide emissions of geothermal power plants vary greatly between 0.1 and 1.05 kg/kWheq where coal power plants have an average of 0.9 kg/kWheq (Bonafin et al., 2019). This large difference is caused by the technologies applied. In closed loop systems, gases extracted from the well are not exposed to the atmosphere and are sent back to the reservoir. Therefore, air emissions are minimal. However, open loop systems emit hydrogen sulfide, carbon dioxide, ammonia, methane, and boron to the atmosphere. Considering operational emissions of a geothermal power plants is two thirds of plant lifecycle emissions and very close to coal powered power plants (Hondo, 2005) the extend of this release can be understood more clearly. These gases, especially hydrogen sulfide is needed to be removed from the gas before releasing. This is one of the major problems in regions where hydrogen sulfide and carbon dioxide ratios are higher such as in Western Anatolia. But it must be added it is seen that the CO_2 concentration decreases with the increase of the production in the region because of high number of open loop systems in the region (Dinçer and Ezan, 2020). Another important fact about geothermal power do not contribute for particulate matter and nitrogen oxide (NO_x) emissions. Supplying heat required to generate steam is supplied from earth crust rather than fossil fuels thus there are no combustion related emissions occur for geothermal.

Hydrogen Sulfide (H_2S) is another important emission debated for geothermal facilities. It is possible to convert 99.9% of the H_2S from geothermal noncondensable gases into elemental sulfur, which can then be used as a nonhazardous soil amendment and fertilizer feedstock. And H_2S emissions have declined from 862 kg/hr to 91 kg/hr or less since 1976 (Kagel et al., 2007). But there are still examples of power plants with no or insufficient sulfur removal capacity.

It must be stated that operation of excessive number of geothermal power plants can result in increased environmental effect in the local region (Dinçer and Ezan, 2020). Büyük Menderes and Gediz Basins in Turkey is an example for such increased effect. Both regions have an intense agricultural activity. Büyük Menders region have a total area of 24.873 km^2 and there are 35 operational, nine licensed and one planned geothermal power plants in the region (Baba et al., 2020). Excessive investment and installation of new power plant has caused the environmental impact of geothermal to increase where both sector and region suffers from wrong practices.

Sustainability is one of the critical issues to achieve international financing which is needed for most geothermal development projects in developing countries. This also force developers to follow international best practices and standards from the outset. The Equator Principles and the IFC/World Bank Performance Standards are most widely accepted standards in the sector. Both of these standards include environmental risks for land and habitat loss, water risks, solid discharges and waste, gas emissions (CO_2 and H_2S), dust and noise, occupational health, and safety. Another important factor is social impacts and risks where large-scale projects in geothermal power generation can impose on local communities. All these issues are also decision points in financial instruments such as risk sharing mechanism applied in Turkey. There are cases where these mechanisms also make mediation by third parties and additional funding for local communities in case there are complaints issued such as in Kenya. Local community expresses their dissatisfaction on plans and their potential impacts on community and issue a complaint in 2014. This led to an agreement which includes wider measures to improve identification of affected individuals, to improve the physical infrastructure of the resettlement site, and to support livelihood restoration in 2016 (World Bank, 2018). Compared to other power generation methods, especially other power plants geothermal plants require much land use. Photovoltaic power generation requires 50 times and other fossil plants require 30 times more land use for the same power generation (Soltani et al., 2021). This advantage is especially important for developing countries where geothermal resources are in rural areas and farming is the main income for local communities. This land cover can be further decreased by directional drilling which can help in geothermal areas with volcanic activities such as Indonesia. Yet this technology is costly and rarely applied in developing countries where alternatives are economically more beneficial.

Beyond these effects new technologies are also bringing their own challenges as in micro-seismic activities observed during hydraulic fracturing works for enhanced geothermal reservoirs. These micro-earthquakes have a magnitude up to around 2 in the Richter scale and pose no danger to existing buildings or structures. But there are unknowns on how they will evolve overtime for the reservoir and geological structures. Yet it must be added that induced seismic events have been observed in a number of geothermal projects such as Basel (Sweden) or Pohang (South Korea). Incident in Basel has a 3.4 magnitude, and in Pohang 5.5 which causes widespread damage and injuring 145 people (Buijze et al., 2019).

9.6 Future prospect

Technological developments in geothermal can be grouped into above the ground and under the ground technologies. Underground technologies focus on exploitation of reserves where above the ground technologies are focused on adapting variable demand changes and capacity improvement.

One of the most widely applied trends in geothermal power generation is cascading, which is pairing of multiple power generation systems is gaining importance in geothermal sector.

Solar, biomass, thermoelectric generators (TEG), hydrogen production, trigeneration, are all paired with geothermal power generation and direct use. Geothermal sources can be paired up stream or downstream base on applications. Most basic example is use of outlet geothermal brine for heat pump heating applications (Anderson, 2019). Power generation on the other hand focus on superheating the working fluid of the power cycle or increasing the quality of vapor in a dry steam application. Concentrated solar or biomass utilization and thermal energy storage (TES) are the most widely invested technologies to improve power generation performance. These hybrid systems aimed reducing adverse effects of diurnal temperature, preheating configurations, or reinjection. For instance, geothermal/solar-thermal hybrid power plants have the ability to operate at higher output levels than relatively equipped stand-alone air-cooled geothermal power plants during periods of solar energy (Enel, 2016). Stillwater power plant in Nevada USA carry this approach even further by combining concentrated solar, geothermal, and photovoltaic (PV) together. A 2 MW thermal solar power plant was commissioned to work with the existing geothermal power plant, and the plant's geothermal power generation capacity of 33.1 MW increased by 3.6%. Further 26.4 MW solar PV unit was added to the geothermal plant to support electricity production to respond to the high demand experienced during the summer days. Another hybrid approach is using biomass directly or after conversion to fuels to increase turbine entry steam temperature. Tuscany hybrid power plant increase its existing 13 MW geothermal power capacity by 5 MW by using locally sourced forest biomass to increase the temperature of the steam generated from the geothermal power plant from 155 to 375°C (Shumkov, 2015).

Current research and investment are mostly aiming renewable heating and cooling in buildings, and applications in the agriculture and food industry. Overall share of renewable energy in district heating worldwide remains negligible to date even these applications require low to medium temperature geothermal sources. The share of renewable energy in district heating may increase from 8% in 2017 to 77% in 2050 (IRENA, 2020). Currently there are notable examples for district heating from both developed world; USA or more notably from Iceland, developing countries such as Turkey. But beyond using geothermal wells there are studies focusing on using abandoned mines, oil, and gas wells together with heat pumps for district heating. These studies are centralized large scale district heating or cooling systems aiming several hundred homes or larger areas (Lettenbichler and Provaggi, 2019).

The low global weighted-average cost of geothermal power generation has been around USD 0.07/kWh since 2016 (IRENA, 2020). And this comparatively low cost and emissions of geothermal energy attract interest beyond utilization of power generation or direct use. Geothermal energy is gaining attention for production of energy carriers in remote areas have included national parks, such as Cerro Pabellón geothermal plant in Chile's Atacama Desert, where hydrogen is used as a long-duration storage for geothermal energy (World Bank, 2020).

Geothermal resource conditions and risks main factors to influence investment decisions. These risks and high up-front investment capital requirements in the initial stage of geothermal development has been a leading factor in the slow pace of geothermal development throughout developing countries (World Bank, 2018). Forty-three percent of the capital cost is related to exploration and drilling operations in geothermal resource utilization (Allahvirdizadeh, 2020). Thus, new drilling and well technologies has been under development for both improved economy and reaching geothermal reservoirs which is not possible until now. European Commission's SET-Plan and its Integrated Roadmap states deep geothermal energy as a promising technology with large innovation potential to supply a large portion of baseload electricity. And to reach this aim large geothermal power plants that tap into ultra-hot, supercritical heat reservoirs becomes a priority. Deep drilling technologies can help tectonically inactive geographies to reach high enthalpy hydrothermal reservoirs. There are significant challenges to realize deep drilling and hard dry rock (HDR) applications. As stated, before enhanced geothermal systems cold fluid was injected to HDR and was heated to high temperature so that we can extract heat energy by heat exchange. There are a few EGS practices executed such as Fenton Hill and Rosemanowes in United Kingdom, Hijori and Ogachi in Japan, and Copper Basin in Australia. All these projects are in developed countries because of high costs and risks associated with new technologies. These projects shows that fracture extension is unpredictable, and it is hard to get good hydraulic conductivity between wellbores and there is very significant water loss in the system (Zhang, 2020).

Beyond technology, utilization of geothermal resources is also evolving. Increased use of heat pumps is also increasing the demand for direct use of geothermal energy (Allahvirdizadeh, 2020). Unlike geothermal resources required to generate electricity, low and medium temperature geothermal resources suitable for direct use applications. To supply these markets 4th generation low temperature direct heating and cooling systems are under development to utilize 55 to 25°C or lower temperature sources (Pinzuti, 2019). This trend is also observed for direct use of geothermal applications in food processing and agriculture even not as rapid as heating and cooling applications. Both urban and agricultural applications do not require high temperatures and can help to reduce emissions from fossil fuels and protect against price volatility (IRENA, 2019). Geothermal energy applications can be integrated to food value chain to reduce already high emissions and environmental effects of agricultural activities.

Public support in the form of feed-in tariffs (FIT), incentives; risk guarantees, longer concession periods and efficient permitting and licensing processes are needed to support geothermal development especially in highly volatile economic conditions in developing countries. One of the examples of such support is risk sharing mechanism (RSM) provided by a contingent grant from the Clean Technology Fund (CTF), provided through the World Bank, to developing countries such as Turkey, Dominic, and Indonesia with the aim of to meet the risks that geothermal energy investors will encounter in well drilling activities for resource exploration and verification. This support covers 40% to 60% of the cost of failed wells will be paid by the RSM to the Beneficiary, up to a total of US$ 4 million under a scheduled program. And significant focus is given to projects targeting less exploited regions (Wietze and Thorbjörnsson, 2018). Cost sharing schemes in developing countries help to reach additional installed capacities of over 1900 MW. Yet a similar but private mechanism, geothermal resource risk insurance, does not show a similar success and can only be applied several tens of MWs (World Bank, 2018). The main reasons for this is geothermal development is a small sector globally where risky small portfolio and the typically high premiums is found not be affordable especially in developing countries. Because of these two attempts at using a geothermal insurance scheme in Turkey and in Mexico were not successful where cost sharing schemes successfully applied.

There is a large variation in fiscal incentives for geothermal. African developing countries such as Nicaragua, Ethiopia, and Kenya waive duties on the import of geothermal equipment and income tax breaks for the first years of geothermal operations. South Asian developing countries such as Indonesia have tax exemptions or reductions in taxable income and exemption of taxes on machinery imported for geothermal development. Philippines focusses on geothermal project developers in these exemptions. Among South American countries Mexico have 100% tax deduction on investment on renewable power which also includes geothermal investments (World Bank, 2018).

Beyond all these issues there are economic barriers for using geothermal power generation in developing countries. One of these is the size of the electrical market compared to required investment. Small markets gain less interest due to high investment requirements, low institutional capacity of governments and the limited local technical capacity and services (World Bank, 2018) which is the case in most of the developing countries. Another factor is fluctuating currency rates where geothermal energy is financed by foreign investment. Asian financial crisis in 1990s is one of the most known examples for this where Indonesian government had to suspend and renegotiate all contracts and devaluate prices almost 50% below pre-crisis contracts after renegotiations took years to complete (Gehringer and Victor, 2012). Current global Covid economic crises may give rise to similar countries in developing world. One such case is already realized where currency fluctuations force Turkish government to change tariff incentives from USD to local currency in 2021.

9.7 Thermodynamic analysis of geothermal energy systems

Geothermal resources are local energy sources. Especially hydrothermal geothermal resources may have different properties even if they are in the same area. For this reason, geothermal resources and systems are primarily evaluated with energy and exergy analyses made in the light of thermodynamic laws. Although geothermal resources are mostly considered as pure water in these analyses, especially in regions with high NCG ratio, the geothermal fluid should be considered as a mixture of water and NCG and their thermodynamic properties should be determined and analyzed in this way.

The fundamental mass, energy and exergy balance equations are applied to the entire system and each component, taking into account that the system is at steady state condition with steady-flow process.

The general mass and energy balance equations are explained in the rate form as (Cengel and Holes, 2015):

$$\sum \dot{m}_{in} - \sum \dot{m}_{out} = \frac{dm_{CV}}{dt} \quad (9.1)$$

The general energy balance can be expressed as the total energy input equals the total energy output:

$$\left[\sum (\dot{m} \cdot h) + \sum \dot{Q} + \sum \dot{W}\right]_{in} - \left[\sum (\dot{m} \cdot h) + \sum \dot{Q} + \sum \dot{W}\right]_{out} = \left[\frac{d(m \cdot e)}{dt}\right]_{CV} \quad (9.2)$$

where, "e" is the specific energy of the nonflowing system including internal, kinetic and potential energies:

$$e = u + \frac{1}{2}V^2 + g \cdot z \quad (9.3)$$

The equivalence must be established for entropy, which can be defined as a quantitative measure of microscopic disorder for a system, before exergy balancing. Because exergy destruction is a function of entropy generation. Since entropy can be transferred to or from a system by heat transfer and mass flow, entropy balance relation is written as follows (Cengel and Boles, 2015):

$$\left[\sum(\dot{m}\cdot s)+\sum\int\frac{d\dot{Q}}{T}\right]_{in}-\left[\sum(\dot{m}\cdot s)+\sum\int\frac{d\dot{Q}}{T}\right]_{out}+\dot{S}_{gen}=\left[\frac{d(m\cdot s)}{dt}\right]_{CV} \quad (9.4)$$

The general exergy balance relations are given in Eq. (9.5).

$$\left[\sum(\dot{m}\cdot\psi)+\sum\left(1-\frac{T_0}{T}\right)\cdot\dot{Q}+\sum\dot{W}\right]_{in}-\left[\sum(\dot{m}\cdot\psi)+\sum\left(1-\frac{T_0}{T}\right)\cdot\dot{Q}+\sum\dot{W}\right]_{out}$$
$$-\dot{X}_{destroyed}=\left[\frac{d(m\cdot\phi)}{dt}\right]_{CV} \quad (9.5)$$

where, T_0, ψ and ϕ are dead state temperature, flow, and nonflow specific exergy,

$$\psi=(h-h_0)-T_0\cdot(s-s_0)+\frac{V^2}{2}+g\cdot z \quad (9.6)$$

$$\phi=(e-e_0)+P_0\cdot(v-v_0)-T_0\cdot(s-s_0) \quad (9.7)$$

respectively. Exergy can be classified as physical and chemical exergy. Physical exergy is mostly taken into consideration in the evaluation of geothermal systems.

$$\dot{X}_{destroyed}=T_0\cdot\dot{S}_{gen} \quad (9.8)$$

In order to calculate the efficiency of a system, it is necessary to know or determine the inputs and outputs of the system. In the simplest way, efficiency is the ratio of how much input is used for the desired output. In this context, the energy efficiency of a system:

$$\eta_{en}=\frac{\text{Desired energy output}}{\text{Required energy input}} \quad (9.9)$$

Similarly, exergy efficiency of a system:

$$\eta_{ex}=\frac{\dot{X}_{useful}}{\dot{X}_{in}}=1-\frac{\dot{X}_{destroyed}}{\dot{X}_{in}} \quad (9.10)$$

The coefficient of performance (COP) is used instead of efficiency for refrigerators, heat pumps, and vapor absorption cycle (VAC). The exergetic efficiency of those devices can be calculated as:

$$\varepsilon=\frac{COP_{act}}{COP_{rev}} \quad (9.11)$$

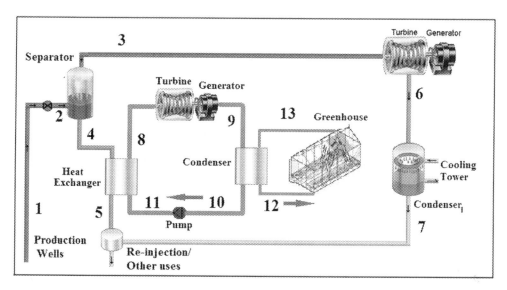

FIGURE 9.13 Schematic view of a combined geothermal system.

9.8 Case study

Here, energy and exergy analyses of a combined geothermal system in which electrical and thermal energy are provided together are discussed as a case study.

A combined geothermal system includes two-stage power generation system and a greenhouse heating system is illustrated in Fig. 9.13. A geothermal resource at 190°C (State 1) with 280 kg/s mass flow rate is flashed to 1/3 to its pressure isenthalpically (State 2). The vapor phase of geothermal fluid (State 3), separated in a separator of single-flash GPP, flows to a steam turbine to produce the electricity, expands to 10 kPa condenser pressure. A binary cycle with R134a is entegrated to single-flash GPP to produce more electricity. The brine (liquid phase of geothermal fluid (State 4)) transfers its energy to R134a by a heat exchanger. The saturated R134a vapor at 2400 kPa is expanded to 800 kPa. After the turbine when R134a is condensed in a condenser of Binary GPP, the release energy is used to heat a greenhouse with a water at 110 kPa. Calculate the total electricity production of the GPP, energy and exergy efficiency of this combined geothermal system. Assume that, the turbines and pump has isentropic efficiency of 85%, the effectiveness of heat exchangers is 85% and temperature approach in the heat exchangers is 10°C. The fluids at the condenser exits are at saturated liquid phase. The dead state properties are 25°C and 100 kPa.

Solution:
Assumptions:

(a) All of processes are under steady state/flow conditions.
(b) Kinetic and potential and energetic influences are excluded while no nuclear/chemical reactions occur.
(c) Due to short lengths of the piping system, the pressure losses in it and the connecting elements used for the components are neglected.

Thermodynamic state point properties are summarized in Table 9.4.

TABLE 9.4 Thermodynamic properties.

State	Fluid	T (°C)	P (kPa)	$\dot{m}$ (kg/s)	h (kJ/kg)	s (kJ/kg.K)	$\dot{E}$ (kW)	$\dot{X}$ (kW)
0	Water	25	100		104.9	0.3672		
00	R134a	25	100		276.9	1.106		
1	Geo. fluid	190	1255	280	807.4	2.235	226,079	40,813
2	Geo. fluid	145.2	418.4	280	807.4	2.261	226,079	38,676
3	Geo. fluid	145.2	418.4	25.75	2740	6.880	70,566	17,879
4	Geo. fluid	145.2	418.4	254.2	611.7	1.793	155,513	20,797
5	Geo. fluid	42.4	418.4	254.2	178.1	0.605	45,281	612
6	Geo. fluid	45.8	10	25.75	2263	7.144	58,290	3,579
7	Geo. fluid	45.8	10	25.75	191.8	0.649	4,940	73
8	R134a	135.2	2400	357.5	359.2	1.103	128,394	29,951
9	R134a	96.1	800	357.5	333.3	1.115	119,159	19,392
10	R134a	31.3	800	357.5	95.47	0.354	34,127	15,438
11	R134a	32.4	2400	357.5	97.06	0.355	34,697	15,922
12	Water	86.1	110	385	360.8	1.148	138,922	8,958
13	Water	41.3	110	385	173.1	0.590	66,645	709

Mass flow rate of geothermal vapor at the steam turbine inlet is determined with quality of the geothermal fluid at State 2.

$$\dot{m}_3 = x_2 \cdot \dot{m}_2 = 0.0919 \cdot 280 \text{ kg/s} = 25.75 \text{ kg/s}$$

Energy balance is applied to the heat exchanger to determine the mass flow rate of R134a in Binary GPP.

$$\dot{m}_8 \cdot (h_8 - h_{11}) = \dot{m}_4 \cdot (h_4 - h_5) \cdot \eta_{HEX}$$

$$\dot{m}_8 \cdot (359.2 - 97.06) \text{ kJ/kg} = 254.25 \text{ kg/s} \cdot (611.7 - 178.1) \text{ kJ/kg} \cdot 0.85$$

$$\dot{m}_8 = 357.5 \text{ kg/s}$$

The net power of the geothermal system is:

$$\dot{W}_{net} = \dot{W}_{s,tur} + \dot{W}_{R134a,tur} - \dot{W}_{pump}$$

9.8 Case study

Turbine electricity production and net power production of the GPPs:

$$\dot{W}_{s,tur} = \dot{m}_3(h_3 - h_6) = 25.75 \cdot (2740 - 2263) = 12283 \text{ kW}$$

$$\dot{W}_{R134a,tur} = \dot{m}_8(h_8 - h_9) = 357.5 \cdot (359.2 - 333.3) = 9259 \text{ kW}$$

$$\dot{W}_{pump} = \dot{m}_{10}(h_{11} - h_{10}) = 357.5 \cdot (97.06 - 95.47) = 568 \text{ kW}$$

$$\dot{W}_{net} = \dot{W}_{s,tur} + \dot{W}_{R134a,tur} - \dot{W}_{pump} = 12283 + 9259 - 568 = 20974 \text{ kW}$$

The mass flow rate of the water used to heat the greenhouse is determined to be 385 kg/s, similar to the calculation of the R134a mass flow rate. Consequently, the transferred heating energy to the greenhouse:

$$\dot{Q}_{gh} = \dot{m}_{12}(h_{12} - h_{13}) = 385 \cdot (360.8 - 173.1) = 72265 \text{ kW}$$

Energy efficiency of the whole system is determined as follow

$$\eta_{en} = \frac{\dot{W}_{net} + \dot{Q}_{gh}}{\dot{E}_1 - \dot{E}_5 - \dot{E}_7}$$

$$\eta_{en} = \frac{20974 + 72265}{226079 - 45281 - 4940} = 53.01\%$$

Exergy efficiency of the whole system:

$$\eta_{ex} = \frac{\dot{W}_{net} + \dot{X}_{12} - \dot{X}_{13}}{\dot{X}_1 - \dot{X}_5 - \dot{X}_7}$$

$$\eta_{ex} = \frac{20974 + 8958 - 709.2}{40813 - 612 - 73} = 72.75\%$$

If only a single flash GPP was used in the system energy and exergy efficiencies:

$$\eta_{en} = \frac{\dot{W}_{s,tur}}{\dot{E}_1 - \dot{E}_4 - \dot{E}_7}$$

$$\eta_{en} = \frac{12283}{226079 - 155513 - 4940} = 18.71\%$$

$$\eta_{ex} = \frac{\dot{W}_{s,tur}}{\dot{X}_1 - \dot{X}_4 - \dot{X}_7}$$

$$\eta_{ex} = \frac{12283}{40813 - 20797 - 73} = 61.56\%$$

Instead of using only a single flash GPP, it is seen that the energy efficiency of the plant can be increased from 18.71% to 53.01% and the exergy efficiency from 61.56% to 72.75% by adding a binary GPP and greenhouse heating. Therefore, combined systems should be considered in designs in order to obtain maximum benefit from geothermal resources.

9.9 Self-evaluation questions and numerical problems

1. Why geothermal resources could be treated as sustainable sources?
2. Write the names of two main groups of geothermal energy utilizations.
3. Draw the Lindal diagram showing the distribution of geothermal applications according to the source temperatures?
4. Write the names of 5 direct use applications of geothermal energy.
5. What is the range of COP for heat pump systems?
6. Write the names of the plant cycles mainly used in geothermal electricity generation.
7. What is the difference between a double-flash geothermal power plant and a single-flash geothermal power plant? Explain it.
8. What is the name of the equipment in flash steam plants that is different from dry steam plants? For what purpose it is used - Explain.
9. Write the names of the major equipment in a geothermal power plant.
10. What is the name of GPP, in which geothermal fluid comes from wells are directly flows to the turbine?

9.9.1 Numerical problems

1. A house is cooled by a ground source geothermal heat pump. The design temperatures of condenser and evaporator of the heat pump in which R-134a is used in its closed loop, are 45°C and 0°C. R134a leaves from the condenser as saturated liquid and from the evaporator as saturated vapor. The isentropic efficiency of the compressor is 80%, and the effectiveness of the condenser and evaporator is 80%. The water circulated in the pipes buried horizontally at 1.5 m deep into the ground enters the condenser of the heat pump at 20°C and leaves at 25°C. What is the COP of this heat pump used to meet the 5 kW cooling requirement of the house and what should be the mass flow rate of the water circulated in the pipes installed into the ground and R134a? Assume that there is no pressure drops in the condenser and evaporator. (Answers: 3.762 (COP), 0.1937 kg/s (water), 0.0297 kg/s (R134a)).
2. What would be the results if R245fa was used in the heat pump as the working fluid under exactly the same operating conditions in the previous question? (Answers: 3.972 (COP), 0.1915 kg/s (water), 0.02757 kg/s (R245fa)).
3. The geothermal fluid, which is extracted from the well as a saturated liquid at 240°C with 50 kg/s mass flow rate, becomes two-phase by decreasing the pressure at constant enthalpy to 400 kPa. The geothermal fluid, which will be used to generate electricity in a single flash GPP, is separated into vapor and liquid phases at constant pressure with the help of a separator. The vapor directed to the steam turbine with 85% isentropic efficiency is expanded to 15 kPa condenser pressure. The geothermal fluid, which comes to the saturated liquid phase at the condenser outlet, is re-injected to the ground together with the liquid part separated in the separator. Calculate the electrical energy to be generated in this plant and the energy efficiency of the power plant taking into account the re-injections. (Answers: 4340 kW, 17%).

4. If the system given in the case study is operated under exactly the same conditions and only R600a is used as the working fluid in Binary cycle, what will be the net electricity generation, energy and exergy efficiency of the whole system? (Answers: 18810 kW, 49.3% (energy), 71.5% (exergy)).

5. Reconsider Problem 13. It is now proposed that the geothermal fluid rejected from the primary separator be flashed to 200 kPa with constant enthalpy and directed to the secondary separator to generate additional geothermal steam. The additional saturated steam is directed to the low pressure steam turbine with 85% isentropic efficiency, then it is expanded to the condenser pressure of 15 kPa. The geothermal fluid leaves the condenser as the saturated liquid and is reinjected into the ground. Calculate the total electricity generation of the power plant that is converted to a double-flash GPP in this way. (Answers: 4949 kW)

Nomenclature

COP	Coefficient of performance (-)
c_p	Specific heat (kJ/kg.K)
e	Specific energy (kJ)
g	Gravitational constant (m/s^2)
h	Specific enthalpy (kJ/kg)
$\dot{m}$	Mass flow rate (kg/s)
P	Pressure (kPa)
$\dot{Q}$	Rate of heat transfer (kW)
s	Specific entropy (kJ/kgK)
$\dot{S}$	Rate of entropy (kW/K)
t	Time (s)
T	Temperature (K)
u	Specific internal energy (kJ/kg)
$\dot{X}$	Rate of exergy (kW)
V	Velocity (m/s)
$\dot{W}$	Rate of work (kW)
z	Elevation (m)

Greek letters

ε	Exergetic efficiency (-)
η	Efficiency (%)
υ	Specific volume (m^3/kg)
φ	Nonflow specific exergy (kJ/kg)
ψ	Flow specific exergy (kJ/kg)

Subscripts

0	Refers to the dead state
act	Actual
comp	Compressor
cond	Condenser
CV	Control volume
en	Energy

ex	Exergy
evap	Evaporator
gen	Generation
geo	Geothermal
gh	Greenhouse
HEX	Heat exchanger
in	Input/ inlet
out	Output/ outlet
rev	Reversible
tur	Turbine
wf	Working fluid

References

Akrouch, G.A., Sánchez, M., Briaud, J.L., 2016. An experimental, analytical and numerical study on the thermal efficiency of energy piles in unsaturated soils. Comput. Geotech. 71, 207–220.

Allahvirdizadeh, P., 2020. A review on geothermal wells: well integrity issues. J. Cleaner Prod., 124009.

Anderson, A., Rezaie, B., 2019. Geothermal technology: trends and potential role in a sustainable future. Appl. Energy 248, 18–34.

Axelsson, G., Gunnlaugsson, E., 2000. Long-term monitoring of high- and lowenthalpy fields under exploitation. In: World Geothermal Congress 2000, Pre-Congress Course, Kokonoe, Japan, p. 226.

Baba, A., Avcı, S., Blank, L., Bozkurt, C., Daylan, B., Evci Kiraz, E.D., Kilicozlu, E., Gissurarson, L., Gunnarsson, G., Okdemir, S., Sener, G.D., Sozbilir, H., Surmeli, S., Top, B.M., Velibeyoglu, K., Yazdani, H., 2020. Türkiye'de Jeotermal Kaynakların Kümülatif Etki Değerlendirmesi Projesi Nihai Rapor. European Bank of Reconstruction and Development, Turkey.

Bertani, R., 2016. Geothermal power generation in the world 2010–2014 update report. Geothermics 60, 31–43.

Bombarda, P., Invernizzi, C.M., Pietra, C., 2010. Heat recovery from diesel engines: a thermodynamic comparison between Kalina and ORC cycles. Appl. Thermal Eng. 30, 212–219.

Bonafin, J., Pietra, C., Bonzanini, A., Bombarda, P., 2019. CO_2 emissions from geothermal power plants: evaluation of technical solutions for CO_2 reinjection. In: Proceedings of the European Geothermal Congress.

Bošnjaković, M., Stojkov, M., Jurjević, M., 2019. Environmental impact of geothermal power plants. Tehnički vjesnik 26 (5), 1515–1522.

Boyd, T., Rafferty, K., 2008. Geothermal greenhouse information package. Goe-Heat Institute, Oregon Institute of Technology, Klamath Falls, OR.

Boyd, T.L., Lund, JohnW., 2006. Geothermal heating of greenhouses and aquaculture facilities. In: 2006 ASAE Annual Meeting. American Society of Agricultural and Biological Engineers.

Bödvarsson, G., 1961. Physical characteristics of natural heat sources in Iceland. In: Proc. UN Conf. on New Sources of Energy, Volume 2: Geothermal Energy, Rome, August 1961. United Nations, New York, pp. 82–89.

BP. 2020. "Statistical Review of World Energy 2020, The Statistical Review of World Energy analyses data on world energy markets from the prior year." 69th edition. https://www.bp.com/content/dam/bp/business-sites/en/global/corporate/pdfs/energy-economics/statistical-review/bp-stats-review-2020-full-report.pdf.

Buijze, L., et al., 2019. Review of induced seismicity in geothermal systems worldwide and implications for geothermal systems in the Netherlands. Neth. J. Geosci. 98.

Burgess, W.G., 1989. Geothermal energy. Geol. Today 5 (3), 88–92. https://doi.org/10.1111/j.1365-2451.1989.tb00630.x.

CanGEA, (2014). Suitable Applications and Opportunities for Canada, https://www.cangea.ca/uploads/3/0/9/7/30973335/directuse_pressrelease2.pdf (accessed 28.02.2021).

Cengel, Y.A., Boles, M.A., 2015. Thermodynamics: An Engineering Approach. McGraw-Hill Education, New York.

Coury, G., Guillen, H.V., Cruz, D.H., 1996. Geothermal noncondensable gas removal from turbine inlet steam. In: Proceedings of the 31st Intersociety 3 Energy Conversion Engineering Conference.

Dickson, M.H., and Fanelli M. "Geothermal energy: utilization and technology." (2013).

Dincer, I., Ozcan, H., 2018. Geothermal Energy. Compr. Energy Syst. 1 (5), 702–732. https://doi.org/10.1016/B978-0-12-809597-3.00119-X.

Dinçer, I., and Ezan M.A. "Tüba-Jeotermal Enerji Teknolojileri Raporu" Türkiye Bilimler Akademisi Yayınları, TÜBA Raporları No: 41, ISBN: 978-605-2249-54-3, 2020.

DiPippo, R., 2005. Geothermal Power Plants Principles, Applications and Case Studies. Elsevier Science, Oxford, UK ISBN-10: 1856174743.

DiPippo, R., 2004. Second law assessment of binary plants generating power from low-temperature geothermal fluids. Geothermics 33, 565–586.

Dumas, P., 2017. Geothermal energy in Europe. Persp. Geothermal Energy Eur. 11–40.

Edenhofer O., Sokona Y., Seyboth K., Eickemeier P., Matschoss P., Hansen G., Kadner S., Schlömer S., Zwickel T., von Stechow C., Change, Climate. "IPCC Special Report on Renewable Energy Sources and Climate Change Mitigation (SRREN), 2010

EIA, U.S. Energy Information Administration. Geothermal power explained, http://www.eia.gov/energyexplained/index.cfm?page=geothermal_power_plants (Accessed 27.02.2021).

Enel Green Power SpA, "Enel Green Power Inaugurates Triple Renewable Hybrid Plant in The Us, Press Release", March, 29 2016

Fridriksson, T., Mateos, A., Audinet, P., Orucu, Y., 2016. Greenhouse Gases from Geothermal Power Production. World Bank, Switzerland.

G. Systems, (2016). Sustainable Energy Handbook, Module 4.7, Geothermal Energy, https://europa.eu/capacity4dev/file/29375/download?token=YTetyYNP (accessed 28.02.2021).

Garabetian T., Dumas P., Serrano C., Mazzagatti V., Kumar S., Dimitrisina R., Ruaud J., Truong C., EGEC Geothermal Market Report, EGEC – European Geothermal Energy Council 2019.

Gehringer M., and Victor L. "Geothermal Handbook: Planning and Financing Power Generation A Pre-launch." (2012).

Ghoreishi-Madiseh, S.A., Kuyuk, A.F., 2017. A techno-economic model for application of geothermal heat pump systems. Energy Procedia 142, 2611–2616.

Gupta H., Roy, S.Worldwide Status of Geothermal Resource Utilization, Geotherm. Energy, Elsevier-The Netherlands, ISBN-13: 978-0-444-52875-9, 2007, pp. 199–229.

Hall, N.R., 1996. Gas extraction system. In: Dunstall, M.G. (Ed.), Lecture notes of Geothermal Utilisation Engineering Geothermal Institute. The University of Auckland, New Zealand.

Hamza, V.M., Cardoso, R.R., Neto, C.F.P., 2008. Spherical harmonic analysis of earth's conductive heat flow. Int. J. Earth Sci. 97 (2), 205–226.

Helvaci, H.U., Gökçen, G., Korel, F., Aydemir, L.Y., 2013. Bir Jeotermal Kurutucu Tasarimi Saha Testleri Ve Kurutma Sisteminin Enerji Analizi, Teskon Ulusal Tesisat Mühendisliği Kongresi ve Fuarı.

Hepbasli, A., Balta, M.T., 2007. A study on modeling and performance assessment of a heat pump system for utilizing low temperature geothermal resources in buildings. Build. Environ. 42 (10), 3747–3756.

Hondo, H., 2005. "Life cycle GHG emission analysis of power generation systems", Japanese Case. Energy 30, 2042–2056.

Horie, T., Muto, T., Gray, T., 2010. Technical Features of Kawerau Geothermal Power Station. In: New Zealand, Proceedings of World Geothermal Congress. Bali, Indonesia April 25–29.

Huttrer, G.W., 2020. Geothermal power generation in the world 2015-2020 update report. In: Proceedings World Geothermal Congress, 2020.

Ipcc, Chapter 4 Geothermal energy, https://archive.ipcc.ch/pdf/special-reports/srren/drafts/SRREN-FOD-Ch04.pdf (accessed 28.02.2021).

IRENA, 2020. Renewable Power Generation Costs in 2019. International Renewable Energy Agency, Abu Dhabi.

IRENA, 2019. Accelerating Geothermal Heat Adoption in the Agri-Food Sector. International Renewable, Energy Agency, Abu Dhabi.

Kagel, A., Bates, D., Gawell, K., 2007. A Guide to Geothermal Energy and the Environment. Geothermal Energy Association, Apr, United States.

Kanoglu, M., 1999. Design and Optimization of Geothermal Power Generation, Heating and Cooling Ph.D. Thesis. University of Nevada, Reno.

Kwambai, C.B., 2010. Exergy Analysis of Olkaria I Power Plant, Kenya. In: Proceedings of World Geothermal Congress. Bali, Indonesia April 25-29.

Lazard's latest annual Levelized Cost of Energy Analysis (LCOE 14.0), Lazard industry insights, 2020

Lettenbichler, S. and Provaggi A. (2019), 100% renewable energy districts: 2050 vision, www.euroheat.org/wp-content/uploads/2019/08/RHC-ETIP_District-and-DHC-Vision-2050.pdf.

Wietze, L., Thorbjörnsson, D., 2018. Workshop: World Bank's Risk Sharing Mechanism. IGC International Geothermal Congress & Exhibition, Turkey 14 March.

Liu, H., Maghoul, P., Bahari, A., Kavgic, M., 2018. Feasibility study of snow melting system using geothermal energy piles in Canadian Prairies. In: GeoEdmonton 2018 –71st Canadian Geotechnical Conference and the 13th Joint CGS/IAH-CNC Groundwater Conference.

Lund, J.W., Aniko, N.T., 2020. Direct utilization of geothermal energy 2020 worldwide review. Geothermics, 101915.

Lund, J.W., Freeston, D.H., Boyd., T.L., 2010. Direct utilization of geothermal energy 2010 worldwide review. Geothermics (April) 25–29.

Maddah, S., Goodarzi, M., Safaei, M.R., 2020. Comparative study of the performance of air and geothermal sources of heat pumps cycle operating with various refrigerants and vapor injection. Alexandria Eng. J. 59 (6), 4037–4047.

Milenić, D., Vasiljević, P., Vranješ, A., 2010. Criteria for use of groundwater as renewable energy source in geothermal heat pump systems for building heating/cooling purposes. Energy Build. 42 (5), 649–657.

Miller, C.C. 2006. "Chapter Two." Forbes 178 (13): 72–74. https://doi.org/10.7312/zhao12754-004.

Moya, P., DiPippo, R., 2010. Miravalles Unit 3 Single-Flash Plant, Guanacaste, Costa Rica: Technical and Environmental Performance Assessment. In: Proceedings of World Geothermal Congress. Bali, Indonesia April 25-29.

Nemzer, M., and Energy Education Group. 2005. "Renewable Energy Source : Geothermal Excerpt From Chapter 3 Energy for Keeps :" 501 (c).

Nihaj Uddin Shan, A.N.M, 2020. A review of kalina cycle. Int. J. Smart Energy Technol. Environ. Eng. 1 (1) September 2020.

Öz, M., Dikel, S., 2016. Use of geothermal sources in aquaculture. In: 1st International Congress on Advances in Veterinary Sciences and Technics (ICAVST).

Petroski, H., 2013. Geothermal energy. Am. Sci. 101 (4), 251–255. https://doi.org/10.1511/2013.103.251.

Pinzuti, V., Dumas, P., Garabetian, T., Manzella, A., Trumpy, E., Laenen, B., & Lagrou, D. (2019). European Technology and Innovation Platform on Deep Geothermal, A presentation. EGC.

Popovska-Vasilevska, S., 2003. Drying of agricultural products with geothermal energy. In: International Summer School on Direct Application of Geothermal Energy, Doganbey(Izmir), Turkey, pp. 2–15.

Ragnarsson, Á., 2014. Geothermal energy in aquaculture. In: Presented at "Short Course VI on Utilization of Low- and Medium-Enthalpy Geothermal Resources and Financial Aspects of Utilization", organized by UNU-GTP and LaGeo, in Santa Tecla. El Salvador March 23-29, 2014.

Saemundsson, K., Axelsson, G., Steingrímsson, B., 2009. Geothermal systems in global perspective. Short Course on Exploration for Geothermal Resources, UNU GTP 11, 17–30.

Salmon, J. P., Meurice, J., Wobus, N., Stern, F., Duaime, M., 2011. Guidebook to geothermal power finance. Contract 303, 275–3000.

Self, S.J., Reddy, B.V., Rosen, M.A., 2013. Geothermal heat pump systems: status review and comparison with other heating options. Appl. Energy 101, 341–348.

Shumkov I., Enel Green finalises geothermal-biomass hybrid project in Italy, Renewables Now, July, 2015.

Soltani, M., et al., 2021. Environmental, economic, and social impacts of geothermal energy systems. Renew. Sustain. Energy Rev. 140, 110750.

Swandaru, R.B., 2009. Modelling and Optimization of Possible Bottoming Units for General Single-flash Geothermal Power Plants M.Sc. Thesis. Department of Mechanical and Industrial Engineering University of Iceland.

Wallace, K., Dunford, T., Ralph, M., Harvey, W., 2009. Aegean Steam: Germencik Dual Flash Plant, GeoFund, IGA Geothermal Workshop "Turkey 2009". Istanbul, Turkey February 16-19.

WEC, 2010. 2010 Survey of Energy Resources. World Energy Council, United Kingdom https://www.worldenergy.org/assets/downloads/ser_2010_report_1.pdf.

World Bank Group, 2018. Opportunities and Challenges for Scaling Up Geothermal Development in Latin America and the Caribbean. World Bank, United States.

World Bank, Energy Sector Management Assistance Program. "Green Hydrogen in Developing Countries. (2020).

Yıldırım Özcan, N. "Modeling, simulation and optimization of flashed-steam geothermal power plants from the point of view of noncondensable gas removal systems." (2010).

Zhang, Y., Zhao, G.-F., 2020. A global review of deep geothermal energy exploration: from a view of rock mechanics and engineering. Geomech. Geophys. Geo-Energy Geo-Resour. 6 (1), 1–26.

Zhao, S.Y., Su, X.S., Chen, C., 2016. A study on geothermal snow-melting technology based on chemical PMMA pavement. Chem. Eng. Trans. 55, 229–234.

Zhao, W., Zhang, Y., Li, L., Su, W., Li, B., Fu, Z., 2020. Snow melting on the road surface driven by a geothermal system in the severely cold region of China. Sustain. Energy Technol. Assess. 40, 100781.

CHAPTER 10

Ocean renewable energy and its prospect for developing economies

Mahfuz Kabir[a], M.S. Chowdhury[b], Nigar Sultana[c], M.S. Jamal[d] and Kuaanan Techato[b]

[a]Bangladesh Institute of International and Strategic Studies (BIISS), Dhaka, Bangladesh [b]Faculty of Environmental Management, Prince of Songkla University, Songkhla, Thailand [c]Knowledge for Development Management (K4DM) Project, UNDP, Bangladesh [d]Institute of Fuel Research and Development (IFRD), Bangladesh Council of Scientific and Industrial Research (BCSIR), Dhaka, Bangladesh

10.1 Introduction

Ocean Renewable Energy (ORE) is a comparatively new but robust and environment-friendly alternative to nonrenewable fossil fuels. It implies various renewable energy resources that can be generated from oceans or marine sites. This industry has been expanding across the world in order to achieve targets of reversing the adverse impacts of climate change according to the Paris Agreement (Horne et al., 2021). ORE comprises tidal currents, wave, thermal, and salinity gradient energy. It is estimated that nearly half of the global population living in cities that comprise more than 100,000 inhabitants are within 100 km from the coast. Therefore, ORE can significantly benefit this population (Barragán and de Andrés, 2015).

Oceans have enormous potential to produce renewable electricity from their hydrokinetic energy ORE is manifested in three forms: mechanical, thermal, and chemical. Rotation of the earth and gravitation of the moon can be attributed to the origin of the ocean's mechanical energy through waves and tides, respectively. The temperature gradient from the sun's heat is captured by the upper ocean surface while the lower surface remains colder, which can be utilized to produce thermal energy. Mixing the flows of two different salt concentrations can be used to generate ocean chemical energy (Hernández-Fontes et al., 2020). Therefore, energy can be harvested from oceans using tides, tidal currents, waves, thermal and salinity gradients. The ORE technologies, which include floating devices as well as submerged or anchored to the seabed, are undergoing rapid transformation through development, trial, and subsequent improvement (Martínez et al., 2021). Ocean tidal and thermal energy are produced in Europe and Asia; wave energy is generated in Europe, South Africa, and Central and South America; and salinity gradient is found in Europe (IRENA, 2020). ORE can be produced in both small and large scales depending on the location and intended coverage. Many developing economies are leading the global production of renewable energy, and they are exposed to various climatic events. Harnessing the potential of ORE would be a viable means of facilitating their rapid economic growth and development, and efforts of mitigating the climatic change.

This chapter has been organized as follows. Section 10.2 presents the theoretical background of each type of ORE along with their respective conversions. Section 10.3 describes the present status of important ORE projects across the world along with their types and capacities. Section 10.4 analyses the state of ORE adoption and potential in developing economies. Section 10.5 elaborates on the contribution of ORE to sustainability with respect to its environmental effects and linkage with the Sustainable Development Goals (SDGs). Section 10.6 presents some selected country cases of technologies and potential. Section 10.7 suggests recommendations to realize the potential, achieve sustainable adoption with community acceptance, and minimize the environmental effects of the ORE projects. Finally, Section 10.8 concludes the chapter.

10.2 Ocean renewable energy conversion

10.2.1 Tidal energy

The ocean tidal/current energy (mechanical) is produced through the conversion of the kinetic energy available in currents of the large open-ocean geostrophic surface into energy for typical use. It is being researched as a viable source of electric power for nearly five decades (VanZwietena and Tang, 2021). Since 1970s, many devices and projects are in place to generate both wave and tidal stream energy. The first wave-based electricity was connected to the grid by 2000 (Queen's University Belfast, 2002), while the first tidal electricity delivered power to the grid in 2004 (Phillips, 2019). Considerable progress has been registered in the wave and tidal energy sector in 2000s and 2010s. Electricity generated from these two sources accounted from approximately 5 GWh to nearly 45 GWh between 2009 and 2019 (IEA-OES, 2020), while the global installed capacity reached nearly 65 MW in 2020 that increased by more than double from that of 2017 (IEA-OES, 2021). Such impressive progress can be mainly attributed to the widespread research and development (R&D) programs in various parts of the world. Moreover, ORE test centers enabled initial development, testing, improvement, installation and decommissioning of the technology (Barclay et al., 2021).

10.2.1.1 Tidal energy conversion

Tang et al. (2014a, 2014b) proposed a set of formulas from which the extraction of tidal energy can be understood. First of all, kinematic energy E is calculated within a flow domain φ.

$$E = \int_\varphi \frac{1}{2}\rho V^2 d\varphi \qquad (10.1)$$

where ρ is seawater density and V is the depth-averaged velocity.

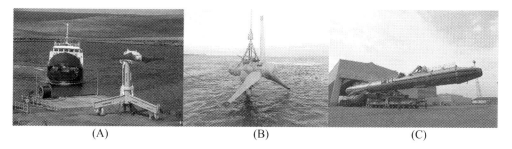

FIGURE 10.1 Examples of tidal stream energy technologies: (A) NovaInnovation M-100 D, (B) Simec Atlantis Energy AR1500, and (C) ScotRenewables SR2000. *Source: Novo and Kyozuka (2021).*

The average kinematic density $\bar{E}$ within the computational domain over the time period of T can be calculated as:

$$\bar{E} = \frac{1}{T} \int_0^T \int_\varphi \frac{1}{2} \rho V^2 d\varphi dt \tag{10.2}$$

The tidal power can be calculated by using the following formula:

$$P = \frac{1}{2} \rho V^3 \tag{10.3}$$

At a specific location the averaged power density $\bar{P}$ over a fixed time period T can be measured by-

$$\bar{P} = \frac{1}{T} \int_0^T \frac{1}{2} \rho V^3 dt \tag{10.4}$$

The total kinematic energy flux $\hat{P}$ along a line ϕ at a certain distance from the coastline can be calculated as-

$$\hat{P} = \int \frac{1}{2} \rho V^3 d\phi \tag{10.5}$$

which can be used to estimate the total tidal power within a specific area, so the averaged total tidal kinematic energy flux is calculated as-

$$\tilde{P} = \frac{1}{T} \int_0^T \int \frac{1}{2} \rho V^3 d\phi dt \tag{10.6}$$

Three tidal energy converters are shown in Fig. 10.1. Fig. 10.1A shows a two-bladed bottom-fixed turbine developed by Nova Innovation (M-100D). It is suitable in areas where tidal current is relatively weak because of its lower cut-in speed. Fig. 10.1B demonstrates a three-bladed bottom-fixed turbine designed by Simec Atlantis Energy (AR1500). This type of converter has larger rotors and lower-rated velocities for efficient functioning in the tidal energy sites are deeper but tidal currents are not much strong. A double-rotor two-bladed floating platform developed by ScotRenewables (SR2000) has been demonstrated in Fig. 10.1C (Novo and Kyozuka, 2021).

10.2.2 Wave energy

There are many types of waves due to different causes and these can be defined in different scales as shown in Fig. 10.2. The wave energy conversion process follows three energy

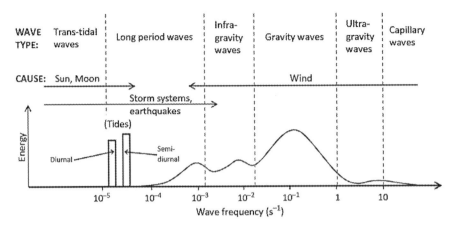

FIGURE 10.2   Scales of surface waves. *Source: Neill (2021).*

conversion stages, that is, primary, secondary, and tertiary. At the stage of primary energy conversion, wave energy transforms into mechanical or pneumatic or potential energy. Secondary energy conversion uses specific power take-off (PTO) to turn absorbed energy into useful mechanical energy. Finally, tertiary energy conversion converts useful mechanical energy into electricity. This final stage takes place with the connection of PTOs to the generators. These three stages are completed with such devices which are interrelated and progressive. The overall power generation efficiency of the WEC is possible to improve if the conversion efficiency among various levels is improved as much as possible (Zhang, 2021).

10.2.2.1 Wave energy conversion

The calculation of wave energy follows a comparatively complex process. At first, a constant overwater wind speed needs to be determined using the following relation:

$$U_A = 0.71(U_{L,10} R_T R_L)^{1.23} \tag{10.7}$$

where $U_{L,10}$ is the conversion of observed wind speed to a reference 10 m height level, R_L is the adjustment for location effects and R_T is the adjustment for stability effects (U.S. Army Engineer Waterways Experiment Station, 1984).

Following formulas are used to estimate the significant wave height (H_m) and significant wave period (T_m):

$$H_m = 0.24 \frac{U_A^2}{g} \left[\tanh\left\{0.49\left(\frac{gd}{U_A^2}\right)^{0.75}\right\} \tanh\left\{\frac{0.0031\left(\frac{gF}{U_A^2}\right)^{0.57}}{\tanh\left\{0.49\left(\frac{gd}{U_A^2}\right)^{0.75}\right\}}\right\} \right] \tag{10.8}$$

$$T_m = 7.54 \frac{U_A}{g} \left[\tanh\left\{0.33\left(\frac{gd}{U_A^2}\right)\right\} \tanh\left\{\frac{0.00052\left(\frac{gF}{U_A^2}\right)^{0.73}}{\tanh\left\{0.33\left(\frac{gd}{U_A^2}\right)\right\}}\right\} \right]^{0.37} \tag{10.9}$$

Here, d denotes depth and F is the fetch length of the wave-generating wind (in m) (Young and Verhagen, 1996; Kamphuis, 2000).

It is necessary to specify a single design wave so that it will not be needed to deal with each wind-casted wave characteristics corresponding to each monthly wind speed data. It can be calculated for both H_m and T_m as follows:

$$\overline{H_m} = \frac{\sum_{j=1}^{12} n_j H_{mj}}{\sum_{j=1}^{12} n_j} \tag{10.10}$$

$$\overline{T_m} = \frac{\sum_{j=1}^{12} n_j T_{mj}}{\sum_{j=1}^{12} n_j} \tag{10.11}$$

Significant wave height and significant wave period for the realistic nonsinusoidal design wave in deep water are used to calculate the wave energy flux (J).

$$J = \frac{\rho g^2}{64\pi} (\overline{H_m})^2 (0.9\overline{T_m}) \tag{10.12}$$

Here, ρ denotes water density (in kg/m^3) and g means gravitational acceleration (in m/s^2) (Su et al., 2018).

When wave energy flux is calculated for finite water depth, it is needed to consider the effect of change in wave celerity with depth. For this, only the depth effect (D_{kd}) needs to be multiplied with (Eq. 10.12) (Falnes, 2002).

$$J = \frac{\rho g^2}{64\pi} (\overline{H_m})^2 (0.9\overline{T_m}) D_{kd} \tag{10.13}$$

However, it is also needed to estimate depth effect (D_{kd}) using the following formula (Falnes, 2002):

$$D_{kd} = \left[1 + \frac{2kd}{\sinh(2kd)}\right] \tanh(kd) \tag{10.14}$$

where k is $2\pi/L$ and L is the wavelength. Wavelength (L) has to be calculated by using the following formula (Sundar, 2015):

$$L = \frac{g(\overline{T_m})^2}{2\pi} \tanh\left(\frac{2\pi d}{L}\right) \tag{10.15}$$

If it seems difficult to deal with L because the equation is implicit, one can directly calculate kd using the following formula (Fenton and Mckee, 1990):

$$kd = \frac{\omega^2 d}{g} \left[\coth\left\{\left(\omega\sqrt{d/g}\right)^{\frac{3}{2}}\right\}\right]^{\frac{2}{3}} \tag{10.16}$$

Wave energy converters (WEC) designed by researchers across the world have a wide range of structures depending on their concepts and application (e.g., Fig. 10.3). Zhang et al. (2021) demonstrated that multi-degree of freedom WEC (MDWEC) has a wide range of applications in the harvesting and development of wave energy. It performs comprehensively

FIGURE 10.3 A submerged wave energy converters functioning on difference of pressure. *Source: Dincer et al. (2018).*

and significantly better than that of the usual WEC with respect to energy production, the economic viability of the technology, reliability, environmental friendliness and adaptability.

Submerged point absorber devices have two parts, one fixed and another movable and using them, the device makes use of the pressure difference between the wave crest and trough. The fixed part is an air-filled cylinder compartment fixed to sea level and the movable part is a moving cylinder. There is a submerged pressure differential device called Archimedes wave swing (see Fig. 10.3) which is a cylindrical buoy anchored to the sea bed. Also, it has to be submerged at least 6 m below the ocean surface and located near the shore of the sea. Archimedes principle is the basis of this pressure differential device. Then mechanical energy is converted into electrical energy through a linear synchronous generator. Slamming forces do not act on this generator as it operates under water. However, its maintenance is a matter of major apprehension (Dincer et al. 2018).

Wave energy convertors are mainly selected according to the location of the power plant (see Fig. 10.4). Oscillating water column and overtopping devices are used at onshore plants. The oscillating water column is the earliest type of convertor that uses the wave-induced oscillatory movement of a column of water in a chamber. Then the water is compressed and forced into the turbine to run the rotor by the water column. At present, this convertor type

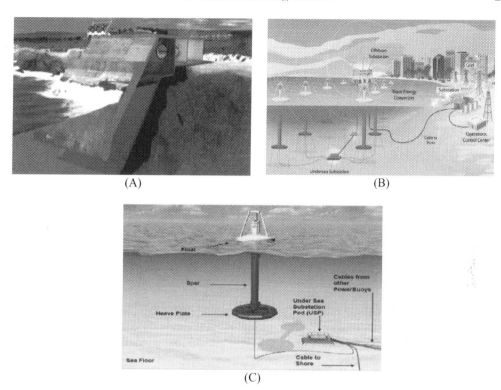

FIGURE 10.4 Common wave energy devices: (A) onshore, (B) near-shore, and (C) offshore. *Source: Wilberforce et al. (2019).*

has many variants which are installed onshore in self-contained structures. Another type of onshore convertor is the overtopping device (Fig. 10.4A). This transforms wave energy into potential energy. Low head turbines are operated with the aid of a reservoir which mainly serves as a storage basin. When the waves break on a ramp, they are directed to the reservoir on the top of the pre-existing water surface. The reservoir stored water generates energy by using low head turbines. At the near shore, mainly three types of convertors are used to convert wave energy, they are, oscillating wave surge convertors, point absorber, and submerged pressure differential devices. The first one can make waves useful by causing an oscillatory motion (Wilberforce et al., 2019).

10.2.3 Ocean thermal gradient energy

Ocean Thermal Energy Conversion (OTEC) is one of the most promising sources of ORE. The average daily heat absorbed by 2.6 km^2 of the ocean surface is greater than that produced by burning 1.1 million liter petroleum (Avery and Wu, 1994). The oceans annual absorption of solar energy is equivalent to nearly 4,000 times greater than global energy consumption per annum (Vega et al., 1999). Thus, the ocean can be conceived as a giant thermal instrument where water can work as working fluid to produce a massive amount of electricity (de Souza et al., 2020). According to Avery and Wu (1994) and Wright (1995), the upper layers of the tropical oceans contain a large amount of hot water with annual mean temperatures of up to 28°C through heat absorption and evaporation. OTEC plants can be installed in the ocean regions with stable vertical thermal gradients of at least 20°C, that is, if the difference of temperature is greater than or equal to 20°C between the surface and deep water (from 600 m to 1 km of depth in the water column).

OTEC is a dependable source of thermal energy for seawater desalination in remote islands that suffer from a lack of access to primary energy or electricity from the nearest grid. More than 100 million tons of CO_2 is being emitted per annum from desalination plants across the world (Shahzad et al., 2017). Therefore, renewable energy sources have been suggested keeping in mind the challenges related to water, energy and environment of traditional desalination plants. Oceans work as the largest collector of solar radiation as it occupies 70.9% of the earth's surface. The temperature gradient formed along the seabed through radiation can be utilized directly for desalination.

OTEC plants provide clean electricity and thus contribute to the inverse anthropogenic effect by withdrawing heat from the oceans and reducing CO_2 emission in the atmosphere. Beside this benefit, a number of by-products are also derived from these plants, such as potable water, cooling water and sea salt (de Souza et al., 2020).

According to Soto and Vergara (2014), for a standard closed Rankine OTEC cycle, the efficiency of the standard OTEC plant (ε_{otec}) can be calculated as-

$$\varepsilon_{otec} = \frac{\dot{W}_{Gnet}}{\dot{Q}_e} \quad (10.17)$$

where $\dot{W}_{Gnet}$ is the generated net power available and $\dot{Q}_e$ is the heat transfer rate at the evaporator.

The efficiency of turbine and generator are calculated based on the following relationships (Ascari et al., 2012 and Martin, 2011):

$$\begin{aligned} h_1 &= Cp_1 T_1 \\ h_2 &= Cp_1 T_2 \\ \dot{W}_{Graw} &= \eta_{TG} * m_{wf}(h_1 - h_2) \\ \dot{W}_{Graw} &= \eta_G W_G \end{aligned} \quad (10.18)$$

where m_{wf} is the mass flow of the working fluid; $h_1, h_2, Cp_1, Cp_2, T_1$ and T_2 are the specific enthalpies (in KJ/kg), entropies or specific heats (in KJ/kgK) and temperatures of NH_3 at the turbine flow rate at inlet 1 and outlet 2 (K); η_{TG} and η_G are the turbine and generator efficiencies; and $\dot{W}_{Graw}$ is raw electrical power available after one operating cycle in MW.

The available net electric power $\dot{W}_{Graw}$ will be-

$$\dot{W}_{Gnet} = \dot{W}_{Graw} - \left(\dot{W}_{WFP} + \dot{W}_{CP} + \dot{W}_{WP} + \dot{\varphi}_{hp}\right) \quad (10.19)$$

here, $\dot{W}_{Graw}$ and $\dot{W}_{Gnet}$ are the generated raw and net power available (in MW), respectively; $\dot{W}_{WFP}, \dot{W}_{CP}$ and $\dot{W}_{WP}$, are the pumping power of working fluid, cold seawater and warm seawater (in MW); and $\dot{\varphi}_{hp}$ is the loss or gain (flow) of heat in the pipes (in MW or MJ/s).

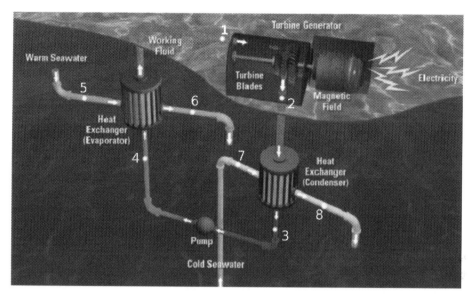

FIGURE 10.5 An OTEC plant and its thermodynamic and power generation devices. *Source: de Souza et al. (2020).*

The equations for each heat exchanger are:

$$\dot{Q}_e = \eta_e \dot{m}_e C_{Pe}[T_{ein} - T_{eout}] \tag{10.20}$$

$$\dot{Q}_c = \eta_c \dot{m}_c C_{Pc}[T_{cin} - T_{cout}] \tag{10.21}$$

where $\dot{Q}_e$ and $\dot{Q}_c$ (point above the terms indicates the temporal variation) are respectively the heat transfer rate to the system through the evaporator and the heat transfer rate of the system to the condenser (J/s), respectively; η_e and η_c are the thermal efficiency of evaporator and condenser, respectively; $\dot{m}_e$ and $\dot{m}_c$ (point above the terms indicates the temporal variation) are respectively the mass flow of hot and cold water (kg/s); T_{ein} (point 5 in Fig. 10.5) and T_{cin} (point 7 in Fig. 10.5) are the input temperature of hot and cold seawater (°C), respectively; T_{eout} (point 6 in Fig. 10.5) and T_{cout} (point 8 in Fig. 10.5) are the output temperature of hot and cold seawater (K), respectively; and C_{Pe} and C_{Pc} are the specific heat of hot and cold water (KJ/kgK), respectively.

The seawater state equations applied to the numerical model through statistical treatment in order to obtain realistic values for the density-

$$\begin{aligned} \rho_e &= f(S_e, T_e, P_0) \\ \rho_c &= f(S_c, T_c, P_{depth}) \\ \dot{m}_e &= \rho_e \dot{v}_e \\ \dot{m}_c &= \rho_c \dot{v}_c \end{aligned} \tag{10.22}$$

here ρ_e and ρ_c are the densities of the hot seawater at the gradient limit depth (kg/m3), depending on the pressure (N/m2), temperature (K) and salinity (psu); $\dot{m}_e$ and $\dot{m}_c$ are the mass flows of oceanic water bodies entering the OTEC plant (kg/s).

According to Sharqawy et al. (2010), the specific heat of hot and cold water can be obtained through a polynomial for the temperature scale in K:

$$C_{Pe} = A + BT_e + CT_e^2 + DT_e^3 \tag{10.23}$$

$$C_{Pc} = A + BT_c + CT_c^2 + DT_c^3 \tag{10.24}$$

where T_e and T_c are the water temperatures at the surface and depth at the gradient limit (K); and A, B, C and D are the coefficients of the specific heat equation (C_{Pe}; C_{Pc}).

However, if the simplest form of thermal gradient energy is to be used, the below-mentioned equation from Nihous (2007) and Rajagopalan and Nihous (2013) can be considered. So, thermal gradient energy (P_t) is:

$$P_t = \frac{3\rho c_p Q_{ww} \gamma \Delta T^2}{16 T_{fs}(1+\gamma)} \tag{10.25}$$

Here, ρ is the water density. Specific heat of water (~4 kj/kg K) is denoted as c_p. Q_{ww} indicates the discharge of warm water, which can be calculated from the formula, $Q_{ww} = \gamma Q_{cw}$. Here, Q_{cw} is the discharge of cold water which has a fixed value of 250 kg/m^3 if an OTEC plant of 100 MW is considered (National Research Council, 2013). T_{fs} is the temperature of water at the free surface and ΔT is the thermal gradient. ΔT can be calculated from the equation as $\Delta T = T_{500} - T_{fs}$, where T_{500} is the water temperature at 500 m depth.

An OTEC system is schematically presented in Fig. 10.5. In the converter, the working fluid utilized in producing mechanical work runs through a circuit. When a low boiling working fluid, such as ammonia, freon, or propylene, passes through an evaporator (a heat exchanger), seawater vaporizes it. Thus, channel-1 in Fig. 10.5 and more energetic stage generate electricity by moderately expanding the steam and driving the turbine. After that, the working fluid releases heat to the ocean when the other heat exchanger (condenser) makes it do so (de Souza et al., 2020).

The main drawback of the OTEC is that the low difference of in temperature reduces the efficiency of electricity output. Therefore, these plants can be installed in tropical belts where a high difference temperature between the surface and deep water is available. However, the electricity generation from traditional OTEC plants is highly expensive because of the average efficiency of 3 to 5% for the Rankine cycles (Yamada et al., 2009). This shortcoming can be addressed by increasing the low difference of temperature between the warmer and cooler side of the system by heating up its operational liquid with solar energy (Assareh et al., 2021).

10.2.4 Salinity gradient energy (SGE)

SGE is one of the least exploited sources of green and clean chemical renewable energy, which is mainly produced in the junction of rivers and the sea (Jianbo et al., 2021). It is the energy that is produced by utilizing the difference between osmotic pressure between two water streams of different salinities (see Fig. 10.6; Abdelkader and Sharqawy, 2021). The theory behind energy production from salinity gradients is the Gibbs free energy gradient. The principal techniques of generating electricity from the SGE are pressure retarded osmosis (PRO) and reverse electrodialysis (RED) (see, Fig. 10.7). The RED technique is advantageous compared with the PRO because of simple devices with a compact system (Jianbo et al., 2021).

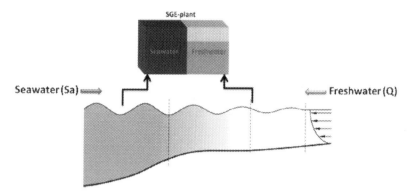

FIGURE 10.6 Salinity gradient energy plant at the entrance of an estuary. *Source: (Haddout and Priya, 2021).*

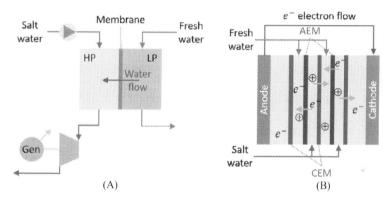

FIGURE 10.7 Processes of salinity gradient energy (A) pressure retarded osmosis, and (B) reverse electrodialysis. *Source: Belyakov (2019).*

The theoretical potential (for the extractable energy from salinity Gradient (G)) is calculated as follows:

$$G = 2RT \times Q \times \left(S_r \ln \frac{2S_{sea}}{S_r + S_{sea}} + S_{sea} \ln \frac{2S_{sea}}{S_r + S_{sea}} \right) \quad (10.26)$$

The estimated values vary with time and seasons, due to the changes in river discharge (Q), temperature (T), sea salinity (S_{sea}), and river salinity (S_r).

The power potential is calculated for energy extraction (G) based on the average values of all factors in (Eq. 10.26; Haddout and Priya, 2021). The released energy is directly proportional to the quantity of the salinity gradient.

The PRO employs the osmosis process in which fresh-water (feed) is separated from saline water of ocean/salty lake (solution) with a semi-permeable membrane. The disparity in salt concentrations causes the flow from the side to the solution side through the membrane for equalization of the concentrations that creates a pressure of fresh water over the membrane (Belyakov, 2019).

According to the solution-diffusion model, the water flux can be expressed as-

$$J_w = -D_w \frac{dc_w}{dx} = -D_w \frac{D_w c_w}{RT} \frac{d\mu_w}{dx} \approx \frac{D_w c_w}{RT} \frac{\Delta \mu_w}{\Delta x} \quad (10.27)$$

where J_w is the water flux; D_w is the coefficient of diffusion of water in the membrane; c_w is the concentration; x is the axis perpendicular to the membrane surface; R is the gas constant; T is the absolute temperature; and μ_w is the chemical potential. If $\bar{V}_w$ is independent of the pressure, then

$$\Delta \mu_w = RT \ln \Delta a_w + \bar{V}_w \Delta P = -\bar{V}_w \Delta \pi + \bar{V}_w \Delta P = \bar{V}_w (\Delta P - \Delta \pi) \quad (10.28)$$

where w a is the chemical activity of water; $\bar{V}_w$ is the partial molar volume of water; P is the hydraulic pressure; and $\Delta \pi$ is the difference of osmotic pressure. Then, the water flux can be expressed as-

$$J_w = \frac{D_w c_w \bar{V}_w}{RT \Delta x} (\Delta P - \Delta \pi) = A(\Delta P - \Delta \pi) \quad (10.29)$$

where A indicates the water permeability. The power density is expressed as the product of the water flux and the hydraulic pressure as an indicator to evaluate the performance of PRO membranes (Jihye et al., 2015):

$$W = J_w \Delta P = A(\Delta \pi - \Delta P)\Delta P \quad (10.30)$$

Conversely, RED is composed of cation exchange membranes (CEM) and anion exchange membranes (AEM). Electric potential is created when the solutions with dissimilar concentration of salinity flow on either side of these ion-exchange membranes. Cations migrate from high to low concentration through CEM, while anions migrate from high to low concentration via AEM. Electrodes are utilized to transform the ionic into the electric current at both sides of the module (Belyakov, 2019).

For a single RED cell,

$$E_{aem} = \alpha_{aem} \frac{RT}{zF} \ln \frac{\gamma_{hc}^- C_{hc}^-}{\gamma_{lc}^- C_{lc}^-} \quad \text{(potential generated on an AEM)}$$

$$E_{cem} = \alpha_{cem} \frac{RT}{zF} \ln \frac{\gamma_{hc}^+ C_{hc}^+}{\gamma_{lc}^+ C_{lc}^+} \quad \text{(potential generated on a CEM)}$$

$$E_{cell} = E_{aem} + E_{cem} \quad \text{(potential generated by a single RED cell)}$$

$$E = NE_{cell} \quad \text{(open-circuit voltage of N RED cells)}$$

$$R_i = N(R_{aem} + R_{cem} + R_{hc} + R_{lc}) + R_{el} \quad \text{(internal resistance in a stack with N RED cells)}$$

where α and γ indicate permeability and activity coefficients, H and L stand for high and low concentrations, R implies gas constant in J/(K·mol), T is temperature, R_{aem} and R_{cem} indicate the resistance of an AEM and a CEM, R_{el} stands for resistance of electrode chamber, and R_{hc} and R_{lc} imply the resistance of concentrated and dilute solution compartment, respectively.

TABLE 10.1 Theoretical reserves of global ocean renewable energy.

Type	Capacity (GW)	Annual generation (TWh)
Marine currents	5000	50,000
Osmotic	20	2000
Ocean thermal	1000	10,000
Tide	90	800
Wave	1000–9000	8000–80,000

Source: Zhang et al. (2021).

And

$$R_{hc} = f_y \frac{d_{hc}}{\Lambda_{hc} C_{hc}}; \; R_{lc} = f_y \frac{d_{lc}}{\Lambda_{lc} C_{lc}}$$

where f_y is the shielding coefficient of the spacer, C is concentration in mol/L, and Λ and d stand for the conductivity of the solution and the width of the spacer, respectively. The output voltage of the stack can be obtained by Ohm's law as follows:

$$U = \frac{E - I}{R_i}$$

$$P = UI \quad \text{(output power of a RED stack)}$$

$$P_D = \frac{P}{A} \quad \text{(power density of a RED stack)}$$

where P is power, A is the area in m^2, U is the output voltage, E is potential, and I is current (Jianbo et al., 2021).

10.3 Present status

The theoretical reserve of various types of ORE is enormous. Wave energy has the biggest capacity with high density and wide distribution across the world even though the capacity varies significantly. Therefore, this type of energy has attracted the attention of coastal countries (Zhang et al., 2021). The other major sources are marine current and thermal energy, which have significant potential (see Table 10.1). There is a significant potential of the production of ORE, which is even greater than the gross demand for electricity of the world (Melikoglu, 2018). It has been estimated that the potential of tidal power is nearly 1,000 TWh, wave power is approximately 93,000 TWh, thermal is about 87,600 TWh, and salinity gradients range from 2,000 to 5,200 TWh that can be increased up to 27,700 TWh (Quitoras et al., 2018). Conversely, according to international energy agency, the total final consumption of electricity was 22,847 TWh in 2019.[1] Thus, the total global demand for electricity can be met with the ORE

---
[1] See, for details, https://www.iea.org/reports/electricity-information-overview/electricity-consumption.

TABLE 10.2 Installed capacity (in MW) of marine energy.

Countries	2010	2011	2012	2013	2014	2015	2016	2017	2018	2019	2020
Republic of Korea	1.00	255.00	255.00	255.00	255.00	255.00	255.00	255.00	255.00	256.00	255.50
France	216.00	214.73	215.91	218.49	220.01	218.33	220.22	218.87	218.00	214.07	214.07
United Kingdom	4.00	4.00	9.00	8.00	9.00	8.94	8.94	13.49	18.40	20.40	22.00
Canada	20.00	20.00	20.00	20.00	20.00	20.00	20.00	20.00	20.00	20.00	20.00
Other Developed	4.55	4.90	5.35	4.23	4.82	5.96	10.19	10.59	10.55	9.97	10.08
Developing											
China	20.00	20.00	20.00	20.00	20.00	20.00	20.00	20.00	20.00	20.00	20.00
Brazil	0.00	0.00	0.10	0.10	0.15	0.05	0.05	0.05	0.05	0.05	0.05
Total	265.55	518.63	525.36	525.82	528.98	528.28	534.41	538.00	542.00	540.48	541.69

Source: Based on IRENA database.

given that the appropriate technologies are developed, the corresponding economic and environmental constraints are addressed (Hernández-Fontes et al., 2020).

However, despite such a considerable theoretical capacity and estimated potential, a meagre amount of ORE is being produced across the world. According to the IRENA database, the cumulative installed capacity of ORE has increased from 265.55 MW in 2010 to 541.69 MW in 2020. The capacity witnessed a sudden increase in 2011 because of a 254 MW capacity added by South Korea. With this, the country has secured the top position in terms of installed capacity (see Table 10.2). In 2020, South Korea, France, and the United Kingdom had about 47.2, 39.5, and 4.1% share in the global installed capacity of ORE. China had the highest installed capacity among developing countries in 2020 (3.7%), which was followed by Brazil (about 0.01%).

Conversely, France generated the highest amount of electricity in 2019 (about 479 GWh), which was followed by South Korea. In fact, these two countries produce nearly the same quantity of marine electricity since 2012. The United Kingdom produced about 14 GWh, which was more than double of China (see Table 10.3).

The ORE technologies are undergoing rapid transformation through R&D, which can be observed from the number of patents for technologies (Fig. 10.8). The cumulative number of patents increased by more than double in the 2010s–from 6,720 in 2010 to 15,240 in 2019 for wave energy and from 5,372 to 11,378 for tidal energy during the same period (IRENA, 2021). It implies that both the technologies are getting momentum in contributing to increasing share of ORE in the global production of renewable energy.

The tidal current ORE plants are a quite common type among all available technology. However, the ones which are at the fully operational stage have a very low capacity compared to the ones at the early planning and concept stage. Almost all high-capacity ones are still being planned or completed, while a few preliminary stages, that is, application submission. The United Kingdom is the dominating country in planning high-capacity plants of this technology. Among the developing countries, only India and Indonesia are currently thinking about constructing tidal current plants with high capacity which indicates they are still at

TABLE 10.3 Generation (in GWh) of marine electricity.

	2010	2011	2012	2013	2014	2015	2016	2017	2018	2019
France	476.397	477.278	457.986	413.572	480.507	487.418	500.516	521.706	479.930	478.964
Republic of Korea	1.039	52.307	465.924	484.000	492.000	496.000	496.000	489.000	485.000	474.000
United Kingdom	1.884	0.940	4.203	4.756	2.221	1.999	0.009	4.194	9.298	13.991
Canada	28.000	26.000	27.000	15.000	16.000	13.000	19.000	6.000	20.000	1.000
Other Developed	0.750	0.880	1.113	1.064	1.192	3.144	1.172	1.033	1.027	1.028
Developing										
China	6.500	6.500	6.500	6.500	6.500	6.500	6.500	6.500	6.500	6.500
Brazil	0.000	0.000	0.001	0.001	0.001	0.001	0.001	0.001	0.001	0.001
Total	514.570	563.905	962.727	924.893	998.421	1008.062	1023.198	1028.434	1001.756	975.484

Source: Based on IRENA database.

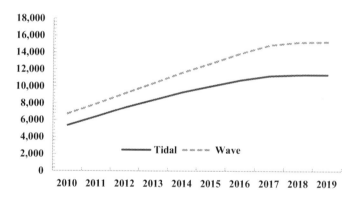

FIGURE 10.8 Number of patents of ocean renewable energy technologies (cumulative). *Source: Based on IRENA database and compilation based on Ocean Energy Systems (OES), 2021.*

the early concept stage (see Table 10.4). Therefore, developed countries dominate in terms of planning and constructing tidal current ORE plants.

Usually, tidal barrage lagoon fence plants have the most capacity to generate electricity than any other type of ORE plant. However, only a few of them are in fully operational status. Among them, Sihwa-Lake Tidal Power Plant in South Korea and Usine marémotrice de la Rance in France has a notable capacity. The first one has a capacity of 254 MW and the latter has a capacity of 240 MW. Some tidal barrage lagoon fence plants, which are among the highest capacity plants of any type, are still at the stage of either early concept or early planning. Also, most of them are projects of developed countries. Developing countries are still at the preliminary stage of adopting this technology. India's Kachchh tidal power project is still at the early concept stage which will have a capacity of 900 MW and China's Dandong city's is also at the same stage with a capacity of 300 MW (see Table 10.5).

Wave ORE plants are the most common ORE plants among all available technologies. However, like tidal current ORE plants, high capacity wave ORE plants are still at the early

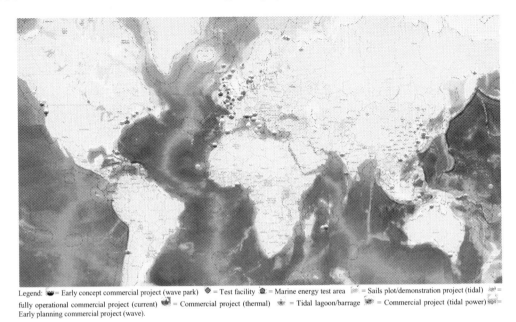

Legend: ⬥ = Early concept commercial project (wave park) ◆ = Test facility ⬢ = Marine energy test area ⬣ = Sails plot/demonstration project (tidal) ⬤ = fully operational commercial project (current) ⬥ = Commercial project (thermal) ⬣ = Tidal lagoon/barrage ⬥ = Commercial project (tidal power) ⬥ = Early planning commercial project (wave).

MAP 10.1 Global offshore installations of ocean renewable energy projects (as of 20 October 2021). *Source: Compilation based on Ocean Energy Systems (OES), 2021.*

TABLE 10.4 Top tidal current power plants.

Name	Country	Status	Capacity (MW)
West Islay Tidal Energy Farm (rest)	United Kingdom	Early concept	370
MeyGen/Inner Sound (rest)	United Kingdom	Early planning	324
SBS @ Indonesia	Indonesia	Early concept	300
Maenggol	South Korea	Early concept	250
Mundra Tidal Power Project (2)	India	Early concept	200
Westray South (rest)	United Kingdom	Early planning	140
Lashy Sound (rest)	United Kingdom	Early planning	120
Brough Ness	United Kingdom	Early planning	100
Palmerah Bridge (2)	Indonesia	Early concept	97
Brims Tidal Array (2A)	United Kingdom	Application submitted	95
Fair Head Tidal Energy Farm (rest)	United Kingdom	Application submitted	90
Brims Tidal Array (rest)	United Kingdom	Application submitted	75

Source: Based on OES (2021).

TABLE 10.5 Major tidal barrage lagoon fence projects.

Name	Country	Status	Capacity (MW)
Cardiff Tidal Lagoon	United Kingdom	Early planning	3,240
Newport Tidal Lagoon	United Kingdom	Early concept	1,800
Incheon Tidal Power Plant	South Korea	Dormant	1,320
Kachchh Tidal Power Project	India	Early concept	900
Mersey Tidal Power	United Kingdom	Early planning	700
Ganghwa Tidal Power Plant	South Korea	Dormant	420
Dandong City	China	Early concept	300
Sihwa-Lake Tidal Power Plant	South Korea	Fully operational	254
Usine marémotrice de la Rance	France	Fully operational	240
Wyre Tidal Gateway	United Kingdom	Early planning	160
Solway Firth Energy Gateway	United Kingdom	Early concept	100
Half Moon Cove Tidal Project	USA	Dormant	90
Penzhin Tidal Power Plant Project (South)	Russia	Early concept	87.1

Source: Based on OES (2021).

stage of conceptualization or planning. The projects which are fully operational are of very low capacity compared to the ones at the early stage. Developed countries are advanced in planning wave ORE plants compared to developing countries. Ghana is currently planning to construct a plant with a capacity of 8.68 MW and thinking about another one with a capacity of 20 MW. Apart from Ghana, no other developing countries are on the list of high-capacity wave ORE plants (see Table 10.6). Also, Ghana's plan for wave ORE plants is not ambitious. The planned capacity is quite low compared to the other planned ones of developed countries.

The least common type of power plant is OTEC, which means only a few countries tried to or planned to adopt this technology. The currently available technology of OTEC cannot support generating a significant amount of technology. This might be a reason for OTEC not being a widely adopted technology. Only France's naval energies onshore OTEC prototype is currently at fully operational status with a capacity of 0.015 MW and Bahamas' Bahama Mar Resort is at dormant status. The other ones are still at a very early stage, which means those plants' construction works have not been started yet. Another noticeable trend is that both developed and developing countries are on the same page in terms of adopting OTEC technology (see Table 10.7).

10.4 ORE in developing economies

ORE is more pertinent to developing economies as a significant part of the remote coastal regions and islands remain out of the electricity connection (Ahuja and Tatsutani, 2009). Developing economies are also yet to significantly explore the ORE technologies because of

TABLE 10.6 Top wave power plants.

Name	Country	Status	Capacity (MW)
Rho-Cee WEC integrated with Floats Pneumatically Stabilised Platform	USA	Early concept	216
Katanes Floating Energy Park (2)	UK	Early concept	104.4
Albany Wave Farm (rest)	Australia	Early concept	81
mWave Wave Farm (2)	Portugal	Early concept	64
Western Star Wave/Project Saoirse (2)	Ireland	Early concept	25
Accra	Ghana	Early concept	20
Mermaid / Bligh Bank	Belgium	Early planning	20
Albany Wave Farm (2)	Australia	Early concept	18
CETO 6 @ Wave Hub (2)	UK	Early planning	14
Audierne Bay Wave Energy Park	France	Early concept	10
Seatricity @ Wave Hub (2)	UK	Early planning	10
GWave @ Wave Hub (1)	UK	Early planning	9
Ada Foah (3)	Ghana	Early planning	8.68

Source: Based on OES (2021).

TABLE 10.7 Major OTEC plants.

Name	Country	Status	Capacity (MW)
Zambales Energy Island	Philippines	Consent authorized	10
Bahama Mar Resort	Bahamas	Dormant	10
Hainan Green Resort	China	Early concept	10
Virgin Islands - St. Croix	USA	Early planning	15
New Energy for Martinique and Overseas (NEMO)	France	Early planning	10.7
Naval Energies onshore OTEC prototype	France	Fully operational	0.015

Source: Based on OES (2021).

their socio-economic backwardness and limited resources that could not be used in generating this energy (Bonar et al., 2015; Felix et al., 2019). It can be generated to provide small and isolated areas adjacent to or in the sea (e.g., coastal islands) without electricity from national or local grids, and ORE can also be catered to coastal larger populations and commercial activities. ORE can also benefit small island developing states (SIDS), even though the cost of producing energy is still high (Langer et al., 2020). However, fiscal and policy support of the government and international development agencies are necessary for installing the devices at the commercial level.

Studies reveal that developing economies have considerable ORE potential. Using numerical simulation, López et al. (2015) found that ORE potential is nearly seven times of the gross demand for electricity of Peru, which can be well harvested because of low temporal variability. Similarly, Kirinus et al. (2018) found considerable potential to produce electricity in Brazil from ocean currents adjacent to the coasts, especially with high current intensity in the southern coasts of the country. de Souza et al. (2020) examined the potential of the Brazilian Ocean Thermal Energy Park, with a potential of up to 376 OTEC plants. They revealed that the utilization of its maximum capacity can produce electricity of 41.36 GW, and absorb 60.16 GJ/s of ocean heat and remove 256.37 Tg/year of CO_2 emission from the atmosphere.

Brazil has nearly 164,000 MW untapped ORE potential in the region between the Pacific and Patagonian and Magellan Trench and the Andes Trench (Huante et al., 2018). Production of wave energy can meet a portion of the demand for electricity in Brazil using the country's advantage of nearly 9000 km coastline. It has the potential to generate approximately 40 GW of ocean wave energy in the southern and southeastern coastline regions, which can be added to the national grid. According to Rusu and Onea (2017), the best energy can be generated between the latitudes of 30° and 60°. According to de Oliveira et al. (2021), Rio Grande do Sul and the southern state of Santa Catarina have the highest potential of producing wave energy because of the presence of greater wave power. However, the cost of installing the conversion technology is significantly high as the devices are at the early commercial stage and the electricity produced from the ocean wave from these sites is not economically viable.

Using data simulation, Huante et al. (2018) estimated ocean thermal energy potential in Mexico and found that the maximum potential is located off the southwestern coasts of the country. The Gulf of Baja California has the potential for tidal energy (Mejia-Olivares et al., 2018) while the areas adjacent to the island of Cozumel have the potential of ocean currents in the country (Alcérreca-huerta et al., 2019).

Langer et al. (2021) estimated the technical and economic potential of OTEC in Indonesia through testing for 100 MWe at provincial and national levels. They demonstrate that the economic potential in four provinces in a range of 0 to 2 GWe, which is equivalent to an annual production of electricity of 0–16 TWh. An OTEC plant with a nominal power output of 100MW at a difference of seawater temperature of 20°C could produce around 1,200 GWh of electricity per annum in the country (ADB, 2014).

ORE can generate nontrivial socio-economic benefits to developing economies through job creation, knowledge diffusion, investment inflow and increasing economic activities in the coastal areas. Nevertheless, reducing the cost of installation, operation and maintenance of the technologies, and minimizing the negative impacts on the ocean environment would help derive maximum benefits of the ORE. A robust coordinated approach among different stakeholders (technology developers, policymakers, and end-users) is necessary to diminish risks and adverse outcomes of the technologies in order to derive optimum benefits of ORE in developing economies. ORE is one of the most promising sources of renewable energy in South East Asia (SEA) since the region is endowed with high tidal intensity in its islands. ORE would be a viable alternative to traditional energy, which would be a strong means of climate change mitigation (Quirapas and Taeihagh, 2021).

A thermocline-based desalination plant with a daily capacity of 100 m^3 piloted at Kavaratti Island in the Indian Ocean. The plant operated successfully for over a decade, which proved that thermocline desalination projects are feasible and dependable in developing regions (Chen et al., 2021).

Because of the high density of power and predictability, tidal current energy has received considerable attention in the energy industry for commercial installation in Europe and North America. China has strived to develop tidal current energy for utilizing its vast ocean in order to supply constant power to its coastal towns or remote islands. Zhoushan Archipelago is identified as the most appropriate site for large-scale commercial installation of tidal current energy devices in China. Moreover, tidal energy technologies can also be installed at Chengshan Cape and Qiongzhou Strait as well as the adjacent waters (Liu et al., 2021).

Located in the Bay of Bengal, Saint Martin's Island is a notable marine tourist destination and the only coral island in Bangladesh. Shahriar et al. (2019) has designed and tested a Wave Energy Converter (WEC) named Searaser at the Island. The estimated capacity of the device was around 144 kW. The results reveal that the technology can be an economically viable and environment-friendly substitute of for existing diesel-based generators that emits a substantial amount of CO_2 into the atmosphere.

The Philippines is actively looking for harnessing the ORE potential to meet the mounting demand for energy with a decent energy mix using both renewable and nonrenewable sources. Quitoras et al. (2018) assessed the potential sites for installing Wave Energy Converters (WECs) in the five coastal areas of the country. AquaBuoy, Pelamis and Wave Dragon converters were used in this assessment to test the technology readiness level. The results demonstrate nearly 10 to 20 kW/m of wave energy flux scattered in its different coastal regions. ORE can greatly help develop isolated islands, supply desalinized water and operate offshore equipment. In addition, it would promote sustainable development of coastal areas and footprint regions in China (Zhang et al., 2019).

Ozturk et al. (2017) assessed the annual potential of the current energy of the Bosphorus Sea strait in Turkey. It demonstrated that late spring to the end of summer is the most suitable period for energy production in the strait. Marta-Almeida et al. (2017) conducted a numerical simulation of tidal stream energy potential for Baía de Todos os Santos, Brazil. They found the considerable potential of renewable energy in the region. Based on numerical simulations, Orhan and Mayerle (2017) found that a significant potential of tidal currents power at the Strait of Larantuka in Indonesia.

10.4.1 ORE projects in developing economies

In Africa, three developing countries have installed ORE projects. Among them, Ghana has a fully operational WEC project titled "Ada Foah" located in the Atlantic Ocean. It was commissioned in 2015 with an installed capacity of 0.4 MW. On the other hand, Mauritius has an early concept WEC project located in the Indian Ocean. Small island-state Cape Verde has an early concept WEC project located in the Atlantic Ocean with a capacity of 0.042 MW.

India has installed three ORE projects. Among them, the OTEC desalination plant is an early concept project, which is located at Kavaratti Island in the Arabian Sea. The country also initiated the Mundra Tidal Power Project (current) in the Gulf of Kutch near Mundra, Kutch district, Gujarat. The installed capacity of this commercial project is 200MW. There is another early concept project titled Kachchh Tidal Power Project (tidal barrage) located in the Gulf of Kutch with a capacity of 900 MW.

China has initiated a number of ORE projects in various parts of the country. Hainan Green Resort is an early concept commercial OTEC project located off the Hainan Island in the South China Sea with a capacity of 10 MW. It also installed the National Wave Energy Test Station coast of Zhuhai in the South China Sea Guangdong province. It has 3 wave energy test berths with 100 kW capacity per test berth and 6 demonstration berths. At the beginning of 2017, China started working on the Wanshan Wave Energy Demonstration Project which is situated at Wanshan Island, Zhuhai (East China Sea). It is a commercial project with a converter capacity of 0.5 MW. HaiShan Tidal Power Station is a fully operational and commercial marine energy (Tidal Barrage) project located at the East China Sea. JiangXia Tidal Power Station is another similar power station located at Zhejiang Province, East China Sea with a converter capacity of 4.1 MW. LHD Tidal Current Energy Demonstration Project is a fully operational marine energy (current) project which has two phases. Phase 1A started operating in around 2016 and Phase 1B was commissioned in 2018. Both are located at Xiushan Island, Zhoushan, Zhejiang Province, the East China Sea. This same project's Phase 1C is still under construction and has a possibility to start operating in 2021. This phase 1C has a converter capacity of 1 MW. This project's rest of the phases are still in the early planning stage. Zhoushan Tidal Current Energy Power Demonstration at East China Sea is a fully installed commercial marine energy project which uses ocean current as its energy source and has a converter capacity of 0.5 MW. A wave energy project with a 1 MW converter capacity located at Wanshan Island, Zhuhai named Wanshan Wave Energy Demonstration Project which is a commercial project and still under construction. China's Dandong City Tidal Barrage Lagoon Fence is still in the early concept phase which will be located at the Yellow Sea. This tidal barrage powered energy plant is expected to have a capacity of 300 MW.

The Philippines Government already gave its consent to construct an OTEC plant named Zambales Energy Island at Cabangan, Zambales, Luzon. It will mainly be located in the South China Sea. This plant's capacity will be 10 MW. The Philippines is also planning to set up a tidal current energy-powered commercial energy plant at Capul Island of San Bernardino Strait named San Bernardino Tidal Energy Farm.

Indonesia's most ORE projects are still at the planning or early concept stage. At Bintuni Bay, Indonesia has a tidal current powered commercial energy plant named BUMWI Wood Chip Factory. Its converter capacity is 0.062 MW. However, it is temporarily offline at present. Buton Tidal Power Plant is still in the early concept and is being planned to be constructed at the Banda Sea. With a capacity of 50 MW, this will be a significant commercial project of Indonesia. Indonesia is also planning to construct the Palmerah Bridge Tidal Current Plant at Larantuka Strait. This commercial project will have two phases. The first phase will have a capacity of 23 MW and the second phase will have a capacity of 97 MW. Indonesia's another early phase tidal current power plant is SBS @ Indonesia. This will be located at the Bali Sea, covering the region Lombok and Bali. This is an ambitious commercial project because it is being planned to have a capacity of 300 MW. Nautilus (2) and Nautilus (3) are two phases of a commercial project which are still at the stage of early concept. Both will be located at the Bali Sea and their source of energy will be the ocean current. The second phase, Nautilus (2) will have a capacity of 57 MW and both phases' converter capacity will be 1.5 MW. Indonesia is also thinking about constructing an ocean wave-powered energy plant with a capacity of 10 MW. This plant will be located at the Bali Sea and named Nusa Penida Wave Farm.

Papua New Guinea is still at the early stage of planning a current powered energy plant named MAKO @ Buka Passage. A Pacific Island country, Kiribati is planning a commercial OTEC at their island of Tarawa in the Pacific Ocean (OES, 2021).

10.5 Contribution of ORE to sustainability

The rapid depletion of traditional fossil fuels is leading to the increasing prices of oil and natural gas. Moreover, the exhaustible nature of fossil fuels would exacerbate the shortage of global energy supply in the coming decades. Countries across the world are now striving to develop new energy businesses and achieve green sustainable development. Oceans contain an enormous volume of energy, which is being received widespread research and attention for developing appropriate technologies for harnessing the potential (Zhang et al., 2021).

Traditional energy production hinges significantly on fossil fuels that accelerate global climate change through CO_2 and equivalent emissions to the atmosphere. Countries, especially developing ones, are exposed to many extreme weather and slow onset events of climate change, which have been posing various threats to their lives, assets, public goods and livelihoods (Eckstein et al., 2021). Three out of seventeen sustainable development goals (SDGs) are related to ORE, which are again linked with sustainability in developing economies. The members of the United Nations (UN) are committed to attaining SDG-7 that entails the production of and access to clean, renewable and affordable energy of which ORE can be a viable complement to solar, wind and hydroelectricity. ORE can help mitigate climate change and reverse anthropogenic effects on the environment. Moreover, ORE sources can be utilized to produce predictable power without much interruption including extreme weather events like cyclones and other natural disasters in the sea compared to solar or wind energy.

Thus, ORE helps reduce CO_2 emission to the atmosphere as an alternative to fossil fuels that helps in achieving SDG-13 through climate change mitigation, it is claimed to have several significant negative consequences on the marine environment. The positive and negative consequences of ORE plants have been illustrated in Fig. 10.9 through interactions among environmental stressors, effects, receptors, and their responses.

ORE can help achieve decarbonization of the oceans (IRENA, 2019). Most commercial and pilot ORE devices are instream turbines that harvest energy from tidal or currents or wave energy converters (WECs) to generate energy from the oscillation of waves. However, ORE devices create risks to marine species and fishes. Specifically, fishes are likely to encounter collision the rotating blades of the turbines as they are unable to detect the safe route toward shelter or food sources through avoiding turbine and related infrastructure in fast-moving waters. There is a high risk of colliding with ORE devices, specifically its moving parts, for example, rotor blades (Horne et al., 2021), which can cause injury or death to fish and other marine species. Onoufriou et al. (2021) found a significant reduction in the number of harbor seals adjacent to a commercial-scale tidal turbine array. They inferred that the potential for an overall decrease in the number of seals in those areas if the turbine array continues to operate for a longer period.

Underwater noise of tidal turbines and WECs is created because of power take-off, rotating turbine blades, and mooring systems of ORE. The noise from ORE devices may mask sounds of fish species and exert an impact on their hearing ability (Polagye and Bassett, 2020), which

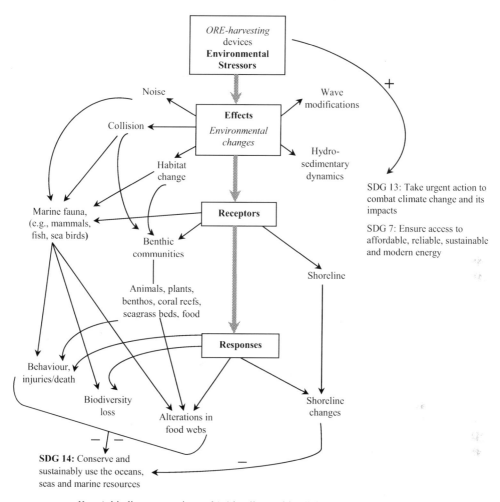

FIGURE 10.9 Environmental impacts of ORE and their linkages with sustainable development. *Source: Authors.*

would distort their communication and usual movement. Auditory masking because of the operation of tidal energy devices has a significantly adverse impact on marine fauna (Pine et al., 2019). Lossent et al. (2018) found that noise produced by a single tidal energy turbine leads to behavioral disturbance and avoidance in a maximal of 3.141 km² area. While fishes and mammals have considerably larger vital areas, such noise results in loss of habitat for a wide range of marine species whose vital area is usually small.

Tidal-stream energy devices (TSEDs) are now examining the possible impacts of commercial installation of large arrays on marine life, especially mammals and the benthic community even though such impact assessments are quite challenging (Fox et al., 2018). Power transmitted from instream tidal turbines and WECs to shore through cables can create electromagnetic fields, which have potentially harmful effects on marine organisms (Gill et al., 2014). Tidal turbines and WECs installed are connected to the seafloor through gravity foundation or with anchors, which may disturb or damage the benthic habitat. The devices, accessories and associated mooring systems may act as artificial reefs and barriers (Hemery, 2020; Bender et al., 2020).

Energy damping because of the presence of devices can lead to erosion and sedimentation (Mendoza et al., 2019). Changed hydro-sedimentary dynamics can lead to shoreline changes through erosion and accumulated sediment, which would again cause erosion of beaches and damage of coastal dunes (Martínez et al., 2021).

The installation of arrays of ORE devices may displace fauna and benthic communities from their habitats if these are partially or fully damaged (Copping et al., 2021). The array of MRE devices can act as barriers to the movement of migratory fish (Copping et al., 2014). The greatest risk to fishes because of the devices is their collision with turbines (Copping, 2018) as instream tidal and river turbines would kill fishes, and create further risk to endangered and stressed fish populations (Sparling et al., 2020). Thus, the ORE plants are likely to cause a threat to the marine ecosystem and physical environment, which would contradict SDG-14: conserve and sustainably use the oceans, seas and marine resources.

Social aspects are intrinsically interwoven with the sustainability of ORE technologies vis-à-vis minimizing their adverse effects on the environment. Sustainability cannot be fully achieved without internalizing the social aspect as it is considered to be a trade-off among social, environmental, and economic aspects of the relevant aspects of ORE (Khan, 2019). Coastal communities of developing economies have scanty information regarding the benefits and disadvantages as the technologies are comparatively new and barely installed in these regions. Inadequate knowledge among the catchment communities might result in ineffective engagement to develop projects, and they are expected to provide better support if they receive timely and precise information concerning the trade-offs (Ramachandran et al., 2020), which would help achieve the sustainability of the projects. However, concerns of the local communities are usually not adhered to the design and installation of the ORE plants because of the lack of social impact assessment, public consultation, and flow of information to the communities. Business-community interface regarding the installation of ORE projects is hardly facilitated and documented.

Catchment communities of the planned ORE project may nurture concerns about the potential adverse impact on, fisheries, tourism, shipping, ocean-based entertainment industry, real-estate price, and employment. Therefore, sustainability of the ORE projects would be achieved through community endorsement if they (especially backward and disadvantaged groups) are convinced about the immediate and long-term positive impacts of the projects and the roadmaps of addressing the negative aspects. Communities can also be involved in other innovative ways, such as designing a framework for sharing project returns with the community (Waldo, 2012), toward achieving greater sustainability.

10.6 ORE in selected developing economies

10.6.1 India

To achieve SDGs and meet the ever-growing energy demand, India needs to explore renewable energy (RE) options more extensively. Its 7,500 km long coastline is spread across

TABLE 10.8 Potential of ORE in India.

Resource	Minimum (GW)	Maximum (GW)
OTEC	180	300
Tidal	9	12.5
Wave	40	60
Total	229	372.5

Source: Chakraborty et al. (2021).
Note: Factoring in 40% losses according to Ministry of New and Renewable Energy.

seven states and it has around 336 islands in the Bay of Bengal and the Arabian Sea. This geographical information makes it easier to visualize how much potential India possesses to adopt ORE technologies. It is a better option for India because its energy demand will reach 4,000 TWh in 2030 and also India has to meet its policy target which states it will meet its energy demand in 2030 by supplying RE five times more than it supplies now. Sen et al. (2016) calculated the maximum (11–8 m) and average (6.77–5.23 m) tidal range in Gujarat. There are several locations in India, that is, Arabian Ocean, the Indian Ocean, the Bay of Bengal, the Sundarbans and Gangetic delta in West Bengal and Lakshadweep, Andaman & Nicobar Islands, which can be worked as potential sites for ORE plants (MNRE, 2021). The most suitable technology for India to adopt would be OTEC because of having a 20°C temperature difference on the Indian coast all year round (Sen et al., 2016).

Though India's potential in the ORE sector is significant as can be observed from Table 10.8, this sector has not been well utilized. India can generate around 229 to 372 GW of energy through ORE (MNRE, 2021) and it can provide India's total energy demand in 2030. The ORE has a capacity utilization factor of 90 to 92% which is higher than any other RE available. Its plant life is more than 50 years, which is also a positive factor in terms of considering ORE as a potential energy source because it is a capital-intensive technology requiring high capital cost and high evacuation costs (Chakraborty et al., 2021).

10.6.2 Bangladesh

The Government of Bangladesh has been adopting various strategies to harness the potential of the Blue Economy in its sovereign territory in the Bay of Bengal, which includes generating energy from the ocean area. Though having plenty of coastal areas to explore the options of deploying WEC technology, the lack of research and development has made the sector quite unapproachable. Kuakata, Sandwip, and St. Martin's Island are some potential locations for implementing wave energy projects. Despite lacking electricity supply, Saint Martin's Island is considered one of the most visited places in Bangladesh because it is the only coral island of Bangladesh located at the Bay of Bengal. Therefore, a recently proposed WEC named Searaser by Shahriar et al. (2019) has explored WEC technology in Saint Martin's Island, Bangladesh. It is very unique in the sense of low cost and ability to implement in remote islands. It is proposed and designed to meet the electricity demand in peak hours. It is a mini hydroelectric power plant (MHPP) which will generate electricity by using a wave energy

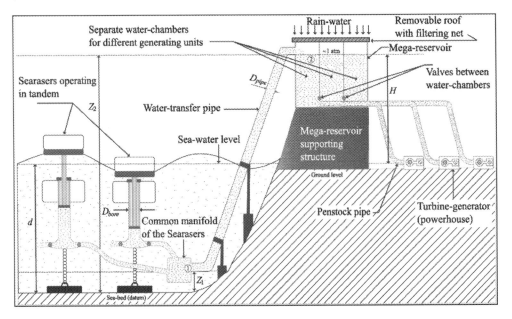

FIGURE 10.10 Searaser WEC piloted in St. Martin's Island, Bangladesh *Source: Shahriar et al. (2019).*

converter (see Fig. 10.10). The MHPP will be a sustainable alternative to currently operating diesel generators in Saint Martin's Island. Its capacity is around 144 kW. Since OREs are not involved in GHG emissions, this plant is a positive factor in promoting eco-tourism.

The estimated wave energy flux data for St. Martin's Island supports that the Searaser WEC can be installed in many coastal regions around the world. It can extract wave energy with such efficiency that it can be installed in regions of low wave power. It needs around US$ 5,278.65 per kW of installed capacity to implement the project. MHPP installation is a prudent decision considering operating costs and end-of-life profit. Nearly US$ 271,478.34 and US$ 705,011.47 profit can be made from this technology for US$ 0.35/kWh and US$ 0.45/kWh, respectively at a 1% rate of interest. It is also possible to get a grip on CO_2 emissions in the Searaser WEC host regions. Around 141,730.45 kg, 102,093.97 kg and 63,658.59 kg of CO_2 emissions can be reduced if it is installed as an alternative to coal-based, petroleum-based and natural gas-based power plants, respectively. Thus, the OER can be a continuous and predictable source of energy in remote coastal areas and islands (Shahriar et al., 2019).

10.6.3 Indonesia

Indonesia can work on its energy security, energy poverty and climate change mitigation by adopting ORE. OTEC would be a viable means to generate clean and steady electricity for both rural and urban areas and to contribute to transitioning to renewable energy in the country. Electricity produced from OTEC up to 16 TWh per year might help Indonesia to reduce reliance on traditional fossil fuels and reduce energy poverty in the coastal and border regions.

Langer et al. (2021) demonstrated that Indonesia is one of the most promising countries in the developing world for OTEC. It would be a viable option for the country to achieve its goals of energy transition in the long run. Therefore, a total of 1021 OTEC project locations have been identified in the country's provincial borders. A pragmatic potential of 102.1 GWe has been calculated for 100 MWe OTEC. It has been revealed that OTEC could be widely installed throughout the country, especially in the northern and eastern side of Sumatera, the southern and western parts of Kalimantan, North Java and the southern part of Papua. For deepwater installations, the Banda Sea is the greatest fit. The country has 57, 52 and 43 GW of theoretical, technical and practical potential of ocean thermal energy, which is considerable compared to its electricity needs. These potentials could be realized with a national feed-in-tariff (FiT) of nearly US$ 0.26 per kWh. It is inexpensive compared to the costs of conventional electricity production. Regions with high electricity demand, that is, Bali and Java, need to set up OTEC plants through FiT (Langer et al., 2021).

10.7 Way forward

Oceans can be the biggest, stable and predictable sources of renewable energy for a large number of developing countries that can access the seas and especially for coastal regions and islands. The potential of generating the energy needs to be harnessed by minimizing the negative impacts on the ocean environment for making development sustainable. However, realizing the potential requires a set of enabling conditions that include technical, economic, environmental, social and policy-induced factors. For example, the Philippines has direct policies among the Asian developing economies to develop and harness ORE. The country has formulated a national energy plan and created agencies to execute ORE projects (Quirapas and Taeihagh, 2021). The renewable energy policies of developing economies must explicitly endorse the ORE, and establish relevant institutions to facilitate public and foreign investments in this sector.

The investment in ORE projects is sensitive to its costs of installation and the per-unit cost of output. Indeed, the ORE technologies are significantly expensive in terms of initial installation and deployment in the ocean water compared to their nonrenewable and renewable energy. In addition, expertise and human resources need to be developed to install and operate the technologies efficiently in cooperation with industries related to research and development on ORE technologies and potential investors. A single device or a small number of devices would be costly in producing ORE. Instead, an array of TEC devices would decrease the average cost of production because of economies of scale. However, it would reduce the mean TEC output and increase adverse impacts on the ocean environment. Overall, even though electricity production from ocean kinetic power is costly in financial terms, it provides constant and predictable energy. Thus it can be a cost-effective means of providing grid energy stability (Rodrigues et al., 2021) along with a significant reduction of CO_2 emission to the atmosphere. ORE technologies would face competition with other baseload renewable energy technologies like hydropower. Therefore, studies should be conducted to determine their profitability for multifunctional use, such as producing freshwater, cooling, mariculture and hydrogen compared to other devices (Langer et al., 2021).

There should be a balance between economies of scale by installing larger plants and increasing the efficiency of producing energy. Hybrid energy systems, such as blending solar and tidal turbines, can also be developed to reduce costs. It would be a financially viable alternative to producing electricity from diesel generators for islands and off-grid locations. Also, the method of cost calculations should be revised to include the life span of the project, operation and maintenance costs during the project life, and benefits of reducing CO_2 emission.

Massive investment is required in R&D where collaboration is imperative among private sector companies, universities and academia, government, international organizations and agencies, and donors in developing cost-effective technologies so that they can compete in the energy market. Cheaper technologies will be able to attract markets of developing economies that are interested to diversify their energy portfolio through tapping the potential of the oceans and supplying energy to the off-grid and isolated regions. In addition, the government and international organizations can create special fund to finance R&D in soft terms to facilitate innovations. It will also encourage the creation of local technology developers and technical manpower in the R&D sector, which would eventually result in a domino effect on the innovation of ORE projects (SEAS, 2018). It is important to formulate policies that encourage R&D to improve the existing technology and introduce new devices to generate ORE from the potential coastline zones. Fiscal and policy incentives would help increase the technical efficiency and commercial viability of these technologies by decreasing the unit costs, thereby making them competitive in the electricity market in developing economies with good potential of ORE.

Detailed environmental information and ecosystem-level data should be generated on the potential adverse impacts because of the production of ORE through conducting in-depth studies (Mendoza et al., 2019). Environmental impact assessments (EIA) are required for ORE projects to compare similar projects in nonrenewable energy to understand the environmental and climatic impacts and suggest measures on how to avoid harming the marine ecosystem due to the installation and operation of the technologies. These would help inform policymakers in developing legal frameworks, formulate policies and plans, create institutions, allocate budgetary resources, and initiate regional and international cooperation to install ORE technologies. For example, India has been successfully operating desalination and tidal energy projects. Regional countries like Bangladesh can learn from their experience and initiate collaboration to establish similar projects in the salinity-prone coastal districts and energy-starved islands. China has developed close collaboration with International Energy Agency (IEA), International Electrotechnical Commission (IEC), Ocean Energy Systems (OES), and European Marine Energy Centre (EMEC) to strengthen its capacity in marine energy (Qiu et al., 2019). However, only China and India are members of OES among developing countries. These agencies can help develop the ORE capacity of developing economies and small-island developing states.

Community-level acceptance of the technologies is important in developing the ORE sector in developing economies as local communities would be the end adopters of such technology. Therefore, a social impact assessment (SIA) can be conducted to understand their perception regarding the potential impact and acceptance of ORE by these communities (Hernández-Fontes et al., 2020). For effective transition in energy, there is a need for involvement of the local communities (Vivoda, 2012) in the planning, installation and commercial operation of the ORE projects. Such a participatory approach and knowledge spillover at the local level regarding various ORE technologies would help ensure adoption at the community level in the long run.

Finally, developing economies need to assess their sustainable blue energy potential for a transition toward ORE, which would include a quantitative target of country-level energy mix and integration of ocean in their respective strategic plans. The corresponding location-specific road map should be prepared for tailoring the ORE devices to local conditions and accelerating their installation. An integrated regional approach is required for realizing the potential of sustainable ORE development (Gilau and Failler, 2020). This kind of approach would target both coastal and adjacent land-locked countries and regions. For example, Bangladesh can undertake a regional approach to include land-locked Nepal, Bhutan and Northeast India in its ocean energy planning. Regional platforms like the Southeast Asian Collaboration on Ocean Energy (SEACORE) (Nachtane et al., 2018) would be an instance for other regional blocs comprising developing countries, such as the Bay of Bengal Initiative for Multi-Sectoral Technical and Economic Cooperation (BIMSTEC) to facilitate collaboration in realizing the potential in which India can take lead from its experience in ORE projects.

10.8 Conclusion

The ocean contains an enormous amount of energy, including marine currents, osmotic, thermal, tidal, and wave energy, which has received considerable attention for investment and research (Zhang et al., 2021). ORE is a scale-independent and viable alternative of fossil fuel for regions especially having oceans in their sovereign territories. It is suitable for power generation for coastal communities and small island developing states. The potential of ORE is enormous, which can be explored by developing economies for addressing power shortages in isolated and backward coastal regions. It would help derive manifold benefits that include climate change mitigation through reduction of CO_2 emission from fossil fuel, investment inflow, development of business and economic activities, creation of jobs and attain parity between catchment communities and other parts of the countries in socio-economic development. However, the negative impacts of the projects on the marine ecosystem and environment must be addressed through improving the design of the technologies, gathering information on ocean ecosystem and biodiversity, and preparing a plan of action. In addition, communities can be made active stakeholders of the projects through the revenue-sharing approach.

Developing economies should introduce policies, create institutions, and allocate budgets to facilitate the installation of the technologies. It would attract regional resources, technological knowledge and expertise, and strengthen cooperation among governments, industry, and beneficiaries. Developing countries are now transforming their energy portfolio through diversifying sources, and ORE would be a viable area of investment in the energy sector of these countries if the technologies become cheaper, cost-effective and affordable to them. At the same time, saving environmental costs, such as reducing carbon emission, can be included in calculating per unit cost of production of energy, which would help attain greater support in formulating policy toward greater adoption of ORE.

Self-evaluation questions

1. What do you mean by ocean renewable energy (ORE)? What are the common sources of ORE?
2. What are the differences between wave and tidal energy-based power plants?
3. What is the present status of ORE in developing economies compared to developed economies?
4. Compare ORE and other renewable energy sources (e.g., solar, wind, hydro) toward sustainable development.
5. Do you think ORE is more potential than other renewable energy sources such as solar, wind, hydro? – Justify your answer.
6. Explain the working principle of ocean thermal energy conversion with a proper diagram.
7. Compare between pressure retarded osmosis (PRO) and reverse electrodialysis (RED) processes for electricity generation from a salinity gradient energy source.
8. Discuss the potential of ocean renewable energy generation in developing economies.
9. How could ocean renewable energy contribute to sustainable energy sector development?
10. How is ORE interlinked with many different sustainable development goals? How can ORE contribute to achieving sustainability?
11. What are the facts that the policymakers should consider to adopt ORE in nations' energy mix?
12. How can developing economies strengthen their technological capacity to adopt ORE widely?
13. Find a developing country that has the potential to exploit ORE. What suggestions would you give them to be more dependent on renewable energy through ORE?
14. Conduct a literature survey and identify recent technology development in the ORE sector toward sustainable development.

Numerical problems

1. (a) Calculate wave energy flux for a significant wave period of 2.5 s and a significant wave height of 2 m (use seawater density). (b) If the wavelength is 8 m, then calculate the wave energy flux at 50 m depth. [Answer: (a) 4384.94 W, (b) 4384.94 W]
2. Using the simplest formula of thermal gradient energy, when the surface water temperature is 25°C, the water temperature at a depth of 500m is 12°C, and the discharge rate of warm water is 1.7 times greater than cold water, calculate the amount of thermal gradient energy. [Answer: 3.51 GW]
3. At a temperature of 17°C, what would be the salinity gradient when sea salinity is 35 g/L, river salinity is 4 mg/L, and river discharge is 30,000 m^3/s. [Answer: 116 MW]

References

Abdelkader, B.A., Sharqawy, M.H., 2021. Temperature effects on salinity gradient energy harvesting and utilized membrane properties – experimental and numerical investigation. Sustain. Energy Technol. Assess. 48 (2021), 101666. https://doi.org/10.1016/j.seta.2021.101666.

Ahuja, D., Tatsutani, M., 2009. Sustainable energy for developing countries. Sapiens 2 (1), 1–16. http://journals.openedition.org/sapiens/823.

References

Alcérreca-huerta, J.C., Encarnacion, J.I., Ordoñez-sánchez, S., Callejas-jiménez, M., Gallegos, G., Barroso, D., Allmark, M., Mariño-tapia, I., Casarín, R.S., Doherty, T.O., Johnstone, C., Carrillo, L., 2019. Energy yield assessment from ocean currents in the insular shelf of Cozumel island. J. Mar. Sci. Eng. 7, 147. https://doi.org/10.3390/jmse7050147.

Ascari, M.B., Hanson, H.P., Rauchenstein, L., Van Zwieten, J., Bharathan, D., Heimiller, D., Langle, N., Scott, G.N., Potemra, J., Nagurny, N.J., et al., 2012. Ocean Thermal Extractable Energy Visualization-Final Technical Report on Award DE-EE0002664, Lockheed Martin Mission Systems and Sensors. October 28, 2012. Technical Report.

Asian Development Bank (ADB), 2014. Wave energy Conversion and Ocean Thermal energy Conversion potential in developing member countries. https://doi.org/10.1007/BF02929925.

Assareh, E., Assareh, M., Alirahmi, S.M., Jalilinasrabady, S., Dejdar, A., Izadi, M., 2021. An extensive thermo-economic evaluation and optimization of an integrated system empowered by solar-wind-ocean energy converter for electricity generation – Case study: Bandar Abas, Iran. Therm. Sci. Eng. Progress 25, 100965. https://doi.org/10.1016/j.tsep.2021.100965.

Avery, W.H., Wu, C., 1994. Renewable Energy from the Ocean: a Guide to OTEC. Oxford University Press, New York.

Barclay, V.M., Culina, J., Neill, S.P., 2021. Ocean renewable energy test centers. Comprehensive Renewable Energy. Elsevier, pp. 1–26. https://doi.org/10.1016/B978-0-12-819727-1.00082-0.

Barragán, J.M., de Andrés, M., 2015. Analysis and trends of the world's coastal cities and agglomerations. Ocean Coastal Manag. 114, 11–20. https://doi.org/10.1016/j.ocecoaman.2015.06.004.

Bonar, P.A.J., Bryden, I.G., Borthwick, A.G.L., 2015. Social and ecological impacts of marine energy development. Renew. Sustain. Energy Rev. 47, 486–495. https://doi.org/10.1016/j.rser.2015.03.068.

Belyakov N., 2019. Sustainable Power Generation: Current Status, Future Challenges, and Perspectives, Elsevier.

Bender, A., Langhamer, O., Sundberg, J., 2020. Colonisation of wave power foundations by mobile mega- and macrofauna – a 12 year study. Mar. Environ. Res. 161, 105053.

Chakraborty, S., Dwivedi, P., Chatterjee, S.K., Gupta, R., 2021. Factors to promote ocean energy in India. Energy Policy 159 (2021), 112641. https://doi.org/10.1016/j.enpol.2021.112641.

Chen, Q., Burhan, M., Ja, M.K., Li, Y., Ng, K.C., 2021. A spray-assisted multi-effect distillation system driven by ocean thermocline energy. Energy Convers. Manag. 245, 114570. https://doi.org/10.1016/j.enconman.2021.114570.

Copping, A., 2018. The State of Knowledge for Environmental Effects: Driving Consenting/Permitting for the Marine Renewable Energy Industry. Ocean Energy Systems, Lisbon, Portugal, p. 25.

Copping A, Battey H, Brown-Saracino J, Massaua M, Smith C. An international assessment of the environmental effects of marine energy development. Ocean Coast Manag 2014; 99:313. DOI:10.1016/J.OCECOAMAN.2014.04.002.

Copping, A.E., Hemery, L.G., Viehman, H., Seitz, A.C., Staines, G.J., Hasselman, D.J., 2021. Are fish in danger? A review of environmental effects of marine renewable energy on fishes. Biol. Conserv. 262, 109297. https://doi.org/10.1016/j.biocon.2021.109297.

de Oliveira, L., Silva dos Santos, I.F., Schmidt, N.L., Filho, G.L.T., Camacho, R.G.R., Barros, R.M., 2021. Economic feasibility study of ocean wave electricity generation in Brazil. Renew. Energy 178, 1279–1290. https://doi.org/10.1016/j.renene.2021.07.009.

de Souza, R.V., Fernandes, E.H.L., de Azevedo, J.L.L., Passos, M.S., Corrêa, R.M., 2020. Potential for conversion of thermal energy in electrical energy: highlighting the Brazilian ocean thermal energy park and the inverse anthropogenic effect. Renew. Energy 161, 1155–1175. https://doi.org/10.1016/j.renene.2020.07.050.

Dincer, I., Rosen, M.A., Khalid, F., 2018. Ocean (marine) energy production. Compr. Energy Syst. 3, 335–379. https://doi.org/10.1016/B978-0-12-809597-3.00316-3.

Eckstein, D., Künzel, V., Schäfer, L., 2021. Global Climate Risk Index 2021: Who Suffers Most from Extreme Weather Events? Weather-Related Loss Events in 2019 and 2000-2019, Germanwatch. Berlin.

Falnes, J., 2002. Wave transport of energy and momentum. Ocean Waves and Oscillating Systems, Linear Interaction Including Wave-Energy Extraction. Cambridge University Press, Cambridge, pp. 75–83.

Felix, A., Hernández-Fontes, J.V., Lithgow, D., Mendoza, E., Posada, G., Ring, M., Silva, R., 2019. Wave energy in tropical regions: deployment challenges, environmental and social perspectives. J. Mar. Sci. Eng. 7. https://doi.org/10.3390/jmse7070219.

Fenton, J.D., Mckee, W.D., 1990. On calculating the lengths of water waves. Coast Eng 14 (6), 499–513.

Fox, C.J., Benjamins, S., Masden, E.A., Miller, R., 2018. Challenges and opportunities in monitoring the impacts of tidal-stream energy devices on marine vertebrates. Renew. Sustain. Energy Rev. 81, 1926–1938. http://dx.doi.org/10.1016/j.rser.2017.06.004.

Gilau, A.M., Failler, P., 2020. Economic assessment of sustainable blue energy and marine mining resources linked to African Large Marine Ecosystems. Environ. Dev. 36, 100548. https://doi.org/10.1016/j.envdev.2020.100548.

Gill, A.B., Gloyne-Philips, I., Kimber, J., Sigray, P., 2014. Marine renewable energy, electromagnetic (EM) fields and EM-sensitive animals. In: Shields, M.A., Payne, A.I.L. (Eds.), Marine Renewable Energy Technology and Environmental Interactions. Springer, Netherlands, Dordrecht, pp. 61–79.

Haddout, S., Priya, K.L., 2021. Preparing for the future: The impact of sea-level rise on salinity gradient energy in estuaries. Energy Climate Change 2 (2021), 100041. https://doi.org/10.1016/j.egycc.2021.100041.

Hemery, L.G., 2020, OES-Environmental 2020 State of the Science Report: Environmental Effects of Marine Renewable Energy Development Around the World. In: Copping, A.E., Hemery, L.G. (Eds.). Report for Ocean Energy Systems (OES), pp. 108–128.

Hernández-Fontes, J.V., Martínez, M.L., Wojtarowski, A., González-Mendoza, J.L., Landgrave, R., Silva, R., 2020. Is ocean energy an alternative in developing regions? A case study in Michoacan, Mexico. J. Cleaner Prod. 266, 121984. https://doi.org/10.1016/j.jclepro.2020.121984.

Horne, N., Culloch, R.M., Schmitt, P., Lieber, L., Wilson, B., Dale, A.C., Houghton, J.D.R., Kregting, L.T., 2021. Collision risk modelling for tidal energy devices: a flexible simulation-based approach. J. Environ. Manage. 278, 111484. https://doi.org/10.1016/j.jenvman.2020.111484.

Huante, A.G., Cueto, Y.R., Silva, R., Mendoza, E., Vega, L., 2018. Determination of the potential thermal gradient for the Mexican Pacific Ocean. J. Mar. Sci. Eng. 6, 20. https://doi.org/10.3390/jmse6010020.

IEA-OES, 2020. Annual Report: An Overview of Ocean Energy Activities in 2019. The Executive Committee of Ocean Energy Systems. https://www.ocean-energy-systems.org/publications/oes-annual-reports/

IEA-OES. 2021, Annual Report: An Overview of Ocean Energy Activities in 2020. The Executive Committee of Ocean Energy Systems. https://www.ocean-energy-systems.org/publications/oes-annual-reports/

IRENA (International Renewable Energy Agency), 2019. Renewable Energy Statistics 2019. Abu Dhabi.

IRENA. 2020. Innovation Outlook. Ocean energy technologies. Abu Dhabi: International Renewable Energy Agency; 2020.

IRENA (International Renewable Energy Agency). 2021. Data & Statistics. https://www.irena.org/Statistics.

Jianbo, L., Chen, Z., Kai, L., Li, Y., Xiangqiang, K., 2021. Experimental study on salinity gradient energy recovery from desalination seawater based on RED. Energy Convers. Manage. 244 (2021), 114475. https://doi.org/10.1016/j.enconman.2021.114475.

Jihye, K., Kwanho, J., Myoung, J.P., Ho, K.S., Joon, HK., 2015. Recent advances in osmotic energy generation via pressure-retarded osmosis (PRO): a review. Energies 8, 11821–11845. https://doi.org/10.3390/en81011821.

Kamphuis, J.W., 2000. Introduction to Coastal Engineering and Management, vol. 16. World Scientific Publishing, New Jersey.

Khan, I., 2019. Power generation expansion plan and sustainability in a developing country: a multi-criteria decision analysis. J. Cleaner Prod. 220, 707–720. https://doi.org/10.1016/j.jclepro.2019.02.161.

Kirinus, E.P., Oleinik, P.H., Costi, J., Marques, W.C., 2018. Long-term simulations for ocean energy off the Brazilian coast. Energy 163, 364–382. https://doi.org/10.1016/j.energy.2018.08.080.

Langer, J., Cahyaningwidi, A.A., Chalkiadakis, C., Quist, J., Hoes, O., Blok, K., 2021. Plant siting and economic potential of ocean thermal energy conversion in Indonesia: a novel GIS-based methodology. Energy 224, 120121. https://doi.org/10.1016/j.energy.2021.120121.

Langer, J., Quist, J., Blok, K., 2020. Recent progress in the economics of ocean thermal energy conversion: critical review and research agenda. Renew. Sustain. Energy Rev. 130, 109960. https://doi.org/10.1016/j.rser.2020.109960.

Langer, J., Cahyaningwidi, A.A., Chalkiadakis, C., Quist, J., Hoes, O., Blok, K., 2021. Plant siting and economic potential of ocean thermal energy conversion in Indonesia a novel GIS-based methodology. Energy 224 (2021), 120121. https://doi.org/10.1016/j.energy.2021.120121.

López, M., Veigas, M., Iglesias, G., 2015. On the wave energy resource of Peru. Energy Convers. Manag. 90, 34–40. https://doi.org/10.1016/j.enconman.2014.11.012.

Liu, X., Chen, Z., Si, Y., Qian, P., Wu, H., Cui, L., Zhang, D., 2021. A review of tidal current energy resource assessment in China. Renew. Sustain. Energy Rev. 145, 111012. https://doi.org/10.1016/j.rser.2021.111012.

Lossent, J., Lejart, M., Folegot, T., Clorennec, D., Di Iorio, L., Gervaise, C., 2018. Underwater operational noise level emitted by a tidal current turbine and its potential impact on marine fauna. Marine Pollut. Bull. 131, 323–334. https://doi.org/10.1016/j.marpolbul.2018.03.024.

Marta-Almeida, M., Cirano, M., Guedes Soares, C., Lessa, G.C., 2017. A numerical tidal stream energy assessment study for Baía de Todos os Santos. Brazil. Renew. Energy 107, 271–287. https://doi.org/10.1016/j.renene.2017.01.047.

Martin, L., 2011. Navfac Ocean Thermal Energy Conversion (Otec) Project, 4, p. 274.

Martínez, M.L., Vázquez, G., Pérez-Maqueo, O., Silva, R., Moreno-Casasola, P., Mendoza-González, G., López-Portillo, J., MacGregor-Fors, I., Heckel, G., Hernández-Santana, J.R., García-Franco, J.G., Castillo-Campos, G., Lara-Domínguez, A.L., 2021. A systemic view of potential environmental impacts of ocean energy production. Renew. Sustain. Energy Rev. 149, 111332. https://doi.org/10.1016/j.rser.2021.111332.

Mejia-Olivares, C.J., Haigh, I.D., Wells, N.C., Coles, D.S., Lewis, M.J., Neill, S.P., 2018. Tidal-stream energy resource characterization for the Gulf of California. México. Energy 156, 481–491. https://doi.org/10.1016/j.energy.2018.04.074.

Melikoglu, M., 2018. Current status and future of ocean energy sources: a global review. Ocean Eng 148, 563–573. https://doi.org/10.1016/j.oceaneng.2017.11.045.

Mendoza, E., Lithgow, D., Flores, P., Felix, A., Simas, T., Silva, R., 2019. A framework to evaluate the environmental impact of OCEAN energy devices. Renew. Sustain. Energy Rev. 112, 440–449. https://doi.org/10.1016/j.rser.2019.05.060.

MNRE, 2021. MNRE declare ocean energy as renewable energy, https://pib.gov.in/Pressreleaseshare.aspx?PRID=1582638

Nachtane, M., Tarfaoui, M., Hilmi, K., Saifaoui, D., el Moumen, A., 2018. Assessment of energy production potential from tidal stream currents in Morocco. Energies 11, 1065. https://doi.org/10.3390/en11051065.

National Research Council. 2013. An Evaluation of the U.S. Department of Energy's Marine and Hydrokinetic Resource Assessments. Washington, DC: The National Academies Press. https://doi.org/10.17226/18278.

Neill, S.P., 2021. Introduction to ocean renewable energy. Compr. Renew. Energy. Elsevier https://doi.org/10.1016/B978-0-12-819727-1.00081-91.

Nihous, G.C., 2007. A preliminary assessment of ocean thermal energy conversion resources. J. Energy Resour. Technol. 129, 10. https://doi.org/10.1115/1.2424965.

Novo, P.G., Kyozuka, Y., 2021. Tidal stream energy as a potential continuous power producer: a case study for West Japan. Energy Convers. Manag. 245 (2021), 114533. https://doi.org/10.1016/j.enconman.2021.114533.

Ocean Energy Systems (OES), 2021. Offshore installations worldwide. Available at: https://www.ocean-energy-systems.org/ocean-energy/gis-map-tool /. Accessed on 20 October 2021.

Onoufriou, J., Russell, D.J.F., Thompson, D., Moss, S.E., Hastie, G.D., 2021. Quantifying the effects of tidal turbine array operations on the distribution of marine mammals: implications for collision risk. Renew. Energy 180, 157–165. https://doi.org/10.1016/j.renene.2021.08.052.

Orhan, K., Mayerle, R., 2017. Assessment of the tidal stream power potential and impacts of tidal current turbines in the Strait of Larantuka, Indonesia. Energy Procedia 76, 7–16. https://doi.org/10.1016/j.egypro.2017.08.199.

Ozturk, M., Sahin, C., Yuksel, Y., 2017. Current power potential of a sea strait: the bosphorus. Renew Energy 114, 191–203. https://doi.org/10.1016/j.renene.2017.04.003.

Phillips, C., 2019. Kvalsund Tidal Turbine Prototype. https://tethys.pnnl.gov/project-sites/kvalsund-tidal-turbine-prototype.

Pine, M.K., Schmitt, P., Culloch, R.M., Lieber, L., Kregting, L.T., 2019. Providing ecological context to anthropogenic subsea noise: assessing listening space reductions of marine mammals from tidal energy devices. Renew. Sustain. Energy Rev. 103, 49–57.

Polagye, B., Bassett, C., 2020, OES-Environmental 2020 State of the Science Report: Environmental Effects of Marine Renewable Energy Development Around the World. In: Copping, A.E., Hemery, L.G. (Eds.). Report for Ocean Energy Systems (OES), pp. 70–89.

Qiu, S., Liu, K., Wang, D., Ye, J., Liang, F., 2019. A comprehensive review of ocean wave energy research and development in China. Renew. Sustain. Energy Rev. 113, 109271. https://doi.org/10.1016/j.rser.2019.109271.

Queen's University Belfast, 2002. Islay Limpet Wave Power Plant. Publishable Report, contract JOR-CT98–0312. Technical Report European Commission. https://tethys.pnnl.gov/sites/default/files/publications/Islay_LIMPET_Report.pdf.

Quirapas, M.A.J.R., Taeihagh, A., 2021. Ocean renewable energy development in Southeast Asia: opportunities, risks and unintended consequences. Renew. Sustain. Energy Rev. 137, 110403. https://doi.org/10.1016/j.rser.2020.110403.

Quitoras, M.R.D., Abundo, M.L.S., Danao, L.A.M., 2018. A techno-economic assessment of wave energy resources in the Philippines. Renew. Sustain. Energy Rev. 88, 68–81. https://doi.org/10.1016/j.rser.2018.02.016.

Rajagopalan, K., Nihous, G.C., 2013. An assessment of global ocean thermal energy conversion resources with a high-resolution ocean general circulation model. J. Energy Resour. Technol. 135, 041202. https://doi.org/10.1115/1.4023868.

Ramachandran, R., Takagi, K., Matsuda, H., 2020. Enhancing local support for tidal energy projects in developing countries: Case study in Flores Timur Regency, Indonesia. Bus. Strategy Dev. 3 (4). 543-533 https://doi.org/10.1002/bsd2.120.

Rodrigues, N., Pintassilgo, P., Calhau, F., González-Gorbeña, E., Pacheco, A., 2021. Cost-benefit analysis of tidal energy production in a coastal lagoon: the case of Ria Formosa e Portugal. Energy 229, 120812. https://doi.org/10.1016/j.energy.2021.120812.

Rusu, L., Onea, F., 2017. The performance of some state-of-the-art wave energy converters in locations with the worldwide highest wave power. Renew. Sustain. Energy Rev. 75, 1348–1362. https://doi.org/10.1016/j.rser.2016.11.123.

Sen, S., Ganguly, S., Das, A., Sen, J., Dey, S., 2016. Renewable energy scenario in India: opportunities and challenges. J. Afr. Earth Sci. 122 (2016), 25–31. 10.1016/j.jafrearsci.2015.06.002.

Shahriar, T., Habib, M.A., Hasanuzzaman, M., Shahrear-Bin-Zaman, A., 2019. Modelling and optimization of Searaser wave energy converter based hydroelectric power generation for Saint Martin's Island in Bangladesh. Ocean Eng. 192, 106289. https://doi.org/10.1016/j.oceaneng.2019.106289.

Shahzad, M.W., Burhan, M., Li, A., Ng, K.C., 2017. Energy-water-environment nexus underpinning future desalination sustainability. Desalination 413, 52–64. https://doi.org/10.1016/j.desal.2017.03.009.

Sharqawy, M.H., Lienhard, J.H., Zubair, S.M., 2010. Thermophysical properties of seawater: a review of existing correlations and data. Desalination Water Treat. 16, 354–380.

Soto, R., Vergara, J., 2014. Thermal power plant efficiency enhancement with ocean thermal energy conversion. Appl. Therm. Eng. 62, 105–112.

Sparling, C.E., Seitz, A.C., Masden, E., Smith, K., 2020. Collision risk for animals around turbines. In: Copping, A.E., Hemery, L.G. (Eds.), OES-Environmental 2020 State of the Science Report: Environmental Effects of Marine Renewable Energy Development Around the World. Report for Ocean Energy Systems (OES), Lisbon, Portugal, pp. 3269

Su, W.R., Chen, H., Chen, W.B., Chang, C.H., Lin, L.Y., 2018. Numerical investigation of wave energy resources and hotspots in the surrounding waters of Taiwan. Renew. Energy 118, 814–824.

Sundar, V., 2015. Ocean Wave Mechanics: Applications in Marine Structures. John Wiley & Sons Limited, Chichester.

Sustainable Energy Association of Singapore (SEAS). A Position Paper by the Sustainable Energy Association of Singapore Marine Renewables Working Group. Singapore: SEAS; 2018.

Tang, H.S., Kraatz, S., Qu, K., Chen, G.Q., Aboobaker, N., Jiang, C.B., 2014a. High-resolution survey of tidal energy towards power generation and influence of sea-level-rise: a case study at coast of New Jersey, USA. Renew. Sustain. Energy Rev. 32, 960–982.

Tang, H.S., Qu, K., Chen, G.Q., Kraatz, S., Aboobaker, N., 2014b. Potential sites for tidal power generation: A thorough search at coast of New Jersey, USA. Renew. Sustain. Energy Rev. 39, 412–425.

U.S. Army Engineer Waterways Experiment Station, 1984. Shore Protection Manual, 4th ed. Coastal Engineering Research Centre, Vicksburg, Mississippi.

VanZwietena, J., Tang, Y., 2021. Ocean Current Energy. Comprehensive Renewable Energy, 2nd ed. Elsevier, pp. 1–15. https://doi.org/10.1016/B978-0-12-819727-1.00167-9.

Vega, L. et al., 1999. Ocean Thermal Energy Conversion (OTEC), OTEC News-Clean Energy, Water and Food.

Vivoda, V., 2012. Japan's energy security predicament post-Fukushima. Energy Pol 46, 135–143. https://doi.org/10.1016/j.enpol.2012.03.044.

Waldo, Å., 2012. Offshore wind power in Sweden—a qualitative analysis of attitudes with particular focus on opponents. Energy Policy 41, 692–702. https://doi.org/10.1016/j.enpol.2011.11.033.

Wilberforce, T., el Hassan, Z., Durrant, A., Thompson, J., Soudan, B., Olabi, A.G., 2019. Overview of ocean power technology. Energy 175, 165–181. https://doi.org/10.1016/j.energy.2019.03.068.

Wright, J., 1995. Seawater: Its Composition, Properties, and Behaviour, vol. 2, Pergamon, 1995.

Yamada, N., Hoshi, A., Ikegami, Y., 2009. Performance simulation of solar-boosted ocean thermal energy conversion plant. Renew. Energy 34 (7), 1752–1758. https://doi.org/10.1016/j.renene.2008.12.028.

Young, I.R., Verhagen, L.A., 1996. The growth of fetch limited waves in water of finite depth. Part 1. Total energy and peak frequency. Coast Eng 29 (1-2), 47–78.

Zhang, J., Xu, C., Song, Z., Huang, Y., Wu, Y., 2019. Decision framework for ocean thermal energy plant site selection from a sustainability perspective: the case of China. J. Cleaner Prod. 225, 771–784. https://doi.org/10.1016/j.jclepro.2019.04.032.

Zhang, Y., Zhao, Y., Sun, W., Li, J., 2021. Ocean wave energy converters: technical principle, device realization, and performance evaluation. Renew. Sustain. Energy Rev. 141, 110764. https://doi.org/10.1016/j.rser.2021.110764.

CHAPTER 11

Hydrogen energy–Potential in developing countries

Minhaj Uddin Monir[a], Azrina Abd Aziz[b], Mohammad Tofayal Ahmed[a] and Md. Yeasir Hasan[a]

[a]Department of Petroleum and Mining Engineering, Jashore University of Science and Technology, Jashore, Bangladesh [b]Faculty of Civil Engineering Technology, Universiti Malaysia Pahang, Gambang, Malaysia

11.1 Introduction

Energy sustainability has recently become a national issue due to the global depletion of fossil-based resources (Nazir et al., 2019). These worries regarding the energy shortage have necessitated the creation of complementary, clean, and affordable energy options. In general, electricity is produced from both renewable and nonrenewable resources (Azam et al., 2021). In terms of economic policy, policymakers could implement certain changes such as tax incentives, structural quality change, and foreign collaboration to profit from trade transparency, enhance green energy demand, and increase nonrenewable energy use (Amri, 2019). Moreover, electricity is closely linked to a country's economic growth in the face of today's global challenges. Nathaniel and Bekun (2021) reported that electricity demand, urbanization, and economic development in developing countries have a new lesson from combined cointegration in the face of systemic breaks. The sustainable development agenda, by programs such as Agenda 21 and 2030, includes a set of priorities and objectives to promote decarbonization and increase living standards (Relva et al., 2021). In this emergency scenario, the world, especially developing countries, needs more electricity from the most readily accessible energy supplies. Hydrogen is an important energy carrier that is being used to support decarbonize the global energy and manufacturing sectors; therefore, generating hydrogen from clean energy sources is of great concern nowadays (Dawood et al., 2020). The hydrogen issue has evolved over the last two decades, with a change of focus away from automotive applications and toward difficult-to-decarbonize sectors like energy-intensive industries, trucks, aircraft, shipping, and heating (Gielen et al., 2019). The need for developing countries to produce hydrogen from renewable resources (solar, biomass and municipal solid waste) is being researched in Pakistan; however, vast reserves of unreliable renewable energy must be stored for later use (Gondal et al., 2018). The study also found that biomass is the most viable feedstock for building a hydrogen supply chain in Pakistan, with a potential to produce 6.6 million tons of hydrogen yearly. By 2030, India wants to reach a combined electric power installed capacity of 40% from nonfossil fuel dependent electricity supplies (Bisht and Thakur, 2019). They reported that the Indian bioenergy market has grown by 5.42 times in the last 10 years. However, in the new Indian power production situation, coal meets the majority of the country's energy requirements. India's traditional electricity demands are: 222.91 GW from thermal (coal, natural gas, and diesel), 6.780 GW from nuclear, 45.29 GW from broad hydro, and 69.78 GW from renewable energy sources. Khan (2020) performed a case study in Bangladesh for the potential of hydrogen production from waste to biogas through the anaerobic digestion process. Ayodele et al. (2020) also studied the latest developments in thermo-catalytic conversion of biomass-derived glycerol hydrogen production. In their research, they explain the potential prospects and challenges of thermo-catalytic conversion, a process for transforming green biofuels into synthetic glycerol. China has made major advances in the hydrogen economy, such as fuel cell vehicles (FCVs), during the last few decades, but it still faces many obstacles in the move to a hydrogen economy (Ren et al., 2020). Moreover, hydrogen has the ability to significantly increase growth rate of the green energy sector and expand the scope of renewable solutions, for example, in industry. Hydrogen is a light gas with a density of 0.089886 kg/m^3 (Ayodele et al., 2019). It may be used to store electricity over the course of a year. As a result, low-cost hydrogen is needed to bring these synergies into action (Gielen et al., 2019).

Hydrogen energy is the practice of utilizing hydrogen and/or hydrogen-containing compounds to produce energy that can be used for all practical purposes while obtaining high intensity production, major environmental and social advantages, and economic viability (Qyyum et al., 2021). This fuel energy is the earth's easiest and most plentiful chemical (Baykara, 2018). It quickly interacts with other chemical components and is often present as part of some materials like water, hydrocarbons, or alcohol. Biomass, like plants and animals, often contains hydrogen. It is thus called an energy transporter, not an energy source (Santhanam et al., 2017). It is generated utilizing a variety of domestic resources such as nuclear, gas, coal, biomass, and other renewable energy sources such as solar, wind, hydroelectric, and geothermal power. Because of the availability of domestic energy supplies, hydrogen is a promising energy carrier for energy security. Four global issues are making the search for nonfossil-based energy sources inevitable: a) depleting supplies, b) rising costs c) current environmental crisis triggered by greenhouse gas emissions, and d) clean, sustainable or organic consumerism (Creutzig et al., 2015). Because the current energy supply is unsustainable, the use of green energy generation for heat, hydrogen, biofuels, and electricity is the only feasible global energy system (Ainas et al., 2017). There are alternative energy sources, such as hydrogen generated from synthetic gas that can be used in transportation and electricity generation (Monir et al., 2018a; Monir et al., 2020a). They also reported that hydrogen production could be enhanced by adding charcoal with biomass feedstocks during gasification through downdraft gasifier. Therefore, hydrogen generation could be used as a solution for present and future energy crisis (Tsai and Chen, 2009).

Hydrogen has the potential for storing and providing resources, and electricity generation. It is three times the energy per unit mass of liquid fuels (Mazloomi and Gomes, 2012).

Although hydrogen has the ability to be used as a fuel cell energy, it is not clean and sustainable yet. Ayodele et al. (2019) worked on measuring electricity generation and the environmental potentials of hydrogen derived from biogas. They reported that 0.334 million tons of hydrogen gas could be converted into 16 million kWh of electricity per year. Approximately, 120 million tons of hydrogen are generated annually, with two-thirds being pure hydrogen and one-third being combined with other gases. According to the International Energy Agency (IEA), this amounts to 14.4 exajoules (EJ), or around 4% of global final energy and nonenergy intake. Gas and coal account for about 95% of all hydrogen production. As a by-product of chlorine production, about 5% is provided by electrolysis (Gielen et al., 2019). Coke oven gas also contributes a significant amount of hydrogen to the iron and steel industry, some of which is recovered. At present, renewable energy sources do not contain a sufficient amount of hydrogen (Gielen et al., 2019). This gas is generated through a number of steps (Pereira et al., 2017). It is important for numerous industrial applications, including ammonia production, petroleum processing, and the water-gas shift reaction (Zhang et al., 2021). The industrialized world's hydrogen supply is of primary importance and in the developing world, progress of hydrogen is deficient. Moreover, hydrogen could be used as an alternative safe cooking fuel. The others most common cooking fuels are wood, crop wastes, tar, dung, and charcoal. However, it can serve as an alternative source of electricity for developing countries since it can be derived from renewable energy resources.

11.2 Status of global energy sector

From the literatures, it is found that energy deficiency is a major problem, stifling a country's rapid development. Bangladesh is one of the developing countries that lacks of power, but the country's energy demand is increasing on a regular basis (Kabir and Khan, 2020; Monir et al., 2020). For this reason, several nations are shifting toward renewable energy as a result of natural resource scarcity and environmental issues. A significant portion of electricity is currently generated from nonrenewable energy sources, but storage capacity is restricted. The main suppliers of these resources are crude oil, gas, and coal, which are usually found in the soil or subsurface (Furuoka, 2017). Over the past two decades, more open attractions about hydrogen have risen up all over the world. The primary goal of this commendable initiative is to reduce emissions and the carbon footprint. Despite the presence of carbon-based hydrogen manufacturing technologies, new hydrogen development methods based on renewable and alternative energy sources have emerged (Arat et al., 2020). As oil or gas is extracted from the earth, it must be processed into fuel and distributed to gas stations where it can be used by customers (Clews, 2016). According to Capuano (2018), the world's energy demand will rise on average over 2050 for all types of fuels shown in Fig. 11.1. From this projection it is clearly found that global energy consumption will rise nearly 50% by 2050 (Capuano, 2020). Fig. 11.2 shows the projected global energy demand of India, China, Africa, other non-OECD Asia, and the rest of the world (Capuano, 2020).

This comparative projection shows that energy demand is growing quickly in the developing world including India, China, and Africa, although it is increasing steadily in other countries before 2050 (Fig. 11.2).

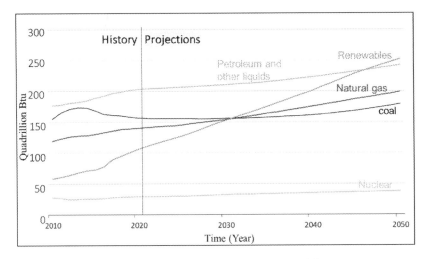

FIGURE 11.1 Projection of global energy demand up to 2050 (Capuano, 2020).

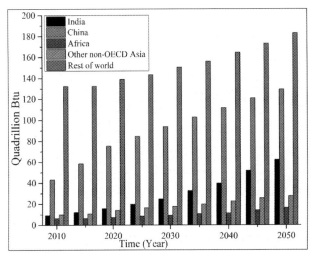

FIGURE 11.2 Projection of energy demand for selected developing countries, and non-OECD Asia, and the rest of the world (Capuano, 2020).

The global power generation and other uses have been presented in Fig. 11.3. In this projection it is shown that the power generation in present time is around 260 Quadrillion Btu, where as it will more than 400 Quadrillion Btu by 2050 (Capuano, 2020). Moreover, global average energy use in 2000 was 12,116 TWh, 15,105 TWh in 2005, 16,503 TWh in 2008, and 19,504 TWh in 2013. According to Birol (2017), total installed electricity generating capacity at the end of 2014 was roughly 6.142 TW, which only included generation linked to local electricity grids. Furthermore, off-grid businesses and cities use a massive amount of heat and electricity. Natural gas and coal have become the two commonly used energy sources. The global energy consumption in 2012 was 18,608 TWh (Birol, 2017). Although, around 80% of global energy in 2016 came from fossil fuels, 10% from biofuels, 5% from nuclear,

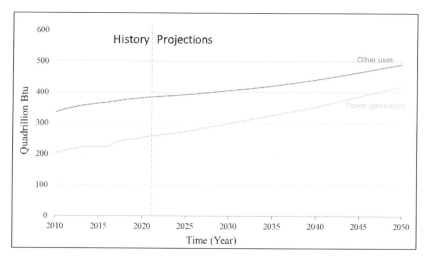

FIGURE 11.3 Global primary energy consumption trends for electric power generation and other uses (Capuano, 2020).

and 5% from renewables (hydro, wind, solar, and geothermal), just 18% of the total global energy came in the form of electricity (Janda and Tan, 2017). In a sense of declining economic development, growth in global energy consumption declined in 2019 (0.6%) compared to an average of 2%/year from 2000 to 2018. China (+3.2%), the world's the largest consumer since 2009, Russia (+1.8%) and India (+0.8%) increased their energy consumption slowly than in previous years. World primary energy consumption more than doubled, from 270.5 EJ in 1978 to 580 EJ in 2018 (Kober et al., 2020). According to the Stated Policies Scenario (STEPS), global electricity demand recovers and surpasses pre-Covid-19 peaks in 2021 (IEA, 2020). By 2030, Indian electricity demand growth will outpace that of other countries, with growth in Southeast Asia and Africa becoming more pronounced. China has the largest real increase in production, accounting for more than 40% of global development by 2030. Electricity demand growth outpaces that of all other energies on a global scale.

Khan (2020) reported that anaerobic digestion is an appropriate technology for producing biogas from organic waste in Bangladesh. There is an enormous scope for generating green hydrogen (hydrogen generated by renewable sources) from biogas helps to reduce greenhouse gas emissions (Alves et al., 2013). With 16 million tons of hydrogen generated in 2012 and 20–22 million tons in 2018, China currently has the world's largest hydrogen production potential reported by Arat et al. (2020). Hydrogen innovation activities in developing countries increased significantly during the last three decades (Arat et al., 2020; Kovač et al., 2021).

11.3 Sources of hydrogen production

Fossil energy (oil, coal, and natural gas) and renewable energy (wind, solar, geothermal, and hydropower) are the examples of the main energy sources. These primary energy sources are transformed into electricity, a secondary energy source that flows to the houses and

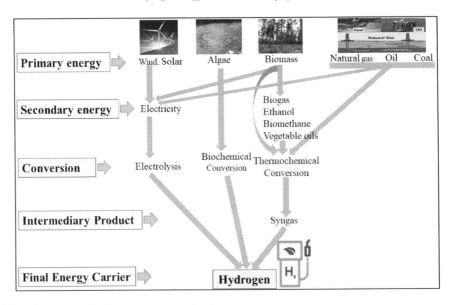

FIGURE 11.4 Sources of hydrogen energy production.

industries through power lines and other transmission infrastructure. The most common source of industrial hydrogen is fossil fuels. During the conversion process, carbon dioxide can be isolated from natural gas with 70 to 85% efficiency for hydrogen productivity and with different degrees of efficiency from other hydrocarbons. The bulk hydrogen is normally created by the steam reforming of methane or natural gas.

11.3.1 Fossil fuel source

A fossil fuel is a fuel that is produced naturally, such as the anaerobic decomposition of buried dead animals, and consists of organic molecules derived from ancient photosynthesis that emit energy when burnt. Millions of years are needed for the formation of fossil fuels (Barber, 2018). The most common fossil fuels are natural gas, oil, and coal; they are used to generate hydrogen (Fig. 11.4) (Muradov, 1993). Hydrogen emits zero pollution during conversion and can be driven by either low-carbon energy or carbon-abated fossil fuels. Annual demand for hydrogen could be more than tenfold by 2050, rising from 8 EJ in 2015 to nearly 80 EJ in 2050 (Denton, 2018; Halloran, 2007). Due to the large market, along with the worsening efficiency of crude oil, the development of fuel-extraction techniques, and advances in hydrogen technology, there have been benefits and challenges to fuel-cell powered vehicles. From the perspective of chemistry, coal is a dynamic and highly variable substance. Coal gasification is the technique to produce electricity, liquid fuels, and hydrogen (Ahamed et al., 2016; Mansur et al., 2020). Precisely, it is done by combining coal with high-pressure oxygen and steam to make synthetic gas (Mansur et al., 2020), which yields mostly carbon monoxide and hydrogen. Subsequently, the hydrogen is removed through pressure swing adsorption, which gives purification (Franco, 2014). Liu et al. (2020) proposed a supercritical water gasification of coal that produces hydrogen according to a reaction kinetic model which includes nitrogen

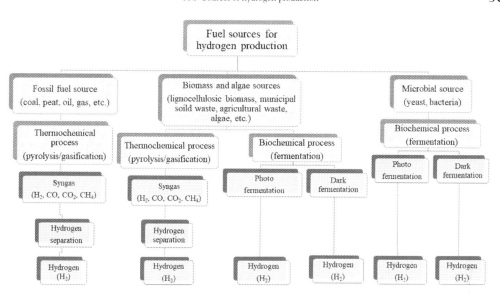

FIGURE 11.5 Hydrogen production from fossil fuel, biomass, algae, and microbial sources.

and sulfur elements. This kinetic model can predict, under different operating conditions not only the production of hydrogen, methane, carbon dioxide, carbon monoxide, ammonia, and hydrogen sulfide, but also products for various forms of coal, which can provide a theoretical foundation for a targeted nitrogen and sulfide factor control in supercritical water. Steam reforming and partial oxidation are primary means of hydrogen extraction from natural gas. Liemberger et al. (2019) reported a hybrid separation process for hydrogen co-transported with natural gas. They found in their research that to extract hydrogen from the natural gas grid, a specific energy demand of 0.5 to 1.0 kWhm^{-3} is required. Petroleum or oil is a complex mixture of hydrocarbons found in the surface of the earth. Heavy oil not only contains carbon, but also contains a lot of hydrogen. Until now, hydrogen extraction from hydrocarbons is not thought to be cost-effective. Therefore, fossil fuels are the main contributor for the production of hydrogen energy. The flowchart of hydrogen production from the sources of fossil fuel, biomass, algae and microbial are shown in Fig. 11.5.

11.3.2 Biomass and algae source

Biomass is a renewable source of energy such as plant residues, grasses, energy crops, woodland residues, wood residues, and urban paper wastes, that is used for the production of bioenergy (Monir, et al., 2018b; Singh et al., 2017). It is more attractive fuel for hydrogen production due to strong reserves and predictable year-round performance. The conversion of biomass into usable sources of energy such as heat, electricity, hydrogen, and liquid fuels are known as biomass-based energy (Kundu et al., 2018; Monir et al., 2018a). Mohammed et al. (2011) studied on the hydrogen sources from biomass. They reported that it is one of the most abundant renewable resources and it is produced by plants fixing atmospheric CO_2 during photosynthesis, making it carbon neutral throughout its life cycle. It is a clean, renewable

TABLE 11.1 Hydrogen production from microorganisms through fermentation process.

Microorganisms	Inoculum	Molar yield (mol H_2/mol hexose)	References
Scenedesmus obliquus	Clostridium butyricum	2.9	Ferreira et al. (2013)
Native consortium	Treated anaerobic sludge	3	Carrillo-Reyes and Buitrón (2016)
Native consortium	Treated anaerobic sludge	2.77	Barragán-Trinidad and Buitrón (2020)
Native consortium	Treated anaerobic sludge	1.02	Barragán-Trinidad and Buitrón (2020)
Chlorella vulgaris ESP6	Clostridium butyricum	1.15	Liu et al. (2012)
Scenedesmus obliquus	Clostridium butyricum	2.23	Batista et al. (2014)

energy source that has the potential to significantly improve the environment, economy, and energy security. Moreover, hydrogen (H_2) gas is extracted from algae, which is promising due to its sustainability, lack of greenhouse gas emissions during hydrogen combustion, and protection of supply even in remote areas (Vijayaraghavan et al., 2010). Sharma and Arya (2017) also reported that algal biomass is responsible for hydrogen production. Algae studies have also focused on ethanol processing of hydrogen since it is derived from fossil or sustainable biomass content. The plasma membrane reactor has also been used to separate hydrogen from ammonia, with a maximum H_2 conversion rate of 24.4% (Hayakawa et al., 2020).

11.3.3 Microbial source

Microorganisms such as bacteria break down organic matter to generate hydrogen through this fermentation-based method. Microbial biomass conversion processes take advantage of microorganisms' ability to consume and digest biomass while also producing hydrogen. *Clostridium, Enterobacter, Klebsiella, Citrobacter*, and *Bacillus* are the most common hydrogen-producing bacteria (Monir et al., 2020c; Monir et al., 2020d), while *Methanorix, Methanosarcina, Methanoculleus, Methanobrevibacter, Methanobacterium, Methanofollis*, and *Methanomassiliicoccus* are the most common methanogens bacteria (Misiukiewicz et al., 2021). *Clostridium butyricum* is an important bacterium because of its hydrogen productivity (Monir, et al., 2020e). *Anaerobes, facultative anaerobes, aerobes, methylotrophs*, and photosynthetic bacteria have all been known to produce hydrogen. *Anaerobica Clostridia* are potential producers that have been immobilized with 50% efficiency and *C. butyricum* produces 2 mol H_2/mol glucose. Recently, Kanchanasuta et al. (2017) suggested that hydrogen is produced by the addition of *C. butyricum* into nonsterile food waste. This research could contribute to commercial-scale systems in the mid- to long-term, depending on the feedstock used, which could be suitable for localized, semicentral, or central hydrogen output. These microbes may degrade complex molecules through a variety of pathways, and the by-products of several of these pathways can be combined by enzymes to generate hydrogen. Mostly used microorganisms with hydrogen yields are shown in Table 11.1.

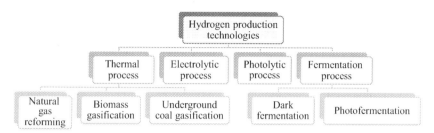

FIGURE 11.6 Hydrogen production technologies.

11.4 Technologies for hydrogen production

Hydrogen is produced using a variety of process technologies (Fig. 11.6), including thermal (natural gas reforming, biomass gasification, and underground coal gasification), electrolytic (water splitting using a variety of energy resources), photolytic (splitting of water using sunlight through biological and electrochemical materials), and fermentation (photo fermentation, dark fermentation).

11.4.1 Thermal process

In this process, heat and chemical reactions are needed to extract hydrogen from organic products such as carbon fuels and biomass (Monir et al., 2018b). Various types of gasifiers are used for thermal processes for the conversion of solid fuel to energy fuel. The intermediate product of syngas is produced through this thermal process. Solid biomass feedstocks are usually used for this conversion process. The feedstock size depends on the types of gasifiers used (Fig. 11.7).

When solid feedstocks are subjected to a thermal action, produces gaseous compounds that can be combined with hydrogen in an oxygen-rich environment, the biomass is usually converted into syngas or synthetic gas-rich syngas mixtures. The main components of syngas are carbon monoxide and hydrogen (Monir et al., 2017). Between polymerization and full combustion, there is a state of polymerization known as pyrolysis. All popular gasifiers used for gasification fall into one of three categories: fixed bed, fluidized bed, or entrained flow system (Fig. 11.7).

The released gas is used as a source for compounding additives and catalysts, in addition to its main uses in microturbines, gas generators, gasoline batteries, and fuel-cell systems. Monir et al. (2020b) reported that when charcoal is added to an empty fruit bunch, it makes a greater flow of hydrogen into the bun, producing a better amount of vapor in the flame, which can then be sent into the resulting extra hydrogen. A simulation process was performed using an Aspen Plus simulator by Monir et al. (2017) to monitor the experimental optimization. In this study, it was found that hydrogen rate was increased by the addition of charcoal with biomass. The energy in various sources, such as natural gas, coal, or biomass, is used in some thermal methods to liberate hydrogen from its molecular structure. Heat, in accordance with closed-chemical cycles, is used in other processes to generate hydrogen from feedstocks including water. Water is decomposed into hydrogen and oxygen by integrating chemical

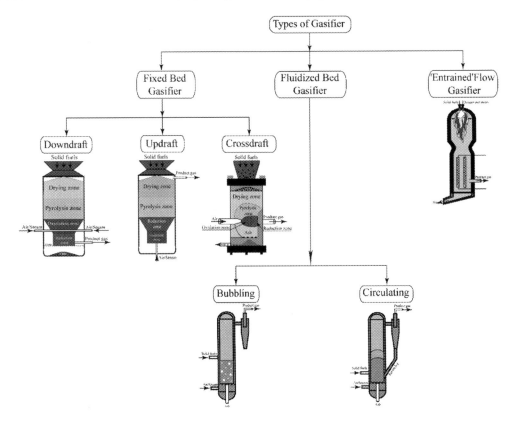

FIGURE 11.7 Types of gasifiers used for thermal process.

reactions. Chemical reactions are performed using heat to move them. The elements and waste heat are released as water and heat are fed into a combination of thermochemical reactions (Eqs. 11.3–11.12). This method is one of the energy conversion methods that convert thermal energy into hydrogen energy, that is, chemical potential, or hydrogen heat.

11.4.1.1 *Natural gas reforming to hydrogen*

Natural gas reforming is the most effective method of producing hydrogen around the world. The technology is now mature and widely used in industry, especially in power plants (Andrews, 2020). Natural gas is used as a raw material for this conversion (Fig. 11.8). Natural gas includes methane (CH_4), which is converted to hydrogen by thermal processes including steam-methane reformation and partial oxidation. Natural gas is a cost-effective hydrogen feed compared to other fossil fuels, partly because it is readily available, simple to handle and has a high hydrogen-to-carbon ratio that minimizes the formation of carbon dioxide by-products (CO_2). It is produced by steam-methane reforming; a mature processing method in which the high-temperature steam is used to extract hydrogen from a methane source. Steam reforming is endothermic, which implies that the process must be heated in order for the reaction to continue. The carbon monoxide and steam are then reacted with a catalyst to

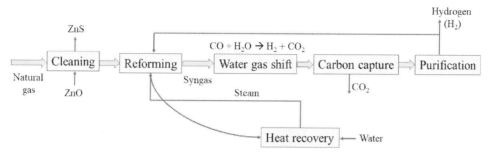

FIGURE 11.8 Hydrogen production steps through natural gas reforming.

create carbon dioxide and more hydrogen in a process known as the "water-gas shift reaction." Carbon dioxide and other impurities are separated from the gas stream in a final process stage named "pressure-swing adsorption," leaving simply pure hydrogen. Some liquids, such as ethanol, propane, or even diesel, are converted into hydrogen through steam reforming.

Steam-methane reforming reaction

$$CH_4 + H_2O \; (+\text{heat}) \rightarrow CO + 3H_2 \tag{11.1}$$

Water-gas shift reaction

$$CO + H_2O \rightarrow CO_2 + H_2 (+ \text{small amount of heat}) \tag{11.2}$$

11.4.1.2 Biomass gasification to hydrogen

Biomass gasification is the incomplete burning of biomass that results in the emission of combustible gases such as hydrogen (H_2), carbon monoxide (CO) and trace amounts of methane (CH_4) as reported by Basu (2018). Biomass includes primarily plant residues, grasses, agricultural residues, wood by-products and municipal solid waste used to produce bioenergy (Zakir Hossain et al., 2014). H_2, CO_2, CO, CH_4, and other hydrocarbons are the ultimate results of biomass gasification (Basu, 2018). Sometimes the fuel mixture through co-gasification approach enhances the quality of syngas. The higher biomass ratio gives lower H_2 content but higher CO, CO_2, CH_4 and hydrocarbons. Whereas it gives lower charcoal, NH_3, and H_2S, although it generates higher tar compounds (Monir, et al., 2020a; Monir et al., 2020f).

Gasification is the thermal process of transforming biomass to a gaseous form of synthetic gas, or syngas, in an oxygen-deficient environment. Syngas is usually composed of carbon monoxide and hydrogen. This process is the intermediate stage between pyrolysis and combustion. The most popular forms of gasifiers used for gasification are fixed bed gasifiers, fluidized bed gasifiers, and entrained flow gasifiers (Fig. 11.7). This process of biomass gasification is achieved by a sequence of complex and well-organized chemical reactions (Eqs. 11.3–11.12). Sometimes, the co-gasification process increased the gasification rate, and the production of hydrogen increased as well. The co-gasification process is also run for the production of hydrogen containing syngas by using lignocellulosic biomass and charcoal (Monir et al., 2018a; Monir et al., 2018). Recently, an experiment was investigated by Anniwaer et al. (2021) for the co-gasification process. In their experiment they used banana peel with rice husk and/or cedarwood. The following chemical reactions are involved during biomass gasification (Monir et al., 2018a).

$$\text{Partial combustion: } C + \frac{1}{2}O_2 = CO \tag{11.3}$$

$$\text{Total combustion: } C + O_2 = CO_2 \tag{11.4}$$

$$\text{Boudouard equilibrium: } C + CO_2 = 2CO \tag{11.5}$$

$$\text{Water-gas reaction: } C + H_2O = CO + H_2 \tag{11.6}$$

$$\text{Methanation reaction: } C + 2H_2 = CH_4 \tag{11.7}$$

$$\text{CO oxidation: } CO + \frac{1}{2}O_2 = CO_2 \tag{11.8}$$

$$H_2 \text{ oxidation: } H_2 + \frac{1}{2}O_2 = H_2O \tag{11.9}$$

$$CH_4 \text{ oxidation: } CH_4 + 2O_2 = CO_2 + 2H_2O \tag{11.10}$$

$$\text{Water-gas shift reaction: } CO + H_2O = CO_2 + H_2 \tag{11.11}$$

$$\text{Methanation: } CO + 3H_2 = CH_4 + H_2O \tag{11.12}$$

11.4.1.3 *Underground coal gasification to hydrogen*

Underground coal gasification (UCG) is a method that converts coal into product gas. UCG is an in-situ gasification procedure that uses oxidants and steam injection to gasify nonmined coal seams. Hydrogen, methane, carbon monoxide, and carbon dioxide are the most common product gases. Coal and peat are mainly consisting of carbon, with varying quantities of other elements such as H_2, S, O, and N. Peat is the first stage in the formation of coal, and it gradually transforms into lignite as pressure and temperature rise as sediment is piled on top of partially decaying organic matter. Peat must be buried 4 to 10 km deep in sand before it can be converted into coal. Coal is produced as dead plant matter dies into peat and is converted into coal through heat and burial pressure over millions of years (Monir and Hossain, 2012). These energy containing materials are also used for the conversion process to generate clean hydrogen gas (Fig. 11.9). Hydrogen is formed by a chemical reaction where coal or peat is reacted with pressurized oxygen and steam at high temperatures to create synthesis gas, which is a combination of carbon monoxide and hydrogen. Gasifiers are used for the conversion of peat or coal to hydrogen gas. If the coal is not feasible to extract through surface or underground mines due to its existence at greater depth; alternative method of underground coal gasification (UCG) is needed to produce hydrogen from these coal deposits. In this process, it requires inserting gasifying medium in an underground coal seam to produce syngas, which can be used to produce electricity, hydrogen, liquid fuels, etc. This is a new technique that uses unmined coal (Gür and Canbaz, 2020). Coal is burnt and gasified underground during the UCG operation. Syngas is generated and extracted from drilled wells to be refined or used directly (Fig. 11.10). Syngas is mostly composed of CO_2, CO, H_2, and CH_4, and its hydrogen content is mostly determined by the gasification agents

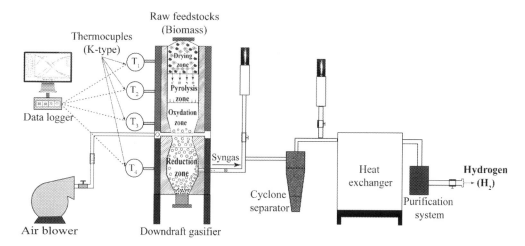

FIGURE 11.9 Hydrogen production through biomass gasification.

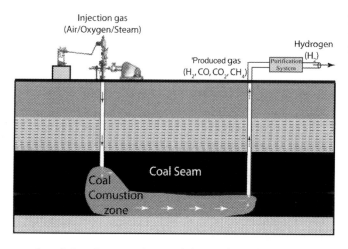

FIGURE 11.10 Hydrogen production through underground coal gasification process.

used and the characteristics of the coal gasified. The combustible components of syngas, such as CO and H_2, are assumed to be metabolites of heterogeneous gasification reactions, with homogeneous combustion reactions of these combustible components absorbing nearly all of the oxygen. There are additional environmental advantages to be gained from coalification, including gasification, carbon sequestration, and carbon capture and storage that are built into the coal production process. Many lab-scale laboratory experiments have been conducted to investigate the impact of various approaches on the processing of hydrogen-rich syngas.

11.4.2 Electrolytic process

Electrolysis is known as the method of decomposing ionic compounds into their constituent elements by passing a direct electric current through the compound in a fluid state. The cathode reduces cations, while the anode oxidizes anions. This is a promising method of

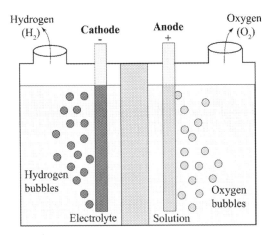

FIGURE 11.11 Hydrogen production through electrolytes.

producing hydrogen from renewable energy (Shiva Kumar and Himabindu, 2019). In this process, water is separated into hydrogen and oxygen using electricity. Chemical reactions (Eqs. 11.13 and 11.14) occur as an electric current pass through a substrate. Electrolytic cell is used for experimental run that consists of positive and negative electrodes separated by a solution comprising positively and negatively charged ions (Fig. 11.11). Water is split into hydrogen and oxygen utilizing energy in electrolysers in this step.

Oxygen evolution reaction (OER):

$$2OH \rightarrow 0.5\,O_2 + H_2 + 2e^- \qquad (11.13)$$

Hydrogen evolution reaction (HER):

$$2H_2O + 2e^- \rightarrow 2OH + H_2g \qquad (11.14)$$

This is the process by which electrical energy is converted into chemicals by trapping electrons as stable chemical bonds. The freshly produced chemical energy are used as fuel or converted back to electricity as desired. An electrolyser is made up of two electrodes known as the cathode and anode. A cathode is a negatively charged electrode, while an anode is a positively charged electrode (Fig. 11.11). Cathodes are separated by electrolytes and covered in water. Because of the use of a particular kind of electrolyte material, different types of electrolysers act in slightly different ways. The electrochemical water splitting is a crucial stage in generating hydrogen from sunlight, whether utilizing a photovoltaic cell to produce an electrostatic potential, then electrolysis, or in a hybrid phase like the Graetzel cell. The resulting hydrogen can be stored as chemical energy through this process.

11.4.3 Photolytic process

Photolysis is a chemical reaction that breaks down molecules into smaller units by absorbing light. Photolysis is less popular method for hydrogen production. In this process, light is used to split water into hydrogen and oxygen (Fig. 11.12). Photolytic and photocatalytic separation techniques are used for the generation of hydrogen from H_2S in aqueous and gas phase media under ultraviolet and/or visible light irradiation (Oladipo et al., 2021).

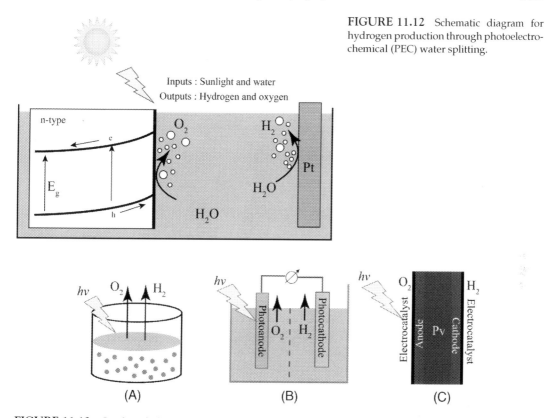

FIGURE 11.12 Schematic diagram for hydrogen production through photoelectrochemical (PEC) water splitting.

FIGURE 11.13 Combined photovoltaic and electrolysis cells used for hydrogen production.

Photocatalytic process is also compared to an electrochemical process in which the electricity supply is substituted with an illumination source. This is a successful method for producing renewable and green hydrogen from sunlight using photocatalysts. The schematic diagram for hydrogen production through photoelectrochemical (PEC) water-splitting is shown in Fig. 11.12.

Light energy can be converted into hydrogen fuel by combining photovoltaic and electrolysis cells (Fig. 11.13). In this type of photoelectrochemical device, photocathodes, and photoanodes could produce 20 kW from electrolysis. In this combined process, photocatalyst is needed to achieve hydrogen production. The development of highly-oriented, affordable, and environmentally sustainable photocatalysts with a relatively long wavelength response is essential for photocatalysis.

11.4.4 Fermentation process

Fermentative hydrogen processing is the process of fermenting organic substrates to produce hydrogen. This kind of hydrogen is often referred to as biohydrogen. Various types of yeast and bacteria are usually used to complete the conversion process. Various forms of

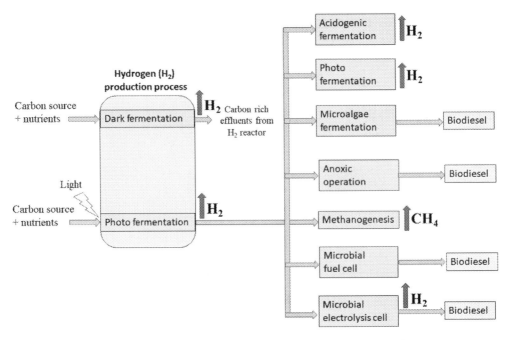

FIGURE 11.14 Hydrogen production processes through photo fermentation and dark fermentation.

fungi and enzymes for hydrocarbon synthesis have been discovered. The current focus of this technology's research is on breeding bacterial origins with anti-inhibition and increased hydrogen production. The production of fermentative hydrogen from glycerol by immobilized microbes has been recently investigated by Chen et al. (2021). From their study, they found that immobilized cells became more tolerant of glycerol and has a greater capacity to produce hydrogen. There are two microbial pathways for the production of hydrogen are photo fermentation and dark fermentation. According to Zhang et al. (2020), the three types of biohydrogen production currently possible from lignocellulosic biomass are photo fermentation (PF), dark fermentation (DF), and integrated dark-photo fermentation (IDPF). The most commonly used bioreactors for fermentation processes are continuous stirred-tank reactor (CSTR), upflow anaerobic sludge blanket reactor (UASB), anaerobic fluidized bed reactor (AFBR), and membrane bioreactor (MBR) (Łukajtis et al., 2018; Monir et al., 2020d; Monir et al., 2020c). Because of the reactors' easy and low-cost nature, fermentation parameters, especially temperature and pH, can be easily regulated (Monir et al., 2020g). Large-scale hydrogen production requires the use of continuous reactors owing to higher expected process efficiencies. To ensure sufficient inoculum preparation and pretreatment, continuous reactors are typically used in batch type. The flowchart of hydrogen production through fermentation is shown in Fig. 11.14.

11.4.4.1 Photo fermentation process

Photo fermentation is the process by which anaerobic bacteria, such as *Rhodobacter, Rhodobium, Rhodopseudomonas,* and *Rhodospirillum,* convert organic acids into hydrogen and carbon dioxide by photosynthesis process. The reduction of molecular nitrogen produces hydrogen in the presence of nitrogenase, which also reduces protons to molecular hydrogen. During these changes, no oxygen is released, which acts as a nitrogenase inhibitor. The yield of hydrogen production is comparable to that of bio-photolysis and is affected by microorganism nature, medium, photo fermenter design, and light intensity (Łukajtis et al., 2018).

11.4.4.2 Dark fermentation process

The fermentative transfer of organic substrate to hydrogen is known as dark fermentation. It is a mechanism through which anaerobic bacteria feed on complex carbohydrates to create hydrogen with a significant number of organic acids as by-products. Dark fermentation produces hydrogen and methane, as well as volatile fatty acids and alcohols. Such compounds can be used by sulfur-free photosynthetic bacteria as a supply of electrons and hydrogen, as well as a carbon source. This process is a complicated process involving a series of biochemical reactions in three stages, similar to anaerobic conversion that is manifested by various bacteria. Dark fermentation is distinct from photo fermentation in that it takes place in the absence of light.

11.5 Current status of hydrogen production

Steam reforming of hydrocarbons, water-electrolysis, noncatalytic partial oxidation, autothermal reforming, gasification, and thermocatalytic processes of fossil fuels including natural gas, naphtha, and other hydrocarbons are all traditional methods for generating hydrogen (Sharma et al., 2020). They introduced a new advance in bio-hydrogen production for the waste-to-energy cycle. Geothermal energy is an effective and promising technology for producing hydrogen. The intensity of hydrogen output, the cost of hydrogen production, energetic performance, exergetic efficiency, exergetic cost, and the amount of electricity generated are all performed by Mahmoud et al. (2021). They also mentioned that the hydrogen production rate with this method ranges from 5.439 kg/h to 13958 kg/h. In terms of unit cost and greenhouse gas pollution, battery-assisted hydrogen production is equivalent to other forms of hydrogen production or fossil fuel (Sako et al., 2021). They reported that the environmental impact of this system on abiotic resource depletion is too high for it to be environmentally compliant with other alternatives. Qureshy and Dincer (2021) implemented a new solar photoelectrochemical design for hydrogen processing. They suggested future research on the impact of solar incident flux, illuminated photoelectrode area, and solar quantity performance on hydrogen efficiency and hydrogen output volume.

As technology advances, electrolysers are rapidly scaling up from megawatt (MW) to gigawatt (GW) capacity. From USD 840 per kilowatt (kW) today, electrolyser prices are expected to halve by 2040 to 2050, although renewable energy rates are expected to continue to decline. For several greenfield applications, renewable hydrogen would soon be the most cost-effective source of sustainable energy (Gielen et al., 2019). The hydrogen production cost and capital cost are shown in Table 11.2.

Producing hydrogen from ethanol steam reforming would not only be environmentally sustainable, but it would also open up new possibilities for renewable, globally accessible

TABLE 11.2 Costs of hydrogen production through various methods (Kayfeci et al., 2019; Zhang et al., 2021).

Energy source	Feedstock	Method	Hydrogen cost ($/kg)	Capital cost (M$)
Fossil fuel	Natural gas	Autothermal reforming of methane	1.48	183.8
Fossil fuel	Natural gas	Steam methane reforming	2.08–2.27	180.7–226.4
Fossil fuel	Coal	Coal gasification	1.34–1.63	435.9–545.6
Internally generated steam	Natural gas	Methane pyrolysis	1.59–1.70	–
Internally generated steam	Woody biomass	Biomass pyrolysis	1.25–2.20	3.1–53.4
Internally generated steam	Woody biomass	Biomass gasification	1.77–2.05	6.4–149.3
Solar	Water + algae	Direct bio-photolysis	2.13	50 $/m^2
Solar	Water + algae	Indirect bio-photolysis	1.42	135 $/m^2
–	Organic biomass	Dark fermentation	2.57	–
Solar	Organic biomass	Photo fermentation	2.83	–
Solar	Water	Solar thermal electrolysis	5.10–10.49	22.1–421.0
Wind	Water	Wind electrolysis	5.89–6.03	499.6–504.8
Nuclear	Water	Nuclear thermolysis	2.17–2.63	39.6–2107.6
Solar	Water	Solar thermolysis	7.89 to 8.40	5.7–16.0

energy. Hydrogen has a wide range of uses in a wide range of industries (Mohammed et al., 2011), including-

- Petroleum and chemical industries, such as fossil fuel processing, petrochemicals
- Hydrogenation agent used to increase the saturation level of unsaturated fats and oils
- Metal fabrication and production
- As a shielding gas in welding processes such as atomic hydrogen welding
- Rotor coolant in power plants' electrical generators
- Float glass manufacturing
- Filling balloons and airships with gas
- Technology for storing energy
- The electronic industry
- Silicon production and processing
- Propulsion fuel for rockets
- Generation of electricity via fuel cells
- Transportation industry

11.6 Challenges of hydrogen conversion

The main challenge in producing hydrogen from oxygenated hydrocarbons through steam reforming is to develop low-cost catalysts with high conversion efficiencies. This is also true for alkaline enhanced reforming, with the added complexity of sequestering the carbon formed by the reforming processes as a sodium carbonate precipitate, which causes concerns with another big challenge to the hydrogen economy: environmental conservation. In southwest China, the hydrogen production from hydropower water abandonment control has an environmental impact (Bamisile et al., 2021). In the future, ASEAN countries should concern about the current and possible economic feasibility and environmental effects of hydrogen energy. The other challenges for hydrogen production are:

- Tackling financial issue
- Large reactor or gasifier construction costs
- Overcoming high initial capital costs
- High threats and complexities
- Inadequate understanding of demand opportunities
- Inadequate government financial assistance
- Technical difficulties
- Inadequate local production of specific machinery
- Lack of uniform technologies

11.7 Hydrogen potentiality in developing countries

Energy has been at the core of many social, political, and economic crises in developing countries. This situation has worsened over time as a result of population expansion and the increasing energy consumption caused by urbanization (Gondal et al., 2018b). Developing countries have a high potential for hydrogen generation because of the abundance of its sources (Khan, 2020). In this regard, the future expected fuel, hydrogen (H_2), has been particularly involved in the energy problem, as well as the negative consequences of carbon-based fuel (Islam et al., 2021a), which have been so destructive to whole countries around the world (Islam et al., 2021b; Monir et al., 2021, Khan, 2018). As a result, there is an increasing demand for low- to zero-emissions fuels, like hydrogen, to be found, evaluated, and implemented to assist developing countries in meeting their climate and development goals. This hydrogen energy may benefit developing countries with a variety of crucial issues associated with energy crises. It provides possibilities for decarbonization in many sectors, including long-distance transportation, chemical properties, and iron and steel, where emissions cannot be substantially reduced. It may also contribute in the improvement of air quality and the conservation of energy resources (Thapa et al., 2021). Although hydrogen fuel technology has been progressed to potential commercial model applications for several years, a complete plan for the development of inexpensive, dependable, low or zero-emission energy for developing countries is needed through deep decarbonization of economic activity (Virji et al., 2020).

A clean hydrogen source has the potential to offer developing countries with strong technologies to help them meet their national sustainable energy and decarbonization objectives.

TABLE 11.3 Potential application of hydrogen and fuel cell technologies in developing countries (ESMAP, 2020).

Energy sources	Methods of green hydrogen production	Hydrogen fuel	Application
Wind energy	Hydrogen electrolysis	Hydrogen direct use	*Transportation* - Light duty vehicles
Solar energy			- Buses
Hydropower			- Trucks
			- Trains
Geothermal energy			- Ships
			Power and heat
			- Grid balancing
			- Cofiring thermal power plant
			- Baseload power
			- Industrial heat and steam
Organic/biowaste	Biogas reforming	Ammonia plant	- Fertilizer for agricultural and export
		Methanol production	- Uninterruptible power supply
			- Telecommunication energy access
			- Natural disaster warming systems
			- Blackstart capabilities

In this regards, green hydrogen (hydrogen generated completely from renewable sources) has the opportunity to enhance national energy security in developing countries. It may also progressively reduce energy prices in nations where diesel is widely utilized. It is also feasible to provide a number of decentralized services capable of meeting all energy needs in buildings, transportation, and industry, as well as assisting in the protection of vital infrastructure from power outages, thus increasing climate and severe weather resilience (ESMAP, 2020). Furthermore, hydrogen and hydrogen-derived fuels are easier to store, transport, and reuse for a variety of energy needs than electrical energy (Table 11.3). This energy has the potential to improve food and energy security while also saving money. Green hydrogen could alleviate some of the main challenges associated with bringing high-quality renewable resources to market. Moreover, electrolyzer cost reductions are required to meet current hydrogen demand, particularly in countries with access to natural gas or cheap coal.

Green hydrogen prices in high-quality renewable resource nations such as Bangladesh, China, India, Egypt, Morocco, Kenya, Nepal, Mexico, Pakistan, South Africa, Turkey, and Somalia may decrease below $2 per kilogram by 2030 if long-term forecasts are fulfilled. For instance, long-term forecasts are fulfilled in China, Bangladesh, Egypt, India, Kenya, Mexico, Morocco, South Africa, and Turkey (ESMAP, 2020). Depending on the price point, the cost of green hydrogen production may be comparable, if not less costly, than the cost of natural gas on-site hydrogen generation. The average hydrogen production costs in some countries have been shown in Fig. 11.15.

There are now near-term opportunities in developing countries to test both green hydrogen for low-emission transportation solutions and remote fuel cells. India, China, Philippines, South Africa, and Indonesia are beginning to acquire experience with ammonia and methanol-based telecommunications fuel cell systems. Smaller stationary cell systems are being tested

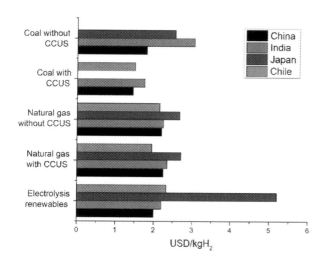

FIGURE 11.15 Hydrogen production costs (average) in some selected developed and developing countries (IEA, 2019).

for residential and tourist customers in Thailand and Namibia. Other significant hydrogen or fuel cell initiatives for stationary energy alternatives are being considered in Mali, Argentina, Uganda, and Martinique. Fuel cell buses have been tested in Malaysia, China and Costa Rica with orders coming in from Indonesia, Bulgaria, and India. South Africa and China have also started to test hydrogen and fuel cell forklift systems (ESMAP, 2020). The aims of these initiatives vary depending on the context, but they all share the notion that, due to hydrogen's higher energy density, it may be more appealing for certain applications than electricity-based storage devices. Green hydrogen represents a significant opportunity for shareholders and policymakers. Island communities, remote cities, countries with existing gas infrastructure, inadequate air quality areas may be attractive locations for near-term green hydrogen and fuel cell projects (ESMAP, 2020). Green hydrogen opportunities may be investigated in these situations.

11.8 Conclusions

Demand for electricity has risen over the past decade as the world's population has grown and the economy has prospered. Clean, low-carbon, and sustainable hydrogen may play a major role. To this end, governments of both developed and developing nations must support hydrogen research and reduce the costs associated with it. Hydrogen may serve various markets, such as power production, transport, and industries. Because of this, politicians should enforce similar policies in these fields. Additionally, hydrogen can be seen as a long-term national electricity transition. National energy policies can be structured such that the manufacturing and commercial industries utilize a minimum percentage of hydrogen in their overall energy production for their own use. Long-distance energy transport is feasible with hydrogen-based fuels. As a result, it may be used as an off-grid power production source for a rural community (particulalry in developing countries), such as on an island, where access to energy is difficult due to geography. As a consequence, researchers will run experiments to see whether these methods are feasible in developing countries. By 2020, the maximum of hydrogen (95%) is generated from fossil fuels through steam reforming of natural gas, partial oxidation of methane, and coal gasification. Alternative renewable fuels, such as waste-to-biogas and biogas-to-hydrogen, should be studied further because producing hydrogen from low-carbon sources is now extremely expensive.

Self-evaluation questions

1. Write a list of the hydrogen energy sources. Why is this energy essential for a country's development?
2. Describe the processes that used to convert hydrogen energy.
3. What are the challenges for the production of hydrogen energy.
4. How will hydrogen help to decarbonize end-uses in different industries?
5. Hydrogen energy could contribute to the sustainable energy sector development in developing countries. Is it possible or not? – Explain.
6. Conduct a literature review and compare the sources, production processes, characteristics, advantages, and disadvantages of grey, blue, and green hydrogen.

References

Ahamed, S., Monir, M.U., Biswas, P.K., Khan, A.A., 2016. Investigation the risk of spontaneous combustion in Barapukuria Coal Mine, Dinajpur, Bangladesh. J. Geosci. Environ. Protect. 4 (04), 74.

Ainas, M., Hasnaoui, S., Bouarab, R., Abdi, N., Drouiche, N., Mameri, N., 2017. Hydrogen production with the cyanobacterium Spirulina platensis. Int. J. Hydrogen Energy 42 (8), 4902–4907.

Alves, H.J., Bley Junior, C., Niklevicz, R.R., Frigo, E.P., Frigo, M.S., Coimbra-Araújo, C.H., 2013. Overview of hydrogen production technologies from biogas and the applications in fuel cells. Int. J. Hydrogen Energy 38 (13), 5215–5225. https://doi.org/10.1016/j.ijhydene.2013.02.057.

Amri, F., 2019. Renewable and non-renewable energy and trade into developed and developing countries. Qual. Quant. 53 (1), 377–387.

Andrews, J.W., 2020. Hydrogen production and carbon sequestration by steam methane reforming and fracking with carbon dioxide. Int. J. Hydrogen Energy 45 (16), 9279–9284.

Anniwaer, A., Chaihad, N., Zhang, M., Wang, C., Yu, T., Kasai, Y., Guan, G., 2021. Hydrogen-rich gas production from steam co-gasification of banana peel with agricultural residues and woody biomass. Waste Manage. (Oxford) 125, 204–214. https://doi.org/10.1016/j.wasman.2021.02.042.

Arat, H.T., Baltacioglu, M.K., Tanç, B., Sürer, M.G., Dincer, I., 2020. A perspective on hydrogen energy research, development and innovation activities in Turkey. Int. J. Energy Res. 44 (2), 588–593. https://doi.org/10.1002/er.5031.

Ayodele, B.V., Abdullah, T.A.R.B.T., Alsaffar, M.A., Mustapa, S.I., Salleh, S.F., 2020. Recent advances in renewable hydrogen production by thermo-catalytic conversion of biomass-derived glycerol: overview of prospects and challenges. Int. J. Hydrogen Energy 45 (36), 18160–18185. https://doi.org/10.1016/j.ijhydene.2019.08.002.

Ayodele, T.R., Alao, M.A., Ogunjuyigbe, A.S.O., Munda, J.L., 2019. Electricity generation prospective of hydrogen derived from biogas using food waste in south-western Nigeria. Biomass Bioenergy 127, 105291. https://doi.org/10.1016/j.biombioe.2019.105291.

Azam, A., Rafiq, M., Shafique, M., Zhang, H., Ateeq, M., Yuan, J., 2021. Analyzing the relationship between economic growth and electricity consumption from renewable and non-renewable sources: fresh evidence from newly industrialized countries. Sustain. Energy Technol. Assess. 44, 100991.

Bamisile, O., Babatunde, A., Adun, H., Yimen, N., Mukhtar, M., Huang, Q., Hu, W., 2021. Electrification and renewable energy nexus in developing countries; an overarching analysis of hydrogen production and electric vehicles integrality in renewable energy penetration. Energy Convers. Manage. 236, 114023. https://doi.org/10.1016/j.enconman.2021.114023.

References

Barber, J., 2018. Hydrogen derived from water as a sustainable solar fuel: learning from biology. Sustain. Energy Fuels 2 (5), 927–935.

Barragán-Trinidad, M., Buitrón, G., 2020. Hydrogen and methane production from microalgal biomass hydrolyzed in a discontinuous reactor inoculated with ruminal microorganisms. Biomass Bioenergy 143, 105825. https://doi.org/10.1016/j.biombioe.2020.105825.

Basu, P. (2018). Biomass Gasification, Pyrolysis and Torrefaction: Practical Design and Theory (3rd ed.): Academic press

Batista, A.P., Moura, P., Marques, P.A., Ortigueira, J., Alves, L., Gouveia, L., 2014. Scenedesmus obliquus as feedstock for biohydrogen production by Enterobacter aerogenes and Clostridium butyricum. Fuel 117, 537–543.

Baykara, S.Z., 2018. Hydrogen: a brief overview on its sources, production and environmental impact. Int. J. Hydrogen Energy 43 (23), 10605–10614.

Birol, F., 2017. Key world energy statistics. International Energy Agency, rue de la Federation. IEA Publications, Paris, France.

Bisht, A.S., Thakur, N.S., 2019. Small scale biomass gasification plants for electricity generation in India: resources, installation, technical aspects, sustainability criteria & policy. Renew. Energy Focus 28, 112–126. https://doi.org/10.1016/j.ref.2018.12.004.

Capuano, L. (2018). International Energy Outlook 2018 (IEO2018). *Available from:*.

Capuano, L. (2020). International Energy Outlook 2020. US Energy Information Administration (EIA). Retrieved from

Carrillo-Reyes, J., Buitrón, G., 2016. Biohydrogen and methane production via a two-step process using an acid pretreated native microalgae consortium. Bioresour. Technol. 221, 324–330.

Chen, Y., Yin, Y., Wang, J., 2021. Comparison of fermentative hydrogen production from glycerol using immobilized and suspended mixed cultures. Int. J. Hydrogen Energy 46 (13), 8986–8994. https://doi.org/10.1016/j.ijhydene.2021.01.003.

Clews, R.J., 2016. Chapter 7 - Petroleum Refining. In: Clews, R.J. (Ed.), Project Finance for the International Petroleum Industry. Academic Press, San Diego, pp. 119–136.

Creutzig, F., Ravindranath, N.H., Berndes, G., Bolwig, S., Bright, R., Cherubini, F., Masera, O., 2015. Bioenergy and climate change mitigation: an assessment. GCB Bioenergy 7 (5), 916–944. https://doi.org/10.1111/gcbb.12205.

Dawood, F., Anda, M., Shafiullah, G., 2020. Hydrogen production for energy: an overview. Int. J. Hydrogen Energy 45 (7), 3847–3869.

Denton, T. (2018). Alternative fuel vehicles: Routledge.

ESMAP. (2020). Green Hydrogen in Developing Countries. Retrieved from

Ferreira, A.F., Ortigueira, J., Alves, L., Gouveia, L., Moura, P., Silva, C.M., 2013. Energy requirement and CO2 emissions of bioH2 production from microalgal biomass. Biomass Bioenergy 49, 249–259.

Franco, C.R.S. (2014). Pressure swing adsorption for hydrogen purification.

Furuoka, F., 2017. Renewable electricity consumption and economic development: New findings from the Baltic countries. Renew. Sustain. Energy Rev. 71, 450–463. https://doi.org/10.1016/j.rser.2016.12.074.

Gielen, D., Taibi, E., & Miranda, R. (2019). Hydrogen: a renewable energy perspective. International Renewable Energy Agency (IRENA).

Gondal, I.A., Masood, S.A., Khan, R., 2018a. Green hydrogen production potential for developing a hydrogen economy in Pakistan. Int. J. Hydrogen Energy 43 (12), 6011–6039. https://doi.org/10.1016/j.ijhydene.2018.01.113.

Gondal, I.A., Masood, S.A., Khan, R., 2018b. Green hydrogen production potential for developing a hydrogen economy in Pakistan. Int. J. Hydrogen Energy 43 (12), 6011–6039. https://doi.org/10.1016/j.ijhydene.2018.01.113.

Gür, M., Canbaz, E.D., 2020. Analysis of syngas production and reaction zones in hydrogen oriented underground coal gasification. Fuel 269, 117331.

Halloran, J.W., 2007. Carbon-neutral economy with fossil fuel-based hydrogen energy and carbon materials. Energy Policy 35 (10), 4839–4846. https://doi.org/10.1016/j.enpol.2007.04.016.

Hayakawa, Y., Kambara, S., Miura, T., 2020. Hydrogen production from ammonia by the plasma membrane reactor. Int. J. Hydrogen Energy 45 (56), 32082–32088.

IEA. (2019). The Future of Hydrogen. Retrieved from

IEA. (2020). World Energy Outlook. Retrieved from

Islam, A., Ahmed, M.T., Mondal, M.A.H., Awual, M.R., Monir, M.U., Islam, K., 2021a. A snapshot of coal-fired power generation in Bangladesh: a demand–supply outlook. Nat. Resour. Forum 45 (2), 157–182. https://doi.org/10.1111/1477-8947.12221.

Islam, A., Hossain, M.B., Mondal, M.A.H., Ahmed, M.T., Hossain, M.A., Monir, M.U., Awual, M.R., 2021b. Energy challenges for a clean environment: Bangladesh's experience. Energy Rep. 7, 3373–3389. https://doi.org/10.1016/j.egyr.2021.05.066.

Janda, K., & Tan, T. (2017). Overview of Sustainable Energy in Central Europe and East Asia.

Kabir, Z., Khan, I., 2020. Environmental impact assessment of waste to energy projects in developing countries: General guidelines in the context of Bangladesh. Sustain. Energy Technol. Assess. 37, 100619.

Kanchanasuta, S., Prommeenate, P., Boonapatcharone, N., Pisutpaisal, N., 2017. Stability of Clostridium butyricum in biohydrogen production from non-sterile food waste. Int. J. Hydrogen Energy 42 (5), 3454–3465. https://doi.org/10.1016/j.ijhydene.2016.09.111.

Kayfeci, M., Keçebaş, A., Bayat, M., 2019. Chapter 3 - Hydrogen production. In: Calise, F., D'Accadia, M.D., Santarelli, M., Lanzini, A., Ferrero, D. (Eds.), Solar Hydrogen Production. Academic Press, pp. 45–83.

Khan, I., 2018. Importance of GHG emissions assessment in the electricity grid expansion towards a low-carbon future: a time-varying carbon intensity approach. J. Clean. Prod. 196, 1587–1599. https://doi.org/10.1016/j.jclepro.2018.06.162.

Khan, I., 2020. Waste to biogas through anaerobic digestion: Hydrogen production potential in the developing world - a case of Bangladesh. Int. J. Hydrogen Energy 45 (32), 15951–15962. https://doi.org/10.1016/j.ijhydene.2020.04.038.

Kober, T., Schiffer, H.W., Densing, M., Panos, E., 2020. Global energy perspectives to 2060 – WEC's World Energy Scenarios 2019. Energy Strategy Rev. 31, 100523. https://doi.org/10.1016/j.esr.2020.100523.

Kovač, A., Paranos, M., Marciuš, D., 2021. Hydrogen in energy transition: a review. Int. J. Hydrogen Energy 46 (16), 10016–10035. https://doi.org/10.1016/j.ijhydene.2020.11.256.

Kundu, K., Chatterjee, A., Bhattacharyya, T., Roy, M., Kaur, A., 2018. Thermochemical conversion of biomass to bioenergy: a review. In: Singh, A.P., Agarwal, R.A., Agarwal, A.K., Dhar, A., Shukla, M.K. (Eds.), Prospects of Alternative Transportation Fuels. Springer Singapore, Singapore, pp. 235–268.

Liemberger, W., Halmschlager, D., Miltner, M., Harasek, M., 2019. Efficient extraction of hydrogen transported as co-stream in the natural gas grid – the importance of process design. Appl. Energy 233-234, 747–763. https://doi.org/10.1016/j.apenergy.2018.10.047.

Liu, C.-H., Chang, C.-Y., Cheng, C.-L., Lee, D.-J., Chang, J.-S., 2012. Fermentative hydrogen production by Clostridium butyricum CGS5 using carbohydrate-rich microalgal biomass as feedstock. Int. J. Hydrogen Energy 37 (20), 15458–15464.

Liu, S., Guo, L., Jin, H., Li, L., Li, G., Yu, L., 2020. Hydrogen production by supercritical water gasification of coal: a reaction kinetic model including nitrogen and sulfur elements. Int. J. Hydrogen Energy 45 (56), 31732–31744. https://doi.org/10.1016/j.ijhydene.2020.08.166.

Łukajtis, R., Hołowacz, I., Kucharska, K., Glinka, M., Rybarczyk, P., Przyjazny, A., Kamiński, M., 2018. Hydrogen production from biomass using dark fermentation. Renew. Sustain. Energy Rev. 91, 665–694. https://doi.org/10.1016/j.rser.2018.04.043.

Mahmoud, M., Ramadan, M., Naher, S., Pullen, K., Ali Abdelkareem, M., Olabi, A.-G., 2021. A review of geothermal energy-driven hydrogen production systems. Therm. Sci. Eng. Prog. 22, 100854. https://doi.org/10.1016/j.tsep.2021.100854.

Mansur, F.Z., Faizal, C.K.M., Monir, M.U., Samad, N.A.F.A., Atnaw, S.M., Sulaiman, S.A., 2020a. Co-gasification between coal/sawdust and coal/wood pellet: A parametric study using response surface methodology. Int. J. Hydrogen Energy 45 (32), 15963–15976. https://doi.org/10.1016/j.ijhydene.2020.04.029.

Mansur, F.Z., Faizal, C.K.M., Monir, M.U., Samad, N.A.F.A., Atnaw, S.M., Sulaiman, S.A., 2020b. Performance studies on co-gasification between coal/sawdust and coal/wood pellet using RSM. IOP Conf. Ser. Mater. Sci. Eng. 736, 022085. 10.1088/1757-899x/736/2/022085.

Mazloomi, K., Gomes, C., 2012. Hydrogen as an energy carrier: prospects and challenges. Renew. Sustain. Energy Rev. 16 (5), 3024–3033.

Misiukiewicz, A., Gao, M., Filipiak, W., Cieslak, A., Patra, A.K., Szumacher-Strabel, M., 2021. Review: methanogens and methane production in the digestive systems of nonruminant farm animals. Animal 15 (1), 100060. https://doi.org/10.1016/j.animal.2020.100060.

Mohammed, M.A.A., Salmiaton, A., Wan Azlina, W.A.K.G., Mohammad Amran, M.S., Fakhru'l-Razi, A., Taufiq-Yap, Y.H., 2011. Hydrogen rich gas from oil palm biomass as a potential source of renewable energy in Malaysia. Renew. Sustain. Energy Rev. 15 (2), 1258–1270. https://doi.org/10.1016/j.rser.2010.10.003.

Monir, M.M.U., Hossain, H.Z., 2012. Coal mine accidents in Bangladesh: its causes and remedial measures. Int. J. Econ. Environ. Geol. 3, 33–40.

Monir, M.U., Aziz, A.A., Kristanti, R.A., Yousuf, A., 2020a. Syngas production from co-gasification of forest residue and charcoal in a pilot scale downdraft reactor. Waste and Biomass Valorization 11 (2), 635–651. https://doi.org/10.1007/s12649-018-0513-5.

Monir, M.U., Aziz, A.A., Dai-Viet, N.V., Khatun, F., 2020b. Enhanced hydrogen generation from empty fruit bunches by charcoal addition into a downdraft gasifier. Chem. Eng. Technol. 43 (4), 762–769. https://doi.org/10.1002/ceat.201900547.

Monir, M.U., Aziz, A.A., Khatun, F., Yousuf, A., 2020c. Bioethanol production through syngas fermentation in a tar free bioreactor using Clostridium butyricum. Renew. Energy 157, 1116–1123. https://doi.org/10.1016/j.renene.2020.05.099.

Monir, M.U., Aziz, A.A., Kristanti, R.A., Yousuf, A., 2018a. Co-gasification of empty fruit bunch in a downdraft reactor: a pilot scale approach. Bioresour. Technol. Rep. 1, 39–49. https://doi.org/10.1016/j.biteb.2018.02.001.

Monir, M.U., Aziz, A.A., Yousuf, A., Alam, M.Z., 2020d. Hydrogen-rich syngas fermentation for bioethanol production using Sacharomyces cerevisiea. Int. J. Hydrogen Energy 1–9. https://doi.org/10.1016/j.ijhydene.2019.07.246.

Monir, M.U., Azrina, A.A., Kristanti, R.A., Yousuf, A., 2018b. Gasification of lignocellulosic biomass to produce syngas in a 50 kW downdraft reactor. Biomass Bioenergy 119, 335–345. https://doi.org/10.1016/j.biombioe.2018.10.006.

Monir, M.U., Hasan, M.Y., Ahmed, M.T., Aziz, A., A., H., M., A., Woobaidullah, A., Biorefinery, 2021. Optimization of fuel properties in two different peat reserve areas using surface response methodology and square regression analysis. Biomass Convers. Bioref. 1–21. https://doi.org/10.1007/s13399-021-01656-x.

Monir, M.U., Khatun, F., Aziz, A.A., Vo, D.-V.N., 2020e. Thermal treatment of tar generated during co-gasification of coconut shell and charcoal. J. Cleaner Prod. 256, 1–9. https://doi.org/10.1016/j.jclepro.2020.120305.

Monir, M.U., Khatun, F., Ramzilah, U.R., Aziz, A.A., 2020f. Thermal effect on co-product tar produced with syngas through co-gasification of coconut shell and charcoal. IOP Conf. Ser. Mater. Sci. Eng. 736, 022007. 10.1088/1757-899x/736/2/022007.

Monir, M.U., Yousuf, A., Aziz, A.A, 2020g. Syngas fermentation to bioethanol. In: Yousuf, A., Pirozzi, D., Sannino, F. (Eds.), Lignocellulosic Biomass to Liquid Biofuels. Academic Press, pp. 195–216.

Monir, M.U., Yousuf, A., Aziz, A.A., Atnaw, S.M., 2017. Enhancing co-gasification of coconut shell by reusing char. Indian J. Sci. Technol. 10 (6), 1–4. http://doi.org/10.17485/ijst/2017/v10i6/111217.

Muradov, N., 1993. How to produce hydrogen from fossil fuels without CO_2 emission. Int. J. Hydrogen Energy 18 (3), 211–215.

Nathaniel, S.P., Bekun, F.V., 2021. Electricity consumption, urbanization, and economic growth in Nigeria: new insights from combined cointegration amidst structural breaks. J. Public Affairs 21 (1), e2102.

Nazir, M.S., Mahdi, A.J., Bilal, M., Sohail, H.M., Ali, N., Iqbal, H.M., 2019. Environmental impact and pollution-related challenges of renewable wind energy paradigm–a review. Sci. Total Environ. 683, 436–444.

Oladipo, H., Yusuf, A., Al Jitan, S., Palmisano, G., 2021. Overview and challenges of the photolytic and photocatalytic splitting of H2S. Catal. Today. https://doi.org/10.1016/j.cattod.2021.03.021.

Pereira, C., Coelho, P., Fernandes, J., Gomes, M., 2017. Study of an energy mix for the production of hydrogen. Int. J. Hydrogen Energy 42 (2), 1375–1382.

Qureshy, A.M.M.I., Dincer, I., 2021. Development of a new solar photoelectrochemical reactor design for more efficient hydrogen production. Energy Convers. Manage. 228, 113714. https://doi.org/10.1016/j.enconman.2020.113714.

Qyyum, M.A., Dickson, R., Shah, S.F.A., Niaz, H., Khan, A., Liu, J.J., Lee, M., 2021. Availability, versatility, and viability of feedstocks for hydrogen production: product space perspective. Renew. Sustain. Energy Rev., 110843.

Relva, S.G., da Silva, V.O., Gimenes, A.L.V., Udaeta, M.E.M., Ashworth, P., Peyerl, D., 2021. Enhancing developing countries' transition to a low-carbon electricity sector. Energy 220, 119659.

Ren, X., Dong, L., Xu, D., Hu, B., 2020. Challenges towards hydrogen economy in China. Int. J. Hydrogen Energy 45 (59), 34326–34345. https://doi.org/10.1016/j.ijhydene.2020.01.163.

Sako, N., Koyama, M., Okubo, T., Kikuchi, Y., 2021. Techno-economic and life cycle analyses of battery-assisted hydrogen production systems from photovoltaic power. J. Cleaner Prod. 298, 126809. https://doi.org/10.1016/j.jclepro.2021.126809.

Santhanam, K.S., Press, R.J., Miri, M.J., Bailey, A.V., & Takacs, G.A. (2017). Introduction to Hydrogen Technology: John Wiley & Sons.

Sharma, A., Arya, S.K., 2017. Hydrogen from algal biomass: a review of production process. Biotechnol. Rep. 15, 63–69.

Sharma, S., Basu, S., Shetti, N.P., Aminabhavi, T.M., 2020. Waste-to-energy nexus for circular economy and environmental protection: recent trends in hydrogen energy. Sci. Total Environ. 713, 136633. https://doi.org/10.1016/j.scitotenv.2020.136633.

Shiva Kumar, S., Himabindu, V., 2019. Hydrogen production by PEM water electrolysis – a review. Mater. Sci. Energy Technol. 2 (3), 442–454. https://doi.org/10.1016/j.mset.2019.03.002.

Singh, Y.D., Mahanta, P., Bora, U., 2017. Comprehensive characterization of lignocellulosic biomass through proximate, ultimate and compositional analysis for bioenergy production. Renew. Energy 103, 490–500. https://doi.org/10.1016/j.renene.2016.11.039.

Thapa, B.S., Neupane, B., Yang, H.-s., Lee, Y.-H., 2021. Green hydrogen potentials from surplus hydro energy in Nepal. Int. J. Hydrogen Energy 46 (43), 22256–22267. https://doi.org/10.1016/j.ijhydene.2021.04.096Get.

Tsai, C.-H., Chen, K.-T., 2009. Production of hydrogen and nano carbon powders from direct plasmalysis of methane. Int. J. Hydrogen Energy 34 (2), 833–838.

Vijayaraghavan, K., Karthik, R., Nalini, S.K., 2010. Hydrogen generation from algae: a review. J. Plant Sci. 5 (1), 1–19.

Virji, M., Randolf, G., Ewan, M., Rocheleau, R.J.I.J.o.H.E., 2020. Analyses of hydrogen energy system as a grid management tool for the Hawaiian Isles. Int. J. Hydrogen Energy 45 (15), 8052–8066. https://doi.org/10.1016/j.renene.2020.04.090.

Zakir Hossain, H.M., Hasna Hossain, Q., Uddin Monir, M.M., Ahmed, M.T., 2014. Municipal solid waste (MSW) as a source of renewable energy in Bangladesh: revisited. Renew. Sustain. Energy Rev. 39, 35–41. http://dx.doi.org/10.1016/j.rser.2014.07.007.

Zhang, B., Zhang, S.-X., Yao, R., Wu, Y.-H., Qiu, J.-S., 2021. Progress and prospects of hydrogen production: opportunities and challenges. J. Electron. Sci. Technol., 100080. https://doi.org/10.1016/j.jnlest.2021.100080.

Zhang, T., Jiang, D., Zhang, H., Jing, Y., Tahir, N., Zhang, Y., Zhang, Q., 2020. Comparative study on bio-hydrogen production from corn stover: Photo-fermentation, dark-fermentation and dark-photo co-fermentation. Int. J. Hydrogen Energy 45 (6), 3807–3814. https://doi.org/10.1016/j.ijhydene.2019.04.170.

CHAPTER 12

The role of demand-side management in sustainable energy sector development

Samuel Gyamfi[a,b], Felix Amankwah Diawuo[a,b], Emmanuel Yeboah Asuamah[a,b] and Emmanuel Effah[a,b]

[a]Regional Centre for Energy and Environmental Sustainability (RCEES), University of Energy and Natural Resources (UENR), Sunyani, Ghana [b]Department of Energy and Petroleum Engineering, School of Engineering, University of Energy and Natural Resources (UENR), Sunyani, Ghana

12.1 Introduction

Energy demand has seen a tremendous increase globally. Consequently, it increases greenhouse gas (GHG) emission levels. This is a major concern to world leaders. A report by British Petroleum (BP) showed a rise of 1.3% in primary energy consumption in 2019 (BP, 2020). Though the year 2020 witnessed a drop in energy demand by 5%, energy-related CO_2 emissions by 7%, and energy investment by 18%, these numbers were mainly due to the coronavirus pandemic (IEA, 2020). There has been a decrease in the various indicators with a sharp contrast to a slight rise in the contribution of renewables (IEA, 2020). However, this is not an indication that the global energy demand will continue to decline as the report is uncertain with recovery from coronavirus pandemic.

The known approach to solve the issue of energy demand growth has been to rely on increasing the supply of conventional energy resources, mostly fossil fuels, which accounted for more than 81% of global energy production in 2018, similar to that of 2017 (International Energy Agency (IEA), 2020). The worry is that energy is not an economic resource that must be exploited at all costs. It rather serves as an input to produce goods and services, such as cooling, heating, and lighting. Consequently, any reduction in the input required to provide these goods and services would increase benefits such as the monetary value (International Energy Agency (IEA), 2020).

There has been an urgent need to substitute and reduce the reliance on fossil fuels. This has subsequently called for the prioritization of renewable energy integration into the public electricity grid (Petinrin and Shaaban, 2015). This has come up as a result of challenges related to energy security and environmental concerns. In 2016, the Paris Agreement was compiled by the United Nations Framework Convention on Climate Change (UNFCCC). The agreement was made to mitigate climate change through many different ways such as reduction of greenhouse gas (GHG) emissions. The strategies to reduce emissions in the energy sector include the use of renewable energy and demand-side management (DSM) which includes energy efficiency, boosted by substantial electrification. Sustainable development has received a lot of attention in recent times with emphasis on economic growth, environmental protection, and social inclusion (Hertog and Luciani, 2009).

Demand-side management has proven to be very useful in the management of the electricity network and has a direct link with sustainable development. This can be seen in how it can fully control loads causing the utilization factor to increase with a cost reduction. Again when the demand is high during critical hours, DSM can be executed for optimal grid operation (Hosseini Imani et al., 2018). The United Kingdom for instance has recognized DSM as a strategy to sustainably develop its energy sector as a way to respond to its commitment to reduce carbon emissions and to improve system efficiency (Strbac, 2008). According to the United Nations' 2030 Sustainable Development Goal, developing a sustainable modern electricity sector is a key place emphasis on the role of DSM opportunities for utilities can play in achieving the set goals (Corbett et al., 2020). In recent times business models are required to include sustainability, thus combining economic viability with environmental and social benefits for various stakeholders operating in the energy sector and manufacturing industry. In this manner, a model that can handle volatile energy availability economically is essential and thus a business model innovation through DSM (Khripko et al., 2017). Research has shown that DSM is capable of improving electricity supply status in developing countries significantly (Diawuo et al., 2021). This is possible because of strategies such as designed electricity tariffs, incentives, penalties, government policies as well as power-saving mechanisms (Torriti, 2012; Ma et al., 2014; Dam et al., 2008; Albadi MH, 2008). The next section describes the concept of demand-side management, followed by the techniques that are in DSM. The following sections also look at the contribution of DSM to sustainable electricity supply followed by some selected case studies.

12.2 Concept of demand-side management

The definitions of demand-side management vary depending on what is included or excluded in the context. Some authors define DSM as management of electricity demand and exclude other forms of energy (Prüggler et al., 2011). Others (Gołębiowska et al., 2021; Khan et al., 2021) define DSM as the mechanism that minimizes the consumption or demand of energy at peak times while some authors also use a similar definition but includes the demand response strategies for consumers to respond to price changes by shifting their loads to off-peak times without compromising their comfort (Warren, 2014). Moreover, some publications include the fundamentals of energy efficiency and energy conservation measures (Sioshansi and Vojdani, 2001; Palensky and Dietrich, 2011).

The DSM is a well-established concept that champions the systematic or gradual growth of electricity as well as the inclusion of alternative (e.g., renewable energy) electricity sources (Diawuo et al., 2019). The inclusion of renewable energy to the grid with a smart grid mechanism requires some level of flexibility to match the consumer demand with supply availability (Davito et al., 2010; Maruf et al., 2020; Majeed Butt et al., 2020). The concept of DSM can fundamentally be categorized into two main areas as shown in Table 12.1: (i) Energy Efficiency (EE), and (ii) Demand Response (DR). Energy efficiency consists of using less energy to complete the same tasks which lead to a reduction in the total energy consumption. This leads to a reduction in CO_2 emissions and the cost of utility bills (Warren, 2014; Diawuo et al., 2018; Davito et al., 2010). On the other hand, the concept of demand response also known as demand side response (DSR), include any strategy that minimizes high-energy demand and gives a clear indication of the future energy market in terms of generation, transmission, and distribution as well as the integration of renewable energy to the grid system (Palensky and Dietrich, 2011; Albadi and El-Saadany, 2007).

Generally, the DSR operates at the end-user's level targeted at influencing the consumption pattern of the end-users. This is done by allowing consumers a greater role in shifting their demand for electricity during the period of high energy demand (peak period). Although demand response may often be complementary to energy efficiency, it is mostly about shifting demand rather than reducing it. It is about changing the shape of a demand curve (Wang et al., 2013; Castillo-Cagigal et al., 2011) rather than reducing the area under the curve. An energy efficiency measure reduces the area under the curve, demand response moves the time of demand rather than directly reducing the load and is intended to reduce peak demand or flatten a load (Castillo-Cagigal et al., 2011; Ki et al., 2011; Zong Y et al., 2012).

Another dimension of demand-side management category is energy saving behavior or consumers' energy-use behavior change. This strategy requires little or no cost to implement and is divided into two, namely: investment behavior and curtailment behavior. Investment behavior involves the use of monetary investment to improve energy-saving behavior whiles curtailment behavior strategy requires little or no monetary investment toward energy-saving (Khan, 2019).

12.3 Basic techniques in DSM

DSM measures are designed to influence, and if necessary, change customer behavior to achieve benefits for both the customer and the electricity industry. Utilities take different actions to achieve a particular load shape objective. Fig. 12.1 shows the six typical load shape objectives for employing DSM. For example, utilities want to flatten out their load profiles and get better utilization of their existing equipment, reduce the installation of new equipment, and thereby lower cost to their consumer. Techniques are employed that could reduce peak load through load clipping, load shifting, and energy conservation. Peak clipping involves the reduction of peak load by using direct load control. Conservation considers decreasing the overall load through a reduction in consumption as well as a change in usage pattern. Appliance efficiency improvement is a typical example. Load building also involves increasing the market share of loads that are or can be served by competing fuel, as well as economic development in the service area. Valley filling considers building loads during the off-peak

TABLE 12.1 Fundamental concept of demand response.

Category of DSM	Category of DR/EE	Category of DR/EE	Pricing and load reduction mechanism/	Description
Demand-side management (DSM)	Demand response (DR)	Price based demand response schemes (PBDRS)	Critical peak pricing (CPP)	CPP rates have higher charges for electricity per kWh used during the critical peak periods designated as critical by the utility providers. CPP is a simplified hybrid of the time of use and real-time pricing. Two or three averaged price points are calculated to reflect different market conditions and the consumer is informed in advance of peak, intermediate and off-peak periods for specific critical hours or day (Association of Edison Illuminating Companies, 2013).
			Time of use (ToU)	Time of use is a utility rate structure in which the per kWh of electricity is higher during the peak demand hours and lower during off-peak hours which varies according to the time of the day, seasons, and day type i.e., either weekday or weekend/holiday and hence can reduce the overall cost for both the utility and the consumers (Zhou et al., Dec. 2019).
			Real-time pricing (RTP)	Real-time pricing is a component of the demand response strategy that delivers efficient and effective utilization of power to adjust the power balance between supply and demand. The cost of power (per-kWh) in this utility rate structure varies hourly based on the real-time electricity production cost at the end of the generation side. Therefore, retail electricity rates are higher during peak times than during shoulder and off-peak times (Wang et al., 2021; Luo and Fong, 2019).

(continued on next page)

Incentive-based demand response scheme (IBDRS)	Direct load control (DLC)	Utilities are allowed to control the operation of some equipment such as air-conditioning systems of customers during specific hours of the day and season. This is usually achieved with the provision of some incentive to consumers in order to minimize energy demand to stabilize the grid. However, DLC is voluntary and customers are not penalized for not curtailing their loads (Krarti, 2018).
	Interruptible tariff	Interruptible tariffs are electricity price structure that is normally offered to residential and commercial customers based on a contractual agreement between the utility and the customer on a unit cost of electricity. In this case, consumers receive a discount for agreeing to reduce their consumption or shifting their loads. This normally helps stabilize the grid or handle an emergency especially when the demand is projected to be higher. However, the energy consumed by the customer does not decrease but rather shifted to an off-peak period (Kostková et al., 2013).
	Emergency demand response programs (EDRP)	EDRP is a voluntary emergency program in which customers receive an incentive for responding to reduce their energy consumption upon short notice from the utility providers when there is a shortfall in supply reserves. There is no penalty if customers do not respond to curtail their loads (Daniar and Talaeizadeh, 2016).

(continued on next page)

TABLE 12.1 Fundamental concept of demand response—cont'd

Category of DSM	Category of DR/EE	Pricing and load reduction mechanism/	Description
		Ancillary service market programs (ASMP)	Ancillary services market programs allow customers to bid on load curtailment in the spot market. This is to maintain energy balance, frequency and voltage regulation, voltage support, and constraint management. In this program, the participating consumer is paid the market price for commitment (Pollitt and Anaya, 2020).
		Capacity market programs (CMP)	CMP is seen as a form of insurance where a participant receives a guaranteed payment for customers who can commit to reducing their load when there are contingencies in the electric grid. Customers are not penalized for not curtailing their load and they normally receive signals a day-head (Aalami et al., 2010).
		Demand-bidding programs (DBP)	In the case of demand-bidding program, consumers are allowed to bid on a specific load reduction in the electricity wholesale market. In this arrangement, the bid is accepted if it is less than the market price. However, customers must curtail their loads by the amount specified in the terms otherwise they become liable for penalties (Office of Electricity of USA, 2006).

(continued on next page)

	Energy saving behaviors	Investment behavior	This energy-related behavior involves the use of monetary investment to improve energy-saving behavior. In this scheme, users are motivated or inspired to reduce their electricity consumption through means such as replacing incandescent bulbs with compact fluorescent lamp (Khan, 2019).
		Curtailment behavior	This strategy requires little or no monetary investment toward energy-saving. Here, the user practice and learn possible ways to save energy. They include strategies such as putting off electric loads which are not in use (Khan, 2019).
Energy efficiency and conservation and load reduction	Energy conservation (EC)		EC refers to the overall reduction of energy consumption or demand for electricity by adjusting behavior. It involves a certain degree of sacrifice, such as using clothes dryer less often or turning down the heat in winter, turning off appliances when they are not in use, etc. (Menegaki and Tsani, 2018; Herring, 2006).
	Energy efficiency (EE)		EE reduces the overall demand for electricity while maintaining the same amount or quality of service output with less energy. For example, instead of lowering the temperature of a conventional furnace, you can install an energy-efficient furnace to keep your house at a certain temperature while consuming less energy than you would with a conventional one (Menegaki and Tsani, 2018; Herring, 2006).

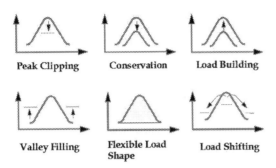

FIGURE 12.1 Basic load shape objectives of demand-side management (Office of Electricity of USA, 2006). Time of day is on the horizontal axis and electricity demand is on the vertical axis.

period whiles flexible load shape looks at specific contracts and tariffs with possibilities to flexibly control consumers' equipment. Load shifting combines the benefits of peak clipping and valley filling by moving existing loads from peak to off-peak hours. The other three load shape objectives are load building, flexible load, and valley filling. The strategies that could result in the load shape objectives in Fig. 12.1 include:

- Actions are taken on the customer side of the electricity meter such as energy efficiency measures
- Arrangements for reducing loads on request, such as interruptible contracts, direct load control and demand response
- Fuel switching, such as changing from electricity to gas for water heating
- Distributed generation, such as standby generators in homes or photovoltaic modules on rooftops
- The different pricing initiatives, such as time of use, real-time pricing, etc.

In this chapter, the basic techniques of demand-side management have been classified into three groups based on the interaction within the electricity network right from the generators to the final consumers. They are namely, flexible load techniques, flexible storage techniques, and demand-side generation techniques.

12.3.1 Flexible load techniques

These techniques are commonly used or applicable in appliances such as air conditioners, water heaters, electric vehicles, and washing machines whose load can be shifted or reduced in demand response events without affecting the comfort of the user.

12.3.2 Flexible storage techniques

Flexible energy-storage devices are increasingly attracting attention as they show unique promising advantages, such as flexibility, shape diversity (Wang et al., 2014; Pushparaj et al., 2007) and it is one of the important technologies that enable the large-scale deployment of renewable energy (Zsiborács et al., 2018). However, flexible energy storage mechanisms with the capacity to charge and discharge energy have been applied to demand-side management in buildings (Groppi et al., 2021; Ren et al., 2021). The most frequently used flexible energy storage technologies in residential facilities for demand-side management are the Thermal Energy Storage System (TESS) and Battery Energy Storage System (BESS).

TESS is divided into three types: sensible heat, latent heat, and thermochemical (Cabeza et al., 2015; Sarbu and Sebarchievici, 2018; Stadler and Sterner, 2018). They can store energy by cooling or heating such that the stored energy can be used later when there is a high demand for cooling, heating, and power generation (Sarbu and Sebarchievici, 2018). Consequently, TESS helps to achieve one or more of the following:

- Increase generation capacity: excess power or generation during off-peak periods or low demand can be used to charge TESS and inject back to the network during high demand periods (Dincer and Rosen, 2013).
- Shift energy consumption to low-cost periods: allows customers to shift their load to a period of low energy demand or low cost of kWh of energy (Dincer and Rosen, 2013).

Similarly, BESS for a residential facility including electric vehicles (EVs) such as plug-in hybrid electric vehicles (PHEVs) and battery electric vehicles (BEVs) can be used to stabilize the grid through a demand-side management program (Vandael et al., 2010). The technology of EVs is such that, the electric power grid can charge the vehicle's battery (Grid-to-Vehicle), or discharge it to a building during high energy demand (Vehicle-to-Grid) (Pang et al., 2012).

12.3.3 Demand-side generation techniques

Demand-side generation techniques play an important role in demand-side management to enhance grid stability and reliability. This technique facilitates the penetration and optimization of renewable energy resources (RERs) at the consumer end and also delivers a timely forecast of intermittent RERs. In this case, the consumers or users own some kind of energy storage and generate their electricity from RERs such as solar and wind into their electricity mix to minimize energy demand from the utility providers (Atzeni et al., 2013).

12.3.3.1 Optimization techniques

Optimization techniques consist of static and dynamic techniques for optimization, such as linear programming, mixed-integer, dynamic programming, etc., for the development and enhancement of good coordination and negotiation in the network (Momoh, 2013). However, Herring (2006) classified optimization techniques in demand-side management based on three main characteristics as 1) optimization architectures; 2) optimization objectives; and 3) optimization algorithms as shown in Fig. 12.2.

12.3.3.2 Game theory techniques

Game theory is a new concept in demand-side management that has contributed massively to making a decision in the energy management arena for efficient coordination and negotiation among players/consumers (Esther and Kumar, 2016; Myerson, 1997; Zhou et al., 2017). The game theory techniques in the demand-side management for the residential microgrid are categorized into two branches, namely cooperative game and noncooperative game (Zhou et al., 2017; Saad et al., 2012; Tushar et al., Jun. 2019) as shown in Fig. 12.3. To exemplify, users are allowed to present how much they would want to buy electricity, then a decision is made on the acceptance price based on the desire to be served. After this, the acquisitive

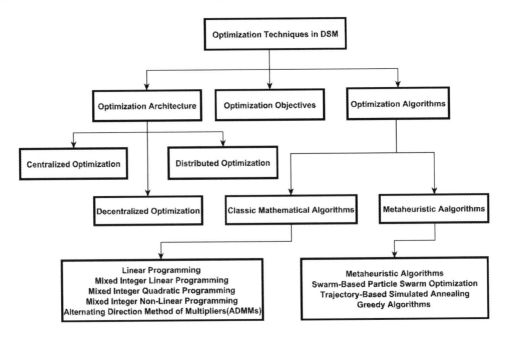

FIGURE 12.2 Optimization techniques in the demand-side management (compiled by authors).

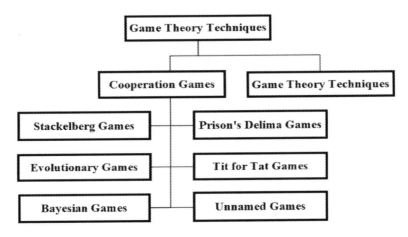

FIGURE 12.3 Classification of game theory techniques in DSM (Zhou et al., 2017).

interaction between operator and users is modeled using for example the Stackelberg game. In such an interaction, the operator aims to maximize profit, while users wish to pay the least price and still be served adequately. After each user states their price, the operator puts out an optimal power price based on the discussions with the user taking into account the source of the electricity (Diamantoulakis et al., 2016).

12.4 DSM contribution to sustainability

The concept of sustainability as well as how it can be realized spans from several viewpoints. In the context of this study, sustainability is defined as the practices and actions through which humankind avoids the depletion of natural resources or a measure to preserve an ecological balance that does not decrease the quality of life of modern societies (Dam et al., 2008; Albadi MH, 2008). The concept of sustainability in the aspect of energy has to do with reducing energy demand, reducing the depletion of fossil fuels, including renewable energies in the energy mix, etc. However, the transition from fossil-based energy sources is accompanied by challenges ranging from volatile availability. Studies have proved that this issue can be lessened through enhanced flexibility of the demand; therefore, DSM becomes critical (Khripko et al., 2017).

Demand-side management is essential for attaining sustainability goals since the objective of implementing DSM is to avert resource production waste, most notably electricity, and increase more efficient use of valuable resources. DSM concept cannot be separated from policies that promote environmental responsiveness and cleaner energy production (Sahin et al., 2019). Demand-side management has recorded significant results over the past years in electricity markets in the area of improving energy efficiency and achieving environmental targets through controlled and reduced consumption (Bergaentzlé et al., 2014).

The energy crisis that occurred in California between the years 2000 and 2001 is an example of the success that DSM can contribute in terms of sustainable development even in advanced and sophisticated economies. The State spent $US 20 billion on establishing new power plants with accompanying emissions and resource depletion only to increase the electricity supply by 2%. In the same instance, $US 1 billion spent on energy efficiency promotion and retrofits resulted in a 10% reduction in electricity demand thereby reducing the use of fossil fuels (Warwick University of and REEEP, 2005).

The discussion so far on DSM has been centered on electrical DSM programs meanwhile in Africa most people utilize other sources of energy; most notably biomass in the form of wood for cooking and heating space. This is common in developing countries and some developed countries as well. In this situation when improved cookstoves are used, it reduces the environmental effects significantly through the reduction in biomass usage (Lugano Wilson, 2006).

Demand-side management in the context of Sustainable Development Goals (SDG) is an alternative to building additional power plants to supply the need of the growing energy demand in any sector due to environmental and economic constraints (Monyei and Adewumi, 2017). Demand-side management has contributed and simulated the transition to a sustainable energy system through the adoption of energy efficiency mechanisms in buildings and other related energy-consuming spaces.

Energy Efficiency, which represents one of the key indicators of DSM contributes to making progress on a range of SDGs, particularly Goals 7 (affordable and clean energy), 11 (sustainable cities and communities), 12 (responsible consumption and production), and 13 (climate action) as shown in Fig. 12.4. Adopting an energy efficiency approach is the most reliable cost-effective alternative for meeting the growing or increasing demand for energy, improving the efficient use of energy, improves the quality of life, securing economic well-being, contributing to a

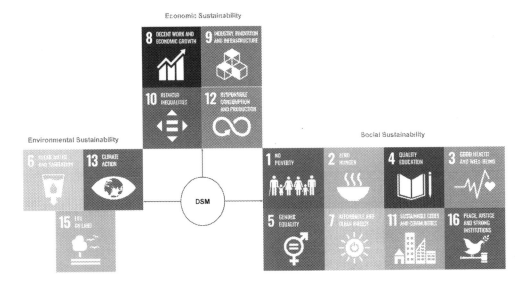

FIGURE 12.4 Contribution of DSM to sustainable development goals.

sustainable environment, and enhancing energy security. The key SDG indicators of DSM as indicated above have impacts on other SDGs, including SDGs 1, 2, 3, 4, 5, 6, 8, 9, 10, 15, and 16.

Again, the inclusion of renewable energy (RE) in demand-side management have a significant reduction in the total electricity cost for household (Sarker et al., 2020) and it enhances access to affordability, reliability, sustainability, and promotion of modern energy. Ensuring access to affordable, reliable, and sustainable energy through RE integration as a result of DSM intervention will open new opportunities for a lot of people across the globe through new economic activities, jobs, better education, and health, empowered women, children, and youth and greater protection and resilience to climate change. Changing energy consumption and production patterns through the implementation of DSM strategies at both the end-use and the generation can help promote economic growth and human well-being.

12.5 Case studies

This section takes a look at the implementation of different DSM programs in developing countries, specifically Ghana, South Africa, China, and India.

12.5.1 Case study 1: DSM in Ghana

Demand-side management in Ghana was started in 1988 by the national energy board. This was introduced by a public utility company, the Volta River Authority (VRA) as a tool to delay investments in putting up new power plants. This initiative however did not survive as it was rejected by consumers and in some cases rejected outright for political and other logistical reasons. Consumers felt it was an attempt for the government to shift blame for supply-side inefficiencies from state-owned utilities. This was attributed to the fact that government officials from the ministry of energy responsible for setting tariffs, energy policy regulations, and similar duties were spearheading the program. Consumers began to accept energy efficiency programs after the removal of subsidies, the separation of functions, and the assignment of regulatory responsibilities to institutions other than the ministry of energy (Ofosu Ahenkorah, 2006).

Ghana has since implemented many DSM programs through the Ghana energy commission which is responsible for developing regulation, planning, and setting policy procedures that address energy issues (Energy Commission of Ghana, 2019; Sakah et al., 2017). The agency that has been instrumental in DSM implementation strategies is the Ghana energy foundation, which is a public-private partnership institution established to promote sustainable development of energy resources and efficient consumption of energy in all of its forms. Their duties have been to educate consumers through public education campaigns, educational programs, and seminars (Energy Foundation, 2021). The offices responsible for the enforcement of energy efficiency standards in the country include the energy commission of Ghana, the Ghana standards authority together with the Customs, Excise, and Preventive Service (CEPS) (Gyam et al., 2017).

The country passed energy efficiency regulations in 2005 for air conditioners and compact fluorescent lights (CFLs) as a step in implementing energy efficiency measures. The energy efficient minimum energy performance standard for these appliances was set up. The appliances had their standards, for instance, the standard for CFLs is to have a minimum service life of 6000 h and a minimum efficacy of 33 lumens per watt (Energy Commision Ghana, 2021).

12.5.1.1 Compact fluorescent lamps (CFL) exchange program

Under this program, the Government of Ghana in 2007 distributed 6 million CFLs to replace incandescent lamps which were seen to be high energy-consuming. This initiative was critical due to the power crisis that occurred during that same period which resulted in load shedding. The distribution and installation of these lamps took place within 3 months. The project brought a reduction in the country's peak electricity demand by 124 MW and peak power consumption by 72.8 GWh/annum. This action delayed the building of a new thermal energy generation plant with an investment of US$105 million. Carbon dioxide savings was estimated at 105,000 tonnes per annum. The project received a global energy efficiency award in 2010 (Energy Commision Ghana, 2021).

12.5.1.2 Refrigerator rebate efficiency project

In September 2012, the Ghana government through the energy commission launched the energy efficient refrigerator rebate and exchange scheme. In this scheme, users of old refrigerators were allowed to exchange them for energy-efficient ones by adding some amount of money. According to the project report, it saved 400 GWh of electricity, 1.1 million tonnes of carbon dioxide, and recovered 1500 kg of chlorofluorocarbon (Energy Commision Ghana, 2021). In the scheme, 10,000 used and inefficient refrigerators in homes were replaced with new energy-efficient ones. The target of the project was to replace 15,000 refrigerators but later discontinued after witnessing a high level of acceptance by Ghanaians.

12.5.1.3 Ghana's new direction toward DSM

The Ghana power compact program which is funded by the United States Government through the Millennium Challenge Corporation (MCC) has created an Energy Efficiency and Demand-Side Management (EEDSM) project to address critical issues facing the electricity sector. The project's objective is to improve building and appliance efficiency thereby reducing wastage of energy. The project has put together four activities, namely:

i) Development and Enforcement of Standards and Labels
ii) Improved Energy Auditing
iii) Education and Public Information
iv) Demand-Side Management

Under this EEDSM project, four interventions are designed to reduce energy waste and upsurge the reserve margin between electricity supply and peak demand. These designed interventions include:

i) The rollout of energy efficiency standards;
ii) Implementation of the pretertiary school curriculum and public information on the use of efficient appliances with standards and labels via different media platforms;
iii) Implementation of an energy auditing and retrofitting activity. In this intervention, three sustainable energy services centers (SESCs) have been established to train and certify energy auditors for the country's installation of energy-efficient LED street lights. These interventions are to be supervised by the Ghana energy commission (Chaplin et al., 2019).

12.5.2 Case study 2: DSM in South Africa

South African energy sector is faced with some level of crisis; this has triggered a widespread blackout as supply falls below demand, threatening to disrupt the national grid. Notwithstanding, the country produces 93% of its electricity from coal making it one of the 15 largest emitters of CO_2 worldwide. This has necessitated the need to focus on DSM initiatives on controlling electrical load between 18:00 and 20:00 hours each day by the electrical utility, ESKOM. To this end, funding has been provided to Energy Service Companies (ESCo's) to implement projects that can achieve load shifting during this period (Rankin and Rousseau, 2008). The programs under this initiative are discussed in following sections.

12.5.2.1 The nondispatchable demand response program

This program is designed to reduce demand periods defined by the system operator during weekdays. This program allows both ESKOM and the municipal top customers to partner with the utility to lessen possible power system constraints. The product is an addition to the demand response (DR) portfolio available to the system operator to maintain adequate daily operating reserve margins to cater for unforeseen circumstances that could affect the stability of the supply. In this program, large consumers are requested to reduce their electricity usage for a defined period to meet supply in return for an incentive. Customers who opt for this portfolio are invited to contribute by decreasing demand on weekdays during predefined periods. The program utilizes both load shifting and peak clipping power management techniques and it is managed via a virtual power station/plant. The demand reduction is monitored from the customer's account, with any adjustments incorporated into the bill. Customers' financial incentive is based on performance therefore they get paid depending on the reduced electricity and it is processed upon verification of demand reduction during evening peaks at R800/MWh for a 14-day notification period or R600/MWh for a 30-day notification (ESKOM).

12.5.2.2 The instantaneous demand response program

The instantaneous demand response program offers the customer an opportunity to partner with ESKOM to mitigate system constraints. The electrical utility, ESKOM compensates customers who have qualified to participate and can reduce load within 6 seconds of notification; and sustain the reduction for a maximum of 10 minutes. The requirement is that the applicant must have a minimum load entry-level of 10MW and should be able to have on average, 200 load reduction events per annum. Other requirements include a maximum number of two load reduction requests of 10 minutes each per scheduled day and the load should be available 24/7 under normal plant conditions. ESKOM is responsible for installing the necessary metering system to enable the customer to participate (ESKOM).

12.5.2.3 Supplemental demand response program

This program is intended to support the system operator to increase daily ancillary reserve and manage system constraints. The system operator can call on participating customers to reduce load when required. Customers have relatively short notification times. Load reduction must be sustained for the contracted number of hours per event. Participants here receive financial compensations for their load reduction. The minimum entry level is 500 kW or 15% of the average MW demand of the plant and they are required to reduce the load for 2 hours when requested (ESKOM).

12.5.3 Case study 3: DSM in China

China is recognized as a major consumer of the world's total energy and this has made management of the national energy security, economic, and social development a major challenge. This is a country that depends so much on coal with a very small proportion of clean energy. Consequently, China has put together requirements to promote the revolution of energy production and consumption and restrain irrational energy consumption to cope with the increasing energy resource and environmental constraints (Xu, 2019; Bambawale and Sovacool, 2011; Gosens et al., 2017). China has put together three main demand response options for both energy consumers and power generating companies to utilize. These include:

i) *Load shifting*: Under this option, peak-hour prices and interruptible load compensations are applied to influence consumers' behavior on energy by using heat and energy storage devices.
ii) *Energy-saving*: Under this strategy, the usage of energy-saving bulbs, high-efficient motors, transformers, and pumps are encouraged.
iii) *Power generation resource replacement*: This was specifically put in place to encourage power generating companies to utilize highly efficient energy resources.

To successfully implement the energy DSM options, the country has put together series of laws, regulations, policies, and legal documents as a first step to promote the country's vision. The implementation of power DSM has recorded from the period 2012 to 2016, a total of 55.3 billion kWh of saved electricity and 12.68 million kW of saved generation capacity, exceeding the target by 13.1 billion kWh and 3.59 million kW respectively (NDRC of China, 2017).

12.5.4 Case study 4: DSM in India

India's step to invest in DSM programs and strategies to promote low carbon transformation was made evident from the launch of the energy conservation act in 2001. The second step was to establish the Bureau of Energy Efficiency (BEE) whose mandate was to specifically focus on energy efficiency and to further initiate the National Mission for Enhanced Energy Efficiency (NMEEE).

Under these initiatives, several schemes to conserve energy were rollout, they included variable speed drives in industry, motor rewinding, downsizing, agricultural pump metering, high-efficiency agricultural pump sets, agricultural pump rectification, improved high-efficiency refrigerators, CFLs, and electronic ballasts. The distribution companies set up time of day (ToD) tariff for the period 6:00 pm to 10:00 pm which represents peak hours. The rate was INR 1.00/kWh (USD 0.02/kWh) more than the energy charges. The electricity rate commission for Andhra Pradesh had a ToD tariff just for night peak hours. Consumption during off-peak hours does not have an incentive.

Assessment of these schemes resulted in an overall electricity savings of 113.16 billion units in 2018-19, which is 9.39% of the net electricity consumption (1,204 billion units). The specific savings for DSM (electrical and thermal) is 16.54 Mtoe, which is 2.84% of the net total energy consumption (581.60 Mtoe) in 2018-19 (Bureau of Energy Efficiency, 2017; Mukhopadhyay et al., 2015)

12.6 Future directions of DSM

In respect of energy sustainability, renewable energies are considered the way forward. Although these sources are intermittent, they will be used to displace a significant amount of energy produced by large conventional plants. To deal with the supply security of the grid network, a significant capacity from conventional plants is always retained. This is where demand-side management will be needed to provide an alternative form of reserve instead of building power plants (Strbac, 2008). Furthermore, with the increasing penetration of Electric Vehicles (EV), its impact concerning system reliability is likely to be felt on the transmission and distribution grid especially at the low voltage level of the distribution network. This is because some of the low voltage networks were not designed or dimensioned to handle a high share of EV charging (Martinenas et al., 2017). This situation presents another opportunity where demand-side management could be utilized to support the penetration of EVs as renewable energies are injected into the electricity grid. Asuamah et al. (2020) proposed demand-side management actions where a real-time tariff scheme was used to manage the charging of EVs contribution to peak demand. The paper suggested that the real-time tariff should only target EV owners when charging and put-up adequate incentives for EV owners to reduce charging during peak demands. Then have an automated EV charger that reads the tariff from the server and decides whether to charge or not, based on the settings defined by the EV owner (Asuamah et al., 2020).

The future seems great for demand-side management schemes for autonomous DC microgrids. In this demand-side management scheme, deferrable loads are shifted from nonsunny hours to sunny hours, and the building's energy demand during nonsunny hours is decreased. In this setup, the charging/discharging cycles of the batteries are reduced while curtailing the power losses in the battery thereby improving the overall system efficiency (Phurailatpam et al., 2016). The call for a smart grid as a means to help integrate renewable energies into the traditional grid makes a new way for demand-side management to automatically manage customers' energy-consuming devices to move shiftable loads from peak hours to off-peak hours. This has been thoroughly discussed in Mahmood et al. (2014) and proposed an autonomous energy scheduling scheme for household appliances in real-time. In this scheme, a smart meter is made available to every user with an embedded Energy Consumption Controlling (ECC) unit. This will be connected with its neighbors through a local area network where consumption information would be shared. Through this, the unit will transfer shiftable loads from peak hours to off-peak hours.

12.7 Conclusion

Demand-side management is considered as one of the key strategies to optimize the utilization of resources thereby promoting energy sustainability. Integration of DSM into energy planning has the potential to increase renewable energy and storage technologies, reduce utility investments and increase consumer benefits. This strategy is particularly important for developing countries where supply deficit, electricity inaccessibility, and climate change vulnerabilities persist. This chapter focused on demand-side management/efficiency strategies in sustainable energy sector development while the basic concepts of DSM were discussed. Cases of demand-side management deployment in developing countries were explored while DSM's future in sustainable energy sector development was presented. The chapter further presented some benefits of DSM in the overall electric system.

Self-evaluation questions

1. What do you mean by demand-side management (DSM)? What are the advantages and disadvantages of DSM?
2. What are the different types of demand-side management techniques- Explain briefly?
3. What are the major differences between demand-side management (DSM) and demand response (DR) programs? Can DR be considered as a part of DSM?
4. To what extent can demand-side management succeed in reshaping the daily load curve to reduce the difference between the maximum and minimum demand?
5. Classify DSM optimization techniques.

6. What accounts for the relatively slow uptake of DSM, particularly in the residential, commercial, and small business sectors?
7. How does the inclusion of DSM in the electricity sector or industry contribute to sustainable development?
8. What role can smart appliances play in the overall DSM concept?
9. What are the available DSM strategies available in some developing countries?
10. Conduct a literature study (research article survey) and identify the challenges or limitations that are responsible for low implementation rate of DSM schemes in developing countries.

References

Aalami, H.A., Moghaddam, M.P., Yousefi, G.R., 2010. Demand response modeling considering interruptible/curtailable loads and capacity market programs. Appl. Energy 87 (1), 243–250.

Albadi MH, E.-S.E., 2008. A summary of demand response in electricity markets. Electr. Power Syst. Res. 78, 1989–1996.

Albadi, M.H., El-Saadany, E.F., 2007. Demand response in electricity markets: an overview. In: 2007 IEEE Power Engineering Society General Meeting. *PES*.

Asuamah E.Y., Kumalo K., Hansen M.R., Jonas M., Bloch M.C., and Løgsted T., "The impact of EV charging on power systems," Aalborg, 2020.

Association of Edison Illuminating Companies. Time-of-Use and Critical Peak Pricing. http://www.aeic.org/load_research/docs/12_Time-of-Use_and_Critical_Peak_Pricing.pdf.

Atzeni, I., Ordóñez, L.G., Scutari, G., Palomar, D.P., Fonollosa, J.R., 2013. Demand-side management via distributed energy generation and storage optimization. IEEE Trans. Smart Grid 4 (2), 866–876.

Bambawale, M.J., Sovacool, B.K., 2011. China's energy security: the perspective of energy users. Appl. Energy 88 (5), 1949–1956.

Bergaentzlé, C., Clastres, C., Khalfallah, H., 2014. Demand-side management and European environmental and energy goals: an optimal complementary approach. Energy Policy 67, 858–869.

BP, "Statistical Review of World Energy globally consistent data on world energy markets. and authoritative publications in the field of energy," p. 66, 2020.

Bureau Of Energy Efficiency, "Impact of energy efficiency measures," vol. 18, 2017.

Cabeza, L.F., Martorell, I., Miró, L., Fernández, A.I., Barreneche, C., 2015. Introduction to thermal energy storage (TES) systems. Advances in Thermal Energy Storage Systems: Methods and Applications. Elsevier Inc., Sawston United Kingdom, pp. 1–28.

Castillo-Cagigal, M.C.-M.E., Gutiérrez, A., Monasterio-Huelin, F., Masa D, J.-L.J., 2011. A semi-distributed electric demand-side management system with PV generation for self-consumption enhancement. Energy Convers. Manag. 52, 2659–2666.

Chaplin D., Ingwersen N., Bos K., and Bernstein D., "Ghana Power Compact: Evaluation Design Report," no. 202, 2019.

Corbett J., Wardle K., and Chen C., "Toward a Sustainable Modern Electricity Grid : The Effects of Smart Metering and Program Investments on Demand-Side Management Performance in the US Electricity Sector 2009–2012," pp. 1–12, 2020.

Dam, Q.B., Mohagheghi, S.M., Stoupis, J., 2008. Intelligent demand response scheme for customer side load management. In: 2008 IEEE Energy 2030 Conference. Energy, p. 2008.

Daniar S. and Talaeizadeh V., "Emergency demand response program modeling on power system reliability Emergency Demand Response Program," no. August, 2016.

Davito, B., Tai, H., Uhlaner, R., 2010. The smart grid and the promise of demand-side management. McKinsey Smart Grid.

Diamantoulakis, P.D., Pappi, K.N., Kong, P.Y., Karagiannidis, G.K., 2016. Game theoretic approach to demand side management in smart grid with user-dependent acceptance prices. IEEE Veh. Technol. Conf. 0 (Dlc).

Diawuo, F.A., De la Rue du Can, S., Baptista, P.C., Silva, C.A.S., 2021. Assessing the impact of demand response on peak demand in a developing country : the case of Ghana. IOP Conf. Ser. Earth Environ. Sci. 642 (012005).

Diawuo, F.A., Pina, A., Baptista, P.C., Silva, C.A., 2018. Energy efficiency deployment: a pathway to sustainable electrification in Ghana. J. Clean. Prod. 186, 544–557.

Diawuo, F.A., Sakah, M., Pina, A., Baptista, P.C., Silva, C.A., 2019. Disaggregation and characterization of residential electricity use: analysis for Ghana. Sustain. Cities Soc. 48 (March), 101586.

Dincer, I., Rosen, M.A., 2013. Exergy analysis of thermal energy storage systems,". Exergy. Elsevier, Amsterdam, Netherlands, pp. 133–166.

Energy Commission Ghana, "Standards and Labelling," Energy Commission,Ghana, 2021.

Energy Commission of Ghana, "National Energy Statistics," 2019.

Energy Foundation, "Energy Foundation – Saving Energy for a Better Tomorrow," 2021.

ESKOM, "Demand Response."

Esther, B.P., Kumar, K.S., 2016. A survey on residential Demand Side Management architecture, approaches, optimization models and methods. Renew. Sustain. Energy Rev. 59, 342–351 Elsevier LtdJun-.

Gołębiowska, B., Bartczak, A., Budziński, W., 2021. Impact of social comparison on preferences for Demand Side Management in Poland. Energy Policy.

Gosens, J., Kåberger, T., Wang, Y., 2017. China's next renewable energy revolution: goals and mechanisms in the 13th five year Plan for energy. Energy Sci. Eng. 5 (3), 141–155.

Groppi, D., Pfeifer, A., Garcia, D.A., Krajačić, G., Duić, N., 2021. A review on energy storage and demand side management solutions in smart energy islands. Renew. Sustain. Energy Rev. 135, 110183.

Gyam, S., Amankwah, F., Nyarko, E., Sika, F., 2017. The energy efficiency situation in Ghana: the energy efficiency situation in Ghana. Renew. Sustain. Energy Rev. (January 2019) 1–9.

Haller, C.R., 2018. Topic-Driven Environmental Rhetoric', in *Sustainability and Sustainable Development : The Evolution and Use of Confused Notions*. Routledge, England, UK, pp. 213–233. doi:10.4324/9781315442044-11.

Handbook of Energy Efficiency and Renewable Energy. CRC Press, 2007.

Herring, H., 2006. Energy efficiency - a critical view. Energy 31 (no. 1 SPEC. ISS.), 10–20 Jan.

Hertog S. and Luciani G., "Energy and sustainability policies in the GCC," 2009.

Hosseini Imani, M., Niknejad, P., Barzegaran, M.R., 2018. The impact of customers' participation level and and various incentive values on implementing emergency demand response program in microgrid operation. Int. J. Elect. Power Energy Syst. 96, 114–125. doi:10.1016/j.ijepes.2017.09.038.

IEA, "Energy and industrial process CO2 emissions and reduction levers in WEO 2020 scenarios, 2015-2030.," 2020.

International Energy Agency (IEA), "World Energy Model 2020 VERSION," no. October, 2020.

Khan, I., 2019. Energy – saving behaviour as a demand – side management strategy in the developing world: the case of Bangladesh. Int. J. Energy Environ. Eng. (0123456789) 493–510. doi:https://doi.org/10.1007/s40095-019-0302-3.

Khan, I., Jack, M.W., Stephenson, J., 2021. Dominant factors for targeted demand side management—an alternate approach for residential demand profiling in developing countries. Sustain. Cities Soc. 67 Apr. doi:https://doi.org/10.1016/j.scs.2020.102693.

Khripko, D., Morioka, S.N., Evans, S., Hesselbach, J., de Carvalho, M.M., 2017. Demand side management within industry: a case study for sustainable business models. Procedia Manuf 8 (June), 270–277.

Kostková K., Ky˘ P., and Jamrich P., "An introduction to load management," vol. 95, pp. 184–191, 2013.

Krarti, M., 2018. Optimal design and retrofit of energy efficient buildings, communities, and Urban centers. Elsevier Inc, Amsterdam, Netherlands.

Lugano Wilson, "Energy And Energy Efficiency, Tanzania Country Report," 2006.

Luo, X.J., Fong, K.F., May 2019. Development of integrated demand and supply side management strategy of multi-energy system for residential building application. Appl. Energy 242, 570–587.

Ma, J., Deng, J., Song, L., Han, Z., 2014. Incentive mechanism for demand side management in smart grid using auction. IEEE Trans. Smart Grid 5, 1379–1388. doi:10.1109/TSG.2014.2302915.

Mahmood, A., et al., 2014. A new scheme for demand side management in future smart grid networks. Procedia Comput. Sci. 32, 477–484.

Majeed Butt, O., Zulqarnain, M., Majeed Butt, T., 2020. Recent advancement in smart grid technology: future prospects in the electrical power network. Ain Shams Eng. J. Ain. Shams. Univ. 12, 687–695.

Martinenas, S., Knezovic, K., Marinelli, M., 2017. Management of power quality issues in low voltage networks using electric vehicles: experimental validation. IEEE Trans. Power Deliv. 32 (2), 971–979.

Maruf, M.H., ul Haq, M.A., Dey, S.K., Al Mansur, A., Shihavuddin, A.S.M., 2020. Adaptation for sustainable implementation of Smart Grid in developing countries like Bangladesh. Energy Rep. 6, 2520–2530 Nov.

Menegaki, A.N., Tsani, S., 2018. Critical issues to be answered in the energy-growth nexus (EGN) research field. Elsevier Inc, Amsterdam, Netherlands.

Momoh, J., Smart grid fundamentals of design and analysis, vol. 53, no. 9. 2013.

Monyei, C.G., Adewumi, A.O., 2017. Demand Side Management potentials for mitigating energy poverty in South Africa. Energy Policy 111 (June), 298–311.

Mukhopadhyay S., Member S., and Rajput A.K., "Demand Side Management and Load Control – An Indian Experience," no. April 2015, 2010.

NDRC of China, 2017. "Notice on deepening the structural reform of supply side and power demand side management under the new situation".

Myerson, RB., 1997. Game Theory: Analysis of Conflicts. Harvard University Press, London.

Ofosu, A., Alfred, S., 2006. Capacity building in energy efficiency and renewable energy regulation and policy-making in Africa. Ghana-Energy Efficiency Country Profile 1–100.

Palensky, P., Dietrich, D., 2011. Demand side management: demand response, intelligent energy systems, and smart loads. IEEE Trans. Ind. Informatics 7, 381–388.

Pang, C., Kezunovic, M., Ehsani, M., 2012. Demand side management by using electric vehicles as distributed energy resources. In: 2012 IEEE International Electric Vehicle Conference, 2012. *IEVC*, pp. 1–7.

Petinrin, J.O., Shaaban, M., 2015. Renewable energy for continuous energy sustainability in Malaysia. Renew. Sustain. Energy Rev.

Phurailatpam, C., Chauhan, R.K., Rajpurohit, B.S., Longatt, F.M.G., Singh, S.N., 2016. Demand side management system for future buildings. In: 2016 1st Int. Conf. Sustain. Green Build. Communities, SGBC 2016.

Pollitt, M.G., Anaya, K.L., 2020. Competition in markets for ancillary services? The implications of rising distributed generation. The Energy Journal 41 (01). doi:10.5547/01956574.41.SI1.mpol.

Prüggler, N., Prüggler, W., Wirl, F., 2011. Storage and Demand Side Management as power generator's strategic instruments to influence demand and prices. Energy 36 (11), 6308–6317 Nov.

Pushparaj, V.L., et al., 2007. Flexible energy storage devices based on nanocomposite paper. Proc. Natl. Acad. Sci. U. S. A. 104 (34), 13574–13577 Aug.

Rankin R. and Rousseau P.G., "Demand side management in South Africa at industrial residence water heating systems using in line water heating methodology," vol. 49, pp. 62–74, 2008.

Ren, H., Sun, Y., Albdoor, A.K., Tyagi, V.V., Pandey, A.K., Ma, Z., 2021. Improving energy flexibility of a net-zero energy house using a solar-assisted air conditioning system with thermal energy storage and demand-side management. Appl. Energy 285, 116433 Mar.

S.-K. Ki, Yuichi I.K., S. Heinen, D. Elzinga, "Impact of smart grid technologies on peak load to 2050," 2011.

Saad, W., Han, Z., Poor, H.V., Başar, T., 2012. Game-theoretic methods for the smart grid: an overview of microgrid systems, demand-side management, and smart grid communications. IEEE Signal Process. Mag. 29 (5), 86–105.

Sahin, E.S., Bayram, I.S., Koc, M., 2019. Demand side management opportunities, framework, and implications for sustainable development in resource-rich countries: case study Qatar. J. Clean. Prod. 241, 118332.

Sakah, M., Diawuo, F.A., Katzenbach, R., Gyamfi, S., 2017. Towards a sustainable electrification in Ghana: a review of renewable energy deployment policies. Renew. Sustain. Energy Rev. 79 (February 2016), 544–557.

Sarbu, I., Sebarchievici, C., 2018. A comprehensive review of thermal energy storage. Sustain 10 (1).

Sarker, E., Seyedmahmoudian, M., Jamei, E., Horan, B., Stojcevski, A., 2020. Optimal management of home loads with renewable energy integration and demand response strategy. Energy 210, 118602.

Sioshansi, F., Vojdani, A., 2001. What could possibly be better than real-time pricing? Demand response. Electr. J. 14 (5), 39–50.

Stadler, I., Sterner, M., 2018. Urban energy storage and sector coupling. Urban Energy Transition. Elsevier, Amsterdam, Netherlands, pp. 225–244.

Strbac, G., 2008. Demand side management: Benefits and challenges. Energy Policy 36 (12), 4419–4426.

Torriti, J., 2012. Price-based demand side management: Assessing the impacts of time-of-use tariffs on residential electricity demand and peak shifting in Northern Italy. Energy 44 (1), 576–583.

Tushar, W., et al., Jun. 2019. A motivational game-theoretic approach for peer-to-peer energy trading in the smart grid. Appl. Energy 243, 10–20.

Office of Electricity of USA, 2006. "Benefits of Demand Response in Electricity Markets and Recommendations for Achieving Them. A report to the United States Congress Pursuant to Section 1252 of the Energy Policy Act of 2005" February 2006.

United Nations Development Programme, World Energy Assessment. Energy and the challenge of Sustainability. 2000.

Vandael, S., Boucke, N., Holvoet, T., Deconinck, G., 2010. Decentralized demand side management of plug-in hybrid vehicles in a Smart Grid. First Int. Work. Agent Technol. Energy Syst. (ATES 2010) (November 2015) 67–74.

Wang, Z.S.H., Gu, C., Li, F., Bale, P., 2013. Active demand response using shared energy storage for household energy management. In: Proceedings of the IEEE Transactions on Smart Grid, pp. 1888–1897.

Wang, L.L., Chen, J.J., Peng, K., Zhao, Y.L., Zhang, X.H., Feb. 2021. Reward fairness-based optimal distributed real-time pricing to enable supply–demand matching. Neurocomputing 427, 1–12.

Wang, X., Lu, X., Liu, B., Chen, D., Tong, Y., Shen, G., 2014. Flexible energy-storage devices: design consideration and recent progress. Adv. Mater. 26 (28), 4763–4782.

Warren, P., 2014. A review of demand-side management policy in the UK. Renew. Sustain. Energy Rev. 29, 941–951.

Warwick University of and REEEP, 2005. Energy Efficiency and Demand Side Management Course Module.

Xu, S., 2019. Energy revolution and the energy demand side management in China. IOP Conference Series: Earth and Environmental Science, 295.

Zhou, Y., Ma, R., Su, Y., Wu, L., Dec. 2019. Too big to change: how heterogeneous firms respond to time-of-use electricity price. China Econ. Rev. 58, 101342.

Zhou, Z., Xiong, F., Xu, C., Jiao, R., 2017. Energy management in microgrids: a combination of game theory and big data-based wind power forecasting. Development and Integration of Microgrids. InTech, United Kingdom.

Zong Y, B.H.M., Mihet-Popa, L., Kullmann, D., Thavlov, A., Gehrrke, O., 2012. Predictive controller for active demand side management with PV selfconsumption in an intelligent building. In: Proceedings of the 3rd IEEE PES International Conference and Exhibition on Innovative Smart Grid Technologies, Europe, pp. 1–8.

Zsiborács, H., usné Baranyai, N.H., Vincze, A., Háber, I., Pintér, G., 2018. Economic and technical aspects of flexible storage photovoltaic systems in Europe. Energies 11 (6), 1–17.

CHAPTER 13

The role of energy storage technologies for sustainability in developing countries

Md Momtazur Rahman[a], Imran Khan[b] and Kamal Alameh[a]

[a]School of Science, Edith Cowan University, Joondalup Drive, WA, Australia
[b]Department of Electrical and Electronic Engineering, Jashore University of Science and Technology, Jashore, Bangladesh

13.1 Introduction

Since the Industrial Revolution, the foundation of the world's economic growth has been dependent on the widespread access to affordable, reliable and modern energy. For a sustainable future, the required energy must be generated from sustainable energy sources, such as solar, geothermal and wind (Magda et al., 2020), which are more compatible with environmental sustainability (Nguyen and Pham, 2020). However, solar and wind energy-based systems require back-up energy generating facilities due to their inherent intermittent nature (Walawalkar et al., 2007). These back-up generating facilities, such as energy storage systems can significantly add flexibility to the renewable energy systems by minimizing the fluctuating nature of solar and wind (Arif et al., 2013). In contrast, nonrenewable resources are limited in supply and cannot be used sustainably (Armaroli and Balzani, 2011). Therefore, renewable energy, particularly solar-based systems, have become a widespread energy generation option in both developed and developing countries.

Electric grids are often highly unreliable in remote areas due to a shortage of generation capacity and other transmission and distribution issues (Levin and Thomas, 2016). Energy storage can help match variable renewable energy supplies to meet the peak electricity demand by storing solar energy during the day and releasing it at night, thus saving money due to storing excess energy. However, the increasing amounts of variable renewable energy penetration, such as solar and wind energy types could negatively impact the energy security process. Therefore, developing countries require a new approach to reliably and cost-effectively manage demand variability and uncertainty (IEA, 2019).

The World Bank group has recently committed $1 billion for developing economies to accelerate investment in 17.5 GWh battery storage systems by 2025, which is more than triple currently installed energy storage systems in all developing countries (Sivaraman, 2019). Thus, renewable energy with storage capability is an excellent alternative to fossil-fuel-based sources, which can bridge the electricity access gap in developing countries cost-effectively and sustainably. For example, a 61 kW peak wind-PV-battery hybrid power system serving a small community in Sitakunda, Bangladesh, generates 169 kWh/day, which corresponds to about USD 0.363/kWh with a negligible amount of generated energy being unused. The analysis also shows that battery energy storage would be an economically and sustainably viable option for a remote community situated within 17 km of the grid (Nandi and Ghosh, 2009). Hence, among the energy storage options available in developing countries, battery is becoming a flexible solution due to its easy deployment and cost-effectiveness manner (Keyes, 2018). However, this is particularly relevant in limited-power grids, and remote locations, where supplying electricity is challenging (Khan, 2019).

The significant advantages for the implementation of energy storage in developing countries include (i) fostering the penetration of renewable energy and backing the deployment of distribution generators (Sedghi et al., 2015); (ii) improving the stability of the grid (Saboori et al., 2015); and (iii) reducing the disparity between the peak and off-peak periods (Barzin et al., 2015).

Many emerging countries have an abundance of renewable energy, including India (Bansal et al., 2019), China (Zhang et al., 2017), Bangladesh (Islam et al., 2014). These countries could benefit from adopting energy storage technologies, especially in remote areas where distributed energy is feasible with traditional options. For example, in China, an optimized hybrid system in a remote area with a battery of 6.94 kWh can decrease the total energy cost by 9 to 11% (Li et al., 2013). However, to sustainably scale up hybrid energy storage deployment in developing countries, associated energy storage technologies, such as maximum power point tracking and battery management system controller need to operate in harsh climatic conditions and sustainably need to manage environmental issues such as reusing and recycling. In addition to that, to open a new market for energy storage systems in developing countries, various barriers need to be resolved beforehand, specifically, (i) the lack of social awareness and reluctance to adopt new technologies; (ii) the lack of knowledge of new technologies and their applications; and (iii) nonproactive regulatory policy and procurement practices that are incapable to pledge cost retrieval.

Therefore, this chapter aims to review the available energy storage system options and their representative technologies, including pumped hydro storage (Deane et al., 2010), compressed air energy storage (Bullough et al., 2004), lead-acid batteries (Hu et al., 2017), redox flow batteries (Li et al., 2011), sensible and latent heat storage (Alva et al., 2018), fuel-cell (A.Kirubakaran and Nema, 2009), and their potential applications in the developing world. Thus, this chapter addresses the advantages and disadvantages of various energy storage technologies, assesses their feasibility for sustainable off-grid electricity systems and their implementation based on future deployment scenarios.

13.1.1 Role of energy storage technologies in energy transitions

Static energy storage was booming at the beginning of the twenty century due to the subsequent expansion of the electricity transmission and distribution networks (Baker and Collinson, 1999). Generally, government policies and public support are crucial to receive benefit from these prospects (Ahuja and Tatsutani, 2009). The energy storage system can play multiple roles in the development of sustainable energy-based power networks for developing countries, including

  (i) Meeting peak load demands (currently, thermal energy storage systems are frequently used to realize peak energy demand in the distributed grid systems) (Desideri and Asdrubali, 2018)
 (ii) Reducing the peak load and shifting it to times of lower demand. For example, at present, North China Power Grid employs pumped hydro storage based on peak load shaving, and currently drives a distributed power system, thus enabling significant savings per unit electricity generation cost (Luan et al., 2018),
(iii) Supporting in peak shrinking, which is another way to reduce the peak demand. For example, a battery bank is employed in a Brazilian microgrid to reduce the peak load, which is a viable option in terms of economic benefits and sustainability, displaying resilience from an electrical point of view (Salles et al., 2020),
 (iv) Peak load shifting, based on the fact that exploratory findings indicate that the thermal energy storage system combined with the effective price-based management techniques could lead to a good peak load shifting (Barzin et al., 2015).
  (v) Alleviating the intermittence of renewable energy sources. For example, it is projected that pumped hydro storage systems can reduce the intermittency and volatility in the power grid network.

The householder can also contribute to the energy transition by incorporating the energy storage system for household utility. Besides, it allows householders to store, share and trade their energy, thus enabling them to be more sustainable in terms of energy uses, reducing their dependence on grid power, and stabilizing the grid (Almehizia et al., 2020).

13.2 Classification of energy storage technologies

The energy storage technologies can be categorized into three major groups depending on the nature of energy stored, as shown in Fig. 13.1. These include (i) mechanical (pumped hydro, compressed air, and flywheels), (ii) electrochemical (lithium-ion battery, vanadium flow battery, lead-acid battery, supercapacitors, hydrogen storage with fuel cells), and (iii) thermal energy storage (sensible heat and latent heat) technologies (Evans et al., 2012).

A detailed description of each type of energy storage technology is discussed in the following sections.

13.2.1 Mechanical energy storage

Mechanical energy storage works in multidimensional systems that use heat, water, air compressors, and turbines. It uses additional apparatus to store energy by exploiting

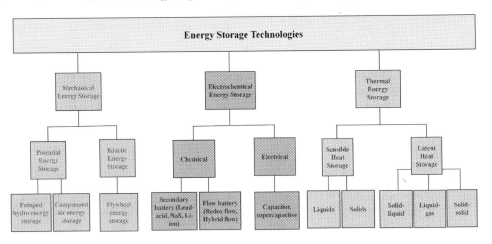

FIGURE 13.1 Classification of major energy storage technologies (Evans et al., 2012).

kinetic or gravitational forces. It primarily consists of pumped hydro storage, compressed-air storage, and flywheel energy storage. The pumped hydro storage is the most durable technology embodied by substantial capacity and long service lifecycle. However, the building of the pumped hydro storage power station is constrained to specific geographical locations. The compressed-air energy storage has the benefits of enormous capacity, long operation time, and long service lifecycle. It can also supply combined heat, coldness, and energy by transforming the compressed air into other energy sources. However, compressed-air energy storage's efficiency is typically low compared to pumped hydro storage, and its system architecture is multifaceted (Xue et al., 2016). On the other hand, the flywheel energy storage has elevated efficiency, rapid response, a long service lifecycle, and a small footprint. However, it has several disadvantages, such as low energy density and susceptibility to self-discharge, making them only appropriate for short-time storage applications (Zhang et al., 2015).

13.2.1.1 Pumped hydro storage

Pumped hydro storage (PHS) is a highly comprehensive and most mature energy storage technology (Rastler, 2010). The first central pumped hydro storage system was built-in in 1929 (Chen et al., 2009). Over 129 GW of pumped hydropower systems are currently in operation globally, making such systems the most widespread energy storage system with higher efficiency and fast response (Chen et al., 2009).

As shown in Fig. 13.2, a PHS station typically consists of an upper and lower water reservoir. During off-peak hours, the water is usually pumped from the lower reservoir to the upper reservoir. Once electricity is required, water is discharged from the upper reservoir across a hydroelectric turbine and stored in the lower reservoir (Rastler, 2010). Then, potential energy turns into kinetic energy to power generators to produce electricity (Hadjipaschalis et al., 2009).

The amount of energy that the PHS system can store in a dam can be calculated according to the following steps (Akour and Al-Garalleh, 2019):

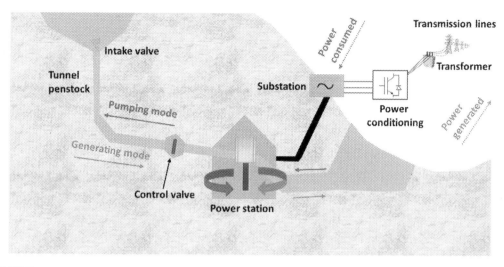

FIGURE 13.2 Schematic diagram of a typical pumped hydro energy storage plant, modified from Abdellatif et al. (2018).

i) Determining the rated pumping head (H);
ii) Calculating the volume flow rate of water (Q) using the following equation:

$$Q = \frac{P \times \eta}{g \times \rho \times H} \quad (13.1)$$

Where P is rated pump power (watt), η is the turbine efficiency, and g is the gravitational acceleration constant (9.81 m/s^2), Q is the fluid flow (m^3/s), H is the hydraulic head height (m), ρ is the fluid density (kg/m^3).

iii) Then, the generated power (P_g) from a hydroelectric turbine can be calculated by using the Eq. (13.2);

$$P_g = Q \times \rho \times g \times H \times \eta \quad (13.2)$$

The PHS efficiency in generating mode ranges from 71.6% to 86.4% (Tronchin et al., 2018), while the pumped hydro's electrical efficiency is around 87% (Noussan and Jarre, 2018).

A pumped hydro storage plant has the lowest investment risk concerning the cost per kilowatt-hour of electricity produced compared to combined cycle gas turbines (Rastler, 2010). However, this technology is correlated with substantial capital cost constraints and long assembly time, making it only suitable for large scale energy storage plants (Walawalkar and Apt, 2007). The majority of reservoirs in the developing countries have been developed for water supply, primarily irrigation, and have received intensive attention in several different countries (Dursun and Alboyaci, 2010). Pumped hydro storage is the most mature technology with the most extended lifespan. Nevertheless, pumped hydro storage's installation costs are highly situational and constrained by geographical locations (Carnegie et al., 2013).

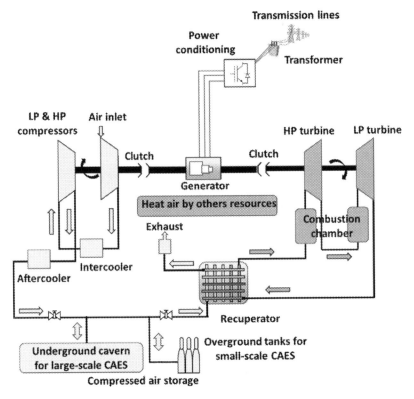

FIGURE 13.3 Schematic diagram of a typical compressed air storage plant, modified from Nikolaidis and Poullikkas (2017) and Luo et al. (2015).

13.2.1.2 Compressed air energy storage

Compressed-air energy storage (CAES) is a commercialized electrical energy storage system that can supply around 50 to 300 MW power output via a single unit (Chen et al., 2013, Pande et al., 2003). It is one of the major energy storage technologies with the maximum economic viability on a utility-scale, which makes it accessible and adaptable modern energy storage systems for developing countries. It consists of five major components (Wang et al., 2017), namely, (i) a driving motor/generator; (ii) an air compressor with intercoolers and aftercoolers, (iii) a turbine train, (iv) a vessel for loading compressed air, (v) apparatus controls and auxiliaries. During peak hours, the loaded compressed air is discharged and heated, or the heat is recovered through the compression process. During discharging, the clutch offers a separation between the motor-generator unit and compressor. Fig. 13.3 shows a schematic diagram of a typical compressed air storage plant (Nikolaidis and Poullikkas, 2017).

According to Sciacovelli et al. (2017), the isentropic air outlet temperature can be calculated using the following equation (Sciacovelli et al., 2017);

$$T_{out}^{ic} = T_{in}^{ic} \beta_i \frac{k-1}{k} \tag{13.3}$$

Where $\beta i = \beta_{HPC} * \beta_{LPC}$ is the compression ratio of the high-pressure compressor (HPC) and low-pressure compressor (LPC) stage. The compressor isentropic efficiency can be defined as-

$$\eta_c = \frac{\text{Isentropic compression work}}{\text{Actual compressor work}} = \frac{T_{out}^{ic} - T_{in}}{T_{out} - T_{in}} \quad (13.4)$$

While in the compressor air tank, the mass and energy balance equation can be written as follows (Sciacovelli et al., 2017; Raju and Khaitan, 2012):

$$\frac{d\rho}{dt} = \frac{m_{in} - m_{out}}{V} \quad (13.5)$$

where ρ is the air density, and m_{in} is the mass flow rate of the incoming air and m_{out} is the mass flow rate of the out-going air, and V is the cavern's volume. Therefore, the rate of increase in the internal energy of the cavern air can be written as follows (Sciacovelli et al., 2017; Raju and Khaitan, 2012):

$$\frac{d(M)}{dt} = m_{in}H_{in} - m_{out}H_{out} - h_{amb}A_{cavern}(T - T_{amb}) \quad (13.6)$$

Where M is the mass of the air in the cavern; H_{in} and H_{out} is the specific enthalpy of the incoming air and outgoing air, respectively; h_{amb} is the heat transfer coefficient; A is the area of the cavern and T and T_{amb} are the temperatures of the air and cavern wall, respectively.

For an ideal gas, the Eq. (13.6) becomes-

$$\rho C_p \frac{dT}{dt} + \frac{m_{in}}{V} C_p (T - T_{in}) - \frac{d\rho}{dt} + \frac{h_{amb} A_{cavern}}{V} (T - T_{amb}) = 0 \quad (13.7)$$

Therefore, Eqs. (13.1) and (13.6) provide the mass and balance energy equations for the compressed air storage in caverns (Sciacovelli et al., 2017; Raju and Khaitan, 2012).

The CAES's efficiency is 70 to 89% (Chen et al., 2009), while electrical efficiency and round trip efficiency are in the range of 70 to 79% and 54% (Tronchin et al., 2018). CAES is considered the leading profitable utility-scale energy storage technology with a high impart of variable energy sources (Lund and Salgi, 2009). This technology presents superior consistency with minimal environmental impacts (Kaldellis, 2010).

13.2.1.3 Flywheel energy storage

The first flywheel energy storage (FES) prototype appeared in the early seventies. They are charged with a motor spin, while energy discharge is achieved throughout the identical motor to deliver electricity. The total energy capacity is a function of the rotor's size and speed, while the power rating is a motor-generator function (Evans et al., 2012). It generally consists of a flywheel, a gigantic cylinder, a motor/generator, and a vacuum pump, as shown in Fig. 13.4. When power is required, electricity is retrieved by the same motor. Then, it works as a generator, which triggers the flywheel to decelerate. Thus the rotational energy is converted back into electricity (Rahman et al., 2012) and can be presented by Eq. (13.8).

$$E = \frac{1}{2} \times I \times \omega^2 \quad (13.8)$$

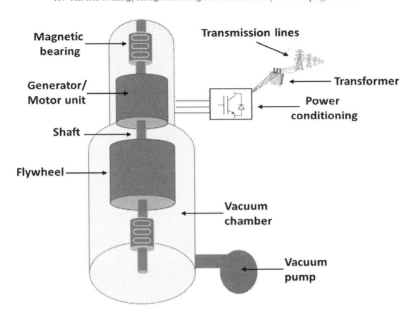

FIGURE 13.4 Schematic diagram of a flywheel energy storage system, modified from Nikolaidis and Poullikkas (2017).

where E is the stored kinetic energy, I is the flywheel moment of inertia [kgm^2], and ω is the angular speed [rad/s].

The rotor is typically installed in an evacuated cylinder, allowing it to accelerate rapidly (Blume, 2015). The main component of a flywheel comprises a rotating mass, which stores rotary kinetic energy according to the Eq. (13.9) (Wheeler, 2010; Berdichevsky et al., 2006); in the case of a hollow cylinder flywheel:

$$E = \frac{1}{2} \times M \times r^2 \times \omega^2 \qquad (13.9)$$

Where M is the mass (kg), and r is the radius (m).

Then the tensile stress of the flywheel can be calculated using Eq. (13.10):

$$\sigma = \rho \times r^2 \times \omega^2 \qquad (13.10)$$

Where ρ is the density, r is the radius, ω is the angular velocity, and σ is the tensile stress.

Finally, substituting the maximum angular velocity from Eq. (13.10) into the Eq. (13.9) yields the maximum attainable energy as-

$$E_{max} = \frac{1}{2} \times \frac{M \times \sigma}{\rho} \qquad (13.11)$$

A flywheel typically requires a small area to install, and it has a higher cycle efficiency in the range of 85 to 95% (Wu et al., 2019). One of the significant drawbacks of the flywheel system is the elevated cost and reasonably high standing losses. Moreover, it creates noise pollution and associates with safety concerns.

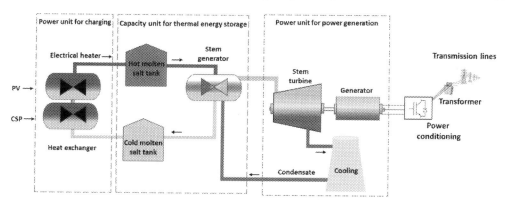

FIGURE 13.5 Schematic diagram of a thermal energy storage system, modified from Olabi et al. (2020).

13.2.2 Thermal energy storage

Thermal energy storage (TES) is a technology that reserves thermal energy by heating or cooling a storage medium and then uses the stored energy later for electricity generation using a heat engine cycle (Sarbu and Sebarchievici, 2018). It can shift the electrical loads, which indicates its ability to operate in demand-side management (Fernandes et al., 2012). The thermal energy storage systems can be used in domestic heating and cooling, as well as in the industrial sector (Olabi et al., 2020). It mainly consists of a thermal storage tank, a medium of transferring heat, and a control system, as shown in Fig. 13.5.

A thermal energy storage (TES) can help rectify the disparity between energy supply and demand (Dincer and Rosen, 2011). Its fundamental principle is similar for all the applications: the power is delivered to the TES during the charging process and collected during the storing process. It is then removed from the TES for later use during the discharging cycle (Kousksou et al., 2010). The energy storage capacity depends on the type of energy storage material used (Dincer and Dost, 1996), namely, latent heat storage, which stores the heat in the phase change material (Wang et al., 2016; Hasan, 2016). In contrast, sensible heat storage stores heat by shifting the storage medium's temperature without changing the phase (Abdin and Khalilpour, 2019).

13.2.2.1 Sensible heat storage

Sensible heat storage (SHS) can increase the temperature of the heat storage material. Generally, water is used as the heat storage medium. This technique converts collected energy into sensible heat into chosen materials and retrieves it while required (Stutz et al., 2017). The governing equation for sensible heat can be expressed as-

$$Q = m \times C_p \times (T_2 - T_1) \tag{13.12}$$

Where Q is the energy (J); C_p is the specific heat at constant pressure (J/k/kg), and T_1 and T_2 are the two temperatures (k); before and after heating, respectively; and m is the mass of heat storage medium (kg).

TABLE 13.1 Properties of some solid materials that can be used for sensible heat storage (Koçak and Paksoy, 2019).

Storage materials	Density (kg/m³)	Thermal conductivity (W/m/K)	Specific heat (J/kg/°C)
Dolerite	2730	–	865
Flint	2650	3.42	945
Quartzite	2600	3.6	877
Gneiss rock	2740	3.0	825
Pebble	2801	–	745
Basalt	2644	2.08	772

TABLE 13.2 Properties of solid-state latent heat storage materials (Tian and Zhao, 2013).

Storage materials	Density (kg/m³)	Thermal conductivity (W/m/K)	Specific heat (J/kg/°C)
Magnesia fire bricks	3000	5.0	1.15
Silica fire bricks	1820	1.5	1.00
NaCl	2160	7.0	0.85
Sand-rock minerals	1700	1.0	1.30
Cast steel	7800	40.0	0.60
Cast iron	7200	37.0	0.56

The SHS is considered as a straightforward, low-cost, and reasonably mature technology. The amount of heat-storing capacity relies on the type of the medium and the volume of storage material (Kumar and Shukla, 2015). The properties of solid material for sensible heat storage is shown in Table 13.1.

13.2.2.2 Latent heat storage

The latent heat storage (LHS) uses phase change materials, such as wax, and salt hydrates, as the storage medium, whereby energy is absorbed or released when a change of phase occurs at a specific temperature. The latent heat storage system's output can be modified corresponding to the grid operation requirement (Zahedi, 2014).

The amount of energy (Q) released or absorbed during a phase change is given by Reinhardt (2010) and Nomura et al. (2010) as-

$$Q = m \times C_p \times dT(s) + m \times L + m \times C_p \times dT \qquad (13.13)$$

Where L is the specific latent heat (J/kg); dT is the temperature difference (k), and m is the mass of the phase change material (kg) (Nomura et al., 2010).

The properties of solid material for (LHS) is shown in Table 13.2.

13.2.3 Electrochemical energy storage

Electrochemical energy storage is a technology used to store electrical energy in a chemical form. The leading electrochemical energy storage technologies consist of a lead-acid battery, lithium-ion battery, redox flow battery, etc. A lead-acid battery comprises a negative electrode made of porous lead and a positive electrode made of lead oxide. Both electrodes are immersed in a sulfuric acid and water solution. The primary benefit of lead-acid batteries is their low capital cost and accessibility. A lithium-ion battery is a rechargeable battery type widely used for electronic mobile and vehicle applications for on and off-grid applications (Yang et al., 2011). The vanadium flow battery (VRB) is also a rechargeable flow battery with a high energy capacity and long lifespan (more than 20 years). However, it has a small energy-to-volume ratio than its counterparts.

13.2.3.1 Lead acid battery

The most commonly used rechargeable battery in the developing world is the lead-acid battery (Ibrahim et al., 2008). It has a rapid response, lower self-discharge rates (<0.3%), high cycle efficiencies (63–90%), and low capital costs (50–60 $/kWh) (Chen et al., 2009). It can be used in multiple areas, including data and telecommunication systems for back-up power supplies. However, it has comparatively lower installation records around the world due to its relatively low cycle (up to ~2000), low energy density (50–90 W h/L), and low specific energy (25–50 Wh/kg) (Baker, 2008). The cathode, anode, and electrolyte of a lead-acid battery consist of PbO_2, Pb, and sulfuric acid. During discharging, both electrodes are converted to lead sulfate while charging; the process is being reversed as shown in Eqs. (13.14) to (13.18) (Yang et al., 2011):

For positive electrode reaction:

$$PbO_2 + 4H^+ + 2e^- \rightarrow Pb^{2+} + 2H_2O \quad (13.14)$$

$$Pb^{2+} + SO_4^{2-} \rightarrow PbSO_4 \quad (13.15)$$

For negative electrode reaction:

$$Pb \rightarrow Pb^{2+} + 2e^- \quad (13.16)$$

$$Pb^{2+} + SO_4^{2-} \rightarrow PbSO_4 \quad (13.17)$$

Overall cell reaction:

$$Pb + PbO_2 + 2H_2SO_4 \rightleftharpoons 2PbSO_4 + 2H_2O \quad (13.18)$$

During overcharging, hydrogen and oxygen gases are released, which are discharged into the atmosphere. Thus, it requires frequent water maintenance (Yang et al., 2011). The cell reaction yields 2.10 V at standard conditions. It has become a widespread and available energy storage option in developing countries; due to its recyclability and low-cost energy storing capacity. However, it is inhibited by toxicity and low energy density (Chen et al., 2009).

13.2.3.2 Lithium-ion battery

In a Li-ion battery, the cathode, anode, and electrolyte are consist of lithium metal oxide [e.g., $LiCoO_2$ (Lithium cobalt oxide), $LiMO_2$ (Lithium transition metal oxides)], graphitic

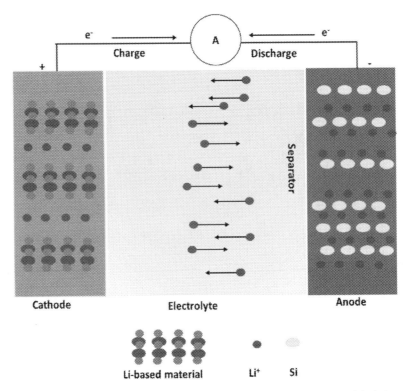

FIGURE 13.6 Schematic diagram illustrating a Li-ion battery cell's operation, modified from Zhang et al. (2011) and David (2020).

carbon, and nonaqueous organic liquid containing dissolved lithium salts ($LiClO_4$), respectively (Díaz-González et al., 2012). During the charging and discharging process, the complete concurrent oxidation and a reduction reaction occur at the two electrodes via lithium ions transfer (Aurbach et al., 2002). Fig. 13.6 demonstrates the operation of a lithium-ion battery in the charging and discharging process.

The Li-ion technologies commenced with realizing intercalation compounds, such as Li_xMO_2 (M = Co, Ni). The energy storage in the cells is achieved via the following chemical reactions (Yang et al., 2011):

Cathode reaction:

$$Li_{1-x}CoO_2 + xLi^+ + xe^- \rightarrow LiCoO_2 \quad (13.19)$$

Anode reaction:

$$Li_xC_6 \rightarrow C_6 + xLi^+ + xe^- \quad (13.20)$$

Overall cell reaction:

$$Li_xC_6 + Li_{1-x}CoO_2 \rightleftarrows LiCoO_2 + C_6 \quad (13.21)$$

$$E = 3.7\,V \text{ at } 25^0\,C \quad (13.22)$$

It has a high energy density, short response time, and high efficiency (Chen et al., 2009). However, the higher cycle depth of discharge can affect its lifetime. Besides, its thermal management becomes more crucial for a bigger battery pack (Viswanathan et al., 2010).

13.2.3.3 Flow battery

A flow battery stores energy in two soluble redox couples, which are comprised of exterior liquid electrolyte containers. During charging, one electrolyte is oxidized at the anode, while during discharging, another electrolyte is reduced at the cathode. In this way, the electrical energy is transferred to the electrolyte. It can be categorized into redox flow batteries and hybrid flow batteries. Its energy capacity can be scaled up independently of the power, with no standby loss.

13.2.3.4 Vanadium redox flow battery

The vanadium redox battery (VRB) is one of the most mature flow battery systems (Divya and Østergaard, 2009). NASA researchers have first studied the chemistry of vanadium redox couples with cyclic voltammetry in the early seventies (Thaller, 1977). The VRB stores energy using vanadium redox couples (V_2^+/V_3^+ and V_4^+/V_5^+) in two electrolyte containers (Yang et al., 2011). During charge/discharge cycles, hydrogen ions (H^+) are exchanged through the ion-selective membrane (see Fig. 13.7).

In a VRB, the energy conversions are achieved via changes in vanadium valence states through the resulting chemical reactions at the electrode (Yang et al., 2011, Lourenssen et al., 2019):

For positive electrode reaction:

$$VO_2^+ + 2H^+ + e^- \rightleftharpoons VO^2 + H_2O \tag{13.23}$$

For negative electrode reaction:

$$V^{2+} \rightleftharpoons V^{3+} + e^- \tag{13.24}$$

$$\text{Overall cell reaction: } VO_2^+ + V^{2+} + 2H^+ \rightleftharpoons VO^{2+} + V^{3+} + H_2O \tag{13.25}$$

$$E = 2.105 \text{ V at } 25°C \tag{13.26}$$

The overall electrochemical reaction delivers a cell voltage of 1.26 V at 25°C. It has a lengthy lifespan with minimal maintenance. It can improve the power quality for stationary applications and compensate the intermittent nature of renewable energy. However, it entails complex construction and suffers from low energy and low power density.

13.2.3.5 Hydrogen fuel cell

Hydrogen can be produced through thermochemical processes, whereby heat and chemical reactions release hydrogen from organic materials (biomass and fossil fuels) (Khan, 2020). Also, electrolysis and solar energy can be applied to split up water (H_2O) into hydrogen (H_2) and oxygen (O_2) molecules. Hydrogen fuel cell technology enables large-scale energy storage in a pollution-free manner. Therefore, storing more than 100 GWh of energy for large-scale applications can be feasible in hydrogen fuel cell technology. However, it has several limitations, such as low energy conversion efficiency (40–50%), high installation cost,

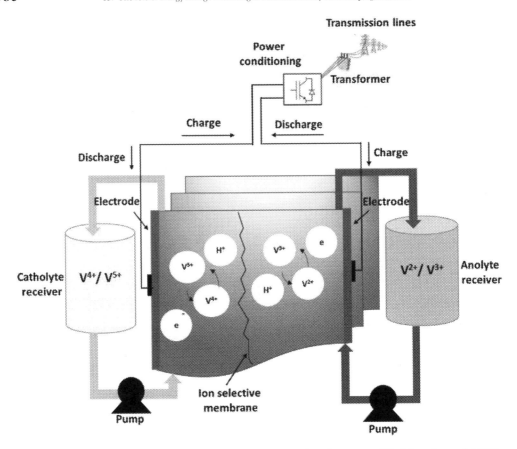

FIGURE 13.7 Schematic diagram illustrating a vanadium redox flow battery, modified from Lu et al. (2020).

and a significant initial investment (Xianxu et al., 2016). In conventional hydrogen energy storage systems, a water electrolysis unit produces hydrogen, which is stored in high-pressure containers and delivered by pipelines for later use (Díaz-González et al., 2012). Therefore, fuel cells can produce electricity using the stored hydrogen and oxygen from the air (Luo et al., 2015). The overall reaction is as follows (Mekhilef et al., 2012):

$$2H_2 + O_2 = 2H_2O + \text{energy} \tag{13.28}$$

Hydrogen can be generated onsite from biomass and can be transferred using a pipeline for any particular application. The construction of a fuel cell is straightforward. A fuel cell comprises an anode, cathode, and an ion-conducting layer (see Fig. 13.8). It receives chemical energy (hydrogen and oxygen are delivered to individual electrodes) and produces electricity. The processes governing energy generation by a fuel cell and the conversion of hydrogen to water are pollution-free, because water is the only by-product resulting from the chemical reactions. Fuel cell types, which are typically categorized by the type of electrolyte used, including proton exchange membrane fuel cell, molten carbonate fuel cell, phosphoric acid fuel cell, alkali fuel cell, and solid oxide fuel cell. It can be integrated into building energy

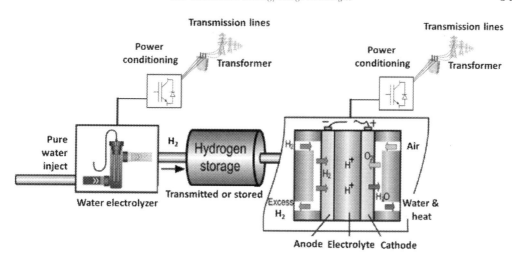

FIGURE 13.8 Typical structure of hydrogen storage and fuel cell, modified from Kleperis et al. (2016).

systems to provide both heat and electricity (Almehizia et al., 2020). It has a longer operating time on a broader temperature range and requires less maintenances. Most importantly, it can run on various fuels. However, a fuel cell has slow power response and low energy and power density (Wilberforce et al., 2016).

13.2.3.6 Capacitor and supercapacitor

A capacitor comprises at least two electrical conductors (metal foils) separated by a thin layer of an insulator (ceramic or glass). During charging, the energy is stored in the electrostatic field of the dielectric material (Chen et al., 2009). It typically has a higher power density and shorter charging time than traditional batteries. However, they are suffered from limited energy storage capacity (Chen et al., 2009). Supercapacitors also comprise two-conductor electrodes, an electrolyte, and a porous membrane separator (see Fig. 13.9) (Díaz-González et al., 2012). It is constructed on nanomaterials to improve the electrode surface area, thus enhancing the capacitance.

A supercapacitor has a higher energy density than a traditional capacitor. It has long cycling times (more than 1×10^5 cycles) and high cycle efficiency (84–97%) (Smith et al., 2008). However, it is prone to self-discharge losses. For example, the typical daily self-discharge rate for a supercapacitor is 5 to 40%. Its capital cost is more than 6000 \$/kWh (Díaz-González et al., 2012). Further, its power and energy densities are typically smaller than those of rechargeable batteries but higher than traditional capacitors (Sharma and Bhatti, 2010). Thus, supercapacitors are reliable for short-term rather than long-term storage applications.

13.2.4 Electromagnetic energy storage

Electromagnetic energy storage primarily encompasses superconducting magnetic energy storage. Its high power density and elevated energy conversion efficiency are found to be crucial for widespread applications (Luo et al., 2015).

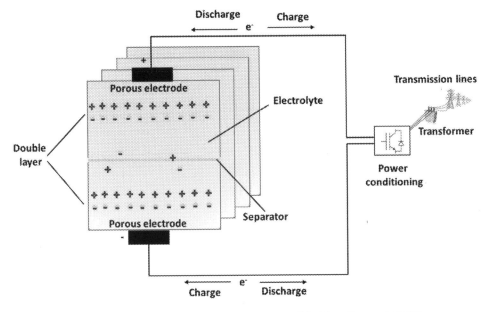

FIGURE 13.9 Schematic diagram of a supercapacitor system, modified from Luo et al. (2015).

13.2.4.1 *Superconducting magnetic energy storage*

A typical superconducting magnetic energy storage (SMES) system consists of a power conditioning unit, a vacuum subsystem, and a superconducting coil unit (Díaz-González et al., 2012). During charging, direct current passes through a superconducting magnetic coil, which stores energy in the magnetic field (Chen et al., 2009). During discharging, the stored energy can be released back to an alternating current system via a power converter module. Note that the amount of stored energy is defined by the self-inductance of the coil and the current flowing through it (Yuan, 2011). A necessary arrangement of a SMES system is shown in Fig. 13.10.

A SMES system has a unique role in intermittent renewable energy systems (Ali et al., 2010) and can be a good choice for short term storage. It is highly reliable for energy storage with immediate response, a high life cycle, and high efficiency. However, the requirement of expensive cryogenic refrigeration limits its application (Chen et al., 2009).

Table 13.3 summarizes the technical characteristics, advantages, disadvantages, and environmental impacts of different energy storage technologies.

13.3 Progress and challenges of energy storage technologies in developing countries

Nowadays, energy storage technologies in developing countries are becoming a crucial part of renewable energy systems. According to the International Energy Agency (IEA), around 2.9 GW of storage capacity is upgraded to power systems in 2019 (Luis, 2020), especially in

13.3 Progress and challenges of energy storage technologies in developing countries

TABLE 13.3 Summary of technical characteristics of different types of energy storage system.

Type	Maturity	Energy density (kWh/m³)	Power rating (MW)	Cycle efficiency	Advantages	Disadvantages	Environment impact and footprint
Pumped hydro storage	Mature	0.5–1.30 (Sabihuddin et al., 2015; Koohi-Fayegh and Rosen, 2020)	100–5000 (Chen et al., 2009; Koohi-Fayegh and Rosen, 2020)	70–85 (Chen et al., 2009)	Higher capacity High upfront cost Variable operational costs	Location constraints Low power and energy density Disturbance to local wildlife	Large negative impact
Compressed; air energy storage	Mature	3–6 (Chen et al., 2009)	Up to 300 (Chen et al., 2009)	60–90 (Sabihuddin et al., 2015)	High energy storage capacity	Safety issues Location constraints	Large negative impact
Li-ion battery	Improving	300–750 (Meishner and Sauer, 2019)	1–100 (Rastler, 2010)	90–97 (Chen et al., 2009)	Higher energy and power density Higher efficiency	High cost Require recycling Life cycle relies on discharge levels	Moderate negative impact
VRB	Mature	20–80 (Chen et al., 2009; Meishner and Sauer, 2019)	~0.03–3 (Chen et al., 2009)	60–90 (Sabihuddin et al., 2015)	High energy storage capacity	Low energy density Low power density	Moderate negative impact
Supercapacitor	Improving	10–30 (Chen et al., 2009; Meishner and Sauer, 2019)	~0.001–0.1 (Farret and Simoes, 2006)	90–100 (Koohi-Fayegh and Rosen, 2020; Meishner and Sauer, 2019)	High power density Long lifetime High efficiency	Low energy density Safety issues Life cycle depends on voltage imbalances Toxic and corrosive	None
Magnetic energy storage	Mature	0.2–14 (Sabihuddin et al., 2015; Koohi-Fayegh and Rosen, 2020)		70–96 (Sabihuddin et al., 2015)	High power Immediate response High efficiency High reliability Long lifetime	Refrigeration energy requirements High cost Requirement for large magnetic fields	Moderate negative impact
Flywheel; energy storage	Mature	20–80 (Chen et al., 2009; Meishner and Sauer, 2019)		60–90 (Sabihuddin et al., 2015)	High-efficiency Small area requirement	Lower energy density Safety issues	Almost none
Hydrogen fuel cell	Research/developing/marketed	500–3000 (Chen et al., 2009)	<50 (Chen et al., 2009)	20–50 (Chen et al., 2009)	Long term storage, Variety applications	Catalyst is expensive	Low
Latent heat storage; sensible heat storage	Relatively mature	100–370 (Sabihuddin et al., 2015) 25–120 (Sabihuddin et al., 2015)	10–30 (Chen et al., 2009)	30–60 (Chen et al., 2009) 75–90 (Sabihuddin et al., 2015)	Larger energy storage capacity Straightforward and low-cost Fast charging and discharging time	Suffers from heat loss	Low

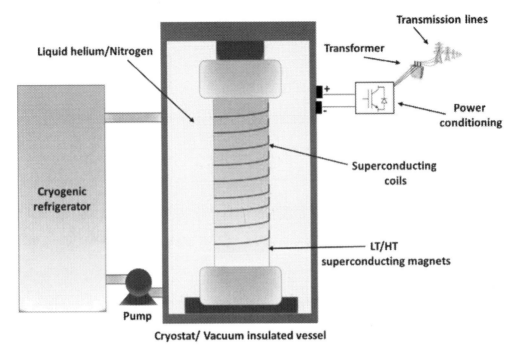

FIGURE 13.10 Schematic diagram of superconducting magnetic energy storage (SMES) system, modified from Molina (2012).

developing countries, including China (Tong et al., 2020), India (Kumar and Shrimali, 2020), Bangladesh (Gulagi et al., 2020), Malaysia (Babu et al., 2020), and Turkey (Çolak and Kaya, 2020).

However, due to urbanization, the landscape in developing countries also changes, and this affects streamflow and hydropower generation. For instance, due to geography and climate changes, the land pattern of Bangladesh changes, and this limits its hydropower potentiality. As a result, the government of Bangladesh has issued an energy policy based on exploiting renewable sources (IEA, 2018), which might open the possibilities to explore proper energy storage technologies in remote areas. On a utility-scale, pumped hydro storage schemes attract more attention in developing countries due to their unparalleled operational flexibility compared to other counterparts. Around 140 GW of pumped hydro storage capacity is mounted worldwide, delivering more than 95% of global energy storage capacity (Buckley, 2019).

In Huizhou, China, the largest PHS power plant, was introduced in 2011 (capacity: 2448 MW). There are currently 24 PHS stations in China with an installed capacity of 16.95 GW, and 95% of its infrastructure owns by the Chinese government (Ming et al., 2013). The government of China also planned to make 50.02 GW by 2020. However, due to COVID-19, the adoption of energy storage systems has been affected severely. Compressed air energy storage systems (CAES) that store energy in the form of high-pressure air can deal with the intermittent nature of renewable energy (Tong et al., 2020). CAES's potentiality is assessed in China's

Northern regions in terms of grid regulation, energy production, and demand-side management (Tong et al., 2020), which indicate their significant commercial viability and ecological benefits. China has completed the integration test of its first 100 MW advanced compressed air energy storage expander (Xinhua, 2020). Additionally, China's engineering and technology institute developed the 1.5 MW, and 10 MW advanced compressed air energy storage systems in 2013 and 2016, respectively (Xinhua, 2020). The flywheel energy storage laboratory of Tsinghua University has set a theoretical and experimental groundwork for China's later research on flywheel technology. While the technology and theory of flywheel energy storage are reasonably mature in China, industrial applications are still at the early stages. The overall growth of electrochemical energy storage in China in 2019 also reached around 519.6 MW (Colthorpe, 2020).

In another developing country, India, the rise in peak power demand requires energy storage schemes to ensure power system stability. Currently, the key constraint of pumped hydro storage systems in India is the scarcity of off-peak power accessibility for pumping. Therefore, fixing an effective operational tariff for pumped hydro storage systems (PHS) throughout India is crucial, and this has recently attracted attention in most states, and proposals are being introduced to the central government for approval (Sivakumar et al., 2013). According to Tim et al. (2019), several projects have been initiated in India over the last few years, including (i) 2 GW PHS projects in Karnataka, (ii) 500 MW PHS proposal for commissioning in Tamil Nadu, (iii) 600 MW PHS unit at its existing 600 MW hydropower plant in Odisha and (iv) another 1GW of PHS in West Bengal (Buckley, 2019). Most importantly, India's Ministry of Power recommended electricity rule modifications in March 2019 to incentivize electricity supply at times of peak demand, which was deemed a vital policy for strengthening the financial guarantee that makes storage projects commercially viable.

On the other hand, some developing countries, such as Bangladesh, face intimidating energy disputes due to the lack of grid expansion in remote locations (Biswas et al., 2011). Therefore, the employment of hybridized energy system with a battery and fuel cell could be economically and technically viable in those areas (Hasan, 2018; Mandal et al., 2018). For example, the battery bank projects (based on the use of 12 SMA Battery inverter) at Cox's Bazar, Bangladesh, have demonstrated commercial viability in terms of meeting the temporary demand of Kutubdia Island cost-effectively (Khan, 2015). Therefore, by realizing the potentiality of renewable energy technologies and employing correlated energy storage technologies in the countryside, Bangladesh might meet its unprecedented energy demand (Khan, 2020).

In Turkey, currently installed pumped hydro storage could only meet around 0.6 to 0.8% of electricity demand (Melikoglu, 2017), and this is too far from country's target of supplying 30% of its electricity from renewable energy sources by 2023 (Melikoglu, 2013). Turkey also intends to integrate various renewable sources, which might impose power grid volatility. Therefore, in order to fulfill Turkey's energy supply targets by 2023, and address the grid volatility, modern energy storage system needs to be implemented (Melikoglu, 2017).

In Malaysia, under the current electricity framework, in addition to the normal consumption charges, commercial and industrial customers are charged the maximum monthly peak power demand charges, which adds up to 20% to their electricity bill (Subramani et al., 2017). Therefore, the distributed small-scale energy storage system, as an alternative option, could potentially be an excellent approach to lowering the peak demand, whereby Malaysian customers can reduce their electricity bills with a reimbursement period of 2.8 years (Chua et al., 2015).

13.4 Sustainability evaluation of energy storage technologies in developing countries

In developing countries, the sustainability of energy storage systems faces significant challenges due to current demography, monetary, societal, and technological developments (Ren and Ren, 2018; Kaygusuz, 2012). Therefore, the policy for sustainable development in economic, social, and environmental dimensions needs to establish an appropriate framework and implementation tools (Khan, 2020). However, implementation of the policy support, reduction of the technology cost and widespread market share are the main barriers to the development of cost-effective energy storage systems. Many developing countries have taken advantage of these factors to restructure their energy sectors, following suitable sustainable goals (Salvarli and Salvarli, 2020).

The situation for adopting energy storage technologies in developing countries is more complicated than the developed ones. For example, due to the resource constraints and social acceptance of new technology, many residents might have difficulties reaching out to essential energy services (Salvarli and Salvarli, 2020). Therefore, sustainable policy, dynamic market share, and cost barriers need to be addressed by introducing more alternatives energy storage systems in developing countries (Salvarli and Salvarli, 2020).

When analyzing the sustainability aspects of electrical energy storage types, it becomes evident that the pumped hydro storage is the most widely used electrical energy storage technology, where electricity is stored in the form of hydraulic potential and can be shifted back to the power grid when required. Hydro pumped storage has commercial benefits over batteries, which use chemicals to store energy. It costs less to convert one unit of energy since they are more resilient than batteries. Because of the low-evaporation property of water stored in reservoirs, hydropower can be stored for months and even years with minimum leakage losses. Hydropower has few limitations though, such as site constraints, and building a large dam enforces environmental sustainability concerns. Though the unit power cost of hydro pump storage is low, its installation cost is much higher. For example, the cost of making hydro pump storage is typically several hundreds of millions of dollars (Tasmania, 2018), which is out of reach for most developing countries' budgets. It also takes a long time to build hydro pump storage with a large storage capacity, and it has social concerns due to its direct impact on landslides and the unwelcoming of the people living in hillside areas.

On the other hand, battery uses chemical reactions to store electricity and then switch it back to appropriate electrical energy. The chemicals used in batteries are sometimes toxic and need a proper recycling plan for environmental sustainability. Some batteries (called primary batteries) are used once and discarded when the power is depleted, while others are recharged (called secondary batteries) and can be used several times (Dimov et al., 2003; Thackeray et al., 2012). However, secondary batteries cannot be used forever because they are susceptible to loss of electrolyte and internal corrosion. For electrical energy storage, batteries are broadly employed in off-grid PV systems in developing countries (Chattopadhyay et al., 2019). Once the cost of the battery starts to follow along with the sharp cost drop of solar and wind energy technologies, batteries hold the immense possibility of transforming the power systems of developing countries. Nevertheless, variation in weather conditions triggers batteries to charge/discharge frequently, which reduces operating life and increasing maintenance expenditures.

In summary, pumped hydro storage is still more sustainable option among all the energy storage technologies. However, the new advances in battery energy storage by eliminating unbalancing factors (Qays, 2020; Qays, 2021) might alter the energy market in developing countries by fostering sustainable development through more effective storage and electric energy transportation. For example, the World Bank group's investments in battery storage systems would allow middle-income countries to move to the next generation of power generation technology (Keyes, 2018).

13.5 The case of China

China's cumulative energy storage capacity is estimated to increase rapidly from 843 MWh in 2017 to 32.1 GWh in 2024 (News-release, 2019). In China, electrochemical energy storage accounts for 4.9% of the country's energy storage capacity. Lead batteries play a leading role in China's energy storage sector, demonstrating 18.6% of its electrochemical storage capacity. For example, Narada's 20 MW/160 MWh project located in Wuxi, China is one of the largest customer-serving battery energy storage projects globally. By giving time shift/storage service, the 20 MW project enhances energy exploitation in the power system by considering peak load.

13.5.1 Summary and benefits of the energy storage project (Narada Power Source, 2020)

- Technical summary: plant power capacity is 20 MW, storage capacity is 160 MWh, plant's designed life-time is 10 years, and architecture consists of 80 sets of 250 kW/20 MW battery.
- Energy storage system includes: power conditioning system, transformer (16000 KVA/1200 KVA), battery management system, monitoring, and control system
- It is a leading commercialized project on battery energy storage system (BESS) and the first augmented distribution and energy storage project in China.
- It supports electricity billing management at the distribution network of the industrial zone serving more than 50,000 people working in industries by shifting the peak load to an off-peak period and partaking in demand response.
- Annual electricity allocation at peak hours: 53,760 MWh and electricity bill savings: $5.43 Million/Year.
- Participating in demand response in Jiangsu province to acquire local government subsidy.

13.6 Challenges and policy implications

Although the potential benefits of installing energy storage systems have been broadly realized in developing countries, several considerable challenges in implementing energy storage systems still exist (Luo et al., 2015). First, how to select the appropriate energy storage systems to match the power system application requirements; second, how to precisely assess the definite values of installed energy storage system facilities; and finally, how to bring the cost down for recently developing technologies. In many circumstances, demand for energy storage and selection of suitable technologies have been considered critical and challenging in developing countries. For example, with the enormous growth of renewable energy sources, energy storage can substantially impact renewable energy integration in developing economies (Rohit et al., 2017). According to the de Sisternes et al. (2019), the challenges that currently hamper the acceptance of sustainable energy storage solutions in developing countries are as follows (de Sisternes et al., 2019);

- Developing country has limited technology access to the needs of sustainable development. For example, battery storage technology is often mismanaged in developing countries and not planned to operate in often-harsh conditions. Therefore, a crucial technical prerequisite needs to be manifested to create a sustainable energy storage market.
- The revenue certainty is impeded in developing countries due to a lack of policy and inappropriate regulatory frameworks, which critically depend on successful project delivery. For example, the lack of clarity for bankable remuneration schemes and obtaining necessary energy storage permits do not exist in most cases.
- Lack of key enabling factors often exists in developing countries, including skilled staff, recycling plans for used batteries, and transportation solutions. For example, trained staff's availability for installing and maintaining systems is commonly quoted as a crucial challenge.

Several enabling factors need to be adopted in developing countries to make the energy storage system cheaper, and more sustainable (de Sisternes et al., 2019). For example, collaboration with the local industries, developing skilled technicians and encouraging the government to initiate awareness program for the people of remote areas. The availability of qualified technicians plays a key role before and after constructing the energy storage system, which also plays a critical role in sustainable economic development in developing countries. The available instrument for energy storage management is not optimized for developing countries' perspectives. An effective policy for recycling energy storage systems is imperative for establishing energy storage deployment in developing countries (de Sisternes et al., 2019). For example, battery technology based on toxic materials needs to be handled appropriately; otherwise, lack of proper recycling systems could hinder environmental sustainability requirements.

Therefore, in power systems, energy storage can effectively improve the reliability of the system and smooth out the fluctuations of intermittent energy (Shi and Luo, 2017). Stakeholders, including power purchaser, operators and regulators have been very conscious since 2006, while adding intermittent power to the national grid, mainly because of the low Capacity Utilization Factor (CUF). CUF is the ratio of actual output produced from the solar power plant over the year to the AC plant capacity multiplied by number of hours in that year. The low CUF was initially assessed by the Regulator to be around 17%, and this factor was used for the calculation of the Feed-in Tariff (FIT) of the solar power plant (Aizad, 2018). The utilization factor of micro/mini-hydro systems in India has been found to be low, due to the unavailability of sufficient water during the dry season versus what is available during monsoons (Stambaugh et al., 2017). The current utilization factor for many pico and micro-hydro systems in Northwest Nepal is surprisingly low, around 10%. This low utilization significantly impacts the systems' benefits, thus making it difficult for them to grow economically and sustainably (Stambaugh et al., 2017). Heat storage systems can effectively enhance power utilization efficiency. Solar collectors can directly heat water, and the hot water of the plants may also be stored (Kalair et al., 2021). Moreover, solid electrode batteries are well suited to compensate for the intermittent nature of renewable energy systems in developing countries (Rugolo and Aziz, 2012).

13.7 Conclusion

Most developing countries do not have adequate grid connections in remote areas for an uninterrupted power supply. In that case, renewable energy has become a popular option in developing countries for electricity generation due to its sustainable nature and cost-effectiveness features. However, due to its oscillation nature, energy storage is likely to play a vital role in energy security in these countries. The primary energy storage types include hydro pumped storage, battery, flywheel, and compressed air storage, which can supply energy during peak-demand hours. However, much research is still needed to fully understand the application of sustainable and cost-effective energy storage technologies in developing economies. This chapter has recommended that government, policymakers, and energy storage technology manufacturers need to invest time and resources in developing effective and efficient sustainable energy storage technologies for the developing world. This has also discussed the various potential and widely used energy storage technologies applications and their effects on the economy and environment, including quality traits and sustainability in various developing countries. As more durable and cost-effective energy storage technologies with diverse applications continue to emerge in the market, adopting renewable energy integration in developing countries will accelerate, leading to better energy safety and reliability. This chapter has also shown that hydro pump storage systems have a longer life than batteries-based systems and are considered the most useful energy storage type. For example, rechargeable batteries can operate for a large number of cycles until the battery stops working efficiently and their energy discharge loss rate is higher than the hydro pump storage evaporation rate, making the latter more appealing for long-term energy storage. While various industries assemble batteries and manufacturing processes take a short time. Further, pump hydro storage facilities take many years and require many construction resources. Additionally, scaling-up sustainable energy storage and the wide variety of alternatives and intricate characteristic patterns make it difficult to evaluate a specific energy storage technology for an application in developing countries. Therefore, it has been recommended that updated energy policies are crucial for the cost-effective implementation of sustainable energy storage systems in developing countries. This should start with the investment in the availability of cost-effective energy storage technologies and support for developing and testing novel energy storage system designs. However, even if these technologies were available, their implementation in developing countries would still be delayed because energy storage is still new in many developing countries. Finally, accomplishing such deployment ecologically, socially, and economically sustainable entails toxic material recycling plans,

13.8 Self-evaluation

13.8.1 Questions

1. What are the most important functionalities of energy storage in unstable power grids?
2. What are the disputes and remedies to scaling-up sustainable energy storage practice in developing countries?
3. What are the energy storage technologies that are well fitted to be implemented in developing countries?
4. What are the leading regulatory challenges to incorporating energy storage systems into power grids?
5. What can we expect to see in innovation in storage technology in the next 5 to 10 years?

13.8.2 Numerical problems

1. A Tesla car is designed with a rechargeable battery to drive its motor. The battery needs 9 hours to charge fully, and can store up to 70 MJ of energy with a potential difference of 320 V.
   a) Calculate the flow of charge during battery charging. [Answer: 218750 C]
   b) How much current is required to charge the battery? [Answer: 6.75 A]
2. A supercapacitor has a full charge of 400 µC at a potential difference of 9 V across its plates. How much energy can be stored in the supercapacitor? [Answer: 1.80 mJ]
3. A government of a developing country has designed an energy-efficient hydropower plant. The water head is positioned at 1.80 m after the dam. The flowing water velocity is 2.20 m/s, while the cross-sectional area of water flow is 8.20 m² with a 1000 kg/m³ density of water.
   a) Calculate the volumetric flow and mass flow rate of a flowing fluid. [Answers: 18.04 m³/s, 18040 kg/s]
   b) What is the maximum theoretical power? [Answer: 318,225.6 W]
   c) What is the actual generated power if the turbine operating efficiency is 92.5%? [Answer: 294,358.68 W]
   d) How much energy would be produced by this turbine in one year? [Answer: 2578582.0368 kWh]
4. A newly installed compressed air energy storage cylinder has a volume of 0.8 m³, which runs at a pressure of 2.8 MPa and a temperature of 25°C. During the discharge mode, the air temperature and pressure fall to 20 °C and 2.5 MPa, respectively. If the air's molecular weight is 28.7 g/mol, what will be the released air mass? [Answer: 2.38 kg]
5. A 10 cm³ methanol fuel cell stack has ten cells that operate at standard conditions. How much electrical energy can be produced per kg of fuel by a methanol fuel cell? [Answer: 21.2 MJ/kg]
6. The rotation speed of a single-piston flywheel is 2000 RPM. It absorbs 1500 J of energy during its half rotation. If the speed fluctuation is ± 50 RPM, what will be the minimum inertia? Assume there is no friction. [Answer: 0.68 kgm²]

7. A piece of aluminum has a mass of 120 g at 40°C. When it is immersed in gases at 100°C temperature, about 0.8 g of steam is condensed. Determine the specific heat capacity of aluminum. Given that the latent heat of steam is 540 cal/g. [Answer: 0.06 cal/g °C]

References

A.Kirubakaran, S.J., Nema, R.K., 2009. A review on fuel cell technologies and power electronic interface. Renew. Sustain. Energy Rev. 13 (9), 2430–2440.

Abdellatif, D., et al., 2018. Conditions for economic competitiveness of pumped storage hydroelectric power plants in Egypt. Renewables Wind Water Solar 5 (1), 2.

Abdin, Z., Khalilpour, K.R., 2019. Single and Polystorage Technologies for Renewable-Based Hybrid Energy Systems. Polygeneration with Polystorage for Chemical and Energy Hubs; Available from: https://opus.lib.uts.edu.au/handle/10453/138640. Elsevier, Melbourne, Australia, pp. 77–131.

Ahuja, D., Tatsutani, M., 2009. Sustainable energy for developing countries. SAPIENS Surv.Persp. Integr. Environ. Soc. 2 (1), 1–16.

Aizad, S. Capacity Utilization Factor & Solar Power Plants. 2018; Available from: https://www.linkedin.com/pulse/capacity-utilization-factor-solar-power-plants-salman-aizad.

Akour, S.N., Al-Garalleh, A.A., 2019. Candidate sites for pumped hydroelectric energy storage system in Jordan. Mod. Appl. Sci. 13 (2).

Ali, M.H., Wu, B., Dougal, R.A., 2010. An overview of SMES applications in power and energy systems. IEEE Trans. Sustain. Energy 1 (1), 38–47.

Almehizia, A.A., et al., 2020. Assessment of battery storage utilization in distribution feeders. Energy Transitions 4 (1), 101–112.

Alva, G., Lin, Y., Fang, G., 2018. An overview of thermal energy storage systems. Energy 144, 341–378.

Arif, M.T., Oo, A.M., Ali, A., 2013. Investigation of energy storage systems, its advantage and requirement in various locations in Australia. J. Renew. Energy 2013.

Armaroli, N., Balzani, V., 2011. Towards an electricity-powered world. Energy Environ. Sci. 4 (9), 3193–3222.

Aurbach, D., et al., 2002. On the use of vinylene carbonate (VC) as an additive to electrolyte solutions for Li-ion batteries. Electrochim. Acta 47 (9), 1423–1439.

Babu, T.S., et al., 2020. A comprehensive review of hybrid energy storage systems: converter topologies, control strategies and future prospects. IEEE Access 8, 148702–148721.

Baker, J., 2008. New technology and possible advances in energy storage. Energy Policy 36 (12), 4368–4373.

Baker, J., Collinson, A., 1999. Electrical energy storage at the turn of the millennium. Power Eng. J. 13 (3), 107–112.

Bansal, N., Srivastava, V., Kheraluwala, J., 2019. Renewable energy in India: Policies to reduce greenhouse gas emissions. Greenhouse Gas Emissions. Springer, Singapore, pp. 161–178.

Barzin, R., et al., 2015. Peak load shifting with energy storage and price-based control system. Energy 92, 505–514.

Berdichevsky, G., et al., 2006. The tesla roadster battery system. Tesla Motors, 1 (5), 1–5.

Biswas, M.M., et al., 2011. Prospects of renewable energy and energy storage systems in Bangladesh and developing economics. Glob. J. Res. Eng. (GJRE) 11 (5), 23–31.

Blume, S., 2015. Global energy storage market overview & regional summary report. Ener. Stor. Coun., Australia.

Buckley, T., 2019. Pumped Hydro Storage in India, EEFA Australasia.

Bullough, C., et al., 2004. Advanced adiabatic compressed air energy storage for the integration of wind energy. In: Proceedings of the European wind energy conference, EWEC. Citeseer.

Carnegie, R., et al., 2013. Utility scale energy storage systems. State Utility Forecasting Group. Tech. Rep. 1 (June), 1–90.

Chattopadhyay, D., et al., 2019. Battery storage in developing countries: key issues to consider. Electr. J. 32 (2), 1–6.

Chen, H., et al., 2009. Progress in electrical energy storage system: a critical review. Prog. Nat. Sci. 19 (3), 291–312.

Chen, H., et al., 2013. Compressed air energy storage. Energy Storage Technol. Appl. 4, 101–112.

Chua, K.H., Lim, Y.S., Morris, S., 2015. Cost-benefit assessment of energy storage for utility and customers: a case study in Malaysia. Energy Convers. Manage. 106, 1071–1081.

Çolak, M., Kaya, İ., 2020. Multi-criteria evaluation of energy storage technologies based on hesitant fuzzy information: a case study for Turkey. J. Energy Storage 28, 101211.

Colthorpe, A., 2020. China deployed 855MWh of electrochemical storage in 2019 despite slowdown, in energy-storage.news.

David, W, 2020. Lithium Ion Secondary Battery Anode Materials Market with (Covid-19) Impact Analysis: Growth, Latest Trend Analysis and Forecast 2026. Galus Australis.

de Sisternes, F.J., et al., 2019. Scaling-up sustainable energy storage in developing countries. J. Sustain. Res. 2 (1), e200002.

Deane, J.P., Gallachóir, B.Ó., McKeogh, E., 2010. Techno-economic review of existing and new pumped hydro energy storage plant. Renew. Sustain. Energy Rev. 14 (4), 1293–1302.

Desideri, U., Asdrubali, F., 2018. Handbook of Energy Efficiency in Buildings: A Life Cycle Approach. Butterworth-Heinemann.

Díaz-González, F., et al., 2012. A review of energy storage technologies for wind power applications. Renew. Sustain. Energy Rev. 16 (4), 2154–2171.

Dimov, N., Kugino, S., Yoshio, M., 2003. Carbon-coated silicon as anode material for lithium ion batteries: advantages and limitations. Electrochim. Acta 48 (11), 1579–1587.

Dincer, I., Dost, S., 1996. A perspective on thermal energy storage systems for solar energy applications. Int. J. Energy Res. 20 (6), 547–557.

Dincer, I., Rosen, M., 2011. Thermal Energy Storage: Systems and Applications. John Wiley & Sons, USA.

Divya, K., Østergaard, J., 2009. Battery energy storage technology for power systems—an overview. Electr. Power Syst. Res. 79 (4), 511–520.

Dursun, B., Alboyaci, B., 2010. The contribution of wind-hydro pumped storage systems in meeting Turkey's electric energy demand. Renew. Sustain. Energy Rev. 14 (7), 1979–1988.

Evans, A., Strezov, V., Evans, T.J., 2012. Assessment of utility energy storage options for increased renewable energy penetration. Renew. Sustain. Energy Rev. 16 (6), 4141–4147.

Farret, F.A., Simoes, M.G., 2006. Integration of Alternative Sources of Energy. John Wiley & Sons, USA.

Fernandes, D., et al., 2012. Thermal energy storage: "How previous findings determine current research priorities". Energy 39 (1), 246–257.

Gulagi, A., et al., 2020. Current energy policies and possible transition scenarios adopting renewable energy: a case study for Bangladesh. Renew. Energy.

Hadjipaschalis, I., Poullikkas, A., Efthimiou, V., 2009. Overview of current and future energy storage technologies for electric power applications. Renew. Sustain. Energy Rev. 13 (6-7), 1513–1522.

Hasan, Mahamudul, et al., 2016. Temperature regulation of photovoltaic module using phase change material: a numerical analysis and experimental investigation. Inter. j. Photo. 2016, 1–8.

Hasan, Md Mahmudul, et al., 2018. Observation of fuel cell technology and upgraded photovoltaic system for rural telecom system in Bangladesh. BAUET J. 1 (2), 71–78.

Hu, X., et al., 2017. Technological developments in batteries: a survey of principal roles, types, and management needs. IEEE Power Energy Mag. 15 (5), 20–31.

Ibrahim, H., Ilinca, A., Perron, J., 2008. Energy storage systems—characteristics and comparisons. Renew. Sustain. Energy Rev. 12 (5), 1221–1250.

IEA, 2018. Renewable energy policy of Bangladesh, Ministry of power. BPDB Dhaka.

IEA, World Energy Outlook 2019. 2019.

Islam, M.T., et al., 2014. Current energy scenario and future prospect of renewable energy in Bangladesh. Renew. Sustain. Energy Rev. 39, 1074–1088.

Kalair, A., et al., 2021. Role of energy storage systems in energy transition from fossil fuels to renewables. Energy Storage 3 (1), e135.

Kaldellis, J.K., 2010. Stand-Alone and Hybrid Wind Energy Systems: Technology, Energy Storage and Applications. Elsevier, Netherland.

Kaygusuz, K., 2012. Energy for sustainable development: A case of developing countries. Renew. Sustain. Energy Rev. 16 (2), 1116–1126.

Keyes, N., 2018. World Bank Group Commits $1 Billion for Battery Storage to Ramp Up Renewable Energy Globally.

Khan, I., 2015. Solar Energy Changes People's Life on Kutubdia Island; Available from: https://www.sma-sunny.com/en/solar-energy-changes-peoples-life-on-kutubdia-island/.

Khan, I., 2019. Drivers, enablers, and barriers to prosumerism in Bangladesh: a sustainable solution to energy poverty? Energy Res. Soc. Sci. 55, 82–92.

References

Khan, I., 2020. Sustainability challenges for the south Asia growth quadrangle: a regional electricity generation sustainability assessment. J. Cleaner Prod. 243, 118639.

Khan, I., 2020. Impacts of energy decentralization viewed through the lens of the energy cultures framework: solar home systems in the developing economies. Renew. Sustain. Energy Rev. 119, 109576.

Khan, I., 2020. Waste to biogas through anaerobic digestion: Hydrogen production potential in the developing world-a case of Bangladesh. Int. J. Hydrogen Energy 45 (32), 15951–15962.

Kleperis, J., et al., 2016. Energy storage solutions for small and medium-sized self-sufficient alternative energy objects. Bulg. Chem. Commun. 48 (E), 290–296.

Koçak, B., Paksoy, H., 2019. Using demolition wastes from urban regeneration as sensible thermal energy storage material. Int. J. Energy Res. 43 (12), 6454–6460.

Koohi-Fayegh, S., Rosen, M., 2020. A review of energy storage types, applications and recent developments. J. Energy Storage 27, 101047.

Kousksou, T., Jamil, A., Bruel, P., 2010. Asymptotic behavior of a storage unit undergoing cyclic melting and solidification processes. J. Thermophys. Heat Transfer 24 (2), 355–363.

Kumar, A., Shukla, S., 2015. A review on thermal energy storage unit for solar thermal power plant application.

Kumar, A.R., Shrimali, G., 2020. Battery storage manufacturing in India: a strategic perspective. J. Energy Storage 32, 101817.

Levin, T., Thomas, V.M., 2016. Can developing countries leapfrog the centralized electrification paradigm? Energy Sustain. Dev. 31, 97–107.

Li, C., et al., 2013. Techno-economic feasibility study of autonomous hybrid wind/PV/battery power system for a household in Urumqi. China Energy 55, 263–272.

Li, X., et al., 2011. Ion exchange membranes for vanadium redox flow battery (VRB) applications. Energy Environ. Sci. 4 (4), 1147–1160.

Lourenssen, K., et al., 2019. Vanadium redox flow batteries: a comprehensive review. J. Energy Storage 25, 100844.

Lu, M., et al., 2020. An optimal electrolyte addition strategy for improving performance of a vanadium redox flow battery. Int. J. Energy Res. 44 (4), 2604–2616.

Luan, F., et al., 2018. Future development and features of pumped storage stations in China. In: IOP Conference Series: Earth and Environmental Science. IOP Publishing, UK.

Luis, M., 2020. Tracking Energy Storage. IEA; Available from: https://www.iea.org/reports/tracking-energy-storage-2020.

Lund, H., Salgi, G., 2009. The role of compressed air energy storage (CAES) in future sustainable energy systems. Energy Convers. Manage. 50 (5), 1172–1179.

Luo, X., et al., 2015. Overview of current development in electrical energy storage technologies and the application potential in power system operation. Appl. Energy 137, 511–536.

Magda, R., Szlovák, S., Tóth, J., 2020. The role of using bioalcohol fuels in sustainable development. Bio-Economy and Agri-Production. Elsevier, Netherland, pp. 133–146.

Mandal, S., Das, B.K., Hoque, N., 2018. Optimum sizing of a stand-alone hybrid energy system for rural electrification in Bangladesh. J. Cleaner Prod. 200, 12–27.

Meishner, F., Sauer, D.U., 2019. Wayside energy recovery systems in dc urban railway grids. eTransportation 1, 100001.

Mekhilef, S., Saidur, R., Safari, A., 2012. Comparative study of different fuel cell technologies. Renew. Sustain. Energy Rev. 16 (1), 981–989.

Melikoglu, M., 2013. Vision 2023: feasibility analysis of Turkey's renewable energy projection. Renew. Energy 50, 570–575.

Melikoglu, M., 2017. Pumped hydroelectric energy storage: Analysing global development and assessing potential applications in Turkey based on Vision 2023 hydroelectricity wind and solar energy targets. Renew. Sustain. Energy Rev. 72, 146–153.

Ming, Z., Kun, Z., Daoxin, L., 2013. Overall review of pumped-hydro energy storage in China: status quo, operation mechanism and policy barriers. Renew. Sustain. Energy Rev. 17, 35–43.

Molina, M.G., 2012. Distributed energy storage systems for applications in future smart grids. In: 2012 Sixth IEEE/PES Transmission and Distribution: Latin America Conference and Exposition (T&D-LA). IEEE.

Nandi, S.K., Ghosh, H.R., 2009. A wind–PV-battery hybrid power system at Sitakunda in Bangladesh. Energy Policy 37 (9), 3659–3664.

Narada Power Source Co., L., 2020. Energy Hybrid Solutions; Available from: http://naradaintl.com/index.php/solutions/energy.html.

News-release, 2019. China to become largest energy storage market in Asia Pacific by 2024; Available from: https://www.woodmac.com/press-releases/china-to-become-largest-energy-storage-market-in-asia-pacific-by-2024/. Wood Mack.

Nguyen, D.C., Pham, V.V., 2020. Pathway of sustainable fuel development with novel generation biofuels. AIP Conference Proceedings, 1st ed., 2235. AIP Publishing LLC, USA.

Nikolaidis, P., Poullikkas, A., 2017. A comparative review of electrical energy storage systems for better sustainability. J. Power Technol. 97 (3), 220–245.

Nomura, T., Okinaka, N., Akiyama, T., 2010. Technology of latent heat storage for high temperature application: a review. ISIJ Int. 50 (9), 1229–1239.

Noussan, M., Jarre, M., 2018. Multicarrier energy systems: optimization model based on real data and application to a case study. Int. J. Energy Res. 42 (3), 1338–1351.

Olabi, A., et al., 2020. Critical review of energy storage systems. Energy 214, 118987.

Pande, P., et al., 2003. Design development and testing of a solar PV pump based drip system for orchards. Renew. Energy 28 (3), 385–396.

Qays, Md Ohirul, et al., 2020. Active cell balancing control strategy for parallelly connected LiFePO$_4$ batteries. CSEE J. Power Energy Syst. 7 (1), 86–92.

Qays, Md Ohirul, et al., 2021. An intelligent controlling method for battery lifetime increment using state of charge estimation in PV-battery hybrid system. Appl. Sci. 10 (24), 8799.

Rahman, F., Rehman, S., Abdul-Majeed, M.A., 2012. Overview of energy storage systems for storing electricity from renewable energy sources in Saudi Arabia. Renew. Sustain. Energy Rev. 16 (1), 274–283.

Raju, M., Khaitan, S.K., 2012. Modeling and simulation of compressed air storage in caverns: a case study of the Huntorf plant. Appl. Energy 89 (1), 474–481.

Rastler, D., 2010. Electricity Energy Storage Technology Options: A White Paper Primer on Applications, Costs and Benefits. Electric Power Research Institute, USA.

Reinhardt, B., 2010. Thermal Energy Storag. Available from: http://large.stanford.edu/courses/2010/ph240/reinhardt1/.

Ren, J., Ren, X., 2018. Sustainability ranking of energy storage technologies under uncertainties. J. Cleaner Prod. 170, 1387–1398.

Rohit, A.K., Devi, K.P., Rangnekar, S., 2017. An overview of energy storage and its importance in Indian renewable energy sector: part I–technologies and comparison. J. Energy Storage 13, 10–23.

Rugolo, J., Aziz, M.J., 2012. Electricity storage for intermittent renewable sources. Energy Environ. Sci. 5 (5), 7151–7160.

Sabihuddin, S., Kiprakis, A.E., Mueller, M., 2015. A numerical and graphical review of energy storage technologies. Energies 8 (1), 172–216.

Saboori, H., Hemmati, R., Jirdehi, M.A., 2015. Reliability improvement in radial electrical distribution network by optimal planning of energy storage systems. Energy 93, 2299–2312.

Salles, R.S., Souza, A., Ribeiro, P.F., 2020. Energy storage for peak shaving in a microgrid in the context of Brazilian time-of-use rate. In: Multidisciplinary Digital Publishing Institute Proceedings.

Salvarli, M.S., Salvarli, H., 2020. For sustainable development: future trends in renewable energy and enabling technologies; Available from: https://www.intechopen.com/chapters/71531. Resour Challenges Appl. 3.

Sarbu, I., Sebarchievici, C., 2018. A comprehensive review of thermal energy storage. Sustainability 10 (1), 191.

Sciacovelli, A., et al., 2017. Dynamic simulation of Adiabatic Compressed Air Energy Storage (A-CAES) plant with integrated thermal storage–link between components performance and plant performance. Appl. Energy, 185, 16–28.

Sedghi, M., Ahmadian, A., Aliakbar-Golkar, M., 2015. Optimal storage planning in active distribution network considering uncertainty of wind power distributed generation. IEEE Trans. Power Syst. 31 (1), 304–316.

Sharma, P., Bhatti, T., 2010. A review on electrochemical double-layer capacitors. Energy Convers. Manage. 51 (12), 2901–2912.

Shi, N., Luo, Y., 2017. Capacity value of energy storage considering control strategies. PLoS One 12 (5), e0178466.

Sivakumar, N., et al., 2013. Status of pumped hydro-storage schemes and its future in India. Renew. Sustain. Energy Rev. 19, 208–213.

Sivaraman, A., 2019. New international partnership established to increase the use of energy storage in developing countries.

Smith, S.C., Sen, P., Kroposki, B., 2008. Advancement of energy storage devices and applications in electrical power system. In: 2008 IEEE Power and Energy Society General Meeting-Conversion and Delivery of Electrical Energy in the 21st Century. IEEE.

Stambaugh, M., et al., 2017. Improving the utilization factor of islanded renewable energy systems. In: Solar World Congress.

Stutz, B., et al., 2017. Storage of thermal solar energy. C.R. Phys. 18 (7-8), 401–414.

Subramani, G., et al., 2017. Grid-tied photovoltaic and battery storage systems with Malaysian electricity tariff—a review on maximum demand shaving. Energies 10 (11), 1884.

Tasmania, H., 2018. Battery of the nation–Tasmanian pumped hydro in Australia's future electricity market.

Thackeray, M.M., Wolverton, C., Isaacs, E.D., 2012. Electrical energy storage for transportation—approaching the limits of, and going beyond, lithium-ion batteries. Energy Environ. Sci. 5 (7), 7854–7863.

Thaller, L., 1977. Redox flow cell development and demonstration project, calendar year 1977. Cleveland: National Aeronautics and Space Administration.

Tian, Y., Zhao, C.-Y., 2013. A review of solar collectors and thermal energy storage in solar thermal applications. Appl. Energy 104, 538–553.

Tong, Z., Cheng, Z., Tong, S., 2020. A review on the development of compressed air energy storage in China: technical and economic challenges to commercialization. Renew. Sustain. Energy Rev. 135, 110178.

Tronchin, L., Manfren, M., Nastasi, B., 2018. Energy efficiency, demand side management and energy storage technologies–a critical analysis of possible paths of integration in the built environment. Renew. Sustain. Energy Rev. 95, 341–353.

Viswanathan, V.V., et al., 2010. Effect of entropy change of lithium intercalation in cathodes and anodes on Li-ion battery thermal management. J. Power Sources 195 (11), 3720–3729.

Walawalkar, R., Apt, J., 2007. Market Analysis of Emerging Electric Energy Storage Systems. National Energy Technology Laboratory. pp. 1–118.

Walawalkar, R., Apt, J., Mancini, R., 2007. Economics of electric energy storage for energy arbitrage and regulation in New York. Energy Policy 35 (4), 2558–2568.

Wang, J., et al., 2017. Current research and development trend of compressed air energy storage. Syst. Sci. Control Eng. 5 (1), 434–448.

Wang, R., Xu, Z., Ge, T., 2016. Introduction to solar heating and cooling systems. Advances in Solar Heating And Cooling. Elsevier, Netherland, pp. 3–12.

Wheeler, B., 2010. Flywheel Energy Storage. Available from: http://large.stanford.edu/courses/2010/ph240/wheeler1/.

Wilberforce, T., et al., 2016. Advances in stationary and portable fuel cell applications. Int. J. Hydrogen Energy 41 (37), 16509–16522.

Wu, F.-B., Yang, B., Ye, J.-L., 2019. Grid-Scale Energy Storage Systems and Applications, 1st ed. Academic Press, Netherland.

Xianxu, H., et al., 2016. Review on key technologies and applications of hydrogen energy storage system. Energy Storage Sci. Technol. 5 (2), 197.

Xinhua, 2020. China completes test on 100 MW compressed air energy storage expander. China Daily.

Xue, X., Mei, S., Lin, Q., 2016. Energy internet oriented non-supplementary fired compressed air energy storage and prospective of application. Power Syst. Technol. 40 (1), 164–171.

Yang, Z., et al., 2011. Electrochemical energy storage for green grid. Chem. Rev. 111 (5), 3577–3613.

Yuan, W., 2011. Second-Generation High-Temperature Superconducting Coils and Their Applications for Energy Storage. Springer Science & Business Media, Germany.

Zahedi, A., 2014. Sustainable power supply using solar energy and wind power combined with energy storage. Energy Procedia 52, 642–650.

Zhang, D., et al., 2017. Present situation and future prospect of renewable energy in China. Renew. Sustain. Energy Rev. 76, 865–871.

Zhang, X., et al., 2011. Electrospun nanofiber-based anodes, cathodes, and separators for advanced lithium-ion batteries. Polym. Rev. 51 (3), 239–264.

Zhang, X., Chu, J., Li, H., 2015. Key technologies of flywheel energy storage systems and current development status. Energy Storage Sci. Technol. 4 (1), 55–60.

CHAPTER 14

Climate change, sustainability, and renewable energy in developing economies

Mahfuz Kabir[a], Zobaidul Kabir[b] and Nigar Sultana[c]

[a]Bangladesh Institute of International and Strategic Studies (BIISS), Dhaka, Bangladesh [b]School of Environmental and Life Sciences, University of Newcastle, Ourimbah Campus, Australia [c]Knowledge for Development Management (K4DM) Project, UNDP, Bangladesh

14.1 Introduction

Climate change is the greatest concern of the nations in today's world. Keeping global warming well below 2°C according to the Paris Agreement is the foremost global target, which requires a robust and rapid transformation of the adoption of low to no-carbon technologies. Environmental hazards and global warming emanating from the predominant use of fossil fuel necessitate the use of alternative energy sources to reverse the negative effects on ecology, environment, and atmosphere with a cumulative stock of greenhouse gases (GHGs). Renewable energy has become important to developing economies since it can help achieve sustainable growth without harming the environment. Economic growth through conventional energy will worsen the quality of the environment. Because of being less efficient than conventional energy, the renewable energy sector needs to focus on its inadequate scale size and operational deficiency to attain environmental sustainability. However, they need to work on their scale size to increase their energy generation capacity by providing enough resources and implementing policies (Ibrahim and Alola, 2020). The reduction of carbon emission from energy production through a decline in using fossil fuels is intrinsically interwoven with sustainable development (Suo et al. 2021). Economic growth has a bidirectional relationship with both CO_2 emission and renewable energy consumption (Radmehr et al. 2021). The most normal practice is to achieve economic development through different processes that degrades the environment. Increasing consumption of renewable energy results in a lesser ecological footprint than nonrenewable energy. Results of the recent empirical studies suggest the adoption of energy-efficient resources to increase the use of renewable energy and reduce the share of consumption of traditional energy to enhance the quality of the environment for sustainable development (Ullah et al., 2021).

Understanding the nexus among climate change, sustainability and renewable energy in the context of developing economies is important for several strong reasons. First, energy consumption is responsible for about three-quarters of total GHGs emissions (OWD, 2020) and nonrenewable sources emit a higher portion of GHGs emissions than renewable sources (IPCC, 2020). Nonrenewable energy, especially fossil fuel, is the prime source of energy in the developing world, which makes production and consumption unsustainable and triggers climate change. Developing economies are leading the global economic growth through massive low-cost production using fossil fuel, which has been creating detrimental effects on the environment and climate through emitting GHGs and hazardous elements in the atmosphere. Therefore, it is important to comprehend how to reduce dependency on nonrenewables and facilitate developing economies to deal with climate change devoid of compromising with economic growth and ongoing developmental efforts. Second, transitioning from nonrenewable to renewable energy is currently a global concern in which the developing economies are taking active interest through accelerated investment in installing the technologies, expanding the market rapidly, and leading the production in many instances. Third, exploring renewable energy sources is the safest and most resilient solution for developing economies to mitigate adverse impacts of climate change and achieve the Paris Agreement and Sustainable Development Goals (SDGs), especially energy-related goals (see, Fig. 14.1) (United Nations, 2018). Some developing economies have expanded their renewable energy capacity while some renewable energy sources are yet to attract required attention because of manifold challenges despite having enormous potential. Fourth, identifying policy options and suitable strategies

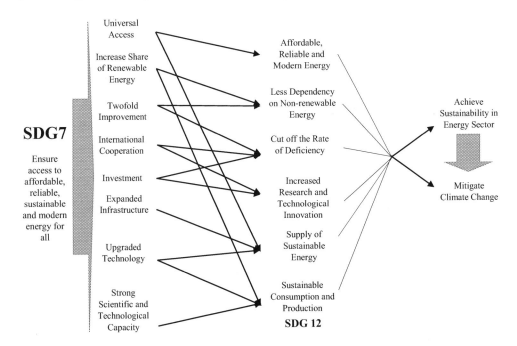

FIGURE 14.1 Energy-related SDGs, climate change, and renewable energy.

is imperative to overcome challenges and harnessing potential toward adopting renewable energy in developing economies for climate change mitigation and achieving sustainable development.

Given this backdrop, the rest of the chapter has been organized as follows. Section 14.2 presents a conceptual link among global climate change, use of energy, and sustainability in the context of the SDGs. Section 14.3 discusses CO_2 emission from fossil fuel and electricity consumption, and climate risks and provides a detailed account of the present state of renewable energy generation of developing economies. It also presents the production of renewable energy of developing economies using maps to demonstrate spatial distribution of production across the world by type of energy. Section 14.4 portrays the adoption of renewable energy technologies by developing economies in the recent period in the context of climate change for mitigation. Section 14.5 discusses the various opportunities and challenges of transformation from nonrenewable to renewables in the backdrop of future scenarios of renewable energy and common policy perspectives in developing economies. Section 14.6 suggests a set of recommendations to promote investment, technological development, and fiscal and monetary policy supports to promote renewable energy technologies. Finally, Section 14.7 concludes the chapter.

14.2 Linkage between climate change, energy, and sustainability

The Environmental Kuznets Curve (inverted U-shaped economic growth-environmental deterioration curve) indicates that economic growth leading to environmental detriments increase at the low-income level of an economy, and the negative environmental standard decreases at the high-income level (Grossman and Krueger, 1991). According to this notion, economic growth triggers environmental pollution of developing economies because of their over-dependence on low-cost fossil fuels in all sectors to accelerate the production of goods and services (Balsalobre et al., 2015; Danish et al., 2019). Thus, economic growth led by traditional energy consumption emits massive CO_2 to the atmosphere, which is the main GHG that accelerates the climate change process (Lau et al., 2014). Though developed countries were mostly responsible in the past for high CO_2 emissions, developing economies are now emitting notable CO_2 because of their rapid urbanization, industrialization, and economic growth (Elum and Momodu, 2017).

Many sustainable development goals (SDGs), such as poverty reduction, decent jobs, income, sustainable cities, sustainable production and consumption, industries, biodiversity, and climate change and ecosystems are directly dependent on the use of energy. Therefore, there are many challenges and opportunities related to various sources of energy. The SDGs cannot be achieved without ensuring a dependable supply of energy. Conversely, the use of fossil fuels creates enormous pressure on the ecosystem, environment, and global climate, which runs the risk of making the development unsustainable. "SDG 7: Ensure access to affordable, reliable, sustainable and modern energy for all" aims to address this challenge (United Nations, 2018). Increasing the share of renewables in the global energy mix by 2030 is one of the key targets (Fuso et al. 2018), which needs massive public and private investment from domestic and international sources toward achieving SDG 7 and other related goals.

Renewable energy is produced directly or indirectly from the sun and other natural processes, which is stocked up within a short period. The sources of this energy include hydro, solar, wind, biofuel (ethanol, biodiesel, green diesel, and biogas), thermal, tidal, ocean wave, geothermal, photochemical, and photoelectric (TREIA, 2015). The largest GHGs emissions come from the use of energy, and energy-related emissions have been estimated to be increased by about 16% by 2040 (OECD/IEA, 2015). As fossil fuel exploration and use cause environmental degradation and discharge of harmful gases to the atmosphere, developing economies need to diversify their energy mix including clean renewable energy to meet the energy requirement. Given the detrimental effects of fossil fuel on the ecosystem, environment, and climate, the global agenda for sustainable development calls for using clean and renewable sources of energy to enhance economic development, human wellbeing, protect the ecosystem and environment, and promote sustainability. Because of recent technological development, renewable energy derived from various cost-effective methods can be utilized to generate heat, power, and fuels for transportation, industry, and commercial and household use.

Renewable energy has been identified as the future for sustainable development. There are fundamental reasons for considering this energy to promote sustainable development: (i) lower adverse impact on the environment than that of nonrenewable energy, (ii) potential to address the environmental apprehensions of the present and future periods, particularly the emission of GHGs, e.g., CO_2, oxides of nitrogen and sulfur, methane, and particulate matters that are generated from using fossil fuel. Nonrenewable energy is an unsustainable energy source because it would be exhausted in the next couple of decades if the current rate of use persists. Conversely, renewable energy sources would not be depleted. Renewable energy is a viable solution of access to power and energy in the geographically vulnerable regions, isolated coastal and riverine islands, and remote hilly areas where energy supply from the national grid, pipeline, network, and storage is physically impossible or financially unviable (Gielen et al., 2019; Saim and Khan, 2021). Thus, it supports system decentralization and facilitates local solutions for meeting the energy needs independently, and augments the flexibility of energy production and supply in isolated communities or areas to facilitate agricultural production, manufacturing, and other economic activities vis-à-vis household needs (Oyedepo, 2012).

The interaction between the use of renewable energy and sustainable development has been displayed in Fig. 14.2 in the context of climate change. It demonstrates that both nonrenewable and renewable sources of energy are used in economic activities. Economic growth is accelerated through energy flow especially in manufacturing, transportation, and other services vis-à-vis irrigation and other activities in agriculture, agro-processing, and storage. However, energy use in these sectors, if dominated by fossil fuel, would cause environmental pollution and excessive GHGs emission leading to climate change. The negative environmental consequences would cause tangible and intangible costs to the society, which would affect the low-income and marginalized segments of the society more adversely. It would make development unsustainable. However, a clean environment and fewer climate-induced hazards through the use of renewable energy would help reduce poverty and inequality, decrease economic and social costs of climate change, and lessen planetary pressure, which would ultimately make development sustainable (UNDP, 2020).

Energy can work as a "golden thread" to promote growth, equity, and sustainability together (IEA, 2017). The use of clean and renewable energy ensures sustainability, increasing energy use would encourage economic growth, and paves to way to attain equity

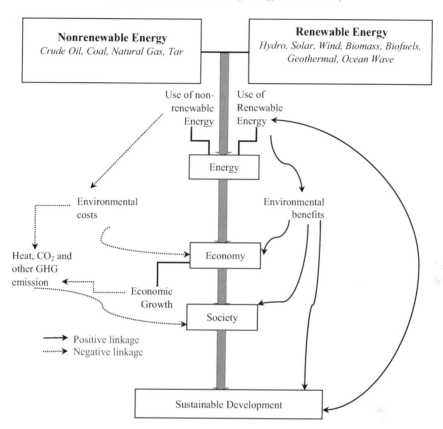

FIGURE 14.2 Energy use, economic growth, and sustainable development.

(Sarkodie et al., 2020). Ensuring environmental sustainability requires reducing the dependence on nonrenewable energy, whereas environmental quality can be maintained through the trio of renewable energy, financial development, and population growth (Riti et al., 2017).

Global warming is related to the loss of biodiversity (Tittensor et al., 2019; Khare et al., 2020). Various social problems such as migration, food security, armed conflicts, and inequalities can also increase (Puaschunder, 2020). Renewable energy can facilitate growth, equity, and sustainability as it has the potential to improve the accessibility of electricity. Employment generation and reduction of poverty and income inequality are the ultimate results of increased accessibility (Solt, 2016). Even social benefits like health, education, and gender equity can be achieved through increased use of renewable energy (Wang et al., 2018).

Continued and affordable access to energy is required for sustainable economic and social development. However, low environmental impact including GHG emissions from the use of energy is a must for development to be sustainable (UNDP, 2002). Notably, renewable energy technologies emit CO_2 much lower than that of fossil fuels (Bilgen et al., 2008; Kaygusuz, 2009).

TABLE 14.1 Net electricity consumption in selected developing countries.

	1995	2000	2005	2010	2015	2017	2018
Total Global (in billion kWh)	11,485.69	13,281.12	15,777.75	18,703.96	21,278.54	22,486.34	23,398.40
Developing countries (% of total)	49.74	48.26	51.33	56.58	61.33	63.05	63.71
Top consumers from the developing world (in % of the world)							
China	7.63	8.87	13.92	19.86	24.68	26.46	27.58
India	2.77	2.89	3.15	3.89	5.10	5.45	5.46
Russia	6.20	5.39	4.91	4.59	4.18	4.09	3.97
Brazil	2.28	2.42	2.33	2.44	2.40	2.30	2.26
Mexico	1.09	1.26	1.22	1.16	1.16	1.16	1.16
Turkey	0.60	0.74	0.82	0.91	1.02	1.10	1.10

Source: Based on US Energy Information Administration (EIA) data (2021).

14.3 Growth of electricity demand and climate change: Impact on developing economies

14.3.1 Electricity consumption in developing economies and climate change

Energy consumption is vital for achieving economic and social development in countries across the world. The crucial role of electricity in economic growth and the improved living standard has been widely recognized (IEA, 2009). However, a major concern in electricity production from fossil fuels is related to GHG emission, which can be reduced through the expansion of alternative power generation technologies (Eskeland et al., 2012).

The amount of electricity consumption has been increasing over the last few decades all over the world. Developing economies also followed the same trend. In 2018, all developing economies cumulatively consumed 63.71% of the world's total electricity consumption and it has been more than 51% since 2005. Among them, China's consumption share is nearly half of the total electricity consumed by the developing economies in the last few years. Before that China's share ranged from one-seventh to one-fourth of the total electricity consumed by the developing economies. India's consumption share also increased throughout the years. However, Russia has experienced a constant decrease in its electricity consumption rate since 1995. Brazil and Mexico's consumption share almost remained the same in the last 25 years. Turkey's change in the rate of consumption over the years is not that significant. Table 14.1 depicts these scenarios.

Over the last two decades, the trend of the production of renewable energy is increasing in nature. However, there was insignificant growth in the share in the 2000s. The share of renewables in the total electricity was around 19-20% at the end of that decade. After that, a steady increase can be noticed in the production of renewable energy, which reached around 27% at the end of the 2010s. Notably, the share of renewable electricity of developing economies has steadily increased over the last two decades, this increase was from around 9 to 16% in the last 20 years.

TABLE 14.2 CO_2 emission from fossil fuel consumption.

	1990	2000	2005	2010	2015	2018	2019
Global Total (million mt)	22,683.30	25,699.80	30,051.44	33,971.15	36,247.49	37,668.11	38,016.57
Developing economies (% of total)	44.05	43.73	49.76	56.28	60.04	61.17	62.16
China (%)	10.60	14.33	20.88	26.97	29.44	29.62	30.34
India (%)	2.64	3.87	4.06	5.18	6.33	6.79	6.83
Sources of emission by developing economies (% of total)							
Coal and coke (%)	13.45	19.72	25.71	32.06	33.81	33.78	
Consumed natural gas (%)	2.34	7.52	7.91	8.55	9.00	9.98	
Petroleum and other liquids (%)	9.76	14.46	14.09	14.41	15.79	16.32	

Sources: Based on EIA (2021) and Crippa et al. (2020).

The recent surge of CO_2 emissions of developing economies is alarming. Over the last decade, the amount of emission increased significantly and ranged between 56 and 62% of the world's total emissions (see Table 14.2). This implies that a large part of developing economies is still heavily dependent on nonrenewable energy sources. Though China's performance in the renewable energy sector is significant, it emitted around 27-30% CO_2 of the global total emission in the last ten years. While India is still at an improving stage in terms of renewable energy consumption, it emitted comparatively low CO_2, around 5 to 7% of the global total emission. Among the sources of fossil fuel, coal and coke have always been responsible for the highest amount of GHG emission (Khan, 2019). Around 32 to 34% of total CO_2 has been emitted from the use of coal and coke in the last ten years. The second responsible source is petroleum and other liquids since 14 to 16% of the global total has come from this source in the last decade. Natural gas is also responsible for a significant amount of CO_2 emission, which was 8 to 10% of the global total in the last ten years (EIA, 2021 and Crippa et al., 2020).

The use of fossil fuels is the primary source of GHG emission leading to global climate change. According to IPCC (2021), the ever-increasing concentrations of GHGs reached the highest amount than ever. The amount for CO_2 is 410 ppm, 1866 ppb for CH_4, and 332 ppb for N_2O in 2019. In the last 120 to 170 years (from 1850-1900 to 2011-2020), the global surface temperature increased by 1.09 (0.95–1.20)°C. The land area experienced a higher temperature rise (1.59 [1.34–1.83]°C) than the ocean (0.88 [0.68–1.01]°C). Clearly, GHGs are responsible for 1°C to 2°C global surface temperature rise. Over the last 270 years (1750–2019), fossil fuel combustion and land-use change released 700 ±75 PgC (1 PgC = 10^{15} g of carbon) to the atmosphere. Unfortunately, the atmosphere still has 41% ± 11% of that total released carbon. Among all the anthropogenic activities, fossil fuels generate 64% ± 15% of CO_2 and over the last decade, there is an 86% ± 14% sharp increase in CO_2 emissions by fossil fuels (IPCC, 2021). Fossil fuel oxidation is the most important reason behind environmental degradation because the world's overwhelming majority of electricity comes from this process (Anand et al., 2015). Currently, fossil fuel is heavily being used in developing economies to produce electricity, which is causing many environmental and climatic risks.

TABLE 14.3 Countries with the highest climate risks.

Year: 2019		Years: 2000-2019	
Country	Losses in billion PPP$	Country	Losses in billion PPP$
India	68.81	Thailand	7.72
Mozambique	4.93	Pakistan	3.77
Zimbabwe	1.84	Philippines	3.18
Bolivia	0.80	Bangladesh	1.86
Afghanistan	0.55	Myanmar	1.51
Malawi	0.45	Haiti	0.39
Niger	0.22	Mozambique	0.30
South Sudan	0.09	Nepal	0.23

Source: Eckstein et al. (2021).

Developing economies are on the top list of global climate risks in both the short and long run (see Table 14.3), which is an outcome of their collective CO_2 emission even though developed economies are also significant emitters through the aggressive use of fossil fuels. Developing countries across continents like India, Thailand, Mozambique and Pakistan are some of the most climate-vulnerable economies in terms of their economic losses (Eckstein et al., 2021). Therefore, delinking the climate-energy nexus can help achieve sustainable development through the rapid adoption of renewable energy technologies as alternatives to fossil fuels, which would help reduce greenhouse gas emissions and the risk of climate-induced disasters.

14.3.2 Production of renewable energy in developing economies

Developing economies have been trying to reduce their dependability on nonrenewable electricity for the last two decades. Consequently, they have been increasing their production capacity of renewable electricity over time, and the surplus electricity has been selling to nearby countries or territories. For example, Bhutan exports about 70% of its hydroelectricity to India[1]. In the 2000s, the share of the production of renewable energy of developing economies increased from 47 to 53% of the world's total renewable electricity. It rose to around 59% in the last decade (c.f., Fig. 14.3). It indicates that developing economies are now collectively leading the production of renewable electricity in the world.

Developing economies are in the process of a gradual transition from fossil fuels to renewables to mitigate environmental pollution and climate change. This transition can be observed from the growth and increasing share of renewable energy in the total energy portfolio. The production of renewable energy has witnessed considerable growth over the last two decades.

---

[1] See, for details, https://www.sasec.asia/index.php?page=news&nid=1178&url=bhu-hydro-export-aug2020 (accessed on 23 September 2021).

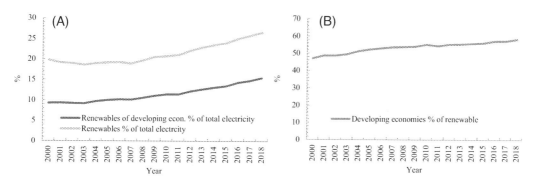

FIGURE 14.3 (A) Total renewable electricity and renewable electricity in developing countries, (B) Percentage of renewable energy in developing economies. *Source: Based on EIA database (2021).*

TABLE 14.4 Global production of renewable energy (billion kWh).

Category	2002	2004	2006	2008	2010	2012	2014	2016	2018
Renewables	2,924.41	3,170.96	3,473.42	3,777.82	4,237.26	4,765.89	5,333.24	5,965.98	6,743.72
Hydroelectricity	2,606.11	2,787.65	3,008.24	3,173.76	3,405.35	3,626.66	3,828.17	3,989.76	4,155.58
Nonhydroelectric	318.29	383.31	465.18	604.06	831.91	1,139.23	1,505.06	1,976.22	2,588.14
Geothermal	51.95	55.85	58.00	63.27	64.56	65.96	71.73	77.27	83.03
Solar, tide, wave, fuel cell	23.74	30.16	32.89	36.28	61.46	132.06	231.09	371.75	601.18
(i) Tide and wave	21.87	27.17	27.21	23.70	27.99	28.26	29.47	27.17	26.25
(ii) Solar	1.88	2.99	5.68	12.59	33.47	103.79	201.62	344.59	574.93
Wind	52.77	84.14	131.83	220.30	339.52	521.27	714.06	956.16	1,265.47
Biomass and waste	189.83	213.16	242.46	284.21	366.37	419.94	488.19	571.04	638.46

Source: US EIA database (2021).

Specifically, between 2002 and 2018, the production of all renewable energy increased by 131%. Hydroelectricity occupies the greatest share of renewable energy. However, its share in the total production of renewable energy has declined from 89% in 2002 to 62% in 2018. Other types of renewable energy, such as wind, biomass and waste, and solar have been gradually occupying a considerable share in the global portfolio of renewable energy (Table 14.4). Among the later technologies, the performance of wind power turned out to be most significant–from only 1.8% in 2002–its share has increased to 18.8% in 2018. Among other types of renewable energy, the share of biomass and waste-to-energy has increased from 6.5 to 9.5% during the same period. Solar energy has experienced the most magnificent growth during this period and secured its share from nearly 0.06% in 2002 to 8.53% in 2018. This technology is being considered to be the most promising source of renewable energy in the long run given its rapid expansion across the world.

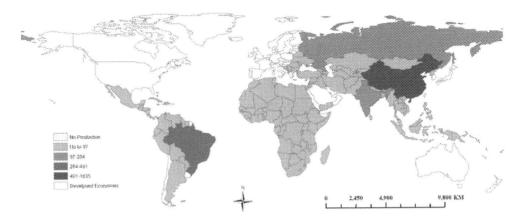

MAP 14.1 Global scenarios of renewable production in developing economies, 2018 (billion kWh). *Source: Prepared by authors based on EIA database.*

Although the transition from fossil fuels to renewables has been growing steadily across the world (Belançon, 2021), three developing economies–China, Brazil and India–secure the first, third, and fourth positions in terms of installed capacity of renewable energy technologies, respectively. The total production of renewable energy has been increasing worldwide led by China–the largest developing economy in the world. The country's capacity is much closer to the capacity of the sum of the other nine economies. China has invested in renewable energy technologies and expanded its capacity rapidly (from 219 GW in 2009 to 925 GW in 2019) for the sustainable transition from nonrenewable to renewable energy as it emits the highest amount of GHGs in the world. The installed capacity of Brazil and India was remarkable, 150 and 139 GW. The other developing countries are also in the process of transition from nonrenewable to renewable energy technologies to reduce the adverse impact of fossil fuels on the environment and climate change.

Among the developing economies, China has produced the highest of renewable energy than all other developing economies and in recent years (2016–2018) they produced more renewable energy than the total production of developed economies. Brazil's amount of share is also significant followed by Turkey and India. Latin American economies have also focused on exploring renewable energy. African and most of the Asian developing economies are still not well-acquainted with these technologies (Map 14.1). Most of the developing economies, such as China, Turkey, India, and Vietnam have been performing well in producing renewable in terms of compounded annual growth rate (CAGR). Their annual growth rate is higher than that of developed economies (Fig. 14.4). It means that developing economies are taking renewable energy as a climate change mitigation measure more seriously than that of developed economies.

14.3.2.1 *Hydroelectricity*

Developing economies across the world opted for massive hydropower projects to meet the energy requirements of their rapid economic growth. Most of the renewable energy used

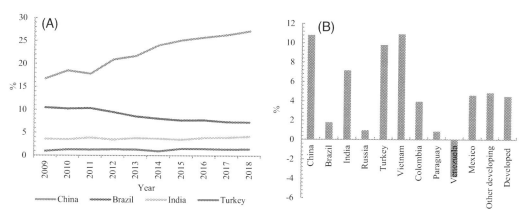

FIGURE 14.4 Share of renewables in selected developing economies (% of global, [A]) and compounded annual growth rate (in %, [B]). *Source: Based on EIA database (2021).*

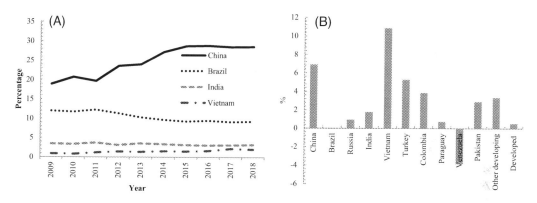

FIGURE 14.5 Share of hydroelectricity in selected developing economies (% of global, [A]) and compounded annual growth rate (in %, [B]). *Source: Based on EIA database (2021).*

came from this source. The consumption of hydroelectricity has been increasing along with its significant effects on sustained economic growth in China. Based on the data over the period 2010 to 2018, Zhang et al. (2021) estimated that the country's consumption of hydropower exceeded 270 million tonnes of oil equivalent (Mtoe) in 2020. Developing economies from the Global South have significantly expanded their installed capacity of hydroelectricity projects to meet the energy requirement of their economic expansion (Mayer et al., 2021). China's share of hydroelectricity reached from around 21 to 29% over the last decade. Brazil also holds a significant amount of share of hydroelectricity production although its share decreased gradually over the last decade. India's share remains almost constant and Vietnam progressed slightly in terms of the percentage of the global share. However, Vietnam's growth rate of hydroelectricity production is the highest among all economies (Fig. 14.5). Most of the Latin American developing economies are also progressing gradually in this energy sector. African and some Asian developing economies are yet to make notable contribution in this

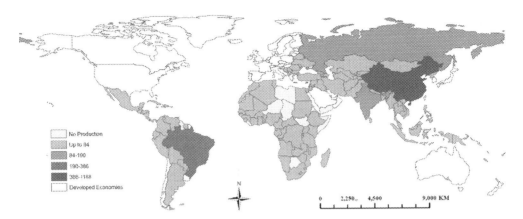

MAP 14.2 Production of hydroelectricity by developing economies, 2018 (billion kWh). *Source: Prepared by authors based on EIA database.*

sector (see, Map 14.2). However, developing economies are experiencing higher growth in hydroelectricity production than that of developed economies.

14.3.2.2 Solar

The rapid growth of solar photovoltaic (PV) installed capacities has been observed in the last two decades. It is mainly because of multiple applications, such as at the household level (solar home system), irrigation pump, office rooftop, mini-grid, and large-scale solar parks. Another major reason is its rapid reduction of cost per unit of electricity generation. Thus, it has become a financially viable solution to reducing GHG emissions from using traditional fossil fuels (Breyer et al., 2015). Solar photovoltaic (PV) systems have expanded rapidly in China over the last ten years (Bai et al., 2021). China emerged as the world's top country in the installed capacity of solar PV (Li et al., 2021). It emerged as the global leader in the solar energy sector through innovation in technology and rapid expansion of production capacity toward a sustainable transition to renewable energy and supplier of cheap solar PV systems in the developing economies. India's contribution to solar energy is also notable followed by Brazil and Turkey (Fig. 14.6). Other developing economies still need to focus properly to perform well in this sector (Map 14.3). All developing economies are exploring this renewable energy sector to some extent (Khan, 2020). A technological breakthrough at the beginning of the 2010s made them adopt this sector more intensely than other ones. Though developed economies performed well in adopting solar power, their growth of production is lower compared to the developing economies.

14.3.2.3 Wind

Wind energy is emerging as an important source of renewable energy. The production of wind energy shows almost similar trends as solar. China has achieved remarkable progress over the years–it occupied around 29% of the world's total wind power in 2018 (Fig. 14.7). According to Soares et al. (2021), China would be one of the top four consumers of this energy

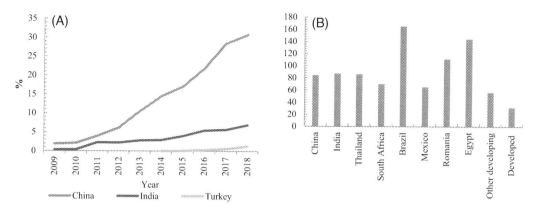

FIGURE 14.6 Share of solar power in selected developing economies (% of global, [A]) and compounded annual growth rate (in %, [B]). *Source: Based on EIA database (2021).*

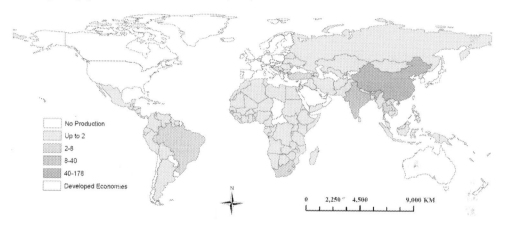

MAP 14.3 Production of solar energy by developing economies, 2018 (billion kWh). *Source: Prepared by authors based on EIA database.*

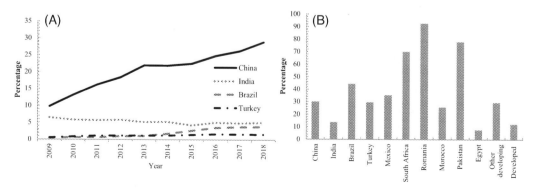

FIGURE 14.7 Share of wind power in selected developing economies (% of global, [A]) and compounded annual growth rate (in %, [B]). *Source: Based on EIA database (2021).*

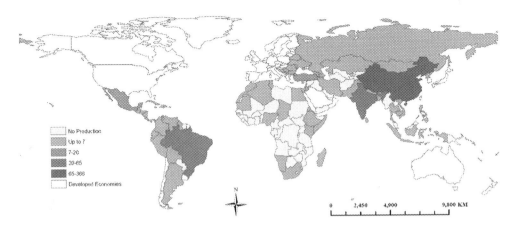

MAP 14.4 Production of wind energy by developing economies, 2018 (billion kWh). *Source: Prepared by authors based on EIA database.*

based on the projected consumption in the business-as-usual scenario for the period 2018 to 2025 along with Canada, Sweden, and Germany. China's rise in this renewable energy is becoming enormous because of its rapid expansion of the installed capacity and technological innovation. India holds the second-highest share among the developing countries, followed by Brazil and Turkey. Other developing economies are also working on wind energy. Asian, Latin American, and African economies performed well in expanding wind power (Map 14.4). Again, the contribution of developed economies to this power is lower than that of the developing economies. Except for China, the individual amount of production of wind energy of all the other developing economies is higher than solar and biomass- and waste-to-energy production. However, considering the volume of production, there is still room to make progress in this energy.

14.3.2.4 Biomass- and waste-to-energy

Biomass is a sustainable source of renewable energy in both developed and developing countries. It occupies a significant share in the energy mix albeit of varying economic feasibility and competitiveness of solid biofuels across developing economies (Angulo-Mosquera et al., 2021). Waste to energy (WtE) is a source of safe, renewable and economically viable, and environmentally sustainable energy, which is a potential alternative to fossil energy. Ever-growing municipal solid waste (MSW) is a major feedstock of WtE, which would partially meet the demand for energy and help manage MSW efficiently and sustainably (Rasheed et al., 2021). The utilization of biomass has grown notably in the energy portfolio and as an input of power generation in Brazil (Sampaio and Sampaio, 2021). China's position is the top among developing economies in this energy, which is followed by Brazil, India, and Thailand (Fig. 14.8 and Map 14.5). Indonesia's growth rate is the highest among developing economies, which indicates its promising future in this energy source. The growth of developed economies in biomass- and waste-to-energy sector is quite lower than developing economies.

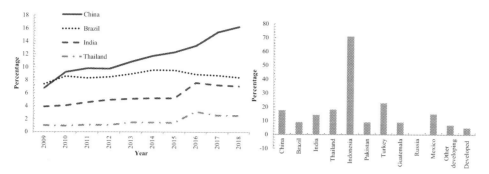

FIGURE 14.8 Share of biomass- and waste-to-energy in selected developing economies (% of global, [A]) and compounded annual growth rate (in %, [B]). *Source: Based on EIA database.*

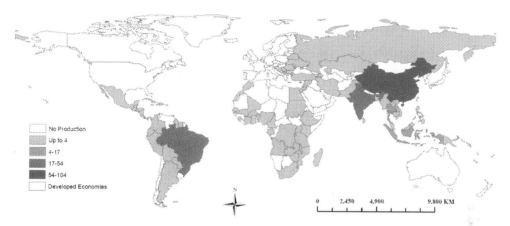

MAP 14.5 Production of biomass- and waste-to-energy by developing economies, 2018 (billion kWh). *Source: Prepared by authors based on EIA database.*

Considering the critical technology behind energy production from biomass and waste, developing economies have demonstrated significant performance in the last ten years. Except for China, energy production from biomass and waste of all developing economies is higher than their production of solar energy.

Biogas generated from organic waste can be used to meet fuel needs at the household level, decrease financial and environmental costs of waste disposal, and help maintain environmental sustainability (Bekchanov et al., 2019). It would also help achieve the SDGs related to increased access to clean fuel, improve the environmental standard, and decrease carbon footprint (Surendra et al., 2014; Bekchanov and Evia, 2018).

14.3.2.5 Bioenergy

Production of biofuel has attracted significant attention at the global level over the past one and a half decades, which is mainly because it is renewable, biodegradable, and harmless fuel. Bioenergy is regarded as a promising source of renewable energy because of its advantages

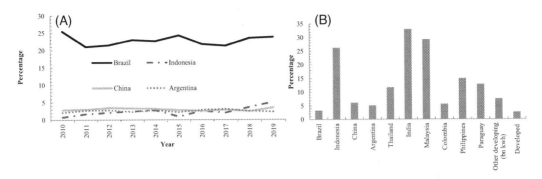

FIGURE 14.9 Share of bioenergy in selected developing economies (% of global, [A]) and compounded annual growth rate (in %, [B]). *Source: Based on EIA database (2021).*

of balanced CO$_2$ emission along with sustainable waste management (Kung and Mu, 2019). Brazil has become the top producer of biofuels during this period because of favorable policies and technologies promoted by the government (Aguilar-Rivera et al., 2021). The biodiesel industry experienced phenomenal growth from nearly no production in 2005 to greater than 4 billion liters by the end of 2017 in Brazil, which made the country the top producer in the developing world. Nearly 90% of biodiesel is produced from Soybean oil in the country. The other inputs include sunflower, peanut, castor, palm, and used oil. The demand for biodiesel has been projected to be 9 billion liters in 2024 in the country (Rodrigues, 2021).

Brazil pays more attention to its bioenergy sector than any other developing economies. Their amount of share ranged between 21 and 26% of the world's total bioenergy in the 2010s. At the end of the last decade, Indonesia had the second-highest share of bioenergy (more than 5 per cent) among developing economies. The other Latin American country performing well in this sector is Argentina (Fig. 14.9). Developing economies, such as India, Malaysia, and Indonesia are putting emphasis on expanding their bioenergy sector. Latin American and some other Asian economies are also trying to boost up their bioenergy production by exploring new sources (Swaraz et al., 2019). However, most of the African economies have not yet explored this type of renewable energy (Map 14.6).

Cumulatively, other developing economies (which are not listed in the top 4) have demonstrated significant progress in each renewable energy sector. Their collective production was more than 43% of the world's total renewable energy in 2018. They jointly produced around 39% of the world's total hydroelectricity in 2018. The next sector that received attention is biofuel, which is followed by biomass and WtE, solar, and wind. They have been experiencing an overall increasing trend although the solar sector showed more progress than others in these economies (Fig. 14.10).

Despite having climate change and sustainability concerns, there is an almost equal increase in energy generation capacity of both renewable and nonrenewable energy sources. Solar technology received the highest attention and a global total of 638GW in 2019 has added to the capacity from merely 25GW in 2009. The wind and gas added 487GW and 438GW to their capacity at the same time. Therefore, if these four sectors are compared, the capacity

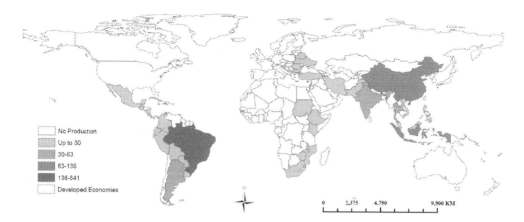

MAP 14.6 Production of biofuel by developing economies, 2019 ('000' barrels per day). Data of 2016 for Costa Rica, Cuba, Swaziland, Ethiopia, Iran, Jamaica, Kenya, Malawi, Mauritius, Mozambique, Nicaragua, North Macedonia, Panama and Sudan. *Source: Prepared by authors based on EIA database.*

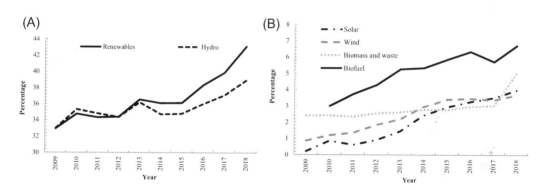

FIGURE 14.10 Share of renewable of other developing economies (% of global). *Source: Based on EIA database (2021)*

of two nonrenewable sources increased almost the same amount as two renewable sources. Hydroelectricity also performed well with some 283 GW added to their capacity in the last decade (Fig. 14.11). A positive factor is that oil and nuclear power registered a downward trend in this case–their worldwide generation capacity has decreased in this timeline (Frankfurt School-UNEP Centre/BNEF, 2019).

Developing economies have huge potential to expand their renewable energy capacity, which is evident from the last decade. According to the data of IRENA (2021) China's performance is distinctive as it added more than 675GW of capacity between 2010 and 2020. Its capacity is higher than the cumulative capacity of all developing economies during the same period. The growth of capacity of Brazil and India is also promising although their capacity is less than one-tenth of China. However, Brazil and India could have engaged themselves more in exploring this sector considering China's performance. Mainly Asian and Latin American developing economies have earnestly been trying to explore renewable energy

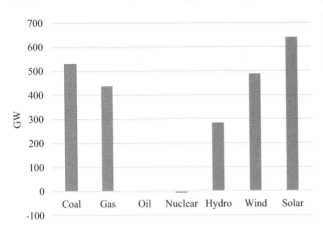

FIGURE 14.11 Net capacity (in GW) added in electricity generation (technology specific) between 2009 and 2019. *Source: Frankfurt School-UNEP Centre/BNEF (2019).*

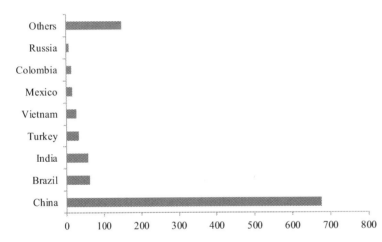

FIGURE 14.12 Capacity (in GW) added in electricity generation in developing economies between 2010 and 2020. *Source: Based on IRENA database (2021).*

technologies (Fig. 14.12). The progress of African economies is very slow in this sector. The possible reason might be their inability to resource utilization and unavailability of renewable energy technologies.

Solar power technologies dominated renewable power technologies in terms of added capacity in the last decade. The solar sector could add around 380GW over the last decade. Wind power technologies also rapidly sharpened their capacity. Around 332GW capacity has been added worldwide over the last decade by this sector. Hydropower technologies also showed considerable progress (around 263GW capacity added) over the same timeline according to IRENA (2021) database. Bioenergy's contribution is insignificant compared to other renewable energy technologies. Only around 41GW more energy could be produced in the last decade than in the 2000s (Fig. 14.13). The least contributing sectors are geothermal and marine energy. They added very little amount in their capacity in the 2010s.

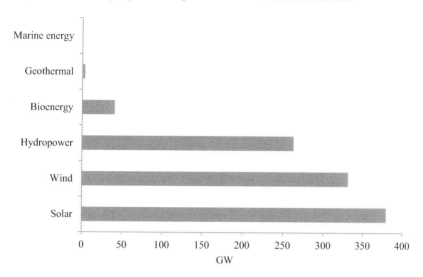

FIGURE 14.13 Capacity (in GW) added in renewable power technologies between 2010 and 2020. *Source: Based on IRENA database (2021).*

14.4 Mitigating climate change: Transition from nonrenewable to renewable

The energy sector needs to be at the front of the targets climate change mitigation as reducing carbon intensity is crucial for climate change mitigation (Gökgöz and Yalçın, 2021). Fostering economic development requires increasing economic activities, which need more energy production and consumption. Therefore, economies need to use a greater amount of energy use on the higher level of development (Tang et al. 2018). To make this development sustainable, economies need to decrease their dependence on nonrenewable energy sources. This would also help mitigate environmental degradation and adverse impacts of climate change (Zubi et al. 2016).

Currently, both developed and developing economies are looking for sustainable development using energy that causes insignificant or no adverse environmental impact. Clean and renewable energy appeared as solutions to mitigate climate change by reducing GHGs emissions (Wang et al., 2018; Urban, 2018). The technologies to generate renewable energy have been undergoing rapid change over time. The first-generation technologies of producing renewable energy that came at the end of the 19th century are the outcomes of the industrial revolution–these included hydroelectricity, biomass combustion, and geothermal power and heat. The second-generation technologies included solar heating and cooling, wind power, bioenergy, and solar photovoltaics (PV), which are major sources of renewable energy in the present time. However, the third-generation technologies are still under development and being made cost-effective for use in economic and business activities vis-à-vis at the household level. These technologies are, inter alia, advanced biomass gasification, biorefinery, solar thermal power, geothermal energy, thermochemical, WtE, and ocean energy (IEA, 2021). These technologies attracted the global attention of the users and witnessed major investment in innovation of technologies since the early 2000s. The rapid evolution of technologies implies that renewable energy would soon be considered as competitive with their nonrenewable counterparts. It can be observed from the growth of the utilization of renewable energy at the global level.

Therefore, in the context of the growing global population, it is believed that the increasing demand for energy based on fossil fuel would lead to unsustainable development because of the cumulative adverse impact on the environment and increased climate variability, especially in developing economies. Since the supply of fossil fuel is exhaustible, exploitation technological advancement in renewable energy would serve as a reliable and long-term supply of energy.

Variable renewable energy (VRE) sources, e.g., solar and wind energy, are exposed to enormous challenges of energy planning and operation of electricity supply because of significant variations in their energy flows during day and night, and across seasons (Khan et al. 2018). Characteristics of power plants, geographical remoteness, weather patterns and catchment area of solar panels and wind turbines are some important factors that also influence the investment decisions by both public and private sectors and pattern of electricity production from large-scale VRE plants (Kiviluoma et al., 2014). Therefore, it is essential to assess variability and long-term potential of energy production keeping in mind the location energy matrix to make the VRE sources reliable and economically viable. However, considering only the locational factors, e.g., local average wind speed or average local solar radiation, are insufficient to determine a precise estimate of the future flow of energy from these VRE sources (Viviescas et al., 2019).

GHG emissions and Earth's average temperature are positively correlated according to historical data (IPCC, 2018). Nonrenewable energy sources are mainly responsible for global warming (Montoya et al., 2021). Therefore, it is imperative to undertake mitigation strategies that require the adoption of practices that promote GHG emissions reduction (Cadez et al., 2019). This will ensure sustainable energy production and consumption (Godfray and Garnett, 2014).

The consumption of nonrenewable energy and economic growth exert a harmful effect on climate change in the long run, while consumption of renewable energy has minimal or no impact on climate change. Nevertheless, bidirectional causality has been found between consumption of nonrenewable energy and climate change, and unidirectional causality has been found from climate change to consumption of renewable energy in the short run (Brini, 2021). The results clearly imply that the increasing use and rapid transition toward renewable energy can facilitate climate change mitigation.

Damage to natural environment and adverse climatic effects from GHG emissions can be mitigated significantly by reducing the use of fossil fuels (Umar et al., 2020). Green energy technologies can facilitate achieving the SDGs related to carbon footprint, sustainable production, and consumption sustainable cities, and climate change. Renewable energy is considered to be the most appropriate alternative to fossil fuel to attain sustainable development (Vickers, 2017). Renewable energy technologies are rapidly becoming accessible and cost-effective with the benefits of mitigating environmental pollution and climate change (Gielen et al., 2019). Most studies reveal that the impact of renewable energy on the environment is positive (Mahjabeen et al., 2020).

Adaptation to climate change and mitigation of GHG emissions can be jointly approached by renewable energy technologies. Both developed and developing economies recognized these technologies as adaptation and mitigation tools (Suman, 2021). Renewable energy can be generated from natural sources, food and waste (Ellabban et al., 2014). All these would work as alternatives to conventional nonrenewable energy for adapting to climate change and meeting the energy needs of people who are dependent on natural resources (Sapkota et al., 2014). The living standard is an important element of human development and multidimensional poverty. So, when the living standard is positively influenced by the use of renewable energy (UNDP, 2005), it implies that it can also contribute to human development and multidimensional poverty.

Hydrogen produced from renewable sources is environment-friendly energy, an alternative to the demand for clean and climate-friendly energy especially in developing economies where water is available to produce hydrogen. In fact, hydrogen is a secondary form of energy that can be manufactured as electricity. Given the increasing concerns over environmental degradation and global warming, research for the production of hydrogen from renewable sources especially from water has increased. It would play a significant role in meeting the demand for clean energy to meet the future requirement of climate-friendly energy. Hydrogen can be produced from the electrolysis of alkaline water (pH > 7.5), which would be a financially viable option vis-à-vis the other renewable energy sources, e.g., solar, wind and wave, with low capital cost and operational stability (Nadaleti et al., 2021). Green hydrogen could also be produced from biogas (Khan, 2020a).

Developing economies which are already in the position of transition from nonrenewables to renewables, need to adopt a different mitigation plan. Developing economies, such as the Philippines, Bangladesh and Pakistan are exposed to massive climate-induced extreme events while these are also emitting GHGs through rapid economic expansion and development. These economies are gradually transitioning to renewable but the level of mitigation is rather low. Therefore, they need to adopt a medium-term strategy for rapid transformation of their energy system by increasing shares of renewable resources, which would help mitigate climate vulnerabilities through decrease GHG emissions. Lessons can be learned from the technical and economic feasibility of such transition of Ethiopia, Jordan and Bolivia. While a combination of several renewable technologies led by solar power would be a starting point to the transition of the energy system, other technologies, such as heating and cooling, pumps, biofuel-run vehicles would help mitigate climate change rapidly (Gulagi et al., 2021). In addition, energy supply from VRE technologies, establishing separate grid connectivity and supplying to national grids would facilitate achieving commercial viability of renewable technologies, encourage private investment to the renewable energy sector, and promote mitigation measures.

Many upper middle-income developing economies are in the process of transition toward renewable as a measure of climate change mitigation through the rapid expansion of installed capacity. However, they are facing various challenges, such as intermittent supply, high cost and less reliability (Irsyad et al., 2019). The government of Malaysia has undertaken a comprehensive strategy to promote renewable energy toward making its notable contributions to the country's sustainable growth and socio-economic development. However, since the country is one of the key destinations of global investment and an Asian manufacturing hub, the production cost in its industries needs to be sufficiently low backed by low energy cost and constant supply. In this regard, solar and other renewables are yet to become cost-effective and constant energy sources even though the country's renewable energy has been projected to surpass up to 13GW installed capacity by 2030 (Chachuli et al., 2021). It is also true for even China, which is the top producer of renewable energy and global supplier, is also experiencing problems of grid connectivity and less-efficient technologies (Irsyad et al., 2019). Fossil fuel would remain a vital source of its energy supply in the coming decades–well beyond the end-line of the SDGs in many developing countries. In order to address these drawbacks of similar fast-growing and upper-middle-income developing economies, developing storage capacity, installation of mega renewable energy projects, mainstreaming renewable energy in the medium-term national development plans, subsidized private investment in the renewables sector, and green investments would be some pragmatic initiatives to support the rapid transition.

Renewable energy like solar and wind can play a catalytic role in ensuring energy security and help climate change mitigation in small island developing states (SIDSs). However, SIDSs attach more priority to climate change adaptation than mitigation of GHG emission through reducing nonrenewable energy (Chien et al., 2021). Mitigation measures turn out to be costlier in these states compared to adaptation because fossil fuel is significantly cheaper than the equivalent per unit cost of renewable energy. The feasible nonfossil fuel energy sources in the SIDSs would be wind and solar power because hydroelectricity, geothermal and biomass would not be available or economically viable. However, in order to promote a gradual and sustainable transition to renewable energy, a step-by-step mitigation approach needs to be followed according to IPCC (2011, 2014). The first step would be not to introduce any new fossil fuel project in the energy system even though ongoing nonrenewable plants should not be immediately stopped. In addition, battery energy storage system (BESS) needs to be implemented for continued supply of wind and solar power throughout the year by overcoming diurnal and seasonal variations. In the second step, the share of renewable energy should be at least one-third of the total energy mix by 2030, which is the end-line of the SDGs. This gradual transition would not create significant pressure on the economy. In the third and final step, a long-term plan can be introduced to full adoption of renewable energy to feed the economic activities along with establishing a full grid network. It would help maintain temperature growth to 1.5°C in line with the IPCC guidelines.

Undeniably, developing economies are exposed to two opposing challenges–meeting the growing demand for inexpensive energy to support rapid economic growth and adversities of global climate change that need mitigation measures through adopting expensive renewable energy technologies. Therefore, developing economies can formulate renewable energy plans for climate change mitigation goals keeping in mind the potential tradeoffs between the opposite challenges. Such a plan to facilitate climate change mitigation would entail low emission climate-resilient development strategies, integrated energy, and local climate action plans. Well-coordinated and integrated renewable energy planning is needed to devise policy frameworks to address reduce GHG emissions and climate change mitigation where multiple sectors and levels of government are involved (Cox et al., 2017).

14.5 Opportunities and challenges in adopting renewable energy technology

The consumption of renewable energy including hydroelectricity will continue to increase in the next decades, which is the greatest opportunity for the developing economies to plan for public and private sector investment. However, the improvement rate in this sector will not be the same for all economies. Non-OECD Asian economies' consumption rate will be

TABLE 14.5 World consumption (in quadrillion Btu) of hydroelectricity and other renewable energy by region, reference case.

Region	2020	2025	2030	2035	2040	2045	2050	Annual % change (2020–2050)
Total Non-OECD	52.1	68.9	86.8	104.1	123.1	141.8	160.2	3.8
Non-OECD Europe and Eurasia	3.5	3.6	4.0	4.6	5.3	5.9	6.3	2.0
Russia	2.0	2.0	2.1	2.4	2.6	2.7	2.8	1.2
Other Europe/Eurasia	1.5	1.6	1.9	2.3	2.7	3.2	3.5	2.8
Non-OECD Asia	33.4	47.7	61.9	74.7	88.6	102.6	116.7	4.3
China	22.1	29.0	34.8	39.3	44.8	50.3	53.3	3.0
India	5.2	10.9	17.4	22.8	28.4	35.4	45.2	7.5
Other Asia	6.2	7.8	9.8	12.5	15.3	16.9	18.3	3.7
Africa	4.8	6.1	7.3	8.6	10.6	13.1	15.5	4.0
Non-OECD Americas	10.1	11.0	12.2	13.4	14.5	15.6	16.7	1.7
Brazil	7.2	7.6	8.3	9.2	9.9	10.3	10.7	1.3
Other Non-OECD Americas	2.9	3.3	3.9	4.2	4.6	5.3	6.1	2.5

Source: EIA, International Energy Outlook (2021).

a lot higher than Africa and non-OECD American economies. Among the non-OECD Asian economies, China's consumption rate is higher than any other country in the world and it will keep increasing at a higher rate in the next three decades. Though Brazil's recent (i.e., in 2020) consumption amount is higher than that of India. India's consumption would be 2-3 times more than Brazil's in the upcoming decades. Even India's rate of annual change in consumption would be higher than that of China. A projection of this consumption is illustrated in Table 14.5. The positive change in consumption amount and rate indicates that the world would gradually be able to shift from nonrenewable to renewable energy sources to meet the demand.

China's different types of renewable electricity generation would continue to increase or remain unchanged. Though the amount of hydroelectricity generation will increase in the next five years, the amount would also remain unchanged in the next decades after that change in five years. Electricity generation through solar and wind systems would have a rapid increase. Solar electricity generation would be almost eight times and wind electricity generation will be around three times higher in 2050 than the present amount of generation. Now China's one-third of total generated electricity comes from renewable energy (EIA, 2021). By 2050, China would generate its half of total electricity through renewable sources.

Other non-OECD Asian economies currently use geothermal, hydro, solar, and wind systems for electricity generation. The amount of solar and wind system-generated electricity would increase rapidly till 2050. The solar sector would produce more than 32 times more electricity in 2050 than now and it would be around 15 times more than the present for wind system. The hydroelectricity sector would also have a steady rise in the upcoming decades.

Geothermal electricity would experience a sharp increase by 2025, followed by a steady increase in 2035. However, their amount of electricity generation would remain unchanged from 2035 to 2050 (EIA, 2021). Recent statistics show that around two-fifth of total electricity comes from renewable sources. It would become one-third of total electricity by 2050.

Africa's total renewable electricity generation would be around 3 times more than the present production in 2050 and will have a 4% average annual change. The current trend of electricity generation of African economies states that they would mainly focus on electricity generation by using hydro and solar technologies among all the renewable energy sources. Though now wind generates more electricity than solar systems, electricity generation by solar systems would experience around a 37-fold rise by 2050 (from 15 to 558 billion kWh) and this would result in the highest annual percent change in electricity generation. Electricity generation from wind would also increase in the next decades. However, the amount would be close to the current amount of hydroelectricity. Hydroelectricity would experience a five-fold rise by 2050. Right now it is the main renewable energy source of Africa. Geothermal sources also have good prospects in Africa. With a 16 fold rise by 2050, this sector would have the second-highest average annual percent change in electricity generation. Although the biomass sector has some prospects in Africa, the projection indicates that the waste sector does not have a future in the African economies (EIA, 2021).

Exploring new renewable technologies can open up avenues for low-cost renewable energy, such as green hydrogen. It can be rapidly expanded as economically viable energy in developing countries to use as a raw material for producing almost all types of power and energy (Khan, 2020a). Hydrogen is necessary to produce electricity, methane, gasoline, and diesel. It has importance as a raw material for industrial purposes. All these can be produced by incorporating hydrogen electrolytic systems in specific applications. Among those applications, the most familiar one is water electrolysis. In this process, when electricity flows in the water-containing electrolytic cell, water molecules get divided into hydrogen and oxygen. Renewable generators, i.e., turbines, photovoltaic panels, and hydraulic turbines can be used to supply electricity to continue the water electrolysis process. Developed countries find it expensive to extract hydrogen through the water electrolysis process due to the high cost, whereas, natural resources of developing countries make it easier to produce hydrogen through electrolysis. Thus, considering the distance and demand, it is possible to reduce the cost of water electrolysis in developing countries (Nadaleti et al., 2021).

The increasing affordability of resources and possible strategic assets are turning the current situation of the renewable energy market. Advanced technologies and progress in the power market are also increasing the prospect of renewable energy technologies. Hence, the cost of renewable energy would substantially decrease and affordability would increase (Amir and Khan, 2021). Cogeneration and tri-generation systems ensure reduced emission through maximum resource utilization. To achieve the SDGs regarding this, the integration of renewable energy sources in tri-generation systems can become an opportunity for developing economies (Sonar, 2021).

Business engagement in the private, market-driven renewable energy sector can advance renewable transition. It can be done directly through consumer-driven increases in renewable generation capacity, and indirectly through turning corporate renewable buyers into the beneficiaries and supporters of public policy and regulatory action that promotes decarbonization (Tzankova 2020). There is still an opportunity to explore indigenous renewable resources that can help mitigate economic and environmental vulnerabilities and increase socio-economic benefits (Gulagi et al., 2020).

There is a rising trend of jobs in the renewable energy sector through production, marketing, and forward and backward linkages. The number of jobs in this sector had been increasing since 2012. According to the IRENA database, the worldwide number of jobs in all renewable energy technologies was 11.46 million in 2019, which was merely 7.28 million in 2012. China alone created 38% of jobs in the global renewable energy sector. Among other developing economies, Brazil, India and Bangladesh which are recognized as the locations of the massive growth of renewable energy technologies created considerable employment (10, 7.3, and 1.4 per cent, respectively in 2019). The solar energy sector offers more job opportunities than any other sector through marketing, supply, and services. In 2019, the total employment in solar energy was 40% of total employment in renewable energy, while one-fourth of total global employment in renewable energy came from China's solar energy. It was followed by hydroelectricity (17%), wind (10%), and biomass and waste (7%). However, a promising sector of job creation is biomass and WtE because of the adoption of technology in energy production from massive urban and rural wastes in developing economies in near future. With the increasing population, a huge amount of biomass and waste is being generated every day. Thus, more manpower is needed to work in this growing sector. Thus, this sector will have more job opportunities in the future, which would help achieve SDG related to decent jobs.

Although renewable energy has great potential to achieve SDGs, the adoption process of renewable energy technology faces different kinds of challenges and their cumulative impact makes it difficult to have a smooth transition from nonrenewable to renewable energy sources.

- Lack of proper attention to renewable energy technologies from the government creates a considerable barrier to the expansion of this sector. Without government incentives and properly designed renewable energy policies, it is difficult to convince investors to invest in renewable energy projects. Since not all financial institutions are willing to support renewable energy technologies as these are considered risky initiatives, it is difficult for investors to solely manage a huge amount of money to invest in these technologies. Mainly insufficient and small market-size influences the financial institutions not to invest in this sector (Adelaja, 2020).
- The production of biofuel needs massive use of agricultural land, which competes with land used to produce food. This would create a negative impact on global food production both in commercial and subsistence farming and run the risk of significant increase in global food prices. Thus, biofuel cannot be recommended in developing economies with a high density of population and lack of arable land. In addition, monoculture for the production of the feedstock of biofuel is harmful to biodiversity, which contradicts SDG-14 (biodiversity and conservation) and SDG-15 (life below water and life on land) (Wallimann-Helmer, 2020).
- COVID-19 shifted attention from many sectors by its devastating impact on human lives and economies. Now some short-term policies are needed to address the current situation immediately. Then medium and long-term policies need to be formulated to set the goals and keep the track of progress to promote the transition to renewable energy (Hoang et al., 2021).

- African economies are still facing challenges like restructuring of the power grid, lack of energy storage technologies, and designing parallel mitigation strategies to address environmental factors despite having abundant resources (Amir and Khan, 2021).
- Complexity of energy distribution networks will make it difficult to ensure efficient energy distribution and entry of individual energy producers. It will also increase the risk of unfair pricing and illegal energy use (Yildizbasi, 2021).
- Large-scale hydropower projects resulted in many unaccounted negative environmental externalities that include damage to river ecosystems, biodiversity, and massive human displacement all over the world (Mayer et al., 2021).
- Ensuring participation from stakeholders toward increasing renewable energy generation needs the governments to put efforts into their agenda-setting and preference-shaping power (Mouraviev, 2021).

14.6 Way forward: R&D, market, and fiscal and monetary policy support

14.6.1 Investment in renewable energy capacity

The notable mitigation options of climate change include the transition from fossil fuels to renewable energy and carbon capture and storage. The transformation of the energy system requires a pronounced reallocation of investments toward low-carbon technologies, but without raising significant affordability issues in most countries (Fragkos et al., 2021).

Developing economies are now aware of the necessity of transitioning from nonrenewable to renewable energy. It can be observed from the presence of developing countries in the top 20 of renewable energy capacity investment. China is in the first place in this ranking and also its amount of investment is higher than the cumulative investment of all European economies. From 2009 to the first half of 2019, China's capacity building investment was US$758 billion that is 31% of the global total, whereas all European counties cumulatively invested US$698 billion that is around 28% of the global capacity investment. The other developing economies in the top 20 are India, Brazil, Mexico, South Africa, and Turkey. Among them, India and Brazil are among the top 10 economies. All developing economies (except China) in this list cumulatively invested US$184 billion, which is higher than Europe's top investor's (Germany US$179 billion) investment (Fig. 14.14). A significant cost drop-down of wind and solar technologies in the mid-decade helped developing economies invest in this sector. There is a possibility that their position in the ranking will be higher in the next decade (Frankfurt School-UNEP Centre/BNEF, 2019).

Currently, the global economy has started to undergo a green transformation where multinational corporations (MNCs) are trying to strengthen their capabilities for sustainable green innovation at the global level. Sustainability-oriented MNCs include both environmental pure-players, for example, specialized renewable energy lead firms and multi-technology conglomerates. In this context, green foreign direct investments (FDIs) have enormous potential to contribute to the deepening of sustainability capabilities, especially in the renewable energy sector of developing economies. An analysis of 1,217 green FDI in renewable energy sectors across the world over the period 1997 to 2015 demonstrates that these strengthen the overall orientation to the sustainability of the MNCs. These investments have a greening effect on the overall technology bases of the firms and they augment specialization in specific green

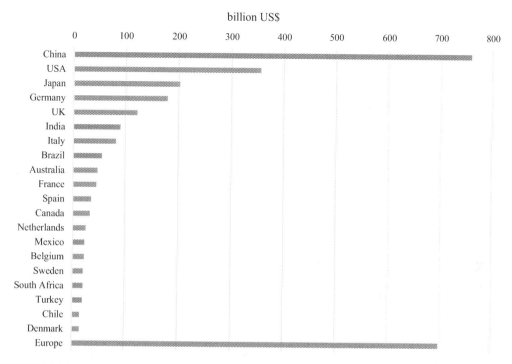

FIGURE 14.14 Investment in renewable energy capacity building between 2010 and first half of 2019 of top 20 countries and Europe. *Source: Frankfurt School-UNEP Centre/BNEF (2019).*

technologies. In addition, green FDIs have a significant positive impact on the degree and quality of innovative capacity of the MNCs in green sustainable technologies. Finally, green FDIs lead newly-established subsidiaries to undertake more innovative and greening activities than the acquisition of foreign firms in the long run (Amendolagine et al., 2021).

As high a US$2.6 trillion was invested in the last decade in the renewable energy sector excluding the large hydropower technology. Solar and wind technologies dominated the other technologies in terms of investment. A total of US$1.349 trillion has been invested in the solar sector which is 52% of total investment. Wind power received around US$1 trillion investment which is 41% of total investment. The biomass and waste sector also gained some importance that resulted in some US$115.5 billion investment (4% of total investment). Small hydro, biofuels, and geothermal technologies received some US$42.7, 27.3, and 19.8 billion investments, respectively (Fig. 14.15). The least investment was received by marine technologies with some US$500 million (Frankfurt School-UNEP Centre/BNEF, 2019).

Renewable energy has the potential to become the source of as much as two-thirds of the world's supply of energy in 2050 (Brini, 2021). Therefore, effective and rapid transition from nonrenewable to renewable energy would depend heavily on recognition of this potential by the policymakers and promote both public and private investments in favor of innovation of technologies and expansion of installed capacity.

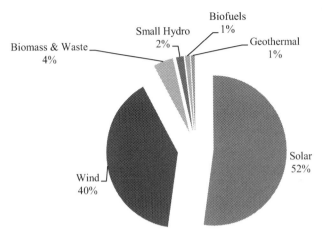

FIGURE 14.15 Investment in renewable energy capacity by energy category between 2010 and 2019. Total investment was US$2.6 trillion during this period. *Source: Frankfurt School-UNEP Centre/BNEF (2019).*

FDI in renewable energy technologies can also foster the environmental sustainability of developing economies. In addition, regulations related to investment in the energy sector would support the transition of developing economies from fossil fuels to renewable energy toward mitigating climate change (Ahmed and Wang, 2019). For this transition, the government of the developing countries could follow suitable sustainable energy infrastructure planning framework (Khan, 2021).

14.6.2 Research and development

Tackling GHG emissions and environmental degradation is possible through the advancement of technological innovation (Ali et al., 2016). Research and development (R&D) would play a significant role in improving the efficiency and reducing the unit cost of renewable technologies compared to their nonrenewable counterparts. It would help expand the rapid adoption of renewable technologies by the public sector and strengthen the market for renewable energy. It would also foster the rapid expansion of installed capacity and enhance the availability of renewable energy, which would reverse environmental degradation at the early and even relatively advanced state of development of the economies. However, renewable energy must be economically affordable up to low-income and disadvantaged groups of the society for becoming sustainable in the long term.

To avoid climate-induced catastrophe and ensuring economic development and environmental sustainability, it is important to invest in public renewable energy research and development (Koçak and Ulucak, 2019). Therefore, the production and consumption of fossil fuel would decrease by the surge of newly invented clean and green technologies. It means that investing in renewable energy research and development can turn out to be a great attempt to perform better in development and sustainability (Lee and Min, 2015). This type of investment will help transition from nonrenewable to renewable energy by introducing new technologies. However, it may require investment from both the public and private sector on R&D because affordability and availability are the major concerns when the issue is to phasing nonrenewable energy out of the energy sector.

The increasing trend of innovation in renewable energy is a positive sign for climate change mitigation and sustainability. The data of IRENA reveals that the number of patents increased from 204,998 in 2009 to 646,416 in 2019. The largest share of patents (57%) was in solar energy in 2019 compared to 54% in 2009. It can be observed that the solar energy sector has got the most attention of the innovators. Wind energy comes next in this case. Bioenergy has also been a potential sector throughout the years that can be understood by the increasing number of patents every year. Third-generation biofuels have emerged as a powerful alternative to decrease reliance on fossil fuels. However, there is a need for investing more in research and development to decrease production costs and achieve commercial viability (Nazari et al., 2021). Besides, more research on waste to energy generation needs particular focus for the developing countries, as it is going to be a sustainable and potential energy generation technology in near future (Khan and Kabir, 2020).

Throughout the years, the unit price of renewable energy has been decreasing with the increasing capacity. Now the highest amount of energy is being generated at the lowest price. Per unit price of solar energy has decreased significantly in recent years (from US$284.4 in 2011 to only US$55.8 in 2019 per MWh). It was possible because of the rapid expansion of its market across the world driven notably by the private sector–the installed capacity of solar energy increased from merely 1GW in 2011 to 59GW in 2019. In addition, the per-unit cost of onshore wind power has decreased from US$76.5 to 49.2 during the same period. This indicates that the world is now experiencing a technological revolution in the renewable energy sector. The reduced unit cost would help expand the technologies and become economically viable, which would pave the way for increasing the installed capacity of various renewable energy technologies in developing economies. Thus, increasing investment in renewable energy and research would result in sustainable development (Ullah et al., 2021).

Supporting research and development on renewable energy by the public sector would help replace fossil fuels and establish a favorable market for green technologies and increase the share of renewable energy that dominates the energy mix (Becker, 2015; Alvarez-Herranz, 2017). Investment in renewable energy research and development would result in a substantial reduction in GHG emission from fossil fuel use. The connection between energy technology budgets and CO_2 emissions was studied by Shahbaz et al. (2018) and they concluded that energy technology budgets mitigate emissions. Environmental sustainability can also be maintained by public budgets on renewable energy (Baloch et al., 2020). Both Jin et al. (2017) and Altintas and Kassouri (2020) agreed that energy budgets help decrease carbon emissions.

14.6.3 Fiscal policy

Fiscal policy support is important to expand renewable energy technology, especially in developing economies, to stimulate capacity expansion, technological innovation, and involvement of the private sector in marketing the technologies among others. Among them, public investment in developing large-scale installed capacity plays a critical role, which may include establishing mega solar parks, wind power plants, and waste-to-energy and hydroelectricity projects. Reduction of domestic and import taxes as well as introducing subsidies would allow and encourage the private sector players in installing micro, small, and medium-sized plants and technologies at the household and commercial levels.

To facilitate the transition from nonrenewable to renewable sources of energy, there is a need for increased investment in innovation and improvement of technologies as well as the rapid expansion of installed capacity. The data of the international renewable energy agency (IRENA, 2021) reveals that four out of the top five recipients of the public investment were developing countries, viz. Brazil, India, Pakistan, and Nigeria. Amongst the technologies, public investment in renewable hydropower is higher than any other power sector in the years from 2009 to 2019 except 2015. Hence, the investment in hydropower throughout the years is almost half of the cumulative investments in major renewable energy sources (US$92.51 billion or 37% of total investment). The wind and solar energy sector also received a significant amount of investments (US$51.55 billion or 20% of total investment). Solar energy received US$39.23 billion during this period, which was 16% of total public investment in the world. Despite being a promising sector, bioenergy and WtE did not receive a noticeable amount of investments. Waste management is a major challenge in developing economies–traditional dumping and landfilling cause considerable environmental hazards and threat to human health. Therefore, expansion of technologies of waste-to-energy would be an important step toward sustainable waste management and at the same time sustainable waste management of developing economies.

It is imperative for the governments to provide subsidies for enabling renewable energy technologies to compete in the electricity market. The subsidy is justifiable to avoid damages to the environment and climate by fossil and traditional fuels. The amount of subsidy on renewable energy would seem to be high because of significantly greater per unit cost than that of fossil fuel, and conventional and nuclear energy. However, Breyer et al. (2015) argued that the trend of subsidy for renewable energy from 1970 to 2012 was inadequate compared to that for fossil fuel and nuclear power. Even unsubsidized renewable power is exposed to adverse completion with subsidized power produced by fossil fuel. In order to facilitate fair completion between them and help mitigate climate change, a significant amount of subsidy is required for renewable energy across the world.

14.6.4 Monetary policy

Monetary policy would play important role in expanding the adoption of renewable technologies in developing economies. The most important barrier to the adoption of renewable energy is related to private investment financing as most of the renewable technologies are costly–higher costs of renewable energy investment, when compared to traditional energy investment, affect the investment decision. However, considering the costs of environmental and climate change for using traditional fossil fuels, it is important to introduce soft financing models for renewable technologies steered by the central banks of developing economies through monetary policies and implemented by formal and quasi-formal financial institutions. Such financial packages would include lower interest, a high grace period, and a longer recovery period compared to the standard packages for business and industrial sectors.

The central banks of developing economies can create special revolving funds and refinancing schemes to channel money through commercial banks for financing technologies on a commercial basis. However, thousands of nongovernmental organizations (NGOs) and microfinance institutions regulated by central banks are working across the developing economies, which are financing many micro, small, and medium enterprises (MSMEs). Special funds can be created for these organizations to finance investment micro and small renewable technologies at the local level. It would facilitate the adoption of these technologies especially in rural, remote, geographically vulnerable, and isolated regions.

Energy policies need to facilitate transformation in the energy system to increasing the share of renewable in the energy mix. In doing so, the government's allocation and tax policies need to facilitate both public and private investment to reduce the cost of renewable energy technologies (Keles and Bilgen, 2012).

14.7 Conclusion

Adoption of low to no-carbon renewable energy technologies will lead the developing economies to keep global warming well below 2°C according to the Paris Agreement. However, achieving this target will be difficult if developmental activities continue using traditional nonrenewable energy. The ecosystem, environment, and global climate are under enormous pressure due to the massive consumption of fossil fuels. Even though adverse impacts of climate change are putting every economy at risk, developing economies are at the highest risk due to their greater exposure, vulnerability and lesser adaptive capacity and resilience. The share of GHG emission of developing economies is more than half of the world's total emission, which implies that they are still struggling to reduce their reliance on nonrenewable energy. However, renewable energy has been recognized as an important means of climate change mitigation and sustainable development. A rapid evaluation of renewable energy technologies has been observed since the end of the 2000s, which significantly helped expand the installed capacity of these technologies in the developing world.

Indeed, developing economies have been experiencing considerable economic growth over the last few decades. The size of the economy of some developing economies has become as large as developed economies, for instance, China, India, Brazil, Turkey, and South Africa. Their contribution to the world's total economic growth is significant. However, consumption of energy, especially fossil fuel, has increased simultaneously to feed their rapid economic growth, industrialization and international trade. Renewable energy is gradually occupying a notable portion of the economic activities of these economies. Large developing economies are leading the transition to renewable energy. For instance, China's production and consumption of renewable energy are significantly higher than that of any other economy. Currently, around two-thirds of the total renewable energy is being produced by developing economies. Their performance in renewable energy capacity addition has demonstrated noticeable progress, which would enable them to mitigate climate change.

To achieve SDG-7, it is necessary to achieve one of its key targets increasing the share of renewables in the global energy mix by 2030. Developing countries are gradually becoming the hub of industrialization and therefore GHG emission rate increased significantly in the last few decades. Some Asian, Latin American and African developing economies are putting efforts to rapidly adopt renewable energy technologies. However, many developing countries are still lagging far behind, which will make it difficult for the economies to achieve sustainability and will adversely impact the environment. China, Brazil, and India are role models for the developing economies to understand how to use renewable energy sources to derive maximum environmental and climatic benefits. The market of renewable energy will expand significantly in developing countries in the near future if it is facilitated by investment, support in innovating low-cost advanced technology, and conducive fiscal and monetary policy. Developing economies need to adopt step-by-step and long-term strategies in line with the SDGs and IPCC guidelines for the expansion of the technologies toward climate change mitigation and sustainable development.

Self-evaluation questions

1. What are the positive and negative linkages among energy use, economic growth, and sustainable development?
2. How does increasing electricity consumption impact climate change and the structure of developing economies?
3. What are the factors that influence developing economies toward the transition from nonrenewable to renewable energy options?
4. Which renewable energy has more potential to expand in the near future in the developing world?
5. In the context of developing countries, which renewable energy needs more focus to develop and why?
6. Which region of the world needs to increase its focus on the renewable energy sector and why?
7. What are the major barriers to adopt renewable energy in developing nations?
8. What are the opportunities of the developing economies to rapidly adopt renewable energy?
9. Explain the links between climate change, sustainability, and renewable energy.
10. Conduct a literature survey and identify the link between negative climate change and fossil fueled electricity generation and explain their mitigation plans.

References

Adelaja, A.O., 2020. Barriers to national renewable energy policy adoption: insights from a case study of Nigeria. Energy Strategy Rev. 30, 100519. https://doi.org/10.1016/j.esr.2020.100519.

Aguilar-Rivera, N., Michel-Cuello, C., Cervantes-Niño, J.J., Gómez-Merino, F.C., Olvera-Vargas, L.A., 2021. Effects of public policies on the sustainability of the biofuels value chain. In: Roy, R C (Ed.), Sustainable Biofuels: Opportunities and Challenges. Elsevier, pp. 345–379. https://doi.org/10.1016/B978-0-12-820297-5.00004-9.

Ahmed, Z., Cary, M., Shahbaz, M., Vo, X.V., 2021. Asymmetric nexus between economic policy uncertainty, renewable energy technology budgets, and environmental sustainability: evidence from the United States. J. Cleaner Prod. 313, 127723. https://doi.org/10.1016/j.jclepro.2021.127723.

Ahmed, Z., Wang, Z., 2019. Investigating the impact of human capital on the ecological footprint in India: an empirical analysis. Environ. Sci. Pollut. Res. 26, 26782–26796. https://doi.org/10.1007/s11356-019-05911-7.

Ali, W., Abdullah, A., Azam, M., 2016. The dynamic linkage between technological innovation and carbon dioxide emissions in Malaysia: an autoregressive distributed lagged bound approach. Int. J. Energy Econ. Policy 6 (3), 389–400. https://www.econjournals.com/index.php/ijeep/article/view/2137.

Altıntas, H., Kassouri, Y., 2020. The impact of energy technology innovations on cleaner energy supply and carbon footprints in Europe: a linear versus nonlinear approach. J. Clean. Prod., 124140. https://doi.org/10.1016/j.jclepro.2020.124140.

References

Alvarez-Herranz, A., Balsalobre-Lorente, D., Shahbaz, M., Cantos, J.M., 2017. Energy innovation and renewable energy consumption in the correction of air pollution levels. Energy Policy 105, 386–397. https://doi.org/10.1016/j.enpol.2017.03.009.

Amendolagine, V., Lema, R., Rabellotti, R., 2021. Green foreign direct investments and the deepening of capabilities for sustainable innovation in multinationals: Insights from renewable energy. J. Cleaner Prod. 310, 127381. https://doi.org/10.1016/j.jclepro.2021.127381.

Amir, M., Khan, S.Z., 2021. Assessment of renewable energy: Status, challenges, COVID-19 impacts, opportunities, and sustainable energy solutions in Africa. Energy Built Environ. https://doi.org/10.1016/j.enbenv.2021.03.002.

Anand, S., Gupta, A., Tyagi, S.K., 2015. Solar cooling systems for climate change mitigation: a review. Renew. Sustain. Energy Rev. 41, 143–161. http://dx.doi.org/10.1016/j.rser.2014.08.042.

Angulo-Mosquera, L.S., Alvarado-Alvarado, A.A., Rivas-Arrieta, M.J., Cattaneo, C.R., Rene, E.R., García-Depraect, O., 2021. Production of solid biofuels from organic waste in developing countries: a review from sustainability and economic feasibility perspectives. Sci. Total Environ. 795, 148816. https://doi.org/10.1016/j.scitotenv.2021.148816.

Bai, B., Wang, Y., Fang, C., Xiong, S., Ma, X., 2021. Efficient deployment of solar photovoltaic stations in China: an economic and environmental perspective. Energy 221, 119834. https://doi.org/10.1016/j.energy.2021.119834.

Baloch, M.A., Ozturk, I., Bekun, F.V., Khan, D., 2020. Modeling the dynamic linkage between financial development, energy innovation, and environmental quality: does globalization matter? Bus. Strategy Environ. 30 (1), 176–184. https://doi.org/10.1002/bse.2615.

Balsalobre, D., Alvarez, A., Cantos, J.M., 2015. Public budgets for energy RD&D and the effects on energy intensity and pollution levels. Environ. Sci. Pollut. Res. 22, 4881–4892. https://doi.org/10.1007/s11356-014-3121-3.

Becker, B., 2015. Public R&D policies and private R&D investment: a survey of the empirical evidence. J. Econ. Survey 29, 917–942. https://doi.org/10.1111/joes.12074.

Bekchanov, M., Evia, P., 2018. Resources Recovery and Reuse in Sanitation and Wastewater Systems: Options and Investment Climate in South and Southeast Asian Countries. In: ZEF Working Paper No 168, ZEF, 2018. Bonn, Germany.

Bekchanov, M., Mondal, M.A.H., de Alwis, A., Mirzabaev, A., 2019. Why adoption is slow despite promising potential of biogas technology for improving energy security and mitigating climate change in Sri Lanka? Renew. Sustain. Energy Rev. 105, 378–390. https://doi.org/10.1016/j.rser.2019.02.010.

Belançon, M.P., 2021. Brazil electricity needs in 2030: trends and challenges. Renew. Energy Focus 36, 89–95. https://doi.org/10.1016/j.ref.2021.01.001.

Bilgen, S., Keles, S., Kaygusuz, A., Sarı, A., Kaygusuz, K., 2008. Global warming and renewable energy sources for sustainable development: a case study in Turkey. Renew. Sustain. Energy Rev. 12 (2), 372–396. http://dx.doi.org/10.1016/j.rser.2006.07.016.

Brini, R., 2021. Renewable and non-renewable electricity consumption, economic growth and climate change: evidence from a panel of selected African countries. Energy 223, 120064. https://doi.org/10.1016/j.energy.2021.120064.

Breyer, C., Koskinen, O., Blechinger, P., 2015. Profitable climate change mitigation: the case of greenhouse gas emission reduction benefits enabled by solar photovoltaic systems. Renew. Sustain. Energy Rev. 49, 610–628. http://dx.doi.org/10.1016/j.rser.2015.04.061.

Cadez, S., Czerny, A., Letmathe, P., 2019. Stakeholder pressures and corporate climate change mitigation strategies. Bus. Strategy Environ. 28, 1–14. https://doi.org/10.1002/bse.2070.

Chachuli, F.S.M., Ludin, N.A., Jedi, M.A.M., Hamid, N.H., 2021. Transition of renewable energy policies in Malaysia: benchmarking with data envelopment analysis. Renew. Sustain. Energy Rev. 150, 111456. http://doi.org/10.1016/j.rser.2021.111456.

Chien, F., Chau, K.Y., Ady, S.U., Zhang, Y., Tran, Q.H., Aldeehani, T.M., 2021. Does the combining effects of energy and consideration of financial development lead to environmental burden: social perspective of energy finance? Environ. Sci. Pollut. Res. 28, 40957–40970. https://doi.org/10.1007/s11356-021-13423-6.

Cox, S., Hotchkiss, E., Bilello, D., Watson, A., Holm, A., 2017. Bridging Climate Change Resilience and Mitigation in the Electricity Sector Through Renewable Energy and Energy Efficiency: Emerging Climate Change and Development Topics for Energy Sector Transformation, Washington, DC: U.S. Agency for International Development.

Crippa, M., Guizzardi, D., Muntean, M., Schaaf, E., Solazzo, E., Monforti-Ferrario, F., Olivier, J.G.J., Vignati, E., 2020. Fossil CO_2 emissions of all world countries–2020 Report, EUR 30358 EN, Publications Office of the European Union, Luxembourg. ISBN 978-92-76-21515-8, http://doi.org/10.2760/143674, JRC121460.

Danish, Baloch, M.A., Mahmood, N., Zhang, J.W., 2019. Effect of natural resources, renewable energy and economic development on CO_2 emissions in BRICS countries. Sci. Total Environ. 678, 632–638. https://doi.org/10.1016/j.scitotenv.2019.05.028.

Eckstein, D., Künzel, V., Schäfer, L., 2021. Global Climate Risk Index 2021: Who Suffers Most from Extreme Weather Events? Weather-Related Loss Events in 2019 and 2000-2019, Germanwatch. Berlin. https://germanwatch.org/sites/default/files/Global%20Climate%20Risk%20Index%202021_1.pdf.

Ellabban, O., Abu-Rub, H., Blaabjerg, F., 2014. Renewable energy resources: current status, future prospects and their enabling technology. Renew. Sustain. Energy Rev. 39, 748–764. https://doi.org/10.1016/j.rser.2014.07.113.

Elum, Z.A., Momodu, A.S., 2017. Climate change mitigation and renewable energy for sustainable development in Nigeria: a discourse approach. Renew. Sustain. Energy Rev. 76, 72–80. http://dx.doi.org/10.1016/j.rser.2017.03.040.

Eskeland, G.S., Rive, N.A., Mideksa, T.K., 2012. Europe's climate goals and the electricity sector. Energy Policy 41, 200–211. https://doi.org/10.1016/j.enpol.2011.10.038.

Fragkos, P., van Soest, H.L., Schaeffer, R., Reedman, L., Köberle, A.C., Macaluso, N., Evangelopoulou, S., De Vita, A., Sha, F., Qimin, C., Kejun, J., Mathur, R., Shekhar, S., Dewi, R.G., Diego, S.H., Oshiro, K., Fujimori, S., Park, C., Safonov, G., Iyer, G., 2021. Energy system transitions and low-carbon pathways in Australia, Brazil, Canada, China, EU-28, India, Indonesia, Japan, Republic of Korea, Russia and the United States. Energy 216, 119385. https://doi.org/10.1016/j.energy.2020.119385.

Frankfurt School-UNEP Centre/BNEF, 2019. Global Trends in Renewable Energy Investment 2019, http://www.fs-unep-centre.org (Frankfurt am Main).

Fuso Nerini, F., Tomei, J., To, L.S., Bisaga, I., Parikh, P., Black, M., Borrion, A., Spataru, C., Broto, V.C., Ananharajah, G., Milligan, B., Mulugetta, Y., 2018. Mapping synergies and trade-offs between energy and the sustainable development goals. Nat. Energy 3, 10–15. https://doi.org/10.1038/s41560-017-0036-5.

Gielen, D., Boshell, F., Saygin, D., Bazilian, M.D., Wagner, N., Gorini, R., 2019. The role of renewable energy in the global energy transformation. Energy Strategy Rev. 24, 38–50. https://doi.org/10.1016/j.esr.2019.01.006.

Godfray, H.C.J., Garnett, T., 2014. Food security and sustainable intensification. Philos. Trans. R. Soc. B, Biol. Sci. 369, 20120273. https://doi.org/10.1098/rstb.2012.0273.

Gökgöz, F., Yalçın, E., 2021. Can renewable energy sources be a viable instrument for climate change mitigation? Evidence from EU countries via MCDM methods. In: Ting, D, Stagner, J (Eds.), Climate Change Science: Causes, Effects and Solutions for Global Warming. Elsevier, pp. 19–39.

Grossman, G.M., Krueger, A.B., 1991. Environmental Impacts of a North American Free Trade Agreement. In: National Bureau of Economic Research Working Paper 3194. National Bureau of Economic Research, Inc.

Gulagi, A., Alcanzare, M., Bogdanov, D., Esparcia, E.Jr., Ocon, J., Breyer, C., 2021. Transition pathway towards 100% renewable energy across the sectors of power, heat, transport, and desalination for the Philippines. Renew. Sustain. Energy Rev. 144, 110934. http://doi.org/10.1016/j.rser.2021.110934.

Gulagi, A., Ram, M., Solomon, A.A., Khan, M., Breyer, C., 2020. Current energy policies and possible transition scenarios adopting renewable energy: A case study for Bangladesh. Renewable Energy 155, 899–920. https://doi.org/10.1016/j.renene.2020.03.119.

Hoang, A.T., Nižetić, S., Olcer, A.I., Ong, H.C., Chen, W.H., Chong, C.T., Thomas, S., Bandh, S.A., Nguyen, X.P., 2021. Impacts of COVID-19 pandemic on the global energy system and the shift progress to renewable energy: Opportunities, challenges, and policy implications. Energy Policy 154, 112322. https://doi.org/10.1016/j.enpol.2021.112322.

Ibrahim, M.D., Alola, A.A., 2020. Integrated analysis of energy-economic development-environmental sustainability nexus: case study of MENA countries. Sci. Total Environ. 737, 139768. https://doi.org/10.1016/j.scitotenv.2020.139768.

IEA. 2009. World energy outlook 2009. International Energy Agency. OECD/IEA, Paris. Available at: https://www.iea.org/reports/world-energy-outlook-2009.

IEA. 2017. Energy Access Outlook 2017. Available at: https://www.iea.org/reports/energy-access-outlook-2017.

IEA. 2020. World Energy Outlook 2020, Paris: IEA.

International Energy Agency (IEA), 2021. Renewable energy explained, https://www.eia.gov/energyexplained/renewable-sources/.

IPCC, 2014. Climate Change 2014: Mitigation of Climate Change. Contribution of Working Group III to the Fifth Assessment Report of the Intergovernmental Panel on Climate Change [Edenhofer, O., Pichs-Madruga, R., Sokona, Y., Farahani, E., Kadner, S., Seyboth, K., Adler, A., Baum, I., Brunner, S., Eickemeier, P., Kriemann,

B., Savolainen, J., Schlömer, S., von Stechow, C., Zwickel, T., Minx, J.C. (eds.)]. Cambridge University Press, Cambridge, United Kingdom and New York, NY, USA. Available at: https://www.ipcc.ch/site/assets/uploads/2018/02/ipcc_wg3_ar5_full.pdf.

IPCC, 2018. Global Warming of 1.5°C.An IPCC Special Report on the impacts of global warming of 1.5°C above pre-industrial levels and related global greenhouse gas emission pathways in The Context of Strengthening the Global Response to the Threat of Climate Change, Sustainable Development, and Efforts to Eradicate Poverty. [Masson-Delmotte, V., Zhai, P., Pörtner, H.O., Roberts, D., Skea, J., Shukla, P.R., Pirani, A., Moufouma-Okia, W., Péan, C., Pidcock, R., Connors, S., Matthews, J.B.R., Chen, Y., Zhou, X., Gomis, M.I., Lonnoy, E., Maycock, T., Tignor, M., Waterfield, T. (eds.)]. Intergovernmental Panel on Climate Change - IPCC. Available at: https://www.ipcc.ch/site/assets/uploads/sites/2/2019/06/SR15_Full_Report_High_Res.pdf.

IPCC, 2020. Emission Factor Database [WWW Document]. Available at: https://www.ipcc-nggip.iges.or.jp/EFDB/main.php.

IPCC, 2021. Climate Change 2021: The Physical Science Basis–Working Group I contribution to the Sixth Assessment Report of the Intergovernmental Panel on Climate Change, Geneva: Intergovernmental Panel on Climate Change.

IRENA (International Renewable Energy Agency), 2021. Data & Statistics. https://www.irena.org/Statistics.

Irsyad, M.I., Halog, A., Nepal, R., 2019. Renewable energy projections for climate change mitigation: An analysis of uncertainty and errors. Renewable Energy 130, 536–546. http://doi.org/10.1016/j.renene.2018.06.082.

Jin, L., Duan, K., Shi, C., Ju, X., 2017. The impact of technological progress in the energy sector on carbon emissions: an empirical analysis from China. Int. J. Environ. Res. Publ. Health 14, 1–14. https://doi.org/10.3390/ijerph14121505.

Kaygusuz, K., 2009. Energy and environmental issues relating to greenhouse gas emissions for sustainable development in Turkey. Renew. Sustain. Energy Rev. 13 (1), 253–270. https://doi.org/10.1016/j.rser.2007.07.009.

Keles, S., Bilgen, S., 2012. Renewable energy sources in Turkey for climate change mitigation and energy sustainability. Renew. Sustain. Energy Rev. 16, 5199–5206. http://dx.doi.org/10.1016/j.rser.2012.05.026.

Khan, I., Jack, M.W., Stephenson, J., 2018. Analysis of greenhouse gas emissions in electricity systems using time-varying carbon intensity. J. Cleaner Prod. 184, 1091–1101. https://doi.org/10.1016/j.jclepro.2018.02.309.

Khan, I., 2019. Temporal carbon intensity analysis: renewable versus fossil fuel dominated electricity systems. Energy Sources A: Recovery, Util. Environ. Eff. 41, 309–323. https://doi.org/10.1080/15567036.2018.1516013.

Khan, I., 2020. Impacts of energy decentralization viewed through the lens of the energy cultures framework: solar home systems in the developing economies. Renew. Sustain. Energy Rev. 119, 1–11. https://doi.org/10.1016/j.rser.2019.109576.

Khan, I., 2020a. Waste to biogas through anaerobic digestion: Hydrogen production potential in the developing world - a case of Bangladesh. Int. J. Hydrogen Energy 45, 15951–15962. https://doi.org/10.1016/j.ijhydene.2020.04.038.

Khan, I., 2021. Sustainable energy infrastructure planning framework: transition to a sustainable electricity generation system in Bangladesh. Chapter 7: Energy and Environmental Security in Developing Countries. *Springer Nature*, pp. 173–198. https://doi.org/10.1007/978-3-030-63654-8_7.

Khan, I., Kabir, Z., 2020. Waste-to-energy generation technologies and the developing economies: a multi-criteria analysis for sustainability assessment. Renew. Energy 150, 320–333. https://doi.org/10.1016/j.renene.2019.12.132.

Khare, N., Singh, D., Kant, R., Khare, P., 2020. Global warming and biodiversity. Current State and Future Impacts of Climate Change on Biodiversity. IGI Global, Pennsylvania, pp. 1–10. http://doi.org/10.4018/978-1-7998-1226-5.ch00.

Kiviluoma, J., Holttinen, H., Scharff, R., Weir, D., Cutululis, N., Litong-Palima, M., Milligan, M., 2014. Index for wind power variability. In: Proceedings of the 13th Int. Work. Large-Scale Integr. Wind Power into Power Syst. as well as Transm. Networks Offshore Wind Plants 2014, Nov 11-13. Berlin, Germany.

Koçak, E., Ulucak, Z.Ş., 2019. The effect of energy R&D expenditures on CO_2 emission reduction: estimation of the STIRPAT model for OECD countries. Environ. Sci. Pollut. Res. 26, 14328–14338. https://doi.org/10.1007/s11356-019-04712-2.

Kung, C.C., Mu, J.E., 2019. Prospect of China's renewable energy development from pyrolysis and biochar applications under climate change. Renew. Sustain. Energy Rev. 114, 109343. https://doi.org/10.1016/j.rser.2019.109343.

Lau, L.S., Choong, C.K., Eng, Y.K., 2014. Investigation of the environmental Kuznets curve for carbon emissions in Malaysia: do foreign direct investment and trade matter? Energy Policy 68, 490–497. https://doi.org/10.1016/j.enpol.2014.01.002.

Lee, K.H., Min, B., 2015. Green R&D for eco-innovation and its impact on carbon emissions and firm performance. J. Cleaner Prod. 108, 534–542 http://dx.doi.org/10.1016/2Fj.jclepro.2015.05.114.

Li, J., Chen, S., Wu, Y., Wang, Q., Liu, X., Qi, L., Lu, X., Gao, L., 2021. How to make better use of intermittent and variable energy? A review of wind and photovoltaic power consumption in China. Renew. Sustain. Energy Rev. 137, 110626. https://doi.org/10.1016/j.rser.2020.110626.

Mahjabeen, Shah, M.Z.A., Chughtai, S., Simonetti, B., 2020. Renewable energy, institutional stability, environment and economic growth nexus of D-8 countries. Energy Strategy Reviews 29, 100484. https://doi.org/10.1016/j.esr.2020.100484.

Mayer, A., Castro-Diaz, L., Lopez, M.C., Leturcq, G., Moran, E.F., 2021. Is hydropower worth it? Exploring amazonian resettlement, human development and environmental costs with the Belo Monte project in Brazil. Energy Res. Soc. Sci. 78, 102129. https://doi.org/10.1016/j.erss.2021.102129.

Montoya, M.A., Allegretti, G., Bertussi, L.A.S., Talamini, E., 2021. Renewable and non-renewable in the energy-emissions-climate nexus: Brazilian contributions to climate change via international trade. J. Cleaner Prod. 312, 127700. https://doi.org/10.1016/j.jclepro.2021.127700.

Mouraviev, N., 2021. Renewable energy in Kazakhstan: challenges to policy and governance. Energy Policy 149, 112051. https://doi.org/10.1016/j.enpol.2020.112051.

Nadaleti, W.C., Lourenc, V.A., Americo, G., 2021. Green hydrogen-based pathways and alternatives: towards the renewable energy transition in South America's regions–Part A. Int. J. Hydrogen Energy 46 (43), 22247–22255. https://doi.org/10.1016/j.ijhydene.2021.03.239.

Nazari, M.T., Mazutti, J., Basso, L.G., Colla, L.M., Brandli, L., 2021. Biofuels and their connections with the sustainable development goals: a bibliometric and systematic review. Environ. Dev. Sustain. 23, 11139–11156. https://doi.org/10.1007/s10668-020-01110-4.

OECD/IEA, 2015. World energy outlook 2015 factsheet: global energy trends to 2040. International Energy Agency. Available at: http://www.worldenergyoutlook.org/media/weowebsite/2015/WEO2015_Factsheets.pdf.

OWD, 2020. Sector by sector: where do global greenhouse gas emissions come from? [WWW Document]. Global Greenhouse Gas Emissions by Sector. Available at: https://ourworldindata.org/ghg-emissions-by-sector.

Oyedepo, S.O., 2012. Energy and sustainable development in Nigeria: the way forward. Energy Sustain. Soc. 2 (15), 1–17. http://dx.doi.org/10.1186/2192-0567-2-15.

Puaschunder, J., 2020. Global climate change-induced migration and financial flows. Governance & Climate Justice: Global South and Developing Nations. Springer International Publishing, pp. 111–143. https://doi.org/10.1007/978-3-319-63281-0_7.

Radmehr, R., Henneberry, S.R., Shayanmehr, S., 2021. Renewable energy consumption, CO_2 emissions, and economic growth nexus: a simultaneity spatial modeling analysis of EU countries. Struct. Change Econ. Dyn. 57, 13–27. https://doi.org/10.1016/j.strueco.2021.01.006.

Rasheed, T., Anwar, M.T., Ahmad, N., Sher, F., Khan, S., Ahmad, A., Khan, R., Wazeer, I., 2021. Valorisation and emerging perspective of biomass based waste-to-energy technologies and their socio-environmental impact: a review. J. Environ. Manage. 287, 112257. https://doi.org/10.1016/j.jenvman.2021.112257.

REN21. Renewable Energy Network. Global renewable energy report for 2010. Available at: www.ren21.netS.

Riti, J.S., Shu, Y., Song, D., Kamah, M., 2017. The contribution of energy use and financial development by source in climate change mitigation process: a global empirical perspective. J. Cleaner Prod. 148, 882–894 http://dx.doi.org/10.1016%2Fj.jclepro.2017.02.037/.

Rodrigues, A.C.C., 2021. Policy, regulation, development and future of biodiesel industry in Brazil. Cleaner Energy Technol. 4, 100197. https://doi.org/10.1016/j.clet.2021.100197.

Saim, M.A., Khan, I., 2021. Problematizing solar energy in Bangladesh: Benefits, burdens, and electricity access through solar home systems in remote islands. Energy Res. Soc. Sci. Vol. 74 (101969), 1–12. https://doi.org/10.1016/j.erss.2021.101969.

Sampaio, R., Sampaio, P., 2021. Promoting sustainable energy in Brazil: the role of biomass. Elgar Encyclopedia of Environmental Law by M Faure (ed.), 498-508. https://doi.org/10.4337/9781788119689.IX.42.

Sapkota, A., Lu, Z., Yang, H., Wang, J., 2014. Role of renewable energy technologies in rural communities' adaptation to climate change in Nepal. Renew. Energy 68, 793–800. https://doi.org/10.1016/j.renene.2014.03.003.

Sarkodie, S.A., Adams, S., Leirvik, T., 2020. Foreign direct investment and renewable energy in climate change mitigation: Does governance matter? J. Cleaner Prod. 263, 121262. https://doi.org/10.1016/j.jclepro.2020.121262/.

Shahbaz, M., Nasir, M.A., Roubaud, D., 2018. Environmental degradation in France: the effects of FDI, financial development, and energy innovations. Energy Econ 74, 843–857. https://doi.org/10.1016/j.eneco.2018.07.020.

Soares, Í.N., Gava, R., de Oliveira, J.A.P., 2021. Political strategies in energy transitions: exploring power dynamics, repertories of interest groups and wind energy pathways in Brazil. Energy Res. Soc. Sci. 76, 102076. https://doi.org/10.1016/j.erss.2021.102076.

Solt, F., 2016. The standardized world income inequality database. Soc. Sci. Q. 97 (5), 1267–1281. https://doi.org/10.1111/ssqu.12295.

Sonar, D., 2021. Renewable energy based trigeneration systems—technologies, challenges and opportunities. In: Ren, J (Ed.), Renewable-Energy-Driven Future. Technologies, Modelling, Applications, Sustainability and Policies. Elsevier, pp. 125–168. https://doi.org/10.1016/B978-0-12-820539-6.00004-2.

Suman, A., 2021. Role of renewable energy technologies in climate change adaptation and mitigation: a brief review from Nepal. Renew. Sustain. Energy Rev. 151, 111524. https://doi.org/10.1016/j.rser.2021.111524.

Suo, C., Li, Y.P., Mei, H., Lv, J., Sun, J., Nie, S., 2021. Towards sustainability for China's energy system through developing an energy-climate-water nexus model. Renew. Sustain. Energy Rev. 135, 110394. https://doi.org/10.1016/j.rser.2020.110394.

Surendra, K.C., Takara, D., Hashimoto, A.G., Khanal, S.K., 2014. Biogas as a sustainable energy source for developing countries: opportunities and challenges. Renew. Sustain. Energy Rev. 31, 846–859. http://dx.doi.org/10.1016/j.rser.2013.12.015.

Swaraz, A.M., Satter, M.A., Rahman, M.M., Asad, M.A., Khan, I., Amin, M.Z., 2019. Bioethanol production potential in Bangladesh from wild date palm (*Phoenix sylvestris* Roxb.): an experimental proof. Ind. Corps Prod. 139 (111507), 1–9. https://doi.org/10.1016/j.indcrop.2019.111507.

Tang, E., Peng, C., Xu, Y., 2018. Changes of energy consumption with economic development when an economy becomes more productive. J. Cleaner Prod. 196, 788–795. https://doi.org/10.1016/j.jclepro.2018.06.101.

Tittensor, D.P., Beger, M., Boerder, K., Boyce, D.G., Cavanagh, R.D., Cosandey-Godin, A., Crespo, G.O., Dunn, D.C., Ghiffary, W., Grant, S.M., Hannah, L., Halpin, P.N., Harfoot, M., Heaslip, S.G., Jeffery, N.W., Kingston, N., Lotze, H.K., McGowan, J., McLeod, E., McOwen, C.J., O'Leary, B.C., Schiller, L., Stanley, R.R.E., Westhead, M., Wilson, K.L., Worm, B., 2019. Integrating climate adaptation and biodiversity conservation in the global ocean. Sci. Adv. 5 (11), eaay9969. http://doi.org/10.1126/sciadv.aay9969.

TREIA, 2015. Definition of renewable energy. Define Renew Energy. http://www.treia.org/renewable-energy-defined.

Tzankova, Z., 2020. Public policy spillovers from private energy governance: new opportunities for the political acceleration of renewable energy transitions. Energy Res. Soc. Sci. 67, 101504. https://doi.org/10.1016/j.erss.2020.101504.

Ullah, A., Ahmed, M., Raza, S.A., Ali, S., 2021. A threshold approach to sustainable development: Nonlinear relationship between renewable energy consumption, natural resource rent, and ecological footprint. J. Environ. Manage. 295, 113073. https://doi.org/10.1016/j.jenvman.2021.113073.

Umar, M., Ji, X., Kirikkaleli, D., Shahbaz, M., Zhou, X., 2020. Environmental cost of natural resources utilization and economic growth: Can China shift some burden through globalization for sustainable development? Sustainable Development 28, 1678–1688. https://doi.org/10.1002/sd.2116.

UNDP, 2002. Energy for sustainable development: a policy agenda. Johansson, T.B., Goldemberg, J. Eds. United Nation Development Program, New York. Available at: http://content-ext.undp.org/aplaws_publications/2101911/Energy%20for%20Sustainable%20Development-PolicyAgenda_2002.pdf.

UNDP, 2020. Human Development Report 2020: The next frontier–Human development and the Anthropocene. New York: UNDP.

United Nations, 2018. Affordable and Clean Energy: Why It Matters, New York: United Nations. Available at: https://www.un.org/sustainabledevelopment/wp-content/uploads/2018/09/Goal-7.pdf.

UNDP, 2005. Energizing the millennium development goals: a guide to energy's role in reducing poverty. UNDP. Available, New York.

Urban, F., 2018. China's rise: challenging the North-South technology transfer paradigm for climate change mitigation and low carbon energy. Energy Policy 113, 320–330. https://doi.org/10.1016/j.enpol.2017.11.007.

US Energy Information Administration (EIA), 2021. International data. https://www.eia.gov/international/data/world.

Vickers, N.J., 2017. Animal communication: when I'm calling you, will you answer too? Curr. Biol. 27 (14), R713–R715. https://doi.org/10.1016/j.cub.2017.05.064.

Viviescas, C., Lima, L., Diuana, F.A., Vasquez, E., Ludovique, C., Silva, G.N., Huback, V., Magalar, L., Szklo, A., Lucena, A.F.P., Schaeffer, R., Paredes, J.R., 2019. Contribution of variable renewable energy to increase energy security in Latin America: complementarity and climate change impacts on wind and solar resources. Renew. Sustain. Energy Rev. 113, 109232. https://doi.org/10.1016/j.rser.2019.06.039.

Wallimann-Helmer, I., 2020. Justice in renewable energy transitions for climate mitigation. Reference Module in Earth Systems and Environmental Sciences. Elsevier http://doi.org/10.1016/B978-0-12-819727-1.00029-7.

Wang, B., Wang, Q., Wei, Y.M., Li, Z.P., 2018. Role of renewable energy in China's energy security and climate change mitigation: an index decomposition analysis. Renew. Sustain. Energy Rev. 90, 187–194. https://doi.org/10.1016/j.rser.2018.03.012.

Yildizbasi, A., 2021. Blockchain and renewable energy: integration challenges in circular economy era. Renew. Energy 176, 183–197. https://doi.org/10.1016/j.renene.2021.05.053.

Zhang, C., Wu, W., Xie, W., Li, Q., Zhang, T., 2021. Forecasting the hydroelectricity consumption of China by using a novel unbiased nonlinear grey Bernoulli model. J. Cleaner Prod. 278, 123903. https://doi.org/10.1016/j.jclepro.2020.123903.

Zubi, G., Dufo-López, R., Pasaoglu, G., Pardo, N., 2016. Techno-economic assessment of an off-grid PV system for developing regions to provide electricity for basic domestic needs: a 2020-2040 scenario. Appl. Energy 176, 309–319. http://dx.doi.org/10.1016/j.apenergy.2016.05.022.

Index

Page numbers followed by "*f*" and "*t*" indicate, figures and tables respectively.

A
Abu Dhabi Fund for Development (ADFD), 215
Additive Ratio Assessment (ARAS), 7
African Development Bank, 98
Alkaline earth materials, 179
Alternative and renewable energies
  policy, 28
Anaerobic Digestion (AD)
  based technology, 34,
    see also (Bioelectricity)
  process, 211
Analytic Hierarchy Process (AHP), 7
Analytic Network Process (ANP), 7
Anion Exchange Membranes (AEM), 274
Ashuganj Power Station Company Limited (APSCL), 10
Asian Development Bank (ADB), 98

B
Bangladesh
  bioelectricity generation capacity, 45*t*
  coal and biomass, 181
Bangladesh Power Development Board (BPDB), 43
Bangladesh Rural Electrification Board (BREB), 10
Battery Energy Storage System (BESS), 333
Betz's constant, 141
Bioelectricity, 29
  anaerobic digestion-based technology, 34
  direct combustion/gasification-based technology, 39
  generation, technologies, 31
  microbial fuel cell, 33
Bioenergy, 69, 391
  economic impacts, 71
  environmental impacts, 69
  social impact, 70
Biofuel, 218
Biogas, 38
Biogas-based power generation and thermal application program (BPGTAP), 42
Biomass, 300, 305
  fired plant, combined heat and power, 46
  gasification, 309
  resources potential, 40
  waste-to-energy, 388

Bio Oxygen Dissolved (BOD), 72
Business engagement, 400

C
Carbon Capture and Sequestration (CCS), 182
Catchment communities, 286
Cation Exchange Membranes (CEM), 274
Chemical reaction, 311
Clean Technology Fund (CTF), 250
Climate change, 377
Coal and biomass, 181
  in Bangladesh, 181
Coastal communities, 286
Coefficient of Performance (COP), 252
Co-gasification, 183
  of biomass and coal, 191
  challenges, 194
  with coal, 186
Commercial gasifiers, 188
COmplex PRoportional ASsessment (COPRAS), 7
Concentrating Solar Power (CSP), 62
  types, 88
Continuous Stirred-Tank Reactor (CSTR), 313
Cumulative Environmental Impact Assessment (CEIA), 116

D
Dark fermentation process, 314
Decision matrix, 13, 11
  normalized, 14
  weighted normalized, 14
Demand Response (DR), 327, 328
Demand-side generation techniques, 333
Demand-side management, 335, 340
Dendro power project, 45
Developing countries
  solar energy, 88
  wind power, 157
Direct combustion/gasification-based technology, 39
Doubly-Fed Induction Generator (DFIG), 147
Drill and Blast Method (DBM), 119
Drying process, 234
Dual-Path Strategy (DPS), 95

E

Economic growth, 377
Electrical energy, 236
Electrical snow melting method, 234
Electric potential, 274
Electrochemical energy storage, 357
Electrodes, 274
Electrolysis, 311
Electrolytic process, 311
Electromagnetic energy storage, 361
Energy, 340, 380
    balance, 251
    consumption, 382
    demand, 325
    efficiency, 335
    efficiency of a system, 252
    intensive industries, 299
    policies, 407
    storage technologies, 349
Energy sustainability, 1, 299
Entropy, 252
Environmental and Social Impact Assessment (ESIA), 115
Environmental Impact Assessments (EIA), 290
Environmental Kuznets Curve, 379
European Marine Energy Centre (EMEC), 290
Exergy, 252

F

Fermentation process, 312, 313
Fiscal policy, 405
Flexible energy-storage devices, 332
Flexible load techniques, 332
Flow battery, 359
Fluidized Bed Gasifier (FBG), 194
Flywheel energy storage prototype, 353
Fossil energy, 303
Fossil fuel, 304
Fuel Cells (FC), 31
    vehicles, 299

G

Game theory techniques, 333
    classification, 334f
Gas combustion turbine, 38
Gasification process, 206
Gasification technology, 186, 206
    commercial gasifiers, 188
    schematic basic view, 207f
Gasifiers
    types, 308f
Gas turbine technology, biogas combustion, 38f
Geothermal aquacultural heating, 234
Geothermal District Heating System (GDHS), 233f
Geothermal energy, 228, 233, 234, 236, 237, 243, 245
    applications, 250
    systems, 251
    utilization, 246
Geothermal fluid, 230
Geothermal greenhouse heating, 233
Geothermal Heat Pump (GHP) systems, 236
Geothermal power generation, 230, 245
    technologies, 231f
Geothermal Power Plant (GPP), 229
Geothermal resources, 228, 229, 230, 251
Geothermal sources, 236, 248
Geothermal steam, 230
Geothermal systems, 227
    classification, 228f
German Corporation for International Cooperation (GIZ), 98
Global warming, 380
Government Backed Subsidies (GBS), 98
Greenhouse Gas (GHG), 1, 57, 182
Grid Price Parity (GPP), 98
Grid Side Converter (GSC), 147
Gross Domestic Product (GDP), 90

H

Heat exchanger, 233
Heat pumps, 235
Heavy oil, 304
High Voltage DC (HVDC), 99
Horizontal Axis Wind Turbines (HAWT), 139
Hot dry rock systems, 237
Human Development Index (HDI), 24
Hydroelectricity, 382, 386
Hydro energy, 65
    economic impacts, 66
    environmental impacts, 65
    social impacts, 66
Hydrogen, 299, 301
    energy, 300
    energy production, 304f
    fuel cell, 359
    production, 305f, 307
    production processes, 314f
    production steps, 309f
    Sulfide, 247
Hydrology, 116
Hydronic snow melting method, 234
Hydropower, 366
    basics, 111
    circuit breakers and relays, 124
    civil works components, 119
    components, 118

cooperation, 127
cost, 126
cumulative impact assessment, 116
desilting structures, 119
in developing countries, 128
diversion structure or reservoir, 119
forebay tank, 120
generators, 123
governing system, 124
historical background, 108
hydraulic turbine, 122
hydroelectrical equipment, 122
industry, 126
penstock, 120
power channel, 119
powerhouse, 120
present day operation strategy, 125
present status, 108
renovation, modernization, and upgrading, 125
reservoir-based, 117
services from, 114
surveys and investigations, 114
tail race channel, 122
technology, 117
transformers, 124
tunnels, 119
Hydropower potential, climate change impact, 117
Hydropower Projects (HPP), 107
   multiplier effects, 130

I
Incineration process, 208, 209f
Infrastructure Development Company Limited (IDCOL), 43, 102
Instantaneous demand response program, 339
Integrated Dark-Photo Fermentation (IDPF), 313
Integrated Gasification Combined Cycle (IGCC), 180, 182
International Electrotechnical Commission (IEC), 290
International Energy Agency (IEA), 107, 301
International Hydropower Association (IHA), 110
Islamic Development Bank (IDB), 98
Island communities, 340

K
Kalina cycle technology, 232
Kenya Off-grid Solar Access Project (KOSAP), 103
Kenya Power and Lighting Company (KPLC), 103
Kenya's Clean Energy Pathway (KCEP), 102
Kinetic model, 304

L
Latent heat storage, 356

Lead acid battery, 357
Least Developed Country (LDC), 219
Levelized Cost of Electricity (LCOE), 98, 153
Light energy, 313
Lignocellulosic biomass, 34
Lindal diagram, 229f
Liquid Ring Vacuum Pumps (LRVP), 231f
Lithium-ion battery, 357
Load building, 327

M
Marine Renewable Energy (MRE) technologies, 263
Mechanical energy storage, 349
Mechanical methods, 234
Microbial biomass conversion processes, 306
Microbial Fuel Cell (MFC), 33
   opportunities and challenges, 35
Monetary policy, 406
Multi-Criteria Decision Analysis (MCDA) methods, 4
Municipal Solid Waste (MSW), 72, 203
   advanced technologies, 204
   collection, 211
   disposal, 213
   generation of waste, 219
   improvement, 204
   methods, 204
   unsustainable management, 204

N
National Climate Change Action Plan (NCCAP), 102
National Electrification Strategy (NES), 102
National energy policies, 340
Nationally Determined Contributions (NDC), 102
Natural Gas Combined Cycle (NGCC), 182
Natural gas reforming, 308
NCG discharge system, 232
Negative electrode reaction, 357
North-West Power Generation Company Limited (NWPGCL), 10

O
Ocean Energy Systems (OES), 290
Ocean Renewable Energy (ORE), 263
   conversion, 264
   technologies, 279
Ocean Thermal Energy Conversion (OTEC), 269, 281
   desalination plant, 282
Ocean thermal gradient energy, 269
Optimization techniques, 333, 334f
Organic products, 307
Organic Rankine Cycle (ORC), 232
Overall cell reaction, 357

P

Parabolic Trough (PT), 88
Paris Agreement and Sustainable Development Goals, 378
Performance score, 15
Phase Change Material (PCM), 234
Photocatalytic separation techniques, 312
Photo fermentation process, 314
Photolytic process, 312
Photolytic techniques, 312
Photovoltaic (PV), 85
Positive electrode reaction, 357
Power Purchase Agreement (PPA), 92, 98, 103
Preference Ranking Organization METHod for Enrichment of Evaluations (PROMETHEE), 7
Pressure Retarded Osmosis (PRO), 272
Private Service Provider (PSP), 103
Probability Density Function (PDF), 143
Producer gas, 39
Proton Exchange Membrane (PEM), 33
Public-Private Partnership (PPP), 103
Pulse Width Modulation (PWM) inverters, 146
Pumped Hydro Storage (PHS), 349
Pumped Storage Power (PSP), 114
Pyrolysis process, 209

R

Rainwater, 227
Renewable energy, 62, 379, 398, 403
    benefits, 59t
    bioenergy, 69, 70, 71
    developing countries, future impacts, 77
    economic impacts, 78
    environmental impacts, 78
    global production, 60f
    hydro energy, 65, 66
    production, global scenario, 60
    scenario by 2040, 61t
    social impacts, 77
    solar energy, 62, 63, 64
    solid waste, 72, 73
    sources, impacts, 75
    technologies, 205
    wind energy, 67, 68, 69
Reverse Electro-Dialysis (RED), 272
Rotor Side Converter (RSC), 147
Rural Electrification and Renewable Energy Corporation (REREC), 103

S

Salinity Gradient Energy (SGE), 272
    processes of, 273f
Sanitary landfilling, 206
Scales of surface waves, 266f
Seawater state equations, 271
Second-generation technologies, 395
Sensible heat storage, 355, 356
Simulation process, 307
Small Island Developing States (SIDS), 280t
Social Impact Assessment (SIA), 290
Solar energy, 62
    contribution to sustainability, 100
    developing countries, 88
    economic impacts, 64
    environmental impacts, 62
    future prospects and challenges, 94
    generation systems, 347
    generation targets, future, 96t
    global and developing world context, 85
    Kenya, 102
    present status, 92
    social impacts, 63
Solar Home System (SHS), 87, 90, 92
    inter connected clusters, 93
Solar light systems, 92
Solar mini-grids, 94
Solar photovoltaic, 86
    grid connected systems, 87
    monocrystalline, 86
    off-grid systems, 87
    polycrystalline, 86
    thin-film, 87
Solar power, 87, 393
Solar technology, 86
    concentrating solar power, 87
    solar photovoltaic, 86
Solar water pump, 94
Solid waste, 72
    economic impacts, 73
    environmental impacts, 72
    social impacts, 73
Solution-diffusion model, 274
South Asian developing countries
    Bangladesh scenario, 43
    bioelectricity generation, 42
    bioelectricity prospects and technological status, 40
    biomass resources potential, 40
    electricity generation mix, 26
    India scenario, 42
    Pakistan scenario, 46
    Sri Lanka scenario, 45
Squirrel Cage Induction Generator (SCIG), 147
Static energy storage, 349
Steam-methane reforming reaction, 308
Steam reforming of hydrocarbons, 315
Steam turbine technology, bioelectricity production, 38f

Stepwise Weight Assessment Ratio Analysis (SWARA), 7
Superconducting Magnetic Energy Storage (SMES) system, 362
Supercritical Pulverized Coal (SCPC), 182
Supplemental demand response program, 339
Suspended Particular Matter (SPM), 57
Sustainability, 248, 335
  analysis, 16
  application, 10
  assessment methods, 5
  concept, 4, 5f
  conduct, 8
  defined, 1, 3t
  economic criteria, 8, 9f
  environmental criteria, 8
  indicators, 10t
  methods, 7
  social criteria, 8
  steps, 7f
  technical criteria, 8
  TOPSIS method, 10
Sustainable and Renewable Energy Development Authority (SREDA), 220
Sustainable Development Goals (SDG), 57, 95, 379, 204
Sustainable energy, 1
  benefits of, 1
  history, 2f
  transition governing process, 4f
Sustainable impact assessment, 6
Syngas, 208

T

Tectonic plate boundaries, 238
Thermal Energy Storage (TES), 87, 248
  system, 333
  technology, 355
Thermal process, 307
Thermochemical process, 206
Thermocline-based desalination plant, 281
Tidal energy, 264
  conversion, 264
  converters, 265
Tidal-Stream Energy Devices (TSEDs), 285
TOPSIS method, 10
Tunnel Boring Machines (TBM), 119

U

Underground Coal Gasification (UCG) method, 310
United Nations Framework Convention on Climate Change (UNFCCC), 325

United Nations Industrial Development Organization (UNIDO), 112
Upflow Anaerobic Sludge Blanket Reactor (UASB), 313

V

Vanadium Redox Battery (VRB), 359
Vapor Absorption Cycle (VAC), 252
Variable Renewable Energy (VRE), 396
Vertical Axis Wind Turbines (VAWT), 139
Vicious Cycle of Energy Poverty (VCEP), 99

W

Waste management tools and techniques, 205
Waste to Energy (WtE), 72, 388
  generation process, 212f
  generation technology, 208
  incineration, 215
  institutional policy, 220
  potential, 220
  reduction and reuse, 206
  schematic basic, 207f
  technology, 204, 214
Wave energy, 265
  conversion, 266
  converter, 282
  converters, 267, 268f
  convertors, 268
  devices, 269f
  flux, 267
  process, 265
Wave power plants, 280t
Weighted Aggregates Sum Product Assessment (WASPAS), 7
Weighted normalized decision matrix, 12
Weighted Product Method (WPM), 7
Weighted Sum Method (WSM), 7
West Zone Power Distribution Company Limited (WZPDCL), 10
Wind-casted wave, 267
Wind energy, 67, 136, 388
  conversion system, 135, 145
    advanced, 146
    conventional, 145
  conversion systems, 141
  cost, 153
  economic impacts, 69
  employment generation, 153
  environmental impacts, 67
  future prospects, 155
  global status, 148
  historical development, 136
  impacts, 167
  power generation, 150

social impacts, 68
  sustainability, 163
  technology manufacturing and supply chain, 151
  wind turbines, 138
Wind speed, 144
Wind turbines
  classification, 138
  components, 141*f*
  design, 144
  historical development, 138*t*
  power, 145
  swept area, 144
  technologies, 145
  WECS, 145, 146
  wind speed, 144
World Bank (WB), 98

Z
Zero-Emission Coal (ZEC), 186
'Zero waste', 206

Printed in the United States
by Baker & Taylor Publisher Services